D0081425

**Noble gases**

Atomic weights are based on carbon-12. Atomic weights in parentheses indicate the most stable or best-known isotope. Slight disagreement exists as to the exact electronic configuration of several of the high-atomic-number elements. Names and symbols for elements 104, 105, and 106 are unofficial.

| | | | IIIA | IVA | VA | VIA | VIIA | 2 Helium **He** 4.00260 |
|---|---|---|---|---|---|---|---|---|
| | | | 5 Boron **B** 10.81 | 6 Carbon **C** 12.011 | 7 Nitrogen **N** 14.0067 | 8 Oxygen **O** 15.9994 | 9 Fluorine **F** 18.99840 | 10 Neon **Ne** 20.179 |
| IB | IIB | | 13 Aluminum **Al** 26.98154 | 14 Silicon **Si** 28.086 | 15 Phosphorus **P** 30.97376 | 16 Sulfur **S** 32.06 | 17 Chlorine **Cl** 35.453 | 18 Argon **Ar** 39.948 |
| 28 Nickel **Ni** 58.71 | 29 Copper **Cu** 63.546 | 30 Zinc **Zn** 65.38 | 31 Gallium **Ga** 69.72 | 32 Germanium **Ge** 72.59 | 33 Arsenic **As** 74.9216 | 34 Selenium **Se** 78.96 | 35 Bromine **Br** 79.904 | 36 Krypton **Kr** 83.80 |
| 46 Palladium **Pd** 106.4 | 47 Silver **Ag** 107.868 | 48 Cadmium **Cd** 112.40 | 49 Indium **In** 114.82 | 50 Tin **Sn** 118.69 | 51 Antimony **Sb** 121.75 | 52 Tellurium **Te** 127.60 | 53 Iodine **I** 126.9045 | 54 Xenon **Xe** 131.30 |
| 78 Platinum **Pt** 195.09 | 79 Gold **Au** 196.9665 | 80 Mercury **Hg** 200.59 | 81 Thallium **Tl** 204.37 | 82 Lead **Pb** 207.2 | 83 Bismuth **Bi** 208.9804 | 84 Polonium **Po** (210)[a] | 85 Astatine **At** (210)[a] | 86 Radon **Rn** (222)[a] |

## Inner transition elements

| 63 Europium **Eu** 151.96 | 64 Gadolinium **Gd** 157.25 | 65 Terbium **Tb** 158.9254 | 66 Dysprosium **Dy** 162.50 | 67 Holmium **Ho** 164.9304 | 68 Erbium **Er** 167.26 | 69 Thulium **Tm** 168.9342 | 70 Ytterbium **Yb** 173.04 | 71 Lutetium **Lu** 174.97 |
|---|---|---|---|---|---|---|---|---|
| 95 Americium **Am** (243)[a] | 96 Curium **Cm** (247)[a] | 97 Berkelium **Bk** (249)[a] | 98 Californium **Cf** (251)[a] | 99 Einsteinium **Es** (254)[a] | 100 Fermium **Fm** (253)[a] | 101 Mendelevium **Md** (256)[a] | 102 Nobelium **No** (254)[a] | 103 Lawrencium **Lr** (257)[a] |

# FUNDAMENTAL TRANSITION METAL ORGANOMETALLIC CHEMISTRY

**Brooks/Cole Series in Inorganic Chemistry**
*Series Editor: Gregory L. Geoffroy*

Inorganic and Organometallic Reaction Mechanisms
*Jim D. Atwood*

Fundamental Transition Metal Organometallic Chemistry
*Charles M. Lukehart*

# FUNDAMENTAL TRANSITION METAL ORGANOMETALLIC CHEMISTRY

*Charles M. Lukehart*
*Vanderbilt University*

Brooks/Cole Publishing Company
Monterey, California

*To Marilyn and our children*

Brooks/Cole Publishing Company
A Division of Wadsworth, Inc.

© 1985 by Wadsworth, Inc., Belmont, California 94002.
All rights reserved.
No part of this book may be reproduced, stored in a retrieval system,
or transcribed, in any form or by any means—
electronic, mechanical, photocopying, recording, or otherwise—
without the prior written permission of the publisher,
Brooks/Cole Publishing Company, Monterey, California 93940,
a division of Wadsworth, Inc.

Printed in the United States of America

10 9 8 7 6 5 4 3 2 1

**Library of Congress Cataloging in Publication Data**

Lukehart, Charles M., [date]
    Fundamental transition metal organometallic chemistry.

    Includes bibliographies and indexes.
    1. Organometallic compounds.   2. Transition
metal compounds.   I. Title.
QD411.L857   1984      547'.05          84-12135
ISBN 0-534-03801-8

Sponsoring Editor: *Mike Needham*
Editorial Assistant: *Lorraine McCloud*
Production Editor: *Gay L. Orr*
Manuscript Editor: *Caroline King*
Permissions Editor: *Carline Haga*
Interior Design: *Victoria Van Deventer*
Cover Design: *Charles Carter Design*
Art Coordinators: *Rebecca Ann Tait and Michele Judge*
Interior Illustration: *Vantage Art, Inc.*
Typesetting: *Syntax International*
Cover Printing: *Lehigh Press—Lithography/Auto Screen*
Printing and Binding: *Maple-Vail Book Manufacturing Group*

# Preface

The rapid growth of transition metal organometallic chemistry since the 1950s has been remarkable. Many new classes of compounds have been prepared, and numerous examples of new chemical reactivity have been discovered. This chemistry has led to the development of new methods of organic synthesis and to a better understanding of catalytic processes. As the importance of this field has grown, there has been an increased need for a book that introduces undergraduate and graduate students to this growing area of chemistry.

This book provides a fundamental introduction to the organometallic chemistry of the transition metals, with the selected inclusion of related organolanthanide and organoactinide chemistry. The core material has been used since 1973 in a course that has been offered annually to upperclassmen and first-year graduate students. Prior knowledge by the students of general and organic chemistry has been assumed, and many of the students have also been exposed to a formal course in inorganic chemistry.

This book is intended to fill the void between simplistic surveys of transition metal organometallic chemistry and the more advanced treatments, which usually exclude much needed discussions of compound preparation, structure, and bonding concepts. It provides a solid foundation in these latter areas. The material presented here is as topically comprehensive as space permits. Students who continue to work in organometallic chemistry are prepared for more advanced texts or concepts as encountered in the literature or in research, while students who continue their study in a peripheral area of chemistry develop an appreciation of how organometallic chemistry can contribute to the resolution of problems within their area of interest.

The book is organized into two parts. Part One provides a general introduction to transition metal organometallic chemistry. Chapter 1 contains a brief historical perspective and a review of selected fundamental concepts. In Chapters 2 through 6, the syntheses, structures, and bonding descriptions of several major classes of compounds are presented. Metal carbonyls are discussed first (in Chapter 2) because these complexes played such an important role in the early development of organometallic chemistry, and because these relatively simple compounds can be used to demonstrate important concepts. Substituted metal carbonyl complexes are covered in Chapter 3. The reader is introduced to a variety of ancillary ligands, to several examples of isomerism in coordination compounds, and to the analysis of the infrared C—O stretching patterns of these molecules. Cyclopentadienyl complexes are presented in Chapter 4. These compounds were also very important in the "re-birth" of organometallic chemistry. The reader is introduced to π-complexes, to molecules having "piano-stool" structures, and to the cyclopentadienyl metal carbonyl complexes that will serve as useful reagents throughout the remainder of the book. Other important classes of molecules having π-ligands are discussed in Chapter 5. These compounds contain other carbocyclic rings, alkene, alkyne, or π-enyl ligands. Chapter 6 is a short chapter that presents complexes containing nitric oxide and other small molecules or monoatomic ligands, and introduces encapsulated and peripheral clusters. Throughout Chapters 2 through 6, discussions involving molecular orbital theory or NMR applications are presented at a qualitative level.

Chapters 7 through 11 present a survey of the fundamental reactions of organometallic compounds, and these reactions are classified according to the overall type of reaction that has occurred rather than according to mechanism. Considerable mechanistic data are presented for each class of reaction. Although proposed reaction mechanisms are frequently revised upon acquiring new data, it is important that the student learn the currently accepted mechanisms for these fundamental reactions. This information is particularly helpful in understanding the chemo-, regio-, or stereospecific reactions that are discussed in Part Two of the text. Chapter 7 focuses on one of the most simple types of reaction—the dynamic intramolecular rearrangements of organometallic complexes. In Chapter 8, hydride, alkyl, acyl, and related compounds are introduced via the protonation, alkylation, or acylation of metal carbonyl anions. Insertion, elimination, and abstraction reactions are discussed in Chapter 9, while oxidative-addition and reductive-elimination reactions are covered in Chapter 10. Reactions classified as either electrophilic or nucleophilic attack on organometallic molecules are treated in Chapter 11. In many of the reactions discussed in Chapters 7—11, the reactants are complexes that were presented earlier in Chapters 2—6. However, several new types of compounds are introduced as products of these reactions.

Part Two is designed to give a "big picture" overview of selected applications of transition metal organometallic chemistry. Applications related to organic synthesis are discussed in Chapter 12. These applications include the use of organometallic moieties as either protecting or activating groups. The compounds and reaction types presented earlier in the text are used at this point for the synthesis of organic compounds. Finally, Chapters 13 and 14 cover applications related to industrial processes and homogeneous catalysis. Chemical catalysis is introduced in Chapter 13, and selected industrial processes are discussed. The mechanisms of these processes are presented as a step-wise sequence involving reactions and compound types that were discussed earlier in the text. Chapter 14 summarizes several selected areas of organometallic chemistry that are now receiving intensive study. These areas are related to the understanding and development of new catalytic processes.

*To the Instructor:* The material covered in this book is presented as a one-semester course (approximately 40 lectures). The background of the students taking the course varies appreciably each year. In some years, the class is composed mainly of undergraduates in pre-professional programs of study, while in other years the class is composed predominantly of undergraduate chemistry majors and first-year graduate students. This course is usually offered in the Spring semester, with a general survey inorganic chemistry course offered in the preceding Fall semester. Lecture presentations focus on explaining concepts presented in the text. Basic coordination chemistry, molecular orbital theory (including other bonding concepts), and a "big picture" overview are continuously stressed in class throughout the study of Part One. The text serves both as "class notes" and as a pedagogical tool. Results from current literature are used in the classroom and as questions on quizzes and hour examinations. The study questions are used likewise, or as assigned homework problems. Several of the study questions require that the student use the available current literature to locate specific data. A short literature project is usually required of the students. At Vanderbilt, undergraduate students are encouraged to use the chemical literature and to submit written work in their courses. However, experience reveals that students prefer to consult the review literature for additional background information rather than the research literature (this is particularly true for undergraduate students). For this reason, the Suggested Reading references are provided for the interested student. This text is not intended to serve as a reference text for research chemists.

*To the Student:* This book serves as an introduction to the exciting new area of transition metal organometallic chemistry. The fundamental knowledge base provided in Part One is utilized in Part Two to illustrate the applications of this chemistry to stoichiometric and catalytic chemistry. In many ways, you will learn a new "chemical language" in studying this material. Just as "benzene" conveys a special meaning in organic chemistry, "metallocene" does likewise in organometallic chemistry. A new nomenclature system is also introduced. Whereas organic chemistry focuses primarily on the chemistry of four-valent carbon, there are metal clusters known where carbon (and even hydrogen) atoms are bonded to six metal atoms. Only by understanding the structural and bonding features of organometallic complexes can you both appreciate and be able to rationalize the observed regio- and stereochemical reactions of these new compounds. Study Questions are provided to enhance your understanding of the material presented. Suggested Reading references will lead you to more detailed reviews of the chemistry discussed in the text.

The author is indebted to several reviewers, including James Atwood, State University of New York at Buffalo; Peter Vollhardt, University of California at Berkeley; Wade Gladfelter, University of Minnesota; William H. Hersh, University of California at Los Angeles; Paul Jones, North Texas State University; Robert C. Kerber, State University of New York at Sunnybrook; Clifford Kubiak, Purdue University; Peter Wolczanski, Cornell University; and J. T. Templeton, University of North Carolina, who have read sections of the manuscript, for their comments and criticism. The author assumes all responsibility for any errors or omissions in the final draft of the text. The help of many students, particularly Randy Willard, in proofreading initial drafts of the manuscript is gratefully acknowledged. The author sincerely appreciates the secretarial assistance of Judi Roy, Jewell Carter, Connie Beasley, Eleanor Davenport, and Lina Zottolla in the preparation of the manuscript.

*Charles M. Lukehart*

# Contents

## INTRODUCTION TO TRANSITION METAL ORGANOMETALLIC CHEMISTRY 1

**PART ONE**

## 1

### Historical Perspective and Fundamental Concepts 3

## 2

### Binary Metal Carbonyl Complexes 20

# 6
## Complexes Containing Nitric Oxide and Other Small-Molecule or Monatomic Ligands    176

# 7
## Dynamic Intramolecular Rearrangements of Organometallic Complexes    196

# 8
## Complexes Containing Hydride, Alkyl, Acyl, or Related Ligands    211

# 9

# Insertion, Elimination, and Abstraction Reactions   233

# 10

# Oxidative-Addition and Reductive-Elimination Reactions   274

# 11

# Electrophilic and Nucleophilic Attack on Organometallic Complexes   300

# APPLICATIONS OF TRANSITION METAL ORGANOMETALLIC CHEMISTRY   345

PART TWO

## 12

## Applications to Organic Synthesis   347

## 13

## Applications to Industrial Homogeneous Catalysis   388

## 14

## Selected Topics Related to Catalysis    413

## Abbreviation List    435
## Index    437
## Index to Structural Data    443

# FUNDAMENTAL TRANSITION METAL ORGANOMETALLIC CHEMISTRY

# INTRODUCTION TO TRANSITION METAL ORGANOMETALLIC CHEMISTRY

## PART ONE

# Historical Perspective and Fundamental Concepts

## CHAPTER 1

## HISTORICAL PERSPECTIVE

In order to appreciate the rapid recent growth of organometallic chemistry and to anticipate future possible extensions, we shall begin here with a historical overview of this area of chemistry and then go on to discuss specific organometallic complexes and their reactions. Table 1.1 presents a selected chronological development of transition metal organometallic chemistry. Throughout the remainder of the book we shall discuss these developments in more detail.

The first transition metal organometallic compounds discovered were very stable and easily prepared complexes, such as Zeise's water-soluble ethylene complex of platinum (**1.1**) and Mond's preparation of $Ni(CO)_4$ from elemental nickel and carbon monoxide gas (see eq. *1.1*).

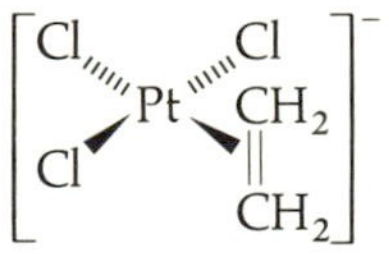

| | |
|:---:|:---:|
| **1.1** | **1.2** |
| $K[(C_2H_4)PtCl_3] \cdot H_2O$ | $(C_5H_5)_2Fe$ |
| Zeise's salt | ferrocene |

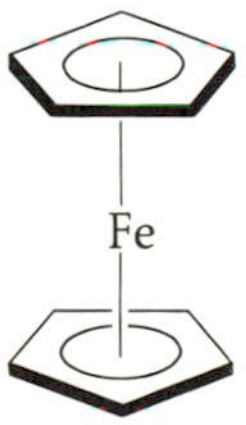

$$\text{Ni(metal)} + \text{CO(g) (excess)} \xrightarrow[25°C]{1 \text{ atm}} \text{Ni(CO)}_4 \qquad (1.1)$$

In the early 1900s, the synthetic importance of alkyl and aryl compounds of the representative metals, such as zinc, magnesium, lithium, and lead, encouraged research of the analogous transition metal compounds. Although alkyl and aryl transition

**TABLE 1.1    Selected chronological developments of transition metal organometallic chemistry.**

| Date | Event |
|---|---|
| 1827 | W. C. Zeise prepares a platinum complex containing ethylene. |
| 1849 | E. Franklin prepares alkyl zinc compounds. |
| 1890 | L. Mond discovers $Ni(CO)_4$ as corrosive agent of nickel valves. |
| 1891 | Berthelot, Mond and Quinche prepare $Fe(CO)_5$. |
| 1925 | F. Fischer and H. Tropsch use metals to prepare hydrocarbon fuels from CO and $H_2$. |
| 1928 | W. Hieber begins research with metal carbonyls. |
| 1938 | O. Roelen develops Oxo process [terminal olefin + CO + $H_2$ $\xrightarrow[\text{cat.}]{\text{M}}$ (C + 1) aldehyde]. |
| 1939 | W. Reppe develops metal-catalyzed acetylene chemistry. |
| 1951 | P. Pauson and S. A. Miller independently discover ferrocene, $(\eta^5\text{-}C_5H_5)_2Fe$. |
| 1952 | G. Wilkinson reports the correct structure of ferrocene. |
| 1955 | K. Ziegler and G. Natta discover metal-catalyzed polymerization of olefins. |
| 1955 | E. O. Fischer reports first $\eta^6$-arene complexes. |
| 1961 | X-ray structure of vitamin $B_{12}$ coenzyme reveals a Co–carbon bond. |
| 1962 | X-ray structures confirm existence of metal–carbonyl cluster complexes. |
| 1963–1966 | Metal–metal bonds of bond orders 1, 2, 3 and 4 are well established. |
| 1964 | First transition metal carbenoid complex isolated. |
| 1968 | Preparation of "uranocene," $(\eta^8\text{-}C_8H_8)_2U$. |
| 1972 | Industrial L-dopa synthesis using a soluble rhodium catalyst. |
| 1973 | E. O. Fischer reports first transition metal carbyne complex. |
| 1973 | G. Wilkinson reports isolation of $W(CH_3)_6$. |
| 1973 | E. O. Fischer and G. Wilkinson receive Nobel Prize for their contributions to transition metal organometallic chemistry. |
| 1974 | Industrial Rh-catalyzed carbonylation of $CH_3OH$ to acetic acid. |
| 1976 | Industrial Rh-catalyzed, low-pressure Oxo process. |
| 1976 | Stoichiometric reduction of a CO ligand by $H_2$. |
| 1979 | Cobalt catalysis used in the synthesis of $d,l$-estrone. |
| 1980 | Industrial use of coal-derived CO and $H_2$ gases announced. |
| 1983 | Reaction between methane and mononuclear complexes reported. |

metal complexes, such as $Et_3Cr$, $Et_3Co$ and $Et_2Mn$, can be prepared, these complexes are very unstable thermally and very air sensitive. This instability, which limited their synthetic utility considerably, was not fully understood until the 1970s.

Industrial applications of metal catalysts beginning in the 1930s (see table 1.1) actually preceded a basic understanding of organometallic chemistry. As various classes of organometallic complexes and reaction types were discovered, a mechanistic interpretation of these catalytic cycles became possible. Today such accumulated knowledge is being used extensively to develop new chemical processes. As we shall discuss in Part Two, specific applications of organometallic chemistry to industrial chemistry are determined by trends in feedstock supplies. After about 1950 acetylene feedstocks were replaced by cheaper olefin supplies derived from inexpensive petroleum resources. In the last decade increasing costs for petroleum and underdevelopment of our natural gas resource has shifted attention to using mixtures of CO and $H_2$ as a feedstock. Such gaseous mixtures can be obtained from water and a carbon source such as coal. Basic understanding of metal-carbonyl chemistry, the reduction of CO by $H_2$ and the catalytic conversion of CO and $H_2$ into larger organic molecules is receiving intensive study by organometallic chemists.

From 1827 to about 1950, transition metal organometallic complexes were considered a curious class of compounds. Most of the alkyl complexes were very unstable thermally; the structure of Zeise's salt (**1.1**) was not yet known. Hieber's research with metal carbonyl complexes revealed classical ligand substitution reactions and the formation of unusual compounds containing a metal-hydrogen bond.

From 1951 and into the 1970s, the preparation of ferrocene (**1.2**), an organometallic complex of unusually high stability, and the availability of x-ray crystallography, IR, and proton NMR spectroscopy for molecular structure elucidation led to a "rebirth" of organometallic chemistry. Many different classes of complexes and reaction types were discovered. Thermally stable complexes were prepared, isolated, and characterized. Basic research flourished (see figure 1.1) as the number of research papers reporting organometallic chemistry increased at an astounding rate.

From the mid-1970s to the present, there has been a distinct shift in interest toward the potential applications of organometallic chemistry and a general systematization of known reaction chemistry. Reaction mechanisms are being studied, and the chemistry of known catalytic cycles and the design of new catalytic processes are being explored. The reactions of coordinate molecules are under study as means for developing useful organic synthetic reagents or for effecting basic chemical conversions relevant to chemical catalysis. However, new classes of compounds are still

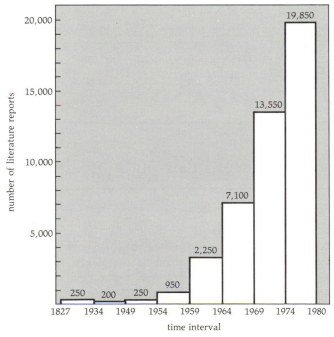

**FIGURE 1.1** The chronological appearance of research papers and patents in transition metal organometallic chemistry.

being discovered. The chemical importance of organometallic compounds is certain to increase in the future.

The chapters of Part One present an introduction to the synthesis, structure, chemical bonding and reaction chemistry of several classes of organometallic complexes. In Part Two, we shall use this basic knowledge to understand the chemistry involved in several applications of organometallic chemistry to stoichiometric and catalytic reactions that illustrate the future importance of this area of chemistry.

## INTRODUCTION TO ORGANOMETALLIC COMPLEXES

Transition metal organometallic complexes are molecules containing at least one chemical bond between a transition metal atom and a carbon atom. A *mononuclear* organometallic complex contains only one metal atom and can be represented by the general formula **1.3**,

$$L_a MR_b$$

**1.3**

where M is a transition metal atom. According to usual convention, the R ligands denote neutral species that would exist independently as radicals, such as methyl or phenyl radicals, while the L ligands signify neutral molecules, like CO, $H_3N$, or $Ph_3P$, which can also coordinate to a metal. Halogen ligands, if present, are usually represented by X. There must be at least one carbon-donor ligand in an organometallic complex. For molecules having more than one L or R ligand, the ligands need not be identical.

We shall focus our attention on organometallic molecules in which M is one of the 21 transition metals of the titanium through nickel groups, as shown below.

| Group IV | Group V | Group VI | Group VII | Group VIII | | |
|----------|---------|----------|-----------|-----------|------|------|
| Ti | V | Cr | Mn | Fe | Co | Ni |
| Zr | Nb | Mo | Tc | Ru | Rh | Pd |
| Hf | Ta | W | Re | Os | Ir | Pt |

Metals of the copper and zinc groups are usually classified as representative metals since they have completely filled *d* subshells in their common lowest oxidation states. Occasional reference to organometallic complexes of the lanthanide and actinide metals will be made. Recent developments in lanthanide and actinide organometallic chemistry forecast exciting and rapid growth in this area.

The total number of ligands coordinated to a metal defines the *coordination number* of the metal atom. In most complexes, the metal has a coordination number of 4, 5 or 6. Four-coordinate complexes have structures based on idealized tetrahedral or square-planar coordination, as shown in **1.4** and **1.5**, respectively.

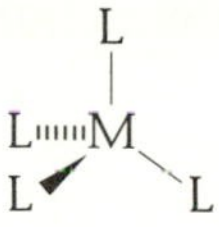

**1.4**

tetrahedral complex

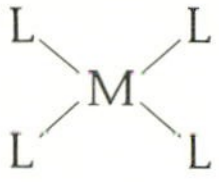

**1.5**

square or square-planar complex

Five-coordinate complexes usually have trigonal-bipyramidal (**1.6**) or square-pyramidal (**1.7**) structures, and the most common idealized structure for six-coordinate complexes is octahedral (**1.8**).

**1.6**

trigonal-bipyramidal
complex

**1.7**

square-pyramidal
complex

**1.8**

octahedral
complex

   When the ligands are not identical, geometrical isomers are possible (except in a tetrahedral complex). Some important examples of such isomerism are shown in **1.9** through **1.12**.

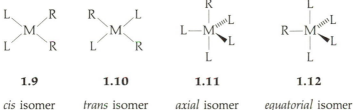

**1.9**

*cis* isomer

**1.10**

*trans* isomer

**1.11**

*axial* isomer

**1.12**

*equatorial* isomer

For square-planar complexes of the type $ML_2R_2$, *cis* and *trans* isomers (**1.9** and **1.10**) are possible. In the trigonal-bipyramidal complexes the unique ligand can occupy either an *axial* or an *equatorial* site as in **1.11** and **1.12**, respectively. Octahedral complexes of composition $ML_4R_2$ can exist as *cis* or *trans* isomers (**1.13** and **1.14**), while a complex of formula $ML_3R_3$, can exist as either *facial* or *meridonal* isomers (**1.15** and **1.16**).

**1.13**

*cis* isomer

**1.14**

*trans* isomer

**1.15**

*facial* (or *fac*)
isomer

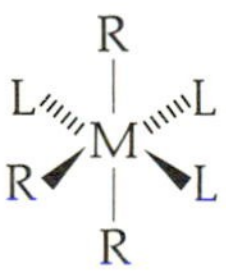

**1.16**

*meridonal* (or *mer*)
isomer

## PRELIMINARY DISCUSSION OF METAL–CARBON BONDING

Specific types of metal–carbon bonding influence the observed chemical reactivity of transition metal organometallic compounds. As we shall discuss in chapters 2 through 6 in some detail, there is a large variety of metal–carbon bonding interaction. For example, if the organic ligand R is a methyl group as shown in **1.17,**

$$L_aM—CH_3 \qquad L_aM\leftarrow \begin{matrix} CH_2 \\ \| \\ CH_2 \end{matrix} \qquad ML_a$$

**1.17**             **1.18**             **1.19**

then we usually regard the M—C bond as a covalent, two-electron, two-center bond similar to other metal–carbon bonds such as the Si—C bond in tetramethyl-silane, $Si(CH_3)_4$. The bonding electron density lies along the M—C internuclear axis; so the bond is referred to as a *σ-bond*. However, M—C bonding is more complex for unsaturated ligands such as ethylene (**1.18**) or benzene (**1.19**).

Structural data show that the carbon atoms in **1.18** or **1.19** are equidistant from the metal atom. We can describe the M—L bonding in such complexes as a *forward donation* of pi-electron density from the occupied *p*-pi molecular orbitals of the organic molecule to the valence shell of the metal atom. Since a highly reduced metal atom usually has several *d* electrons in its valence shell, we can also propose a *back donation* of metal electron density into empty *p*-pi molecular orbitals on the organic molecule. These M—C bonding interactions are depicted for an ethylene ligand in **1.20** and **1.21.**

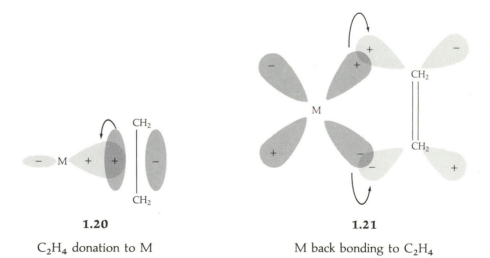

**1.20**                                    **1.21**

$C_2H_4$ donation to M                    M back bonding to $C_2H_4$

However, it is important to mention at this point that the degree of metal–ligand back donation to an unsaturated organic molecule depends on the "chemical accessibility" of the $d$-electron density of the metal atom. If the $d$ electrons reside on a highly electronegative metal atom, then M → L back donation is insignificant, and the complex may be very unstable. This is the primary reason why highly charged metal cations do not generally form stable complexes with unsaturated organic molecules. When significant M → L back donation does occur, it strengthens the M—L bond, and it reduces the amount of negative charge that builds up on the metal atom as a result of the ligand donation of electron density. Therefore, M → L back donation actually stabilizes the lower oxidation states of metal atoms.

Ligand → metal forward donation of electron density varies in degree according to the nature of the ligand. Amines, such as $NH_3$, are strong $\sigma$ donors, but they lack any $\pi$-accepting ability because the nitrogen donor atom does not possess an empty valence shell orbital to accept back-donated electron density. Trialkylphosphines, such as $Et_3P$, are moderately strong $\sigma$-donors and moderately strong $\pi$-acceptors (phosphorus does have empty $3d$ orbitals in its valence shell). Trifluorophosphine and CO are relatively poor $\sigma$-donors, but strong $\pi$-acceptors of electron density. Isocyanides, RNC, are strong $\pi$-acceptors and are also relatively strong $\sigma$-donors. Alkynes are usually stronger donors than alkenes.

## NOMENCLATURE FOR ORGANIC LIGANDS

Clearly, the chemical bonding in complexes **1.17**, **1.18**, and **1.19** is not the same. Metal–ligand bonding extremes range from an M-C $\sigma$ bond to a formal M-C $\pi$ bond. Since the designation of specific metal–ligand bonding interactions becomes increasingly difficult for more complex organometallic compounds, a recently developed nomenclature system that specifies only the number of carbon atoms chemically bonded to a metal atom has been adopted. A metal–carbon bond is recognized by spectroscopic measurements or by the relatively short M-C distances obtained from a molecular structure determination.

According to this system, an organometallic ligand is described by the number of carbon atoms connected to the metal atom by using the appropriate Greek numerical prefix, *mono, di, tri,* etc., followed by the word *hapto* (from the Greek *haptein,* "to fasten, denoting contact or combination"). In this nomenclature, the presence of a metal–ligand interaction is described without specifying the detailed nature of the metal–ligand chemical bond. Complexes **1.17** through **1.19** have *monohapto*-methyl, *dihapto*-ethylene and *hexahapto*-benzene ligands, respectively. In a chemical formula, the *hapto* prefix is abbreviated as "eta," $\eta$, the Greek letter for $h$, and the numerical prefix becomes a superscript. The organometallic ligands in **1.17** through **1.19** could be represented in a chemical formula as $\eta^1$-$CH_3$, $\eta^2$-$C_2H_4$ and $\eta^6$-$C_6H_6$, respectively. However, the hapticity of ligands that contain only *one* carbon atom, such as $CH_3$ or CO, is not shown. Also, when *all* of the carbon atoms of a ligand are bonded to a metal, the numerical superscript can be deleted as required by the accepted IUPAC rules of nomenclature. In subsequent chapters we shall use this nomenclature system, but the numerical superscripts, when greater than one, will always be shown as, for example, in $\eta^2$-$C_2H_4$ and $\eta^6$-$C_6H_6$.

## THE EFFECTIVE ATOMIC NUMBER RULE

For any new class of molecules such as **1.3**, it is helpful to develop simple empirical bonding concepts or rules that rationalize and predict which molecular formulations represent realistic structures of thermally stable compounds. In coordination chemistry, it is useful to recognize the preferred coordination numbers of the various metal ions and the effect of particular $d$-electronic configurations on structure and chemical reactivity as predicted by general bonding theories, such as the *angular overlap model*. In organic chemistry, recognition of the "octet rule" is of paramount importance. By knowing that a carbon atom prefers to enter into chemical bonding to achieve a valence shell electronic configuration similar to that of neon, one can rationalize why the simplest stable hydrocarbon is $CH_4$ and not $CH_5$, $CH_6$, or some other species. The octet rule also predicts that electron deficient species, such as carbenium ions, $R_3C^+$, or carbenes, $R_2C:$, have lower stability and react with other reagents to complete the valence shell electronic configuration of the carbon atom. Fortunately, the concept of the octet rule can be extended to transition metal atoms of organometallic complexes affording a nearly equally powerful "18-electron rule."

A transition metal atom has an inner-core electron configuration identical to that of the next lighter noble gas atom and a valence shell consisting of five $d$ orbitals, one $s$ orbital and three $p$ orbitals, as shown below.

### Electronic Configuration of Transition Metal Atoms

[noble gas core] $(n-1)d^m\, ns^{1,2}\, np^0$     [Ar core] $3d^5\, 4s^1\, 4p^0$

general transition metal atom        gaseous Cr atom

[Ar core] $3d^6\, 4s^0\, 4p^0$

Cr atom in condensed phase

When a gaseous chromium atom enters a condensed phase, such as a liquid or solid, the $4s$ electron, which extends quite far from the nucleus, encounters enough electron–electron repulsion with neighboring atoms that it drops back into the $3d$ orbitals giving a $3d^6$ valence shell electronic configuration. This electrostatic repulsion is sufficient to overcome the energy required to pair two $3d$ electrons. If the chromium atom were to bond to other atoms thus filling the $3d$, $4s$, and $4p$ valence orbitals completely, then its valence shell would contain 18 electrons, and the chromium *in this molecule* would have attained the electronic configuration of krypton. When this happens, the chromium atom is said to obey the "*18-electron rule*." Similarly, transition metals in the second and third transition metal series would adopt the xenon and radon configurations, respectively, when completely filling their valence shells.

Transition metals acquire electrons to fill their valence orbitals by forming bonds to ligand molecules that act as electron donors. The 18 valence electrons are then the sum of the number of electrons initially present on the "bare" metal atom and the electrons donated by the ligands. Although metals in organometallic complexes frequently obey the 18-electron rule, metals in more "classical" coordination complexes, those prepared by early coordination chemists, such as $[Cr(NH_3)_6]^{3+}$, frequently *disobey* the 18-electron rule. The reason lies in the differences between the metals and ligands in each type of complex.

For the more classical complexes, the metal ions are in relatively high oxidation states. This means that there are relatively few electrons on the "bare" metal ion. When ligands coordinate to the metal ion, the preferred coordination number of the metal ion (usually 4 or 6) is reached before 18 valence electrons can be acquired. For example, the $Ti(H_2O)_6^{3+}$ complex is composed of a $Ti^{3+}$ ion and six water molecules, each donating an electron pair to the metal ion. A $Ti^{3+}$ ion has one $d$ electron $(3d^1)$ in its valence shell, and the twelve electrons donated by the $H_2O$ ligands give a valence shell electron count of only 13. More than six water ligands cannot be accommodated in the valence shell of the $Ti^{3+}$ ion because of strong interligand steric repulsions.

Classical coordination complexes also disobey the 18-electron rule because the more electronegative heteroatomic donor atoms of the usual types of ligand molecules impose a weak crystal field upon the metal ion. Two weak-field complexes of $Fe^{2+}$, a $3d^6$ ion, demonstrate this point. The tetrahedral complex, $[FeCl_4]^{2-}$, has six $3d$ electrons from the $Fe^{2+}$ ion and four electron pairs donated by the chloride ligands. The electron configuration of $[FeCl_4]^{2-}$,

$$[FeCl_4]^{2-} \quad \underline{1\!\!\downarrow} \, \underline{1} \, \underline{1} \, \underline{1} \, \underline{1} \qquad \underline{1\!\!\downarrow} \qquad \underline{1\!\!\downarrow} \, \underline{1\!\!\downarrow} \, \underline{1\!\!\downarrow} \qquad (1.2)$$
$$3d \qquad\qquad 4s \qquad 4p$$

which is consistent with magnetic measurements, reveals only 14 electrons in the valence shell of the iron ion. The $Fe^{2+}$ ion disobeys the 18-electron rule because the chloride ligands cannot force the complete pairing of the $3d^6$ electrons, thus freeing two of these orbitals for additional ligand bonding. Even the octahedral aquo complex, $[Fe(H_2O)_6]^{2+}$, where the $H_2O$ ligands are stronger-field ligands than chloride ligands, remains a high-spin complex. The electron configuration for the iron atom in this complex can be represented simplistically as

$$[Fe(H_2O)_6]^{2+} \quad \underline{1\!\!\downarrow} \, \underline{1} \, \underline{1} \, \underline{1} \, \underline{1} \qquad \underline{1\!\!\downarrow} \qquad \underline{1\!\!\downarrow} \, \underline{1\!\!\downarrow} \, \underline{1\!\!\downarrow} \qquad \underline{1\!\!\downarrow} \, \underline{1\!\!\downarrow} \, \underline{\quad} \, \underline{\quad} \, \underline{\quad} \qquad (1.3)$$
$$3d \qquad\qquad 4s \qquad 4p \qquad\qquad 4d$$

Although the $Fe^{2+}$ ion now has 18 electrons in its "valence shell," it still violates a rigorous application of the 18-electron rule since its electron configuration does not resemble that of a noble gas atom. Metals having 18-electron configurations similar to those of the noble gases are said to obey the *effective atomic number (EAN) rule*.

In many types of organometallic complexes, such as metal carbonyl compounds, the metals are in low oxidation states. Furthermore, the more electropositive carbon donor atoms form such strong covalent bonds to a metal that the metal atom nearly always adopts a low-spin electronic configuration. These features ensure first, a large number of valence electrons on the "bare" metal atom and, second, that these electrons are maximally paired in the complex. Electron pairs donated by the ligands then complete the filling of the valence shell.

To apply the EAN rule efficiently, we need a quick method of determining the $d$-electron configurations for the transition metal atoms. It is convenient to remember the number of $d$ electrons for each group of metals as the $M^0$ configurations shown in line $(1.4)$.

| Sc | Ti | V | Cr | Mn | Fe | Co | Ni | |
|----|----|---|----|----|----|----|----|---|
| $3d^3$ | $3d^4$ | $3d^5$ | $3d^6$ | $3d^7$ | $3d^8$ | $3d^9$ | $3d^{10}$ | $(1.4)$ |

The valence $s$ electrons of the metals are formally placed in the $d$ orbitals. To determine the number of $d$ electrons for a given metal atom we just add or subtract electrons from these values as indicated by the charge of the metal atom or ion in the complex.

For example, the thermally stable carbon monoxide complexes of nickel, iron, and chromium are $Ni(CO)_4$, $Fe(CO)_5$, and $Cr(CO)_6$, respectively. As shown below, in each complex the metal atom obeys the EAN rule:

**$Ni(CO)_4$**

| $Ni^0(3d^{10})$ | 10 electrons |
| 4CO ligands | 8 electrons |
| | 18 electrons |

**$Fe(CO)_5$**

| $Fe^0(3d^8)$ | 8 electrons |
| 5CO ligands | 10 electrons |
| | 18 electrons |

**$Cr(CO)_6$**

| $Cr^0(3d^6)$ | 6 electrons |
| 6CO ligands | 12 electrons |
| | 18 electrons |

(*Note:* A CO ligand acts as a two-electron donor as can be shown by the Lewis structure, $:^-C{\equiv}O^+:$.) The EAN rule not only explains the observed stability of these complexes, but it also predicts that species like $Ni(CO)_5$ and $Ni(CO)_4{}^+$ should be difficult, if not nearly impossible, to prepare and should easily undergo loss of carbon monoxide or chemical reduction, respectively.

When the organic ligand can be treated formally as a radical or ionic ligand, two conventions are commonly used for applying the EAN rule. These are shown below for the alkyl complex, $CH_3Mn(CO)_5$:

**$CH_3Mn(CO)_5$**

| $Mn^0(3d^7)$ | 7 electrons |
| $\cdot CH_3$ | 1 electron |
| 5CO ligands | 10 electrons |
| | 18 electrons |

**$CH_3Mn(CO)_5$**

| $Mn^+(3d^6)$ | 6 electrons |
| $:CH_3{}^-$ | 2 electrons |
| 5CO ligands | 10 electrons |
| | 18 electrons |

These conventions differ in assignment of the bonding electron pair of the Mn—$CH_3$ bond to the manganese and carbon atoms. Homolytic cleavage of the Mn—

**TABLE 1.2   A selected listing of ligands according to the number of donated electrons by usual electron counting conventions.**

| Number of electrons donated to M | Ligand type |
|---|---|
| 1 | H, alkyl, aryl and halogen radicals |
| 2 | CO, CS, RNC, $R_3N$, $R_3P$, $R_3As$, RCN, $R_2S$, $H^-$, $R^-$, alkenes, $\eta^3$-allyl$^+$, alkynes, $NO^+$, halide anions |
| 3 | $\eta^3$-allyl radicals, NO |
| 4 | $\eta^3$-allyl$^-$, $\eta^4$-$C_4R_4$(cyclobutadienes), $\eta^4$-dienes (conjugated or unconjugated) |
| 5 | $\eta^5$-$C_5R_5$ radicals |
| 6 | $\eta^5$-$C_5R_5{}^-$, $\eta^6$-$C_6H_6$, $\eta^7$-$C_7H_7{}^+$ (tropylium cation) |
| 7 | $\eta^7$-$C_7R_7$(cycloheptatrienyl radicals) |
| 8 | $\eta^8$-$C_8R_8$(cyclooctatetraenes) |

$CH_3$ bond forms an $Mn^0$ atom and a methyl radical; heterolytic cleavage affords an $Mn^+$ ion and a methyl carbanion. The methyl ligand can be regarded formally as either a one- or a two-electron donor. For electron counting purposes either convention is acceptable provided that the appropriate charge is placed on the metal atom. *It is important to not mix the two conventions*, i.e., a methyl radical ligand and an $Mn^+$ ion.

Table 1.2 lists some common types of ligands and the number of electrons donated by each ligand under these electron counting conventions. The values shown represent the number of electrons donated by a ligand in its normal bonding mode to a single metal atom. In other structures, the same ligand might donate fewer or more electrons than the number shown in table 1.2. A bridging halogen atom acts as a three-, or four-electron donor depending on whether it is regarded as a halogen atom or a halide ion.

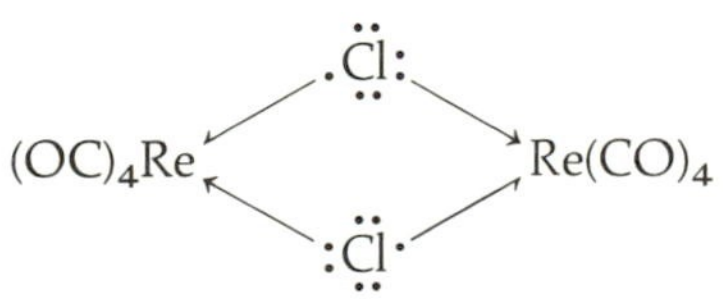

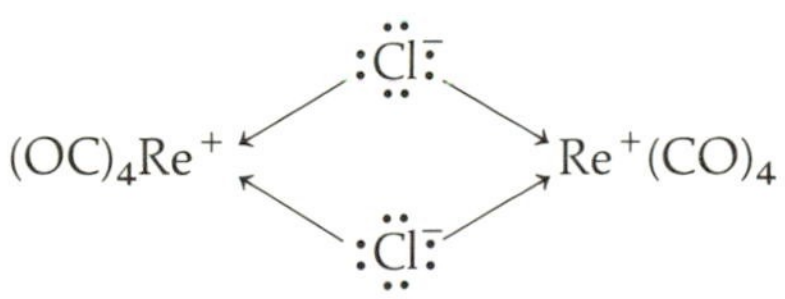

| Re ($5d^7$) | 7 electrons | Re$^+$ ($5d^6$) | 6 electrons |
|---|---|---|---|
| 4CO ligands | 8 electrons | 4CO ligands | 8 electrons |
| Cl· | 1 electron | 2Cl:$^-$ | 4 electrons |
| Cl (lone pair) | 2 electrons | | 18 electrons |
| | 18 electrons | | |

Pi-bonded ligands are usually treated as neutral species, as shown below for manganese and chromium complexes, even though they might be introduced into a chemical synthesis as ionic reagents. Electron counting is only a formalism used to determine whether a metal atom obeys the EAN rule and is, therefore, *electronically saturated*. It should not be seen as representing the actual electron charge distribution within a complex.

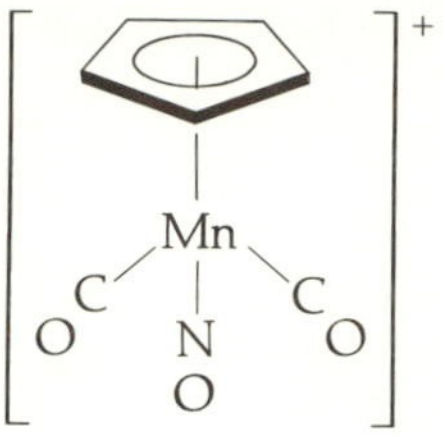

| Mn$^+$ ($3d^6$) | 6 electrons | Cr ($3d^6$) | 6 electrons |
| $C_5H_5\cdot$ | 5 electrons | $C_6H_6$ | 6 electrons |
| 2CO ligands | 4 electrons | 3CO ligands | 6 electrons |
| NO ligand | 3 electrons | | 18 electrons |
| | 18 electrons | | |

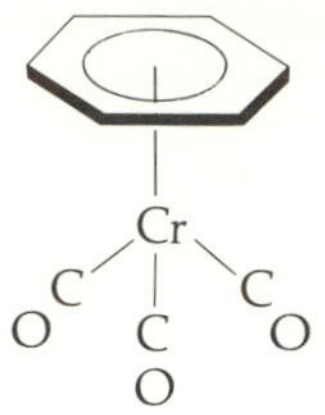

## COMPLEXES THAT DISOBEY THE EAN RULE

Although the vast majority of known organometallic complexes obey the EAN rule, many compounds have been isolated that violate this rule. Some of these complexes, such as **1.22–1.24,** have metals with far fewer than 18 electrons.

$$(\eta^5\text{-}C_5H_5)_2ZrCl_2 \qquad Me_3TaCl_2 \qquad Me_6W$$

**1.22**               **1.23**               **1.24**

16 electrons        10 electrons      12 electrons

In many cases, these compounds either involve metals in the early transition metal groups, which have very few $d$ electrons, or have relatively bulky ligands, a factor that favors low coordination numbers due to ligand–ligand steric repulsion. An example of this latter instance is the 16-electron, three-coordinate complex, $(Ph_3P)_3Pt$. Nickelocene, $(\eta^5\text{-}C_5H_5)_2Ni$, is an example of a 20-electron complex. The stability of this molecule is related to the detailed nature of the metal–ligand bonding, as will be discussed in chap. 4.

Square-planar $d^8$-complexes are another important class of compounds that violate the EAN rule. Examples are Zeise's salt (**1.1**) and structures **1.25** and **1.26**.

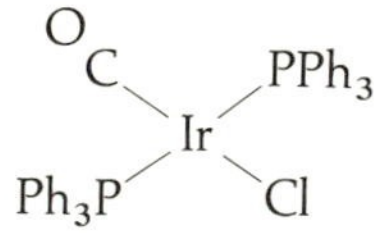

**1.25** 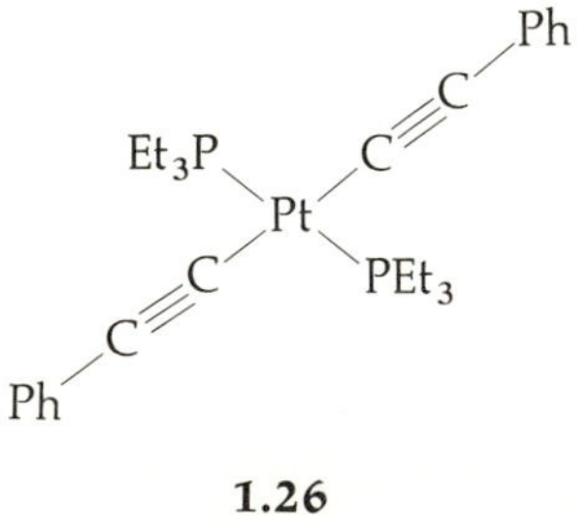                           **1.26**

These complexes are 16-electron compounds. Their stability is apparently related to the relative orbital energies for $d^8$-metals in various coordination geometries.

Detailed derivations of these energy differences using the angular overlap model for metal–ligand bonding are provided in several of the general references listed at the close of this chapter. To summarize these results, a $d^8$-complex has the same electronic stabilization energy in a square-planar geometry as it does in a trigonal-bipyramidal geometry. Therefore, a 16-electron square-planar complex is neither more nor less preferred than its 18-electron trigonal bipyramidal analogue.

For $d^8$-metals, square-planar (low-spin) complexes are also as stable as the corresponding octahedral complex. However, if a $d^8$-metal in a square-planar complex could undergo a two-electron oxidation, the $d^6$-product complex would gain considerable stabilization energy upon forming a six-coordinate complex of octahedral geometry. Square-planar $d^8$-complexes show this type of reactivity. These reactions are called oxidative-additions, as shown in equation 1.5.

$$(1.5)$$

$$\textbf{1.25} \qquad\qquad\qquad \textbf{1.27}$$

| | | | | |
|---|---|---|---|---|
| $Ir^I$ $(5d^8)$ | 8 electrons | | $Ir^{III}$ $(5d^6)$ | 6 electrons |
| $2Ph_3P$ | 4 electrons | | $2Ph_3P$ | 4 electrons |
| $CO$ | 2 electrons | | $CO$ | 2 electrons |
| $Cl^-$ | 2 electrons | | $3Cl^-$ | 6 electrons |
| | 16 electrons | | | 18 electrons |

Compound **1.25** is a 16-electron square-planar complex. *When all ligands are regarded as two-electron donors,* the iridium atom becomes a $d^8$-metal in the *formal $+I$ oxidation state.* Reaction with $Cl_2$ gives **1.27**. Both chlorine atoms have *added* to the metal. Using the *same* electron counting convention, one can show that the iridium atom in **1.27** has undergone a *formal* two-electron *oxidation*. This complex is very stable, because it is now a $d^6$, six-coordinate complex that obeys the EAN rule. A detailed discussion of oxidative-additions and the reverse reaction, reductive-eliminations, is presented in chap. 10.

The angular overlap model for metal-ligand bonding can be used similarly to determine the preferred structure of coordination complexes of any $d^n$ configuration with any coordination number. As an important example related to the above discussion, this bonding model reveals that low-spin complexes having $d^0$-, $d^2$-, or $d^{10}$-metals do not prefer either square-planar or tetrahedral coordination geometries. However, low-spin complexes of the remaining $d^n$-metals prefer a square-planar geometry over a tetrahedral geometry, with the greatest preference occurring for $d^8$-metal complexes. This analysis explains why compounds **1.1**, **1.25**, and **1.26** have square-planar structures. However, four-coordinate organometallic complexes of $d^0$-$Zr^{4+}$, such as **1.22**, or $d^{10}$-metal compounds, such as $Ni(CO)_4$, prefer a tetrahedral structure, because the lack of a structural preference based on M—L bonding will permit steric factors to prevail. A tetrahedral structure minimizes interligand steric repulsion.

## COMMON REACTIONS USED IN COMPLEX FORMATION

In discussing the synthesis of organometallic complexes we shall encounter several general reaction types. Some of these reactions are defined here with a more detailed discussion to follow later. Other specific reaction types will be presented where appropriate. The reactions introduced here are reductive-ligation, ligand substitution, and insertion. The important oxidative-addition reaction was discussed in the previous section (eq. *1.5*).

Metal atoms of transition metal organometallic complexes frequently exist in oxidation states lower than the naturally occurring ones, and these lower oxidation states are stabilized by the M—L chemical bonding. Several types of organometallic complexes can be prepared via a one-step process in which a transition metal salt $MX_n$, a reducing agent such as an alkali metal, and an organic ligand molecule are combined to afford an organometallic complex.

### Reductive-ligation Reaction

$$WCl_6 + 3Zn + 6CO \longrightarrow W(CO)_6 + 3ZnCl_2 \qquad (1.6)$$

Usually the organic reactant is a neutral molecule such as CO or an alkene. The transition metal atom has undergone a formal reduction from the $+n$ oxidation state to a much lower oxidation state with concomitant formation of the M—C bonds. This chemical conversion is called a *reductive-ligation* reaction (eq. *1.6*).

Like other coordination complexes organometallic complexes undergo ligand dissociation and ligand-substitution reactions of the kind shown in equation *1.7*.

### Ligand-substitution Reaction

$$W(CO)_6 + PPh_3 \xrightarrow[\text{or } h\nu]{\Delta} W(CO)_5(PPh_3) + CO \qquad (1.7)$$

Both organic and other types of ligands may participate in these reactions. Ligand dissociation, which is frequently the first step in a ligand-substitution reaction, is initiated by thermal ($\Delta$) or photochemical ($h\nu$) activation. These reactions are discussed in chap. 3.

It is quite common for an M—R bond, where R is usually a *monohapto* organic ligand, such as an alkyl group or a hydrogen atom, to undergo *insertion reactions* where a diatomic or polyatomic molecule is inserted chemically into the M—R bond. The most common insertion reactions are *1,1-* and *1,2-insertion reactions* (eqs. *1.8* and *1.9*, respectively).

### 1,1-Insertion Reaction

$$CH_3Mn(CO)_5 + CO \longrightarrow CH_3\overset{\displaystyle O}{\overset{\displaystyle \|}{C}}Mn(CO)_5 \qquad (1.8)$$

### 1,2-Insertion Reaction

$$HMn(CO)_5 + F_2C{=}CF_2 \longrightarrow HF_2C{-}CF_2{-}Mn(CO)_5 \qquad (1.9)$$

A 1,2-insertion is formally analogous to the addition of boron hydrides or boranes across an olefinic double bond. The reverse reactions are referred to as *1,1-* or *1,2-elimination reactions.* A more detailed discussion of insertion reactions is presented in chap. 9.

With this general introduction to the definition, nomenclature, bonding and reaction chemistry of organometallic compounds, we are now ready to discuss synthesis, structure, and bonding for several major classes of organometallic compounds.

## STUDY QUESTIONS

1. In which of the complexes listed below do the metal atoms obey the EAN rule?
   (a) $Mn(CO)_6^+$
   (b) $Mo(CO)_2Cl_2^-$
   (c) $Fe(CO)_4Cl_2$
   (d) $Cr(CO)_5I^-$
   (e) $(Ph_3P)Fe(CO)_4$
   (f) $(F_3P)_2Mo(CO)_4$
   (g) $Cr(CO)_5^{2-}$
   (h) $Mn(CO)_4Cl_2^{2-}$

2. For the molecular structures shown below, determine if a metal-metal bond should be present to satisfy the EAN rule. If an M—M bond is present, then specify the M—M bond order. *Hint:* With an M—M single bond, each metal acts as a one-electron donor to the other metal.

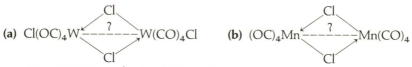

   (a) $Cl(OC)_4W\underset{Cl}{\overset{Cl}{\cdots\cdots\overset{?}{\cdots\cdots}}}W(CO)_4Cl$     (b) $(OC)_4Mn\underset{Cl}{\overset{Cl}{\cdots\cdots\overset{?}{\cdots\cdots}}}Mn(CO)_4$

   (c) $C_5H_5(OC)_2Mo\overset{?}{-\!-\!-\!-}Mo(CO)_2C_5H_5$
   (d) $C_5H_5(OC)Ni\overset{?}{-\!-\!-\!-}Ni(CO)C_5H_5$

3. For molecules of the type $M(CO)_x(NO)_y$, where M is V, Cr, Mn, Fe, Co or Ni, determine which integral values of $x$ and $y$ greater than or equal to zero afford molecular compositions in which M obeys the EAN rule. *Note:* NO is a three-electron donor in these complexes.

4. Write the chemical formula of each complex shown below using the *hapto* nomenclature. *Note:* The CO ligands need *not* be specified as *monohapto;* specify the hapticity of the other organic ligands.

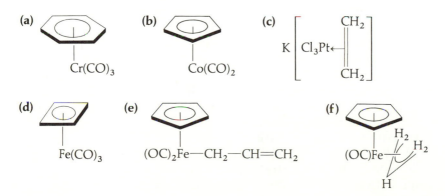

(a) $Cr(CO)_3$     (b) $Co(CO)_2$     (c) $K\begin{bmatrix} Cl_3Pt\leftarrow \|\begin{smallmatrix}CH_2\\CH_2\end{smallmatrix} \end{bmatrix}$

(d) $Fe(CO)_3$     (e) $(OC)_2Fe-CH_2-CH=CH_2$     (f) $(OC)Fe\begin{smallmatrix}H_2\\H_2\\H\end{smallmatrix}$

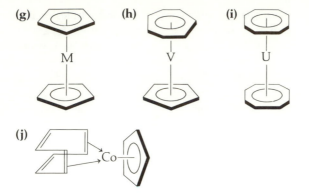

5. Classify each reaction below as one or a combination of the reaction types mentioned in this chapter:

(a) $Mo(CO)_6 + CH_3CN(xs) \xrightarrow{\Delta} Mo(CO)_3(CH_3CN)_3 + 3CO$

(b) $CrCl_3 + Al(xs) \xrightarrow{C_6H_6} [Cr(C_6H_6)_2]^+ AlCl_4^-$

(c) $Ir(Cl)(CO)(PPh_3)_2 + CH_3I \longrightarrow Ir(Cl)(CO)(PPh_3)_2(CH_3)(I)$

(d) $Cr(CO)_6 + R_4N^+I^- \xrightarrow{\Delta} R_4N[Cr(CO)_5I] + CO$

(e) $R-\overset{\overset{\displaystyle O}{\|}}{C}-Rh(PPh_3)_2(CO) + H_2 \longrightarrow R-\overset{\overset{\displaystyle O}{\|}}{C}-Rh(PPh_3)_2(CO)(H)_2$

(f) $R-\overset{\overset{\displaystyle O}{\|}}{C}-Rh(PPh_3)_2(CO)(H)_2 \xrightarrow{\Delta} HRh(PPh_3)_2(CO) + R-\overset{\overset{\displaystyle O}{\|}}{C}-H$

(g) $R-\overset{\overset{\displaystyle O}{\|}}{C}-Rh(PPh_3)_2(CO)(H)_2 \xrightarrow{\Delta} RRh(PPh_3)_2(CO)(H)_2 + CO$

(h) $Pt(Et)(PPh_3)_2Cl \xrightarrow{\Delta} Pt(H)(PPh_3)_2Cl + C_2H_4$

6. For metals of the first transition series, list three different metals or metal ions that would have **(a)** $d^6$, **(b)** $d^8$, or **(c)** $d^{10}$ electronic configurations.
7. A ligand must have a relatively low-energy *empty* orbital in its valence shell if it is to participate significantly in M → L back bonding. For the following ligands, determine whether such an orbital is present, and, if so, identify the orbital: **(a)** $CH_3$; **(b)** $PH_3$; **(c)** $CO$; **(d)** $NH_3$; **(e)** $NO$; **(f)** $SiPh_3$; **(g)** $C_2H_2$; **(h)** $C_6H_6$.
8. For each complex listed below, determine the *formal* oxidation state of the metal atom (i.e., the charge on the metal atom when each ligand is regarded as donating an even number of electrons).
   (a) $Fe(CO)_3(PPh_3)_2$        (b) $Pt(CH_3)_2(PEt_3)_2$
   (c) $Me_3TaCl_2$            (d) $(\eta^5\text{-}C_5H_5)_2ZrCl_2$
   (e) $ClMn(CO)_5$            (f) $Me_6W$
   (g) $Cr(CO)_4{}^{4-}$          (h) $(\eta^5\text{-}C_5H_5)Mo(CO)_3CH_3$

9. Complex **1.25** can be alkylated by oxonium salts, $R_3O^+$, to give square-pyramidal, cationic complexes such as $[Ir(CO)(PPh_3)_2(Cl)R]^+$. Rationalize the formation of these products.

10. Sketch all of the possible geometrical and optical isomers for the octahedral complex, $Ir(CO)(Cl)(PPh_3)_2(Br)_2$.

## SUGGESTED READING

### General References

Abel, E. W., and Stone, F. G. A., eds. *Organometallic Chemistry*. London: The Chemical Society. Periodic specialist reports reviewing the literature of organometallic chemistry.

Collman, J. P., and Hegedus, L. S. *Principles and Applications of Organotransition Metal Chemistry*. Mill Valley, CA: University Science Books, 1980.

Cotton, F. A., and Wilkinson, G. *Advanced Inorganic Chemistry*. 3rd (1972) and 4th (1980) ed. New York: John Wiley & Sons.

Green, M. L. H. *Organometallic Compounds*. Vol. 2, *The Transition Elements*. London: Methuen & Co., 1968.

King, R. B. *Transition Metal Organometallic Chemistry*. New York: Academic Press, 1969.

Pauson, P. L. *Organometallic Chemistry*. New York: St. Martin's Press, 1967.

Purcell, K. F., and Kotz, J. C. *An Introduction to Inorganic Chemistry*. Philadelphia: Saunders College/Holt, Rinehart and Winston, 1980.

Wender, I., and Pino, P., eds. *Organic Synthesis via Metal Carbonyls*, Vol. 1. New York: John Wiley & Sons, 1968.

Wilkinson, G., Stone, F. G. A., and Abel, E. W., eds. *Comprehensive Organometallic Chemistry*. Oxford: Pergamon Press, 1982.

*Advances in Organometallic Chemistry*. New York: Academic Press.

*Annual Surveys*, published in the *J. Organometal. Chem.* Lausanne: Elsevier Sequoia.

*Inorganic Syntheses*. New York: John Wiley & Sons.

*Progress in Inorganic Chemistry*. New York: John Wiley & Sons.

### Data Compilations and Literature Summaries

Bruce, M. I. *Adv. Organometal. Chem.* **10**, 273 (1972); **11**, 447 (1973); and **12**, 379 (1974). Literature summary from 1950 to 1972.

Dub, M., ed. *Organometallic Compounds*. Vol. 1, 2nd ed. (1966), and Vol. 1, 2nd ed., 1st Supplement by K. Bauer and G. Haller (1975). New York: Springer–Verlag. Compound data compilation.

Kennard, O., Allen, F. H., and Watson, D. G., eds. *Molecular Structure and Dimensions*. Bohn, Utrecht, Netherlands: Scheltema and Holkema, 1977. Literature guide to organic and organometallic crystal structures from 1935 to 1976.

### Specific References

Angelici, R. J. *Synthesis and Technique in Inorganic Chemistry*, 2nd ed., Philadelphia: W. B. Saunders Company, 1977.

Jolly, W. L. *The Synthesis and Characterization of Inorganic Compounds*. Englewood Cliffs, NJ: Prentice-Hall, 1970.

King, R. B. *Organometallic Syntheses*. Vol. 1, *Transition Metal Compounds*. New York: Academic Press, 1965. Experimental organometallic chemistry.

Shriver, D. F. *The Manipulations of Air-Sensitive Compounds*. New York: McGraw-Hill, 1969.

Tolman, C. A. *Chem. Soc. Rev.* **1**, 337 (1972). The 16- and 18-electron rules.

# Binary Metal Carbonyl Complexes

## CHAPTER 2

The first major class of organometallic complexes to be studied extensively was the neutral, *binary metal carbonyl compounds*, $M_x(CO)_y$. These complexes contain only metal atoms and carbonyl ligands. A discussion of organometallic compounds logically begins with these complexes because they are common reagents for further reaction chemistry and because they exhibit chemical bonding and special spectroscopic properties of general importance. Table 2.1 lists the known neutral binary metal carbonyls by element. Binary complexes of this type, $(ML_n)_x$, are also referred to as *homoleptic* complexes.

**TABLE 2.1   Selected neutral binary metal carbonyl complexes, $M_x(CO)_y$.**

| Metal | Complex | Metal | Complex | Metal | Complex |
|-------|---------|-------|---------|-------|---------|
| V | $V(CO)_6$ | Cr | $Cr(CO)_6$ | Mn | $Mn_2(CO)_{10}$ |
|   |   | Mo | $Mo(CO)_6$ | Tc | $Tc_2(CO)_{10}$ |
|   |   | W | $W(CO)_6$ | Re | $Re_2(CO)_{10}$ |

| Metal | Complex | Metal | Complex | Metal | Complex |
|-------|---------|-------|---------|-------|---------|
| Fe | $Fe(CO)_5$ $Fe_2(CO)_9$ $Fe_3(CO)_{12}$ | Co | $Co_2(CO)_8$ $Co_4(CO)_{12}$ $Co_6(CO)_{16}$ | Ni | $Ni(CO)_4$ |
| Ru | $Ru(CO)_5$ $Ru_2(CO)_9$[a] $Ru_3(CO)_{12}$ | Rh | $Rh_2(CO)_8$ $Rh_4(CO)_{12}$ $Rh_6(CO)_{16}$ | Pd | $Pd(CO)_4$[b] |
| Os | $Os(CO)_5$ $Os_2(CO)_9$ $Os_3(CO)_{12}$ $Os_5(CO)_{16}$ $Os_6(CO)_{18}$ $Os_7(CO)_{21}$ $Os_8(CO)_{23}$ | Ir | $Ir_4(CO)_{12}$ $Ir_6(CO)_{16}$ | Pt | $Pt_3(CO)_4$[b] |

[a] Existence not well established.
[b] Formed in low-temperature noble gas matrices.

## PREPARATION OF BINARY METAL CARBONYLS

Binary metal carbonyls are usually prepared by direct synthesis, by reductive-carbonylation, or from another metal carbonyl complex. Direct synthesis involves reaction of a transition metal powder with excess carbon monoxide under appropriate temperature and pressure conditions as shown in equations 2.1 through 2.5. This method is most feasible for preparing nickel and iron carbonyls.

$$Ni^0(s) + CO(g) \xrightarrow[\text{1 atm}]{30°C} Ni(CO)_4 \qquad (2.1)$$
$$\text{colorless liquid, high yield}$$
$$\text{bp } 43°C, \text{ mp } -25°C$$

$$Fe^0(s) + CO(g) \xrightarrow[\text{200 atm}]{200°C, \text{ 15 hr}} Fe(CO)_5 \qquad (2.2)$$
$$\text{yellow liquid, 26\%}$$
$$\text{bp } 103°C$$

$$Co^0(s) + CO(g) \xrightarrow[\text{30--40 atm}]{150°C} Co_2(CO)_8 \qquad (2.3)$$
$$\text{orange solid, 90\%}$$
$$\text{mp } 51°C$$

$$Mo^0(s) + CO(g) \xrightarrow[\text{250 atm}]{200°C} Mo(CO)_6 \qquad (2.4)$$
$$\text{white solid}$$
$$\text{mp } 150°C$$

$$Ru^0(s) + CO(g) \xrightarrow[\text{400 atm}]{300°C, \text{ 14 days}} Ru(CO)_5 \qquad (2.5)$$
$$\text{colorless liquid, 10\%}$$
$$\text{mp } -22°C$$

Nickel tetracarbonyl is the only metal carbonyl formed via direct synthesis near ambient conditions of temperature and pressure. Mond recognized that $Ni(CO)_4$ was a corrosion product of CO attack on nickel valves, and he subsequently developed the *Mond process* for purifying nickel metal based on this reaction. This process uses the facile formation of $Ni(CO)_4$ to extract nickel from a crude mixture of metals. Gaseous $Ni(CO)_4$ is removed from the reaction chamber and decomposed thermally to afford very pure nickel metal and CO gas (which is recycled).

Reductive-carbonylation is a specific type of reductive-ligation reaction (eq. 1.6) in which the organic ligand is CO. A transition metal in a high oxidation state is reduced in the presence of excess carbon monoxide to afford a metal carbonyl complex. Some common reducing agents are the following: electropositive metals, such as the alkali metals, Mg, Zn, or Al (eq. 2.6); alkyl compounds of representative metals, such as $R_3Al$, where the anionic alkyl ligands are undergoing a formal oxidation (eqs. 2.7 and 2.8); electron transfer reagents, such as benzophenone ketyl anion (eq. 2.9); and even $H_2$ or CO gas itself (eqs. 2.10–2.12). Because of relatively

$$CrCl_3 + Al^0(s) + CO(g) \xrightarrow[\text{C}_6\text{H}_6]{AlCl_3\text{cat}} Cr(CO)_6 + AlCl_3 \qquad (2.6)$$
$$\text{white solid, 88\%}$$
$$\text{mp } 154°C$$

$$WCl_6 + 3Et_3Al + CO(g) \xrightarrow[C_6H_6]{50°C, 1000 \text{ psi}} W(CO)_6 \qquad (2.7)$$

$$\text{white solid, 92\%}$$
$$\text{mp } 169°C$$

$$Mn(OAc)_2 + (isobutyl)_3Al + CO(g) \xrightarrow[\text{pressure}]{140°C, 20 \text{ hr}} Mn_2(CO)_{10} \qquad (2.8)$$

$$\text{yellow solid, 48\%}$$
$$\text{mp } 152°C$$

$$2MnCl_2 + 4Na^+[Ph_2CO]^- + CO(g) \xrightarrow[\text{THF}]{200°C, 250 \text{ atm}}$$

$$Mn_2(CO)_{10}(35\%) + 4NaCl + 4Ph_2CO \quad (2.9)$$

$$Ru(acac)_3 + H_2(g) + CO(g) \xrightarrow[\text{MeOH}]{150°C, 200 \text{ atm}} Ru_3(CO)_{12} \qquad (2.10)$$

$$\text{orange solid, 82\%}$$
$$\text{mp } 150°C$$

$$2[Co(H_2O)_4](OAc)_2 + 8(CH_3CO)_2O + 2H_2(g) + CO(g) \xrightarrow[160 \text{ atm}]{170°C}$$

$$Co_2(CO)_8 \quad + 20HOAc + 8H_2O \quad (2.11)$$
$$\text{orange solid, 60\%}$$
$$\text{mp } 51°C$$

$$Re_2O_7 + 17CO(g) \xrightarrow[350 \text{ atm}]{250°C, 16 \text{ hr}} Re_2(CO)_{10} + 7CO_2 \qquad (2.12)$$

$$\text{white solid, 95}+\%$$
$$\text{mp } 177°C$$

low product yields, high reagent costs, and the need for long reaction times at high temperatures and pressures, metal carbonyls are rather expensive compounds to produce. The principal exception is $Fe(CO)_5$ for which the cost of the metal is quite low. For example, at 1984 commercial prices, 25-g quantities of $Fe(CO)_5$, $Ni(CO)_4$, $Cr(CO)_6$, $Re_2(CO)_{10}$, $Ru_3(CO)_{12}$, and $Rh_4(CO)_{12}$ cost approximately \$1, \$10, \$23, \$243, \$400 and \$1,875, respectively.

Metal carbonyls can also be prepared from other binary metal carbonyl complexes. Usually chemical conversion is effected by photolysis (eq. 2.13), or by pyrolysis of neutral binary metal carbonyls (eqs. 2.14–2.16), or by oxidation of anionic metal-

$$Fe(CO)_5 \xrightarrow[\text{HOAc}]{hv} Fe_2(CO)_9 \downarrow \qquad (2.13)$$

$$\text{orange solid, 90\%}$$
$$\text{mp } 100°C$$

$$Fe_2(CO)_9 \xrightarrow[\text{toluene}]{95°C} Fe_3(CO)_{12} \qquad (2.14)$$

$$\text{black solid, 10\%}$$
$$\text{mp } 140°C$$

$$Os_3(CO)_{12} \xrightarrow[12 \text{ hrs}]{195-200°C} Os_4(CO)_{13} + Os_5(CO)_{16} + Os_6(CO)_{18}$$

$$+ Os_7(CO)_{21} + Os_8(CO)_{23} + \text{other products} \qquad (2.15)$$

$$Rh_4(CO)_{12} \xrightarrow[\text{solvent}]{60-80^\circ C} Rh_6(CO)_{16} + CO(g) \qquad (2.16)$$
$$\text{black solid}$$

carbonyls (eqs. 2.17–2.20). The preparation of cationic and anionic metal carbonyl complexes is discussed later (chap. 8).

$$Na[V(CO)_6] \xrightarrow{HX} V(CO)_6 + NaX + \tfrac{1}{2}H_2(g) \qquad (2.17)$$
$$\text{blue–green solid, ca. } 50\%$$
$$\text{mp } 60-70^\circ C \text{ (dec)}$$

$$Na[HFe(CO)_4] \xrightarrow{MnO_2} Fe_3(CO)_{12} + NaOH + MnO \qquad (2.18)$$
$$70\%$$

$$[Co_6(CO)_{15}]^{2-} \xrightarrow[H_2O]{Hg^{2+},\ 25^\circ C} Co_6(CO)_{16} + Co_4(CO)_{12} + Co(aq)^{2+} \quad (2.19)$$
$$\text{black solid}$$

$$[Ir_6(CO)_{15}]^{2-} + 2H^+ \xrightarrow[HOAc]{CO} Ir_6(CO)_{16} + H_2(g) \qquad (2.20)$$

## STRUCTURE OF BINARY METAL CARBONYLS

Mononuclear metal carbonyls adopt molecular structures that minimize interligand repulsions. Thus, 4-, 5- and 6-coordinate complexes have tetrahedral, trigonal-bipyramidal, and octahedral structures, respectively, as shown in table 2.2. The metal–carbonyl distances reflect the variation in covalent radii of the metal atoms; the nearly equivalent M—CO distances for second- and third-row transition metals within the same group is an expected result of the *lanthanide contraction.*

When a carbonyl ligand is bonded to only one metal atom, it is referred to as a *terminal carbonyl ligand.* Intraligand C—O distances are usually in the range 1.14–1.16 Å, slightly longer than the C—O distance of 1.128 Å in gaseous carbon monoxide. The M—C—O angles for terminal carbonyl ligands vary from 165° to 180°. The deviations from linearity may result from intermolecular packing forces for those molecular structures that are determined in the solid state or from intramolecular bonding interactions.

Binuclear metal carbonyls have three-dimensional geometries that are described frequently as a condensation of two idealized mononuclear polyhedra (table 2.3). The $M_2(CO)_{10}$ structures show essentially octahedral geometry about the two metal atoms; however, a covalent two-center, two-electron M—M bond defines a vertex common to both octahedra. The two sets of equatorial CO ligands adopt a staggered relative orientation. Since equatorial and apical CO ligands are not equivalent by symmetry, they may exhibit different chemistry. Such a difference is reflected in the slightly longer M—C (equatorial) distances relative to the M—C (apical) distances. An equatorial CO ligand is *trans* to another CO ligand, and an apical CO ligand is *trans* to the M—M bond.

The confacial bioctahedral structure of $Fe_2(CO)_9$ shows two octahedra sharing three facial ligands. Each iron atom has three terminal CO ligands and three *doubly*

**TABLE 2.2** Molecular structures of mononuclear metal carbonyl complexes, $M(CO)_x$.

| $M(CO)_x$ | Structure | Average M—C distance (Å)[a] |
|---|---|---|
| $Ni(CO)_4$ $Pd(CO)_4$ | 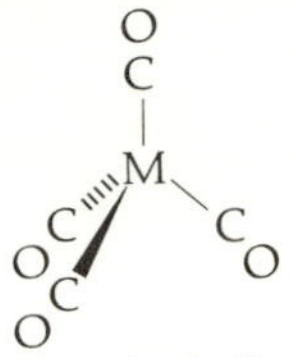 (tetrahedral) | Ni—C  1.838(2) |
| $Fe(CO)_5$ $Ru(CO)_5$ $Os(CO)_5$ | 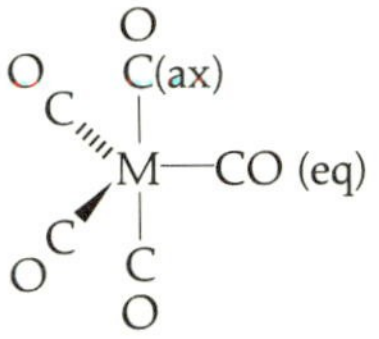 (trigonal bipyramidal) | Fe—C(ax)  1.810(3)[b] Fe—C(eq)  1.833(2) |
| $V(CO)_6$ $Cr(CO)_6$ $Mo(CO)_6$ $W(CO)_6$ | 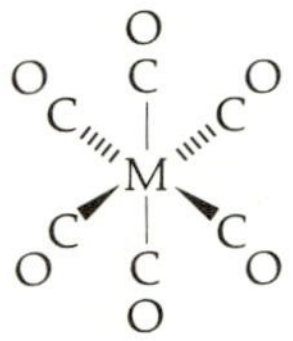 (octahedral) | V—C  2.008(3) Cr—C  1.913(2) Mo—C  2.06(2) W—C  2.06(3) |

[a] Estimated standard deviations are shown in parentheses in all tables, figures and the text. [b] Values obtained from electron diffraction in the vapor phase. Recent single-crystal x-ray diffraction results obtained at $-120°C$ reveal a slightly distorted trigonal bypyramidal geometry in which the average Fe—C(ax) and Fe—C(eq) distances are 1.814(2) Å and 1.812(3) Å, respectively.

*bridging CO ligands.* Both the Fe—C and C—O distances of the bridging CO ligands are slightly longer than the corresponding distances of the terminal CO ligands. An Fe—Fe distance of 2.523 Å is interpreted as a normal Fe—Fe single bond length. The analogous osmium complex is believed to have only one doubly bridging CO ligand although this structure has not been confirmed by x-ray or neutron diffraction.

Two isomers of $Co_2(CO)_8$ have been observed: In the solid state, an isomer having two doubly bridging CO ligands is observed; in solution, a structural isomer having all terminal CO ligands is in equilibrium with the bridged isomer. Both isomers possess a covalent Co—Co single bond. Solution infrared data indicate a very small enthalpy difference between these bridged and terminal isomers. The structural preference for the bridged isomer in the solid state presumably results from a more favorable packing of the molecules in the crystalline lattice.

The structures of selected polynuclear metal carbonyls are shown in table 2.4. The $Fe_3(CO)_{12}$ geometry is derived from $Fe_2(CO)_9$ by replacing one of the doubly

**TABLE 2.3** Molecular structures of binuclear metal carbonyl complexes, $M_2(CO)_y$.

| $M_2(CO)_y$ | Structure | Average M—C distances (Å) | | Average M—M distances (Å) | |
|---|---|---|---|---|---|
| $Mn_2(CO)_{10}$ $Tc_2(CO)_{10}$ $Re_2(CO)_{10}$ | 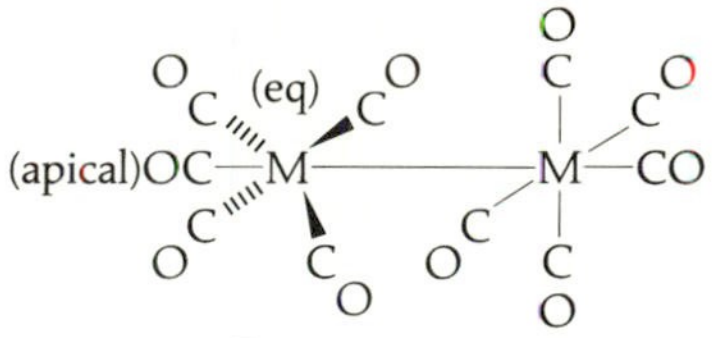 | Mn—C(apical) Mn—C(eq) Tc—C(apical) Tc—C(eq) | 1.79(2) 1.83(2) 1.90(1) 2.00(2) | Mn—Mn Tc—Tc Re—Re | 2.923(3) 3.036(6) 3.02(1) |
| $Fe_2(CO)_9$ | 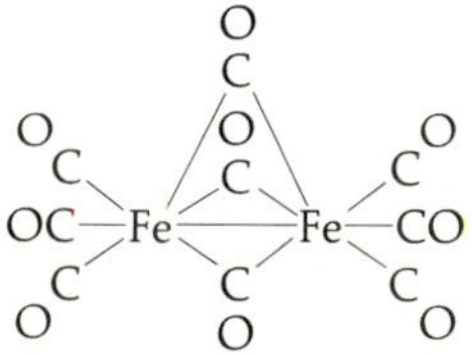 | Fe—C(terminal) Fe—C(bridging) | 1.838(3) 2.016(3) | Fe—Fe | 2.523(1) |
| $Os_2(CO)_9$ | 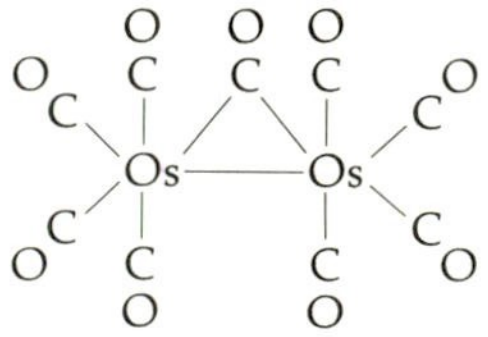 | IR only | | | |
| $Co_2(CO)_8$ (solid) | 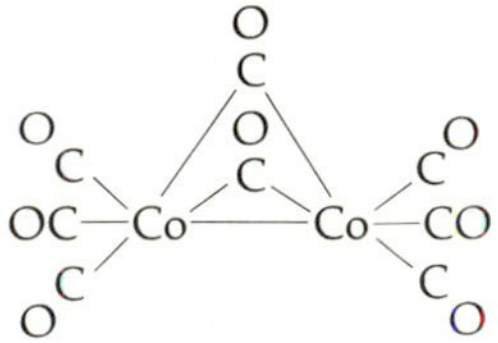 | Co—C(terminal) Co—C(bridging) | 1.80(2) 1.90(2) | Co—Co | 2.522(2) |
| $Co_2(CO)_8$ (soln) | 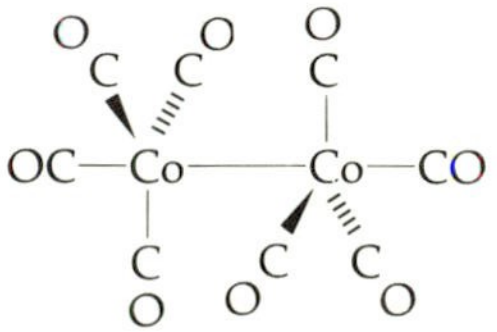 | | | | |

25

**TABLE 2.4** Molecular structures of selected polynuclear metal carbonyls, $M_x(CO)_y$.

| $M_x(CO)_y$ | Structure | Average M—C distances (Å) | | Average M—M distances (Å) | |
|---|---|---|---|---|---|
| $Fe_3(CO)_{12}$ | | Fe—C(terminal) | 1.82 | $(Fe—Fe)^a$ | 2.558(1) |
| | | Fe—C(bridging) | 2.05 | $(Fe—Fe)^b$ | 2.680 |
| $Ru_3(CO)_{12}$ | | Ru—C(axial) | 1.942(4) | Ru—Ru | 2.852(1) |
| | | Ru—C(eq) | 1.921(5) | | |
| $Os_3(CO)_{12}$ | | Os—C(axial) | 1.946(6) | Os—Os | 2.877(3) |
| | | Os—C(eq) | 1.912(7) | | |
| $Co_4(CO)_{12}$ | | Co—C(terminal) | 1.83 | $(Co—Co)^a$ | 2.50 |
| | | Co—C(bridging) | 2.04 | $(Co—Co)^b$ | 2.48 |
| $Rh_4(CO)_{12}$ | | Rh—C(terminal) | 1.96 | $(Rh—Rh)^a$ | 2.72 |
| | | Rh—C(bridging) | 1.99 | $(Rh—Rh)^b$ | 2.75 |

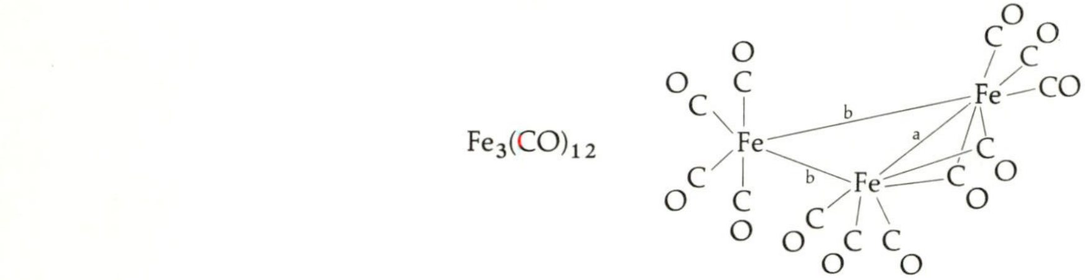

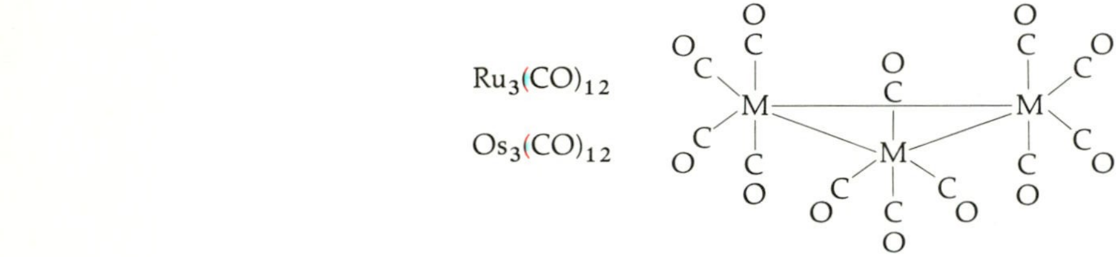

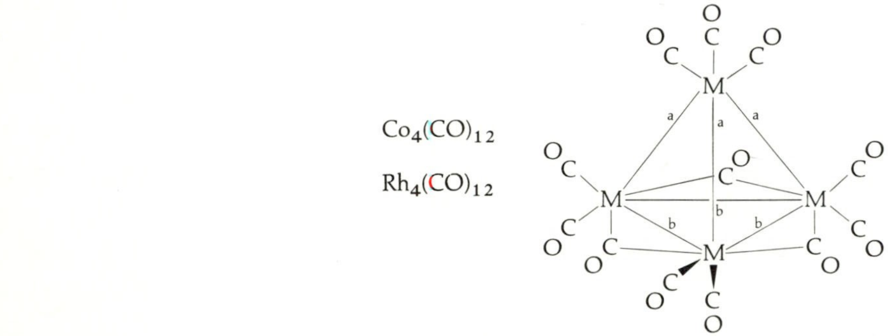

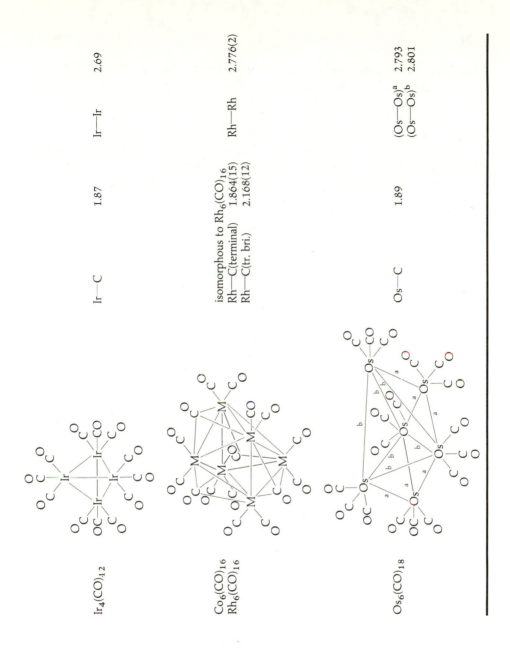

Ir₄(CO)₁₂

$Ir-Ir$    2.69

$Ir-C$    1.87

Co₆(CO)₁₆
Rh₆(CO)₁₆

$Rh-Rh$    2.776(2)

isomorphous to Rh₆(CO)₁₆
$Rh-C$(terminal)    1.864(15)
$Rh-C$(tr. bri.)    2.168(12)

Os₆(CO)₁₈

$(Os-Os)^a$    2.793
$(Os-Os)^b$    2.801

$Os-C$    1.89

bridging CO ligands with a doubly bridging $Fe(CO)_4$ moiety. The $(Fe\!-\!Fe)^a$ distance, 2.558 Å, is very similar to the $Fe\!-\!Fe$ bond length of 2.523 Å in $Fe_2(CO)_9$. However, the $(Fe\!-\!Fe)^b$ distances to the $Fe(CO)_4$ group are significantly longer and may reflect a slight lengthening of $M\!-\!M$ distances when doubly bridging CO ligands are not present. Interestingly, the analogous ruthenium and osmium complexes adopt a triangular $M_3$ core structure in which each metal has four terminal CO ligands and an approximate octahedral coordination geometry. Originally the absence of doubly bridging CO ligands for these heavier metals was attributed to the relatively long $M\!-\!M$ distances, but this conclusion has been seriously questioned recently. The $Ru\!-\!Ru$ and $Os\!-\!Os$ distances are about 0.33 Å longer than the $Fe\!-\!Fe$ bond length in $Fe_2(CO)_9$, an amount consistent with the larger covalent radii for these heavier elements.

Complexes of the type $M_4(CO)_{12}$ have structures based on a tetrahedral arrangement of the four metal atoms with each metal atom covalently bonded to the other three. The cobalt and rhodium compounds have doubly bridging CO ligands that span each $M\!-\!M$ bond in a basal plane with a unique $M(CO)_3$ group in an apical site. Each of the basal metal atoms also has two terminal CO ligands. The iridium complex $Ir_4(CO)_{12}$ has tetrahedral symmetry with three terminal CO ligands on each Ir atom. The absence of doubly bridging CO ligands in the Ir complex reflects the known tendency for heavier metals to prefer terminally coordinated CO ligands.

The six metal atoms of $Rh_6(CO)_{16}$ are located at the vertices of an octahedron with each metal atom formally bonded to four adjacent metal atoms. Each rhodium atom has two terminal CO ligands, but the four remaining CO ligands are centered above a set of alternate triangular faces of the $Rh_6$ octahedron. These CO molecules act as *triply bridging ligands.* Each carbon atom is bonded to three Rh atoms defining a triangular face. The $Rh\!-\!C$ distances to these triply bridging CO ligands are approximately 0.31 Å longer than $Rh\!-\!CO$ (terminal) distances; $C\!-\!O$ bond lengths within these bridging ligands are just slightly longer than $C\!-\!O$ distances within terminal carbonyl ligands.

The related osmium complex, $Os_6(CO)_{18}$, has a metal core structure that defines a bicapped tetrahedron. Each metal atom has three terminal CO ligands. The capping osmium atoms are bonded directly to three osmium atoms defining a face of the $Os_4$ tetrahedron, and the Os atoms of the $Os_4$ tetrahedral core are bonded directly to each other and to either one or both of the capping Os atoms as shown.

Most of the above structural data are obtained from single crystal x-ray diffraction studies. The molecular structures shown represent *idealized* descriptions of the actual molecular geometries. In these idealized molecular geometries, the gross structure of a particular molecule can be described simply. Major structural features can be shown while neglecting any minor, but perhaps statistically significant, deviations of bond distances or angles from the values required in the more symmetrical idealized structure. We have already encountered one such structural simplification regarding the definition of "linear" terminal $M\!-\!CO$ ligands.

Some important structural features are obtained by analyzing the molecular geometries of these neutral binary metal carbonyls:

1. The carbonyl ligand coordinates to metal atoms through the carbon atom.

2. For a first-row transition metal, M—CO distances to terminal CO ligands are approximately 1.85 ± 0.05 Å, and M—C—O angles are usually in the range 165–180°; C—O distances are approximately 1.15 Å.

3. Doubly (**2.1**) and triply (**2.2**) bridging CO ligands, referred to as $\mu_2$-CO and $\mu_3$-CO ligands, respectively, are observed also. The M—C distances to these ligands are about 0.2 Å and 0.3 Å longer, respectively, than M—C distances to terminal CO ligands for a given M atom, and C—O bond lengths in $\mu_2$- and $\mu_3$-CO ligards appear to be slightly longer than C—O bond lengths of terminal CO ligands.

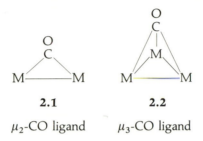

| | |
|:---:|:---:|
| **2.1** | **2.2** |
| $\mu_2$-CO ligand | $\mu_3$-CO ligand |

4. Both mononuclear and polynuclear metal carbonyls are known. Covalent metal–metal bond formation is obvious in structures like $Mn_2(CO)_{10}$, $Ru_3(CO)_{12}$ and $Ir_4(CO)_{12}$ where bridging ligands are not present. Metal–metal bonds are usually assumed to be present in metal carbonyls where the metal atoms are connected via bridging CO ligands. The M—M single bond distances usually fall in the range 2.4–2.9 Å. Polynuclear molecules that possess extensive M—M bonding, such as the $M_3$, $M_4$ and $M_6$ cores of the above carbonyl complexes, are referred to as *clusters*.

Many of these structural features are common to other metal carbonyl compounds also, as we shall discuss in subsequent chapters. Understanding how these structural features arise through chemical bonding is of paramount importance. The next section addresses the following questions raised by these structural data: (1) How can we rationalize the stability of these observed structures? (2) How can we describe M—CO bonding to a terminal CO ligand? (3) How can we describe M—CO bonding to bridging CO ligands? (4) How should we formulate the chemical bonding and rationalize the molecular structures of clusters? The first two of these can be answered rather well, but the answers to the last two questions are quite controversial and require a theoretical interpretation of subtle structural differences and a theoretical definition of a metal–metal bond.

## CHEMICAL BONDING IN METAL CARBONYLS

We rationalize the thermal stability of the metal carbonyls by demonstrating that these structures obey the EAN rule. All of the mononuclear metal carbonyls shown in table 2.2 obey the EAN rule except for $V(CO)_6$, which has only 17 electrons in the valence shell of the vanadium atom. As expected, since $V(CO)_6$ is a radical, it has unusually high chemical reactivity compared to the other mononuclear carbonyls. A more stable form of $V(CO)_6$ is $V(CO)_6^-$, which is an 18-electron complex.

   Although binuclear metal carbonyls exhibit a diversity of structural types, all obey the EAN rule as shown below:

| | Complex | | Electron count on M | | | |
|---|---|---|---|---|---|---|
| Group | Structural isomeric types | $M^0$ | CO (terminal) | CO (bridging) | M—M | Total |
| Mn | $(OC)_5M$—$M(CO)_5$ | 7 | 10 | — | 1 | 18 |
| Fe | $(OC)_3M(\mu_2\text{-}CO)_3M(CO)_3$ | 8 | 6 | 3 | 1 | 18 |
| | $(OC)_4M(\mu_2\text{-}CO)M(CO)_4$ | 8 | 8 | 1 | 1 | 18 |
| Co | $(OC)_3M(\mu_2\text{-}CO)_2M(CO)_3$ | 9 | 6 | 2 | 1 | 18 |
| | $(OC)_4M$—$M(CO)_4$ | 9 | 8 | — | 1 | 18 |

   Each of the above structural isomers has equivalent metal atoms. An M—M single bond is a normal two-center, two-electron covalent bond, so each metal atom is counted as a one-electron donor to the other metal atom. Also, the Fe and Co carbonyls show that an M—M bond having a terminal CO ligand on each metal atom is electronically equivalent to a structural isomer in which the two carbonyl ligands are doubly bridging as shown in **2.3** and **2.4**.

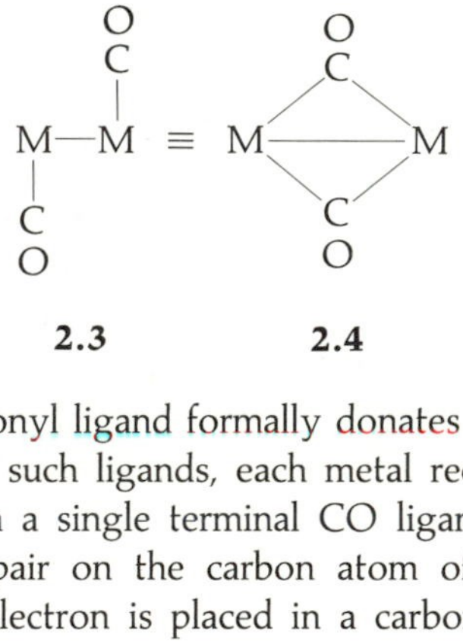

              **2.3**              **2.4**

   A doubly bridging carbonyl ligand formally donates one electron to each metal atom. Since there are two such ligands, each metal receives the same number of electrons as it would from a single terminal CO ligand. In this analysis, the *sp*-hybridized lone electron pair on the carbon atom of a terminal CO ligand is unpaired (**2.5**), and each electron is placed in a carbon $sp^2$ hybrid orbital in the $\mu_2$-CO ligand (**2.6**).

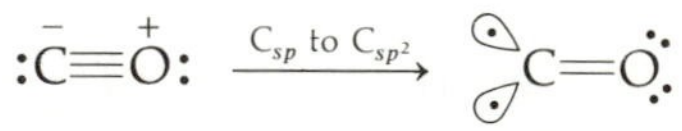

         **2.5**                    **2.6**

   Although this "ketonic" description of a bridging CO ligand is too simplistic, it is adequate for electron counting. The question of whether two metal atoms adopt terminal or doubly bridging CO ligands *cannot* be answered using the EAN rule. Clearly a more sophisticated theory is needed.

Polynuclear metal carbonyls containing up to four metal atoms can also obey the EAN rule:

| | *Complex* | | *Electron count on M* | | | |
|---|---|---|---|---|---|---|
| *Group* | *Structural isomeric types* | $M^0$ | *CO (terminal)* | *CO (bridging)* | *M—M* | *Total* |
| Fe | $\{(OC)_3M(\mu_2\text{-}CO)_2$ | 8(M) | 6 | 2 | 2 | 18 |
| | $[\mu_2\text{-}M'(CO)_4]M(CO)_3\}$ | 8(M') | 8 | — | 2 | 18 |
| | $[M(CO)_4]_3$ | 8 | 8 | — | 2 | 18 |
| Co | $\{[(OC)_2M(\mu_2\text{-}CO)]_3$ | 9(M) | 4 | 2 | 3 | 18 |
| | $[\mu_3\text{-}M'(CO)_3]\}$ | 9(M') | 6 | — | 3 | 18 |
| | $[M(CO)_3]_4$ | 9 | 6 | — | 3 | 18 |

The same basic structural features (M—M bonding, terminal and/or doubly bridging CO ligands) are observed in these complexes. Polynuclear metal carbonyls having more than four metal atoms frequently violate the EAN rule when it is applied to each metal atom. For these molecules an electron counting scheme that recognizes the unique core of metal atoms has been developed. However, our discussion of the electronic and molecular structures of these larger clusters must await a more detailed analysis of M—CO chemical bonding.

At first glance, the high thermal stability of metal carbonyls is unexpected. The relative Brønsted basicity of CO is very low compared to that of classical coordination ligands such as amines, alkoxide or halide anions or cyanide, yet M—CO bonds appear to be quite strong. From calorimetric measurements of the heat of formation of metal carbonyls in the vapor phase, it is possible to calculate the total M—C and M—M bond disruption enthalpy, $\Delta H_D$ (eq. *2.21*),

$$M_x(CO)_y(g) \longrightarrow xM^0(g) + yCO(g) \qquad \Delta H_{rx}^{298} = \Delta H_D \qquad (2.21)$$

where $M^0$ and CO are in their ground electronic states. For a mononuclear complex such as $Cr(CO)_6$, the Cr—CO (terminal) bond enthalpies are the sole contributors to $\Delta H_D$. An average Cr—CO enthalpy is evaluated easily as $\Delta H_D/6$. For polynuclear complexes M—CO (bridging) and M—M bond enthalpies also make a contribution to $\Delta H_D$. By assuming that M—CO (terminal) bond enthalpies are constant for a given metal, we can calculate M—CO (bridging) and M—M bond enthalpies from $\Delta H_D$ data. Table 2.5 lists the individual bond enthalpy contributions for several complexes.

Terminal M—CO bond enthalpies range from 24 to 45 kcal/mole and are greater than the 23 kcal/mole dissociation energy of the B—C bond in $H_3B\cdot CO$. This borane adduct is the most stable known CO complex in which the Lewis acid is not a transition metal. Clearly, we need to develop a bonding theory to explain these large M—CO (terminal) bond enthalpies.

TABLE 2.5   Selected bond enthalpy contributions of metal carbonyls to the total enthalpy of disruption, $\Delta H_D$.

| Complex | *Average bond enthalpy (kcal/mole)* | | |
| --- | --- | --- | --- |
| | M—CO (terminal) | M—CO (doubly bridging) | M—M |
| $Cr(CO)_6$ | 25.8 | | |
| $Mo(CO)_6$ | 36.3 | | |
| $W(CO)_6$ | 42.5 | | |
| $Mn_2(CO)_{10}$ | 23.9 | | 16.0 |
| $Re_2(CO)_{10}$ | 44.7 | | 30.6 |
| $Fe(CO)_5$ | 28.0 | | |
| $Fe_2(CO)_9$ | 28.0 | 15.3 | 19.6 |
| $Fe_3(CO)_{12}$ | 28.0 | 15.3 | 19.6 |
| $Ru_3(CO)_{12}$ | 41.1 | | 28.0 |
| $Os_3(CO)_{12}$ | 45.4 | | 31.1 |
| $Co_2(CO)_8$ | 32.5 | 16.3 | 19.8 |
| $Co_4(CO)_{12}$ | 32.5 | 16.3 | 19.8 |
| $Rh_4(CO)_{12}$ | 39.7 | 19.8 | 27.2 |
| $Ir_4(CO)_{12}$ | 45.4 | | 31.1 |
| $Ni(CO)_4$ | 35.1 | | |

Data taken from J. A. Connor, *Topics Current Chem.* **71**, 71 (1977).

Simple Lewis structures of an M—CO interaction suggest a reason for this unusual M—C bond stability. In this bonding formalism, a metal atom is a Lewis acid since it has empty $d$ orbitals or $d^x sp^y$-hybrid orbitals in its valence shell (**2.7**), and a CO molecule is a Lewis base because of the lone electron pair in the valence shell of the carbon atom (**2.8**).

$$\bigcirc M^0 \bigcirc \qquad\qquad :\overset{-}{C}\!\!\equiv\!\!\overset{+}{O}: \longleftrightarrow :C\!\!=\!\!\overset{..}{\underset{..}{O}}:$$

**2.7**                                                    **2.8**

Carbon monoxide is predicted to act as a carbon donor because the lower electronegativity of carbon makes its lone electron pair more basic. A coordinate-covalent bond between M and CO can be formed (**2.9**) and represented electronically by Lewis structure **2.10**.

$$M^0 \longleftarrow \overset{-}{C}\!\!\equiv\!\!\overset{+}{O}: \;\equiv\; \overset{-}{M}\!\!-\!\!C\!\!\equiv\!\!\overset{+}{O}: \longleftrightarrow M\!\!=\!\!C\!\!=\!\!\overset{..}{\underset{..}{O}}:$$

**2.9**                          **2.10**                          **2.11**

Upon realizing that the metal atom has several $d$ electrons in its valence shell and that oxygen is more electronegative than M, we can write the nonpolar resonance structure **2.11**. These resonance structures represent a formal *back bonding* of electron density on M to the CO ligand. From a resonance hybrid, the M—C and C—O bond orders should be in the ranges 1.0–2.0 and 2.0–3.0, respectively. Unfortunately, such Lewis formulations lack the orbital descriptions of chemical bonding needed to establish specific bonding interactions.

Let us examine metal–carbonyl bonding in more detail using $Cr(CO)_6$ as an example. Figure 2.1 shows a qualitative bonding scheme for a Cr—CO interaction.

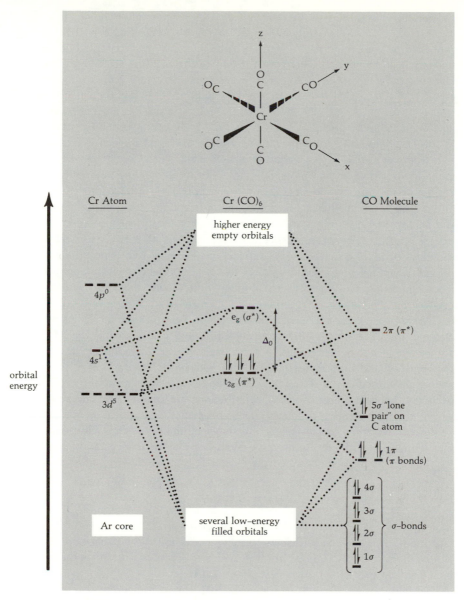

**FIGURE 2.1**   A qualitative bonding scheme for $Cr(CO)_6$.

The orbital energies represent a simplification of the results obtained from a molecular orbital (MO) calculation.

A chromium atom has an argon inner core and a valence shell consisting of $3d$, $4s$ and $4p$ atomic orbitals, which together contain a total of six electrons. A CO molecule has seven filled molecular orbitals. The $1\sigma$-$4\sigma$ orbitals are essentially core orbitals that define both C—O $\sigma$ bonding, and the lone pairs of electrons on the oxygen atom, and the two $1\pi$ orbitals define C—O multiple bonding. The *highest occupied molecular orbital, HOMO,* on CO is the $5\sigma$ MO. This orbital has a large contribution from the carbon $2s$ atomic orbital, and a small contribution from an oxygen $2p$ atomic orbital. Therefore, the two electrons in the $5\sigma$ MO represent the lone electron pair on the carbon atom in resonance structure **2.8.** A rather unusual aspect of the $5\sigma$ MO is that electron density in this MO is slightly *antibonding* with respect to C—O bond formation. The $2\pi$ molecular orbitals are the *lowest-energy unoccupied molecular orbitals, LUMOs,* of CO. They are antibonding counterparts to the $1\pi$ orbitals and are frequently referred to as the $\pi^*$ molecular orbitals.

In forming $Cr(CO)_6$, we permit orbital interactions to occur between a chromium atom and *six* carbon monoxide molecules located at the vertices of an octahedron at the appropriate Cr—C distance. Appreciable orbital interaction occurs only between orbitals having similar spatial distributions (for good orbital overlap) and similar energies (for good energy match). For these reasons, we can neglect orbital interaction between valence and core orbitals or between orbitals having $\sigma$ and $\pi$ spatial distributions. Also, interactions between two filled orbitals do not provide much net bonding energy. In $Cr(CO)_6$, important chemical bonding arises from interacting Cr valence shell atomic orbitals with the CO "valence shell" orbitals ($5\sigma$ and $2\pi$).

By referring to figure 2.1, we realize that the Cr—CO($5\sigma$) interactions *destabilize* and *split* the Cr $d$ orbitals into the expected $t_{2g}$ and $e_g$ levels while the $5\sigma$ orbitals are *lowered* in energy. These interactions result in a *net* lowering of the energy of the complex, and they represent Cr ← CO $\sigma$-*donation.* The $t_{2g}$ and $e_g$ orbitals are mainly composed of Cr $3d_{xz}$, $3d_{yz}$, $3d_{xy}$, and $3d_{z^2}$, $3d_{x^2-y^2}$ atomic orbitals, respectively, but some ligand orbital contributions are present also. Because strong Cr—C covalent bonding produces a large octahedral ligand field splitting, $\Delta_O$, the six valence electrons on chromium completely fill the $t_{2g}$ orbitals. However, notice that the filled $t_{2g}$ orbitals are $\pi$-type orbitals just like the empty $2\pi$ orbitals on the CO ligands. Interactions between these orbitals transfer electron density from the Cr atom to the CO ligands. These interactions represent the Cr → CO $d\pi$-$p\pi^*$ *back bonding.*

Figure 2.2 depicts Cr—CO bonding interactions for *one* Cr—CO bond. *An equivalent $3d_{yz} \rightarrow \pi^*_{yz}$ back bonding interaction occurs perpendicular to the one shown.* In fact, the six $t_{2g}$ electrons participate to some degree in Cr → CO back bonding with all 12 CO ($\pi^*$) orbitals. Figure 2.2 now provides an orbital description of the M → CO back bonding predicted by Lewis structure **2.11.**

Inspection of figure 2.1 also reveals that the chromium atom and carbon monoxide molecules in $Cr(CO)_6$ are in excited electronic states relative to their respective ground state. This realization suggests that the average Cr—CO bond enthalpy in $Cr(CO)_6$ might be represented best by equation 2.22.

$$Cr(CO)_6 \longrightarrow Cr^* + 6CO \qquad \Delta H_{rx} = \Delta H_D^* \qquad (2.22)$$

where CO is in its ground state, but where Cr remains in its excited valence state as though it were still in the complex. The bond enthalpy $\Delta H_D^*$ is $\Delta H_D$ (eq. *2.21*)

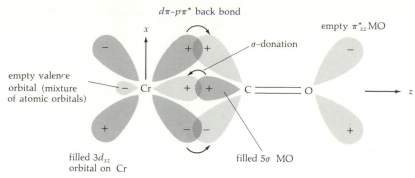

**FIGURE 2.2**   A pictorial orbital description of Cr—CO bonding in Cr(CO)$_6$ in the $xz$ plane.

plus the valence state promotion energy of a Cr atom in Cr(CO)$_6$, which is estimated to be 38.0 kcal/mole. A Cr—CO bond enthalpy is now $\Delta H_D*/6$ or 32.1 kcal/mole. The significance of this calculation will become more evident in chap. 3 when we discuss ligand substitution reactions. Most chemical reactions of Cr(CO)$_6$ and related compounds involve the retention of the metal atom *in a complex.* Therefore thermodynamic analyses of these complexes should incorporate the valence state promotion energy of the metal atom.

The M—CO bonding described above is the first type of *synergistic chemical bonding* we have encountered. This term implies that M ← CO σ-donation of electron density works together with M → CO *d*π-*p*π* back bonding to afford a stronger M—CO chemical bond than would be expected via either bonding interaction by itself. As a CO ligand donates electron density to M, the amount of electron density on M increases with a concomitant decrease in electron density on the CO. This shift of electron density reduces the electronegativity of M and increases the electronegativity of CO, thus facilitating *d*π-*p*π* back bonding. However, *d*π-*p*π* back bonding increases the amount of electron density on the CO ligand and thereby enhances the σ-donation of electron density to M. Eventually electron flow between M and CO "equilibrates" to the proper bond polarity.

Synergistic bonding involving CO ligands explains several observations: that CO appears to act as a "stronger Lewis base" than expected; that multiple bonding can occur between M and CO, and that CO coordinates preferentially to metals in low formal oxidation states (i.e., those metals with *d*-electron density available for back bonding).

The concept of synergistic bonding also allows us to make some predictions. For example, we would predict (1) that M—C distances to CO ligands should be shorter than expected for M—C(*sp*) single bonds; (2) that carbonyl C—O bonds should undergo a net bond weakening due to *d*π-*p*π* back bonding; and (3) that any alteration of the electronegativity of the M atom, such as changing the charge on the complex or introducing more basic ancillary ligands, will affect the synergistic bonding to CO ligands. Much theoretical and physical evidence support these predictions.

Although this qualitative description of M—CO bonding is conceptually simple, a quantitative analysis requires detailed MO calculations. Several MO calculations have been performed for $Cr(CO)_6$. An *ab initio* calculation reported in 1971 revealed very significant $Cr \rightarrow CO$ $\sigma$-donation and $Cr \rightarrow CO$ back bonding. However in 1977 another calculation method indicated that Cr—CO bonding is predominantly sigma in character with only a small $\pi$-contribution. Apparently, $Cr(3d\pi) \rightarrow CO(2\pi)$ back bonding affords $M \rightarrow CO$ *charge transfer*, which strongly affects the C—O bonds but not the Cr—C bonds. In 1980 two calculations refuted this latter claim and reported results indicating that Cr—CO $\pi$ back bonding makes a significant contribution to Cr—C bonding. These calculations estimate that 1.4 to 1.5 electrons are involved in the $Cr \leftarrow CO(5\sigma)$ $\sigma$-donation while 0.3 to 0.5 electrons are back bonded from $Cr(3d)$ orbitals into the $CO(2\pi)$ orbitals of *each* CO ligand. Such apparently diverse results from different methods of calculation are fairly common—although quite disconcerting to the chemist seeking a clear understanding of the chemical bonding.

Experimental values for the electron population of the $5\sigma$ and $2\pi$ molecular orbitals of each CO ligand are 1.65 and 0.38 electrons, respectively, according to x-ray and neutron diffraction studies. Estimated atomic charges indicate that the chromium and carbon atoms are slightly positively charged, while the oxygen atoms are negatively charged; however, Cr—CO back bonding lowers the positive charge on the C atom relative to its charge in free CO. Presumably, the presence of significant $Cr \rightarrow CO$ $d\pi$-$p\pi^*$ back bonding has survived the test of detailed theoretical calculations, and M—CO back bonding is generally accepted by organometallic chemists.


## PHYSICAL EVIDENCE RELATED TO M—CO BONDING

Physical evidence supporting synergistic M—CO bonding is usually obtained from x-ray structural data or from IR stretching frequencies of CO ligands. Significant M—C double bonding should shorten the M—C distance relative to an M—C($sp$) single-bond length, and population of the $CO(2\pi)$ antibonding MO via $M \rightarrow CO$ back bonding should lengthen the C—O bond relative to the C—O distance in free CO. Notice, however, that the extent of C—O bond lengthening due to $d\pi$-$p\pi^*$ back bonding is reduced *slightly* by $M \leftarrow CO$ $\sigma$-donation because the $5\sigma$ MO is slightly antibonding.

Can experimentally measured M—C and C—O distances provide a diagnostic indication of bond order for these bonds? Figure 2.3 shows that Cr—C($sp$) and C($sp$)—O bond lengths change dramatically only for bonds of *low* bond order. For highly multiple bonds, bond length is nearly independent of bond order, and the precision of bond length measurements limits any useful interpretation of bond order. Therefore, bond distances are useful in determining the multiplicity of M—C bonds, but other methods must be found to probe changes in C≡O bond orders.

Since M—C distances are sensitive to the covalent radius of M *in a particular ligand environment*, it is best to compare M—C bond lengths for different types of

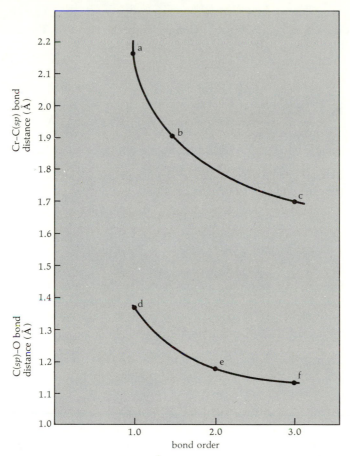

<sup>a</sup>Derived from *fac*–Cr(CO)₃(dien). <sup>b</sup>From Cr(CO)₆. <sup>c</sup>From *trans*–I(OC)₄Cr ≡ C–CH₃.
<sup>d</sup>Derived from CH₃OH. <sup>e</sup>From CO₂. <sup>f</sup>Average of C–O distances in CO(g) and CO⁺(g).

**FIGURE 2.3**  Correlations of Cr—C(*sp*) and C(*sp*)—O bond distance with bond order.

M—C bonding *within the same molecule.* An x-ray structure of the anionic (*n*-propyl)-Fe(CO)₄⁻ complex (**2.12**) reveals two different Fe—C distances.

$$
\begin{array}{c}
CH_3 \\
| \\
CH_2 \\
| \\
CH_2 \\
| \quad \,_{\backslash\backslash\backslash}CO \\
OC-Fe\!\!\blacktriangleleft\!\!CO \\
| \\
C \\
O
\end{array}
$$

**2.12**

The Fe—CO distances of 1.75(2) Å are all equal within experimental error, but the Fe—propyl distance of 2.20(2) Å is much longer. Since the Fe—$CH_2R$ distance *must* represent a single-bond length, a single-bond covalent radius for the Fe atom *in this molecule* can be calculated easily using the known bonding radius of a $C(sp^3)$ atom.

$$
\begin{array}{ll}
\text{Fe—}C(sp^3) \text{ single-bond distance} & = 2.20 \text{ Å} \\
- \text{[Single-bond covalent radius for } C(sp^3) & = 0.77 \text{ Å]} \\
\hline
\text{Single-bond covalent radius of Fe in } \mathbf{2.12} = 1.43 \text{ Å}
\end{array}
$$

An Fe—$C(sp)$ single-bond length for this molecule can be estimated similarly:

$$
\begin{array}{ll}
\text{Single-bond covalent radius of Fe in } \mathbf{2.12} & = 1.43 \text{ Å} \\
+ \text{[Single-bond covalent radius for } C(sp) & = 0.70 \text{ Å]} \\
\hline
\text{Estimated Fe—}C(sp) \text{ single-bond distance in } \mathbf{2.12} = 2.13 \text{ Å}
\end{array}
$$

The observed Fe—CO distance of 1.75 Å in **2.12** is 0.38 Å *shorter* than the calculated Fe—$C(sp)$ single-bond length, thereby substantiating considerable Fe—CO multiple bonding.

Similar analysis can be performed on any complex that contains CO and alkyl ligands. For example, in a series of $(\eta^5\text{-}C_5H_5)Mo(CO)_3R$ complexes (see chap. 4 for a discussion of the $\eta^5\text{-}C_5H_5$ ligand), where R is an alkyl ligand, the Mo—R and Mo—CO distances are 2.38 Å and 1.99 Å, respectively. Repeating the above calculations gives Mo—CO bond lengths 0.32 Å *shorter* than the calculated Mo—$C(sp)$ single-bond length in these complexes.

The observation of short M—CO distances relative to M—$C(sp)$ single-bond lengths is strong evidence of M—CO multiple bonding. A highly polarized $\sigma$ bond would probably not be able to account for the observed shortening of 0.3–0.4 Å. Known M—CO distances are usually interpreted as corresponding to bond orders of about 1.5. In $Cr(CO)_6$ there are six $3d$ electrons in the $t_{2g}$ orbitals that overlap with the two $2\pi$ MOs of each CO ligand. By assuming a maximum back bonding of all six Cr $3d$ electrons to the six CO ligands, the Cr—CO $\pi$-*bond order* would be 0.5 (a total of one $\pi$ electron back bonded to each CO ligand). Since the Cr—CO $\sigma$ bond is assumed to have a bond order of 1.0, the total Cr—CO bond order in $Cr(CO)_6$ is then 1.5. However, we still need physical evidence to support $d\pi$-$p\pi^*$ back bonding into the $2\pi$ orbitals of CO ligands. This evidence should demonstrate a weakening of the C—O bond when such back bonding occurs.

The C—O distances in gaseous CO and $CO^+$ are 1.128(2) Å and 1.115(1) Å, respectively. The $CO^+$ ion is formed by removing an electron from the slightly antibonding $5\sigma$ MO. A C—O triple-bond length is, therefore, about 1.12 Å. The C—O distance of 1.17(1) Å in ketene, $H_2C{=}C{=}O$, approximates a $C(sp)$—O double-bond length. Since essentially all terminal carbonyl C—O distances fall in the narrow range of 1.13(2)–1.16(2) Å, the use of C—O bond lengths for bond order determinations is not possible (see fig. 2.3). Structural data indicate a slight elongation of a carbonyl C—O distance upon coordination, but experimental uncertainty precludes detailed comparisons between different organometallic complexes.

Variations of C—O bond order are conveniently monitored by IR spectroscopy using C—O stretching vibrational frequencies. For terminal CO ligands in neutral complexes, these vibrations occur in the range 2125–1850 cm$^{-1}$, a region of the IR spectrum normally free of absorptions by organic functional groups. Hooke's law, as applied to gaseous CO shows that the C—O stretching frequency, $v_{CO}$, is a simple function of the C—O force constant, $k_{CO}$, and the reduced mass, $\mu_{CO}$, (eq. 2.23).

$$v_{CO} = \frac{1}{2\pi c} \sqrt{\frac{k_{CO}}{\mu_{CO}}} \qquad (2.23)$$

The value of the force constant represents the force required to stretch the C—O bond by a unit distance and, therefore, is an indirect probe of C—O bond order. As C—O bond order decreases, the value of $k_{CO}$ should decrease, and the C—O stretching frequency should appear at lower energies.

For polyatomic molecules such as metal carbonyls, the simple relationship between $v_{CO}$ and $k_{CO}$ shown in equation 2.23 must be modified. A given CO ligand will vibrate anharmonically and may "couple" to other vibrational modes present within the molecule. Strong coupling occurs between oscillating CO ligands, and weaker coupling between M—C and C—O stretching vibrations is observed also. Interactions between CO ligands are helpful in determining molecular geometry as we shall discuss in chap. 3. Although computational methods are available for extracting CO force constants from observed $v_{CO}$ frequencies of complex molecules, we shall find that within a series of similar complexes, the observed C—O stretching frequencies still provide a direct measure of CO bond order.

Figure 2.4 shows an experimental correlation of $v_{CO}$ to CO bond order. Notice that $v_{CO}$ provides a sensitive probe for measuring variations in the C—O bond order even at high bond multiplicities. For example, the frequencies of a CO stretching vibration (which usually appears as an intense and narrow band in the IR spectrum) can be measured conveniently to within $\pm 5$ cm$^{-1}$ using standard IR spectrometers; yet, the $v_{CO}$ range for terminal CO ligands in neutral complexes is very large (ca. 275 cm$^{-1}$). This gives us a sensitive probe for monitoring changes in C—O bond multiplicity as we make minor electronic changes at the metal atom.

The $v_{CO}$ frequencies for CO(g) and CO$^+$(g) are 2170 cm$^{-1}$ and 2214 cm$^{-1}$, respectively. Since these frequencies are higher than those of coordinated CO molecules, the presence of $d\pi$-$p\pi^*$ back bonding is already indicated. This interpretation would be substantiated strongly by observing changes in $v_{CO}$ as predicted from synergistic bonding theory.

Consider three hexacarbonyl metal complexes—Mn(CO)$_6$$^+$, Cr(CO)$_6$, V(CO)$_6$$^-$. These complexes form an isostructural and isoelectronic series since each complex has octahedral geometry and contains a $3d^6$ metal atom. Their M—C distances are similar, and the number and types of M—CO orbital overlaps are identical. By recalling that M ← CO $\sigma$-donation is relatively weak in the absence of $d\pi$-$p\pi^*$ back bonding, we can predict the relative importance of $d\pi$-$p\pi^*$ back bonding in these complexes from simple electrostatic considerations.

The $2\pi$ orbitals of each CO ligand act as Lewis acids (electron density acceptors) when participating in M—CO back bonding. If we assume a constant intrinsic $\pi$-acidity for CO ligands, then the degree of M—CO $d\pi$-$p\pi^*$ back bonding in each

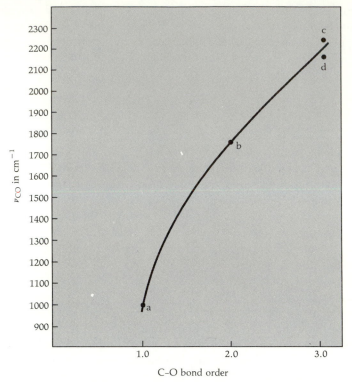

$^a(CH_3)_2O.$ $^b$Acetone. $^cCO^+(g).$ $^dCO(g).$

**FIGURE 2.4**  Experimental correlation between $\nu_{CO}$ and C—O bond order neglecting the formal hybridization of the carbon atom.

complex depends only on the "availability," or orbital electronegativity, of the $3d$ electron density on M. It should be most difficult to remove negative charge from $Mn^+$ and least difficult to remove negative charge from $V^-$. Therefore, we would predict that M—CO $d\pi$-$p\pi^*$ back bonding should *increase* in the direction $Mn(CO)_6^+ < Cr(CO)_6 < V(CO)_6^-$, and the $\nu_{CO}$ frequencies should *decrease* in the direction $Mn(CO)_6^+ > Cr(CO)_6 > V(CO)_6^-$. The following observed $\nu_{CO}$ frequencies support this prediction:

| Complex | $\nu_{CO}(cm^{-1})$ |
|---|---|
| $Mn(CO)_6^+$ | 2090 |
| $Cr(CO)_6$ | 2000 |
| $V(CO)_6^-$ | 1860 |

Similar arguments can be made to explain the observed trend in CO stretching frequencies for the isostructural and isoelectronic series $Ni(CO)_4$, $Co(CO)_4^-$ and

$Fe(CO)_4^{2-}$:

| Complex | $\nu_{CO}(cm^{-1})$ |
|---------|---------------------|
| $Ni(CO)_4$ | 2046 |
| $Co(CO)_4^-$ | 1883 |
| $Fe(CO)_4^{2-}$ | 1788 |

In subsequent chapters we shall use CO stretching frequencies to determine the relative Lewis basicity and $\pi$-acidity of various types of ligands.

## CHEMICAL BONDING TO BRIDGING CO LIGANDS IN METAL CARBONYL CLUSTERS

For electron counting purposes, a $\mu_2$-CO ligand donates one electron to each metal atom to which it is bonded. A simple orbital representation invoking a formal $sp^2$ hybridized carbon atom and an unpairing of the lone electron pair on carbon (**2.13**) is compatible with this electron counting formalism.

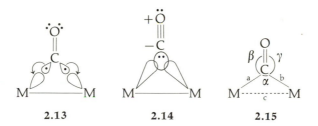

| 2.13 | 2.14 | 2.15 |

These $sp^2$ orbitals would overlap an empty valence orbital on each M atom. Alternatively, a three-center, two-electron orbital overlap description (**2.14**) can be formulated also. Such bonding schemes describe $\sigma$-bonding interactions. The two perpendicular $2\pi$ orbitals of the CO ligand can overlap filled $d$ orbitals on the metal atoms to generate $d\pi$-$p\pi^*$ back bonding as well. The net effect is fairly strong M—C bond formation and a lowering of the C—O bond order. However, a delocalized MO description is probably the only adequate bonding description for $\mu_2$-CO ligands. From chemical reactivity studies and spectroscopic data, the "ketonic" representation of $\mu_2$-CO bonding (**2.13**) is an over simplistic bonding description.

The bond enthalpy for *each* M—C bond to $\mu_2$-CO ligands (table 2.5) is approximately half the M—CO (terminal) bond enthalpy for a given metal atom. Since two M—CO bonds to $\mu_2$-CO ligands are nearly equivalent thermodynamically to a terminal M—CO bond, the energy difference between structures **2.3** and **2.4** is predicted to be slight. With $^{13}$C-NMR spectrometry it is now possible to observe facile exchange reactions between terminal and bridging CO ligands. These observations substantiate the near degeneracy of terminal and $\mu_2$-CO coordination geometries. We might also expect to observe varying degrees of unsymmetrical coordination of $\mu_2$-CO ligands if the potential energy surface for terminal-to-bridge interconversion is nearly flat. Detailed studies of molecular structural data reveal several types of unsymmetrical $\mu_2$-CO coordination.

At present there is considerable controversy over the correct interpretation of these structural data for $\mu_2$-CO ligands. A recent review of solid-state structural data for 120 complexes containing $\mu_2$-CO ligands reveals a nearly continuous range of structures for CO ligands between the terminal and symmetrical $\mu_2$-CO structural extremes. Four types of $\mu_2$-CO structures have been defined based on the relative magnitudes of the bond distances, $a$ and $b$, and bond angles, $\alpha$, $\beta$ and $\gamma$, in **2.15**. These structural types are shown in figure 2.5.

A *symmetrical* $\mu_2$-CO ligand (**2.16**) has equivalent M—C distances, and the C—O vector is perpendicular to the M—M vector. This is the idealized $\mu_2$-CO structure; the angle $\alpha$ is usually in the range 77–90°. In neutral complexes, $\nu_{CO}$ for these ligands is generally in the range 1860–1700 cm$^{-1}$. An *asymmetrical* $\mu_2$-CO ligand (**2.17**) has significantly different M—C distances, yet the C—O vector remains essentially perpendicular to the M—M vector. This structure is generated from (**2.16**) by translating the $\mu_2$-CO ligand toward one of the metal atoms. The *semi-bridging* $\mu_2$-CO ligand (**2.18**) is like an asymmetrical ligand except that the angles $\beta$ and $\gamma$ are sufficiently different that the C—O vector is now not perpendicular to the M—M vector. A *linear* $\mu_2$-CO ligand (**2.19**) has a $\beta$ angle of about

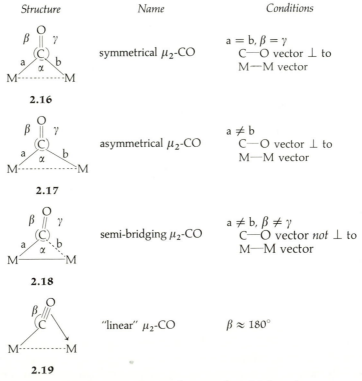

| Structure | Name | Conditions |
|---|---|---|
| **2.16** | symmetrical $\mu_2$-CO | $a = b$, $\beta = \gamma$ <br> C—O vector $\perp$ to M—M vector |
| **2.17** | asymmetrical $\mu_2$-CO | $a \neq b$ <br> C—O vector $\perp$ to M—M vector |
| **2.18** | semi-bridging $\mu_2$-CO | $a \neq b$, $\beta \neq \gamma$ <br> C—O vector *not* $\perp$ to M—M vector |
| **2.19** | "linear" $\mu_2$-CO | $\beta \approx 180°$ |

**FIGURE 2.5**   Structural types of $\mu_2$-CO ligands.

180° and is regarded as a terminal CO ligand that donates electron density from the filled C—O pi molecular orbital system to a second metal atom. This ligand acts as a formal *four-electron donor*, and it represents one of the less controversial structural types of $\mu_2$-CO coordination.

Our understanding of the chemical bonding considerations that favor one structural type over another is very limited. Most differences of opinion center around the relative importance of intra- and intermolecular forces. For example, minor structural effects observed in the solid state might be caused by intermolecular crystal packing forces rather than by intramolecular bonding forces. Such structural features would not be observed in the gas phase. Metal–metal bonding may also contribute significantly to intramolecular forces. However, since the presence of a $\mu_2$-CO ligand ensures that two metal atoms are fairly close together, the question of the actual existence of an M—M bond arises. Usually an M—M bond is assumed to accompany a $\mu_2$-CO ligand. Recently, however, a $\mu_2$-CO ligand has been observed in a palladium complex where a direct Pd—Pd bond is not considered very likely. A further complication is that different $\mu_2$-CO ligand structures are occasionally observed bridging the same two metal atoms. Clearly, the prediction and rationalization of structural preferences involving very small energy differences without an adequate bonding theory is nearly impossible.

In tables 2.3 and 2.4, $Fe_2(CO)_9$, $Co_2(CO)_8$, $Co_4(CO)_{12}$ and $Rh_4(CO)_{12}$ have symmetrical $\mu_2$-CO ligands. The $\mu_2$-CO structure observed in $Fe_3(CO)_{12}$ is shown in more detail in (**2.20**).

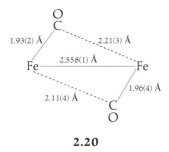

**2.20**

This structure has been described as a *compensating set of asymmetric $\mu_2$-CO ligands* which keep the two iron atoms electronically equivalent.

A semi-bridging $\mu_2$-CO ligand is observed in the anion $[FeCo(\mu_2\text{-CO})(CO)_7]^-$ (**2.21**).

$$\left[ \begin{array}{c} \overset{\displaystyle O}{\overset{\|}{\underset{\diagdown}{C}}} \\ (OC)_4Fe\text{————}Co(CO)_3 \end{array} \right]^-$$

**2.21**

This structure reveals a strong Co—C bonding interaction and a weaker Fe—C bonding interaction. An M—M bond is presumably present in *semi-bridging* $\mu_2$-CO

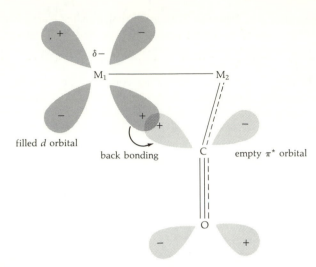

**FIGURE 2.6**   A proposed $d\pi\text{-}p\pi^*$ back bonding model for semi-bridging CO ligands.

structures since the weaker M—C interaction apparently arises from $d\pi\text{-}p\pi^*$ back bonding of excess $d$-electron density on one metal atom to the $\pi^*$ orbitals of a terminal CO ligand on the other metal atom, as shown in figure 2.6. Such back-bonding equalizes the M—M bond polarity that would be present if all CO ligands were terminal. The EAN rule helps in anticipating when these semi-bridging $\mu_2$-CO structures may occur because it predicts the formation of highly polarized M—M bonds.

We demonstrate this analysis using **2.21** as an example. In the absence of the Fe-($\mu_2$-CO) interaction, the anion has the structure

$$(OC)_4Fe \leftarrow \bar{C}o(CO)_4 \quad \text{or, that is,} \quad (OC)_4\bar{F}e\text{—}Co(CO)_4$$

where both metal atoms obey the EAN rule. Since the iron atom bears a full formal unit of negative charge, one of the terminal CO ligands on the cobalt atom accepts some of this electron density via $d\pi\text{-}p\pi^*$ back bonding into its $2\pi$ MO to become a semi-bridging $\mu_2$-CO ligand. The high polarity of the Fe—Co bond is thereby reduced. In subsequent chapters we shall encounter several more examples of asymmetric and semi-bridging $\mu_2$-CO structures.

Inspection of tables 2.3 and 2.4 also reveals that heavier metal complexes, such as $Ru_3(CO)_{12}$, $Os_3(CO)_{12}$, and $Ir_4(CO)_{12}$, prefer a nonbridged structural isomer. The usual explanation—that heavier metal atoms have M—M distances too long for efficient $\mu_2$-CO bonding—is now disputed by considerable data. There is no apparent correlation between M—M separation and the specific type of $\mu_2$-CO structure, either.

One rationalization of the observed structural differences among metal carbonyl clusters centers on the importance of intramolecular repulsive forces between carbonyls. For a given cluster, the carbonyl ligands arrange themselves in a way that minimizes these repulsive interactions and thereby defines a carbonyl polyhedron in three-dimensional space. Insertion of the metal core into the internal cavity of this carbonyl polyhedron gives the ground-state structure. The relative orientation of the metal atoms and the carbon atoms determines whether bridging or terminal carbonyl ligands are present. When a cluster has many carbonyl ligands, several carbonyl polyhedra of similar energy might be possible. In this case, the size of the metal core relative to the size of the interstitial holes of the various carbonyl polyhedra determines which structure is adopted by the ligands. Larger metal cores require large internal cavities; and so, the carbonyl polyhedron having an internal cavity of the appropriate size will be preferred.

As an example, for metal carbonyls having 12 carbonyl ligands, the two most stable carbonyl polyhedra which have 12 vertices are the icosahedron (**2.22**) and the cubooctahedron (**2.23**).

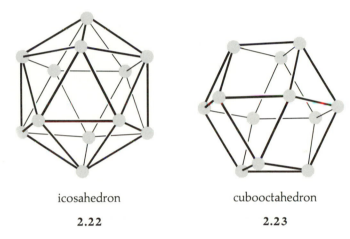

icosahedron                          cubooctahedron

**2.22**                                **2.23**

Each vertex represents a carbonyl oxygen atom. The shapes of the interstitial cavities of these polyhedra are different, and placing different sized $M_3$ or $M_4$ cores in these holes leads to different molecular structures. Large metal cores, such as $Ru_3$, $Os_3$, and $Ir_4$, fit better in the interstitial hole of the cubooctahedron affording only terminal M—CO bonding, as shown in figure 2.7 for $Os_3(CO)_{12}$. Smaller metal cores, like $Fe_3$, $Co_4$, and $Rh_4$, fit into the interstitial hole of the icosahedron, as shown in figure 2.8 for $Fe_3(CO)_{12}$. Inspection of the latter structure reveals that some of the CO ligands are equidistant from two metal atoms and, therefore, represent $\mu_2$-CO ligands. Similar arguments have been used to account for the structural preferences of $Fe_2(CO)_9$, $Os_2(CO)_9$, and other complexes. Figure 2.9 shows the structure of $Fe_2(CO)_9$ derived in this way.

In this analysis, the size of the metal core relative to the size of the interstitial cavities of the *carbonyl polyhedron* determines the observed molecular structure. Heavier metals have large covalent radii and form larger metal cores than do the lighter metals. Heavier metals therefore favor terminal CO coordination because of the larger core size and *not* because they would form especially long M—($\mu_2$-CO) bond distances. As the number of M—M bonds increases, the M—M bond enthalpy contribution may eventually surpass the total M—C bond enthalpy con-

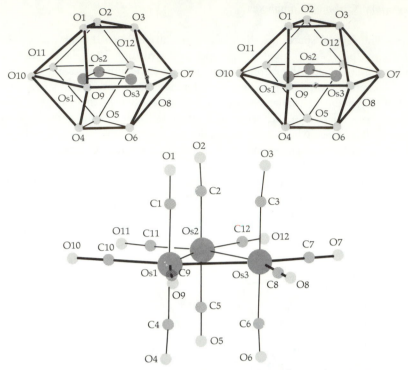

**FIGURE 2.7** Stereoscopic view showing the $Os_3$ metal core of $Os_3(CO)_{12}$ in the interstitial cavity of a cubooctahedral carbonyl polyhedron. *Source: Adapted from B. F. G. Johnson and R. E. Benfield,* Topics Stereochem., **12**, *253 (1981).*

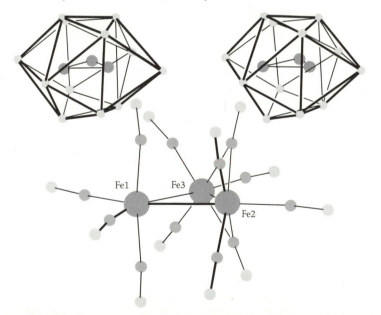

**FIGURE 2.8** Stereoscopic view of the $Fe_3$ metal core of $Fe_3(CO)_{12}$ in the interstitial cavity of an icosahedral carbonyl polyhedron. *Source: Adapted from B. F. G. Johnson and R. E. Benfield,* Topics Stereochem., **12**, *253 (1981).*

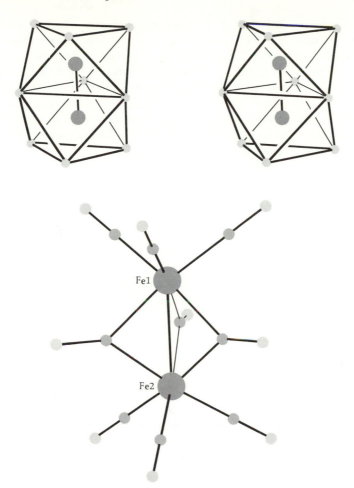

**FIGURE 2.9** The structure of $Fe_2(CO)_9$ as derived by inserting the $Fe_2$ metal core into a tricapped trigonal prismatic carbonyl polyhedron. *Source: Adapted from B. F. G. Johnson and R. E. Benfield,* Topics Stereochem., **12,** *253 (1981).*

tribution. Likewise, intramolecular repulsion between CO ligands becomes less significant, so the molecular structure may be determined by the structural preference of the metal core. We now need to develop a method for predicting the structure of a metal core.

Although a description of chemical bonding for clusters of metal atoms is best approached using MO theory, a simpler bonding theory would be helpful. Unfortunately, for clusters having five or more metal atoms application of the EAN rule frequently breaks down because of extensive M—M bonding. An empirical set of rules, known as *Wade's rules,* have been developed to facilitate electron counting and to predict metal core structures in clusters.

Wade's rules, as applied to organometallic clusters, are based upon an observed correlation between the three-dimensional structure of a metal core and the number of electron pairs available for M—M bonding. All metal core structures are derived from closed polyhedra having triangular faces, such as the tetrahedron, trigonal bipyramid, octahedron, and pentagonal bipyramid, where each metal atom occupies a vertex (figure 2.10). The choice of a particular polyhedral structure for a given metal core is made after determining the number of M—M bonding electron pairs.

Some metal clusters have fewer metal atoms than the number of vertices required by the polyhedron. In these cases, a metal core structure is derived from the predicted polyhedron by removing the appropriate number of vertices. Core structures

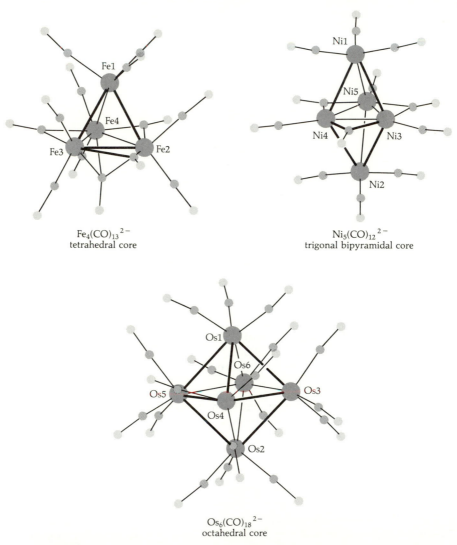

$Fe_4(CO)_{13}^{2-}$
tetrahedral core

$Ni_5(CO)_{12}^{2-}$
trigonal bipyramidal core

$Os_6(CO)_{18}^{2-}$
octahedral core

**FIGURE 2.10**  Examples of metal carbonyl clusters with different metal core geometries. *Source: Adapted from B. F. G. Johnson and R. E. Benfield,* Topics Stereochem., **12,** *253 (1981).*

are classified as *closo*, *nido* or *arachno* depending on whether there are enough metal atoms to fill, respectively, all, all minus one, or all minus two of the vertices of the predicted polyhedron. Table 2.6 illustrates these relationships. If a cluster has seven electron pairs involved in M—M bonding, then the metal core structure is based on a regular octahedron. Such clusters having six, five, or four metal atoms would have, respectively, octahedral (no vertices vacant), square pyramidal (one vertex vacant), or square planar/distorted tetrahedral (two vertices vacant) geometries for the metal core.

**TABLE 2.6   Predicted metal core structures based on number of M—M bonding electron pairs according to Wade's rules.**

| *Number of* M—M *bonding electron pairs* | *Reference polyhedron* | *Core structure for specific core sizes* | | |
|---|---|---|---|---|
| | | *closo* | *nido* | *arachno* |
| 5 | tetrahedron | 4 | 3 | 2 |
| 6 | trigonal bipyramid | 5 | 4 | 3 |
| 7 | octahedron | 6 | 5 | 4 |
| 8 | pentagonal bipyramid | 7 | 6 | 5 |
| 9 | dodecahedron | 8 | 7 | 6 |
| 10 | tricapped trigonal prism | 9 | 8 | 7 |
| 11 | bicapped Archimidean antiprism | 10 | 9 | 8 |
| 12 | octadecahedron | 11 | 10 | 9 |
| 13 | icosahedron | 12 | 11 | 10 |

The number of electron pairs involved in M—M bonding, $P_{MM}$, for a metal carbonyl cluster, $M_n(CO)_x$, is determined from equation 2.24.

$$P_{MM} = \tfrac{1}{2}[n \text{ (number of valence electrons on each M)} + 2x - 12n] \quad (2.24)$$

This formula is derived from an assumption that six of the nine valence orbitals of each M atom are involved in $\sigma$ and $\pi$ bonding to CO ligands and the remaining three orbitals are involved in M—M bonding. Since M—CO orbitals are of lowest energy, these orbitals are filled before any M—M bonding orbitals. The first two terms in equation 2.24 represent the total number of valence electrons. Subtracting the electrons needed to fill six M—CO orbitals on each metal atom from the total number of valence electrons and dividing by two, we obtain the number of valence electron pairs that remain for M—M bonding.

Consider the isoelectronic complexes $Co_6(CO)_{16}$ and $Rh_6(CO)_{16}$. If each metal atom obeyed the EAN rule, there would be $6 \times 18$, or 108, valence electrons involved in M—CO and M—M bonding. From the structures shown in Table 2.4, the octahedral metal core has 12 edges—each of which *might* represent a formal M—M single bond. By subtracting these 24 M—M bonding electrons (12 bonds × 2 electrons/bond) from 108, we calculate 84 valence electrons must be contributed by the metal atoms and the CO ligands for the complexes to obey the EAN rule. However, since each metal atom contributes 9 valence electrons for a total of 54 electrons and the 16 CO ligands contribute a total of 32 electrons, the total number of valence electrons is 86. Therefore, these complexes appear to violate the EAN rule by two electrons. *An obvious difficulty in applying the EAN rule to clusters is the formal specification of M—M bonding within the metal core.* Other difficulties arise in applying the EAN rule to individual metal atoms in clusters when $\mu_3$-CO ligands are present.

Wade's rules can be applied to $Co_6(CO)_{16}$ and $Rh_6(CO)_{16}$ *based solely on the empirical formula.* Six valence orbitals on each metal atom contain all of the M—CO bonding electron density, and these orbitals must be filled first. This means that $6 \times 6 \times 2$, or 72, of the 86 actual valence electrons are assigned to M—CO $\sigma$ and $\pi$ bonding. The remaining 14 electrons are assigned to M—M bonding, thus affording $14/2 = 7$ M—M bonding electron pairs. By consulting Table 2.6, we determine that the presence of seven M—M bonding pairs implies a metal core structure based on a regular octahedron. Since these complexes contain six metal atoms, the metal core adopts the *closo* structure—which is an octahedron. Table 2.4 reveals that this is the observed metal core geometry.

For the Co group clusters, $M_4(CO)_{12}$, where M is Co, Rh or Ir, each complex has four metal atoms. Therefore $4 \times 6 \times 2$, or 48, of the actual valence electrons must be assigned to M—CO bonding. Since the total number of valence electrons is $4 \times 9$ (provided by the metal atoms) + $12 \times 2$ (provided by the CO ligands), or 60, the number of electron pairs involved in M—M bonding is $(60 - 48)/2$ or 6. By consulting Table 2.6, we determine that the metal core structure should be derived from the trigonal bipyramid with one vertex missing. A nearly tetrahedral core geometry is predicted. The observed metal core geometries have idealized tetrahedral geometry.

Similarly, for $M_3(CO)_{12}$ clusters, where M is Fe, Ru or Os, the number of valence electrons assigned to M—CO bonding is $3 \times 6 \times 2$, or 36, electrons. Since there are $3 \times 8$, or 24, electrons provided by the metal atoms and $12 \times 2$, or 24, electrons provided by the CO ligands, the number of M—M bonding electron pairs is $(24 + 24 - 36)/2$, or 6. Table 2.6 predicts that the metal core structure will be derived from the trigonal bipyramid where two vertices are absent. By removing the two axial vertices, we generate a triangular metal core structure, which is the observed structure.

Wade's rules provide a method for predicting the metal core structure of many clusters. The method does not prescribe an unambiguous choice between various *nido* and *arachno* structures, and it does not specify which type of M—CO bonding (terminal, $\mu_2$-CO or $\mu_3$-CO) is expected. However, these rules can be extended to include other types of ligands, and recent refinements of Wade's rules have correctly rationalized very unusual metal core structures, such as that found for $Os_6(CO)_{18}$. The discovery of these applications is left to the interested reader.

## HETERONUCLEAR METAL CARBONYL COMPLEXES

Binary metal carbonyls of the type $M_a{}^1M_b{}^2 \ldots M_n{}^m(CO)_x$ which contain two or more different metals are called *heteronuclear metal carbonyls.* Some examples are $ReMn(CO)_{10}$, $FeRu_2(CO)_{12}$, $Fe_2Os(CO)_{12}$, $Co_3Rh(CO)_{12}$, $Co_2Rh_2(CO)_{12}$,

$Co_2Ir_2(CO)_{12}$, $Rh_3Ir(CO)_{12}$, and $Rh_2Ir_2(CO)_{12}$. These complexes are usually prepared via designed synthetic routes (some of which are discussed later) to avoid
difficult separation procedures that would be required if a random mixture of heteronuclear complexes were formed. As might be expected, molecular structures of these
complexes are derivatives of the related homonuclear structures. Chemical interest
in heteronuclear metal carbonyl clusters centers about the potential catalytic activity
of these compounds. Complexes having different metal atoms within the same molecule may have unusual chemical reactivity. Different metals can exhibit unique
chemistry independently, and different metals in the same molecule may produce
a cooperative effect. For example, one metal might preferentially bond to the carbon
atom of carbon monoxide, and a different metal might coordinate to the oxygen
atom, thereby doubly activating CO to subsequent chemical attack.

## STUDY QUESTIONS

1. Symmetrical and linear $\mu_2$-CO ligands have CO stretching frequencies usually
   in the ranges $1860-1700$ cm$^{-1}$ and $1640-1500$ cm$^{-1}$, respectively. Qualitatively describe the metal–CO bonding for these ligand types by constructing
   a pictorial representation of major orbital interactions. Why are the C—O frequencies of linear $\mu_2$-CO ligands lower than those of symmetrical $\mu_2$-CO
   ligands?
2. Describe the differences in chemical bonding that can be used to distinguish
   a semi-bridging $\mu_2$-CO ligand from a symmetrical $\mu_2$-CO ligand?
3. Within a transition metal group, how do the M—C bond enthalpies of the
   metal carbonyls vary as a function of M?
4. As mentioned in the text, Wade's rules can be adapted easily to include any
   type of ligand provided we know how many electrons are donated by each
   ligand type. These rules can be applied to ionic complexes as well.
   (a) Using Wade's rules, predict the metal core structure in each of the fol-
       lowing complexes. (Assume that the $C_5H_5$ ligand acts as a neutral five-
       electron donor.)
       (1) $Ni_6(CO)_{12}{}^{2-}$        (2) $(C_5H_5)Fe_3Rh(CO)_{11}$
       (3) $(C_5H_5)_3Rh_3(CO)_3$        (4) $(C_5H_5)_3Co_2Mn(CO)_4$
       (5) $[(C_5H_5)Fe(CO)]_4$
   (b) The metal core structures in $Ni_5(CO)_{12}{}^{2-}$ and $Mo_2Ni_3(CO)_{16}{}^{2-}$ are es-
       sentially trigonal bipyramidal. Show how Wade's rules can be used to
       derive these metal core structures.
5. Devise a qualitative pictorial representation of the M—CO bonding associated
   with a $\mu_3$-CO ligand.
6. Explain the difference between $\Delta H_D$, as defined in equation 2.21, and $\Delta H_D{}^*$
   as defined in equation 2.22.
7. How does the *average* Cr—CO bond enthalpy calculated from equation 2.22
   relate to the bond enthalpy required to remove only the *first* CO ligand from
   $Cr(CO)_6$?
8. From figure 2.2, predict other organic molecules that might be expected to
   bind to transition metals in low oxidation states.

9. Sketch the accepted structures for the following three-dimensional objects: **(a)** pentagonal bipyramid; **(b)** dodecahedron; **(c)** trigonal prism; **(d)** tricapped trigonal prism.
10. Given that the polarization within a terminal carbonyl ligand is

$$M\text{---}\overset{\delta+}{C}\equiv\overset{\delta-}{O}$$

predict the sites of nucleophilic and electrophilic attack. Assuming that a terminal carbonyl ligand can be regarded as a metalla-ketene, $M\text{=}C\text{=}O$, give one example of an expected reaction for this ligand type.

## SUGGESTED READING

### Synthesis and Structure of Metal Carbonyls

Abel, E. W., and Stone, F. G. A. *Quart. Rev.* **24**, 498 (1970).

Abel, E. W., and Stone, F. G. A. *Quart. Rev.* **23**, 325 (1969).

Calderazzo, F., Ercoli, R., and Natta, G. In *Organic Syntheses via Metal Carbonyls*, Vol. 1. I. Wender and P. Pino, eds. p. 1. New York: Interscience, 1968.

Chini, P. *Inorg. Chim. Acta. Rev.* **2**, 31 (1968).

Churchill, M. R. *Perspective Struct. Chem.* **3**, 123 (1970).

Churchill, M. R., and DeBoer, B. G. *Inorg. Chem.* **16**, 878 (1977).

Churchill, M. R., Hollander, F. J., and Hutchinson, J. P. *Inorg. Chem.* **16**, 2655 (1977).

Churchill, M. R., and Hutchinson, J. P. *Inorg. Chem.* **17**, 3528 (1978).

Colton, R., and McCormick, M. J. *Coord. Chem. Rev.* **3**, 1 (1980).

Gladfelter, W. L., and Geoffrey, G. L. *Adv. Organometal. Chem.* **18**, 207 (1980).

King, R. B. *Organometallic Syntheses.* Vol. 1. New York: Academic Press, 1965.

Penfold, B. R. *Perspectives Struct. Chem.* **2**, 71 (1968).

Roberts, D. A., and Geoffrey, G. L. *Comprehensive Organometallic Chemistry.* Vol. 6. G. Wilkinson, F. G. A. Stone and E. W. Abel, eds. 763−877. Oxford: Pergamon Press, 1982.

### Chemical Bonding and Thermochemical Data for Metal Carbonyls

Benfield, R. E., and Johnson, B. F. G. *J. Chem. Soc., Dalton Trans.* **1980**, 1743.

Bursten, B. E., Freier, D. G., and Fenske, R. F. *Inorg. Chem.* **19**, 1810 (1980).

Connor, J. A. *Topics Current Chem.* **71**, 71 (1977).

Eady, C. R., Johnson, B. F. G., and Lewis, J. *J. Chem. Soc., Dalton Trans.* **1975**, 2606.

Hillier, I. H., and Saunders, V. R. *Mole. Phys.* **22**, 1025 (1971).

Johnson, B. F. G., and Benfield, R. E. *Topics Stereochem.* **12**, 253 (1981).

Johnson, B. F. G. *Transition Metal Clusters.* New York: Wiley-Interscience, 1980.

Sherwood, D. E., Jr., and Hall, M. B. *Inorg. Chem.* **19**, 1805 (1980).

Wade, K. *Adv. Inorg. Chem. Radiochem.* **18**, 1 (1980).

# Substituted Metal Carbonyl Complexes and Related Compounds

## CHAPTER 3

There are many metal carbonyl complexes of type **3.1** where the ancillary ligands L are other than CO.

$$L_aM(CO)_b$$

**3.1**

In this chapter, we focus on those complexes in which L is a two-electron ligand containing C, N, P, As, Sb, O, S, Te or halide donor atoms. Our selection of common ancillary ligands includes isocyanides (RNC), amines (including $NH_3$), nitriles (RCN), trivalent phosphorus ligands ($PZ_3$, where Z = H, X, R, OR, $NR_2$), arsines ($R_3As$), stibines ($R_3Sb$) and less frequently, ethers, thioethers, tellurides or alcohols. These complexes are usually prepared via ligand substitution reactions of metal carbonyls (eq. *1.7*). In some instances, a low-valent metal complex that lacks CO ligands (**3.1** with $b = 0$) can be prepared directly. These complexes are formal derivatives of metal carbonyls in which all of the CO ligands have undergone substitution. Several specific types of complexes—for example those in which L is NO, H, $Me_3S^+$, CS, CSe, CTe or other carbon donors—are discussed more fully in chapters 6, 8 and 11. Much of the ligand substitution chemistry presented in this chapter is generally applicable to other organometallic complexes as well.

## PREPARATION OF SUBSTITUTED METAL CARBONYL AND METAL PHOSPHINE COMPLEXES

Substituted metal carbonyl complexes are usually prepared by (1) direct synthesis, (2) reductive-ligation (eq. *1.6*), (3) ligand substitution (eq. *1.7*), or (4) chemical conversion of one kind of ligand into another. The first three of these methods are discussed here; examples of the last synthetic route are presented in chapters 9 and 11.

## Direct Synthesis

Until recently, the only direct synthesis of a metal phosphine complex was that of $Ni(PF_3)_4$:

$$Ni^0 + 4PF_3 \xrightarrow[100°C]{70\ atm} \underset{99\%}{Ni(PF_3)_4} \qquad (3.1)$$

Equation 3.1 suggests a possible similarity in M—$PF_3$ and M—CO bonding, for $Ni(CO)_4$ is also prepared by direct synthesis.

Many new complexes have been prepared and isolated by a recently developed synthetic technique in which metal vapor is deposited on a cold probe coated with a layer of ligand molecules (figure 3.1). The complexes form when the reaction probe is warmed after condensation of the metal vapor and ligand molecules. This technique has been used to prepare several phosphine complexes:

$$\underset{\substack{1,2\text{-}bis(\text{dimethylphosphino})\text{ethane}\\(\text{dmpe})}}{Me_2PCH_2CH_2PMe_2} \xrightarrow{\text{M vapor}} M(\text{dmpe})_3 \qquad (3.2)$$

where M = Cr, Mo, W, V, Nb, or Ta.

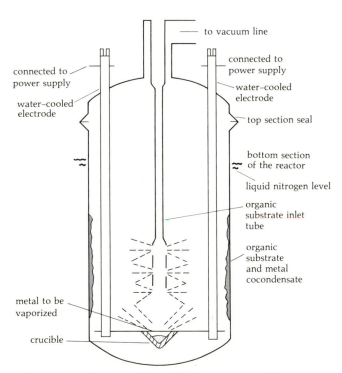

**FIGURE 3.1**  Experimental apparatus for the cocondensation of metal vapor and an organic substrate. *Source: Adapted from K. J. Klabunde,* Acc. Chem. Res., **8,** *393 (1975).*

## Reductive-Ligation

Several examples of low-valent metal complexation via reductive-ligations are shown in equations 3.3–3.9.

$$Ni(OH)_2 + 5CNR \xrightarrow{R'OH} \underset{\substack{\text{yellow solids} \\ 10-30\%}}{Ni(CNR)_4} + RNCO + H_2O \qquad (3.3)$$

R = aryl

$$FeBr_2 + Me_3CNC(xs) + Na/Hg(xs) \xrightarrow[\text{THF}]{25°C, 24 \text{ hr}} \underset{\substack{\text{yellow solid} \\ 70\%}}{Fe(CNCMe_3)_5} \qquad (3.4)$$

$$NiI_2 + Cu(xs) + PF_3(xs) \xrightarrow[\text{135 atm}]{100°C} \underset{100\%}{Ni(PF_3)_4} \qquad (3.5)$$

$$Ni(NO_3)_2 \cdot 6H_2O + 4P(OPh)_3 \xrightarrow[\text{EtOH}]{NaBH_4} \underset{92\%}{Ni[P(OPh)_3]_4} \qquad (3.6)$$

$$K_2PtCl_4 + 2KOH + 4PPh_3 + EtOH \xrightarrow[\text{EtOH}]{65°C} \underset{\substack{\textit{yellow solid, 79\%} \\ \textit{dec. } 118-120°C}}{Pt(PPh_3)_4} + 4KCl + CH_3CH(O) \quad (3.7)$$

$$2RhCl_3 \cdot 3H_2O + 6CO \xrightarrow{100°C} \underset{\substack{\textit{red solid, 96\%} \\ \text{mp } 124-125°C}}{[Rh(CO)_2Cl]_2} + 6H_2O + 2OCCl_2 \quad (3.8)$$

$$IrCl_3 \cdot 3H_2O + Ph_3P(xs) \xrightarrow[\text{reflux, 12 hr}]{DMF} \underset{\substack{\textit{yellow- green solid, 88\%} \\ \text{Vaska's complex}}}{trans\text{-}(Ph_3P)_2Ir(CO)Cl} \qquad (3.9)$$

(Rh analog also)

Alkali metals or electropositive representative metals are commonly used as reducing agents, but $NaBH_4$, EtOH, CO, and isocyanides can also reduce transition metals. The formation of Vaska's complex (eq. 3.9) is particularly interesting since the CO ligand is presumably extracted from dimethylformamide.

    Several of the above products are not organometallic complexes according to the definition of chap. 1. However, these low-valent metal compounds are formal derivatives of organometallic complexes, and several of them can be prepared via complete ligand substitution of the analogous metal carbonyl. Some of these complexes are useful reagents for synthesizing other organometallic compounds.

## Ligand Substitution

The most general reaction in organometallic chemistry is ligand substitution. As shown in equations 3.10 and 3.11 for metal carbonyl complexes, an incoming ligand L' can displace either CO or an ancillary ligand L.

$$M(CO)_xL_y + L' \xrightarrow[\Delta]{hv \text{ or}} M(CO)_{x-1}L_yL' + CO \qquad (3.10)$$

or

$$M(CO)_xL_y + L' \xrightarrow[\Delta]{h\nu \text{ or}} M(CO)_xL_{y-1}L' + L \tag{3.11}$$

Usually L′ is also a two-electron donor, and therefore it replaces only one CO or L ligand; however, we shall encounter several examples where L′ is a polydentate ligand that can displace more than one ligand of the original complex. The reactant complex may contain only CO ligands ($y = 0$), both CO and ancillary ligands ($x$ and $y \geq 1$), or no CO ligands ($x = 0$). These substitution reactions are initiated either thermally or photochemically.

Some specific examples of ligand substitution reactions are shown in equations 3.12–3.27. The isocyanide, amine, and acetonitrile reactions (eqs. 3.12–3.17) illustrate CO and $PR_3$ substitution reactions.

$$Ni[P(OR)_3]_4 + 2C\equiv N-C_6H_{11} \longrightarrow Ni[P(OR)_3]_2(CNC_6H_{11})_2 + 2P(OR)_3$$

$$R = Me, Et \tag{3.12}$$

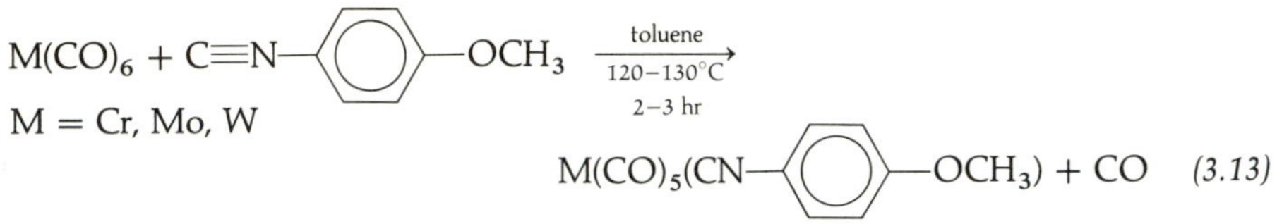

$$M(CO)_6 + C\equiv N-\!\!\left\langle\bigcirc\right\rangle\!\!-OCH_3 \xrightarrow[120-130°C]{\text{toluene}}$$

$$M = Cr, Mo, W \qquad\qquad 2-3 \text{ hr}$$

$$M(CO)_5(CN-\!\!\left\langle\bigcirc\right\rangle\!\!-OCH_3) + CO \tag{3.13}$$

colorless solids
mp 105–125°C

$$Fe(CO)_5 + C\equiv N-C_6H_{11} \xrightarrow[C_6H_6]{h\nu} Fe(CO)_4(CNC_6H_{11}) + CO \tag{3.14}$$
$$\text{yellow oil, 67\%}$$
$$\text{dec. 167°C}$$

$$Cr(CO)_6 + C_5H_{10}NH \xrightarrow[THF]{h\nu, 30°C} Cr(CO)_5(NHC_5H_{10}) + CO \tag{3.15}$$
$$\text{piperidine} \qquad\qquad \text{yellow solid, 60\%}$$
$$\text{mp 69.5°C}$$

$$M(CO)_6 + \overset{\frown}{N\quad N}(\text{en or bipy}) \xrightarrow{120-130°C} cis\text{-}M(CO)_4(\overset{\frown}{N\quad N}) + 2CO \tag{3.16}$$
$$\text{yellow solids, 30-50\%}$$

$$M = Cr, Mo, W; \text{ en} = \text{ethylenediamine};$$
$$\text{bipy} = 2,2'\text{-bipyridine}$$

$$Mo(CO)_6 + \text{dien} \xrightarrow{150°C} fac\text{-}Mo(CO)_3(\text{dien}) \tag{3.17}$$
$$\text{yellow solid, 80\%}$$
$$\text{dien} = diethylenetriamine \qquad\qquad \text{dec.} > 300°C$$

Many of the product complexes serve as useful reagents for further ligand substitution. For example, direct photochemical substitution of CO by phosphines (eq. 3.18) frequently occurs in low yield and gives complex mixtures of products.

$$M(CO)_6 + PH_3 \xrightarrow[C_6H_6]{mild\ \Delta} M(CO)_5PH_3 \quad + CO \qquad (3.18)$$

M = Cr, Mo, W

colorless or pale yellow,
2–10%
dec. 112°–116°C

However, thermal substitution of N-donor ligands by phosphines (eq. 3.19) proceeds in high yield.

$$fac\text{-}M(CO)_3(NCCH_3)_3 + PH_3 \xrightarrow[THF]{0°C \rightarrow 25°C} fac\text{-}M(CO)_3(PH_3)_3 \quad + 3CH_3CN \quad (3.19)$$

M = Cr, Mo, W

yellow solids, ~60%
dec. 130°–150°C

Ligand substitution of polynuclear metal carbonyls gives products resulting from loss of CO (eq. 3.20) or loss of a smaller metal carbonyl molecule (eq. 3.21).

$$Mn_2(CO)_{10} + PH_3 \xrightarrow[C_6H_6]{hv} Mn_2(CO)_9PH_3 \qquad (3.20)$$

orange solid, 17%
dec. 113°C

$$Fe_2(CO)_9 + PH_3 \xrightarrow{\Delta} Fe(CO)_4PH_3 + Fe(CO)_5 \qquad (3.21)$$

yellow solid, 70%
mp 36°C

When iron carbonyls are treated with isocyanides, the *mononuclear* complexes $Fe(CO)_4$ (CNR) and *trans*-$Fe(CO)_3(CNR)_2$ are formed. Similarly, triphenylphosphine or $Ph_3As$ cleaves the dimeric complex $[Rh(CO)_2Cl]_2$ to afford mononuclear rhodium complexes (eq. 3.22).

$$[Rh(CO)_2Cl]_2 + 4Ph_3M \xrightarrow[C_6H_6]{\Delta} 2\ trans\text{-}(Ph_3M)_2Rh(CO)Cl + 2CO \quad (3.22)$$

M = P, As

yellow solids, 88%
dec. 195°–242°C

Ligands having oxygen donor atoms form relatively unstable complexes that are usually prepared photochemically (eq. 3.23). Ligands with sulfur and tellurium donor atoms form metal carbonyl compounds that are more stable thermally (eq. 3.24).

$$M(CO)_6 + L(as\ solvent) \xrightarrow[N_2\ flush]{hv} M(CO)_5L + CO \qquad (3.23)$$

L = DMF, acetone, $i$-$Pr_2O$, EtOH

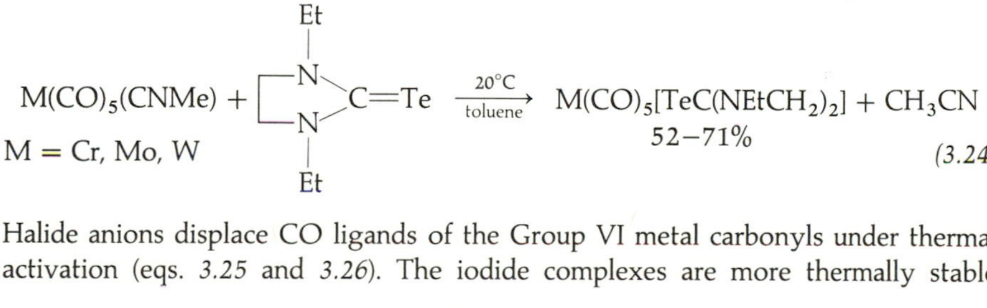

$$M(CO)_5(CNMe) + \begin{bmatrix} Et \\ | \\ -N \\ \diagdown \\ C=Te \\ \diagup \\ -N \\ | \\ Et \end{bmatrix} \xrightarrow[toluene]{20°C} M(CO)_5[TeC(NEtCH_2)_2] + CH_3CN$$

M = Cr, Mo, W      52–71%

$$(3.24)$$

Halide anions displace CO ligands of the Group VI metal carbonyls under thermal activation (eqs. 3.25 and 3.26). The iodide complexes are more thermally stable than the chloride or bromide compounds. Fluoride, which is a poor ligand, can be introduced via silver ion-assisted halide exchange of a corresponding chloride complex, as shown in equation 3.27.

$$M(CO)_6 + Et_4N^+X^- \xrightarrow[\text{diglyme}]{120°C} [Et_4N][M(CO)_5Cl] + CO \qquad (3.25)$$
$$\text{yellow solids}$$

M = Cr, Mo, W; X = Cl, Br;
diglyme = *diethyleneglycol dimethyl ether*

$$M(CO)_6 + [C_5H_5NCH_3]I \xrightarrow[\text{THF}]{\text{reflux}} [C_5H_5NCH_3][M(CO)_5I] \quad + CO \quad (3.26)$$
$$\text{yellow-orange solids, } 40-70\%$$

M = Cr, Mo, W

$$2 \text{ } trans\text{-}(Ph_3P)_2M(CO)Cl + Ag_2CO_3 + 2NH_4F \xrightarrow[\text{MeOH}]{\Delta}$$

$$2 \text{ } trans\text{-}(Ph_3P)_2M(CO)F + 2AgCl + CO_2 + 2NH_3 + H_2O \quad (3.27)$$
$$\text{yellow solids, } 80-90\%$$

M = Rh, Ir

   The types of substituted carbonyl complexes that a metal forms is generally characteristic of its group. Group VI metal hexacarbonyls undergo extensive ligand substitution according to the following equation:

$$M(CO)_6 + nL \xrightarrow[\text{solvent}]{\Delta \text{ or } hv} M(CO)_{6-n}L_n + nCO \qquad (3.28)$$

M = Cr, Mo, W

Totally substituted complexes are known for L = RNC, $PF_3$, $RN(PF_2)_2$, dmpe, and some $P(OR)_3$ ligands. Several general observations concerning these substitution reactions can be summarized:

   1. Ligands having phosphorus or arsenic donor atoms afford more thermally stable complexes than do ligands with nitrogen, oxygen, or halide donor atoms.
   2. As the degree of substitution increases, it becomes increasingly difficult to displace additional CO ligands.
   3. Most of the $M(CO)_4L_2$ and $M(CO)_3L_3$ complexes exist as the *cis* and *facial* geometrical isomers, respectively, although geometrical isomerization to *trans* isomers occurs frequently.
   4. Chelating ligands such as en, bipy, dien, and di- or tri(tertiary)phosphines and arsines form more stable complexes than do their monodentate analogs. Two common bidentate ligands are 1,2-*bis*(diphenylphosphino)ethane, diphos (**3.2**) and *o*-phenylenebisdimethylarsine, diars (**3.3**).

$$Ph_2PCH_2CH_2PPh_2$$

**3.2**                              **3.3**

diphos                           diars

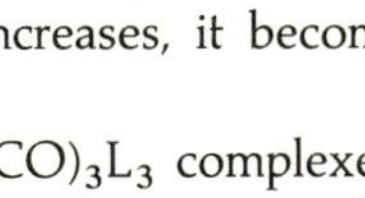

5. Steric factors influence the degree of substitution. For example, when M is Cr and L is $PH_3$, a tetrasubstituted complex, $Cr(CO)_2(PH_3)_4$, can be prepared, but with the larger $Ph_3P$ ligand, only disubstitution occurs affording trans-$Cr(CO)_4$ $(PPh_3)_2$. The ligand $PEt_3$ has an intermediate steric influence; the direct reaction of $Cr(CO)_6$ and $PEt_3$ affords both cis- and trans-$Cr(CO)_4$ $(PEt_3)_2$. Recently, fac-$Cr(CO)_3$ $(PEt_3)_3$ was prepared from $CrCl_2$ via reductive-ligation. However, an x-ray structure revealed elongated Cr—P bond distances due to steric crowding.

Group VII metal carbonyls form three types of ligand substitution complexes: 17-electron mononuclear complexes, $M(CO)_4L$, and mono- or disubstituted bimetallic complexes, $M_2(CO)_9L$ and $M_2(CO)_8L_2$. The mononuclear complexes have trigonal bipyramidal geometry with L occupying an axial site. With monodentate ligands, the disubstituted, dinuclear complexes have an L ligand on each metal atom, usually in an equatorial site. Complexes of the form $M_2(CO)_9L$ normally exist as axial isomers. Specific substitution behavior depends on the metal atom. For example, direct reaction of $Mn_2(CO)_{10}$ with $Ph_3P$ gives $Mn(CO)_4PPh_3$ and $[Mn(CO)_4PPh_3]_2$, but with $Re_2(CO)_{10}$ only the dimeric product forms.

Ligands having phosphorus, arsenic, or antimony donor atoms react directly with the three iron carbonyls to afford axial-$Fe(CO)_4L$ and trans-$Fe(CO)_3L_2$ complexes. Chelating ligands, such as diphos (**3.2**) and diars (**3.3**) form cis-$Fe(CO)_3$(L L) compounds in which the donor atoms of the chelate occupy axial and equatorial sites. Similar substitution reactions occur with isocyanides.

Phosphine substitution reactions of $Fe_3(CO)_{12}$ occasionally afford minor products that retain the $Fe_3$ core:

$$Fe_3(CO)_{12} + PPh_3 \xrightarrow[CHCl_3]{45°, 45 \text{ min}}$$

$$\underbrace{Fe(CO)_5 + Fe(CO)_4PPh_3 + Fe(CO)_3(PPh_3)_2}_{\text{major products}} + \underset{\text{green-black solid, 2\%}}{Fe_3(CO)_{11}PPh_3} \qquad (3.29)$$

$$\textbf{3.4}$$

Complex **3.4** is a structural derivative of $Fe_3(CO)_{12}$.

An interesting example of the importance of an O-donor ligand is the enhanced reactivity of $Fe_2(CO)_9$ in THF. Synthetic chemists have used the increased "solubility" of $Fe_2(CO)_9$ in THF to prepare π-olefin complexes for many years. Recently, however, this $Fe_2(CO)_9$ "solution" has been studied in more detail. Presumably, $Fe_2(CO)_9$ reacts with THF under a CO atmosphere to give the THF complex **3.5**:

$$Fe_2(CO)_9 + THF \xrightarrow{CO} Fe(CO)_4(THF) + Fe(CO)_5 \qquad (3.30)$$

$$\textbf{3.5}$$

Although complex **3.5** has not been isolated, it does react with many other heteroatomic donor ligands to afford isolable complexes. When CO is purged from the solution, di- and trinuclear substituted iron complexes are obtained as shown in equations *3.31* and *3.32*.

$$\text{Fe}_2(\text{CO})_9 + \text{bipy} \xrightarrow{\text{THF}} \text{Fe}_2(\text{CO})_7(\text{bipy}) \qquad (3.31)$$

**3.6**

$$\text{Fe}_2(\text{CO})_9 + \text{diars} \xrightarrow{\text{THF}} \text{Fe}_2(\text{CO})_8(\text{diars}) + \text{Fe}_3(\text{CO})_{10}(\text{diars}) \qquad (3.32)$$

**3.7**

Complexes **3.6** and **3.7** are stoichiometric but not structural derivatives of $\text{Fe}_2(\text{CO})_9$ and $\text{Fe}_3(\text{CO})_{12}$, respectively. Complex **3.6** has only *two* $\mu_2$-CO ligands, and its structure is discussed later. Complex **3.7** has two semi-bridging $\mu_2$-CO ligands that do *not* bridge the same Fe—Fe bond. In both complexes, the hetero-atomic ligands act as bidentate, chelating ligands.

The cobalt carbonyl $\text{Co}_2(\text{CO})_8$ reacts with phosphines, $(\text{PhO})_3\text{P}$, $\text{Ph}_3\text{As}$, and $\text{Ph}_3\text{Sb}$ in nonplanar solvents to give $[\text{Co}(\text{CO})_3\text{L}]_2$ dimeric complexes having all terminal CO ligands and a Co—Co single bond. At $0°\text{C}$ a monosubstituted complex, $\text{Co}_2(\text{CO})_7\text{PPh}_3$, can be prepared by direct reaction.

Substituted derivatives of rhodium and iridium carbonyl complexes can be prepared also. The Rh complexes, $\text{Rh}_4(\text{CO})_8[\text{P}(\text{OPh})_3]_4$ (**3.8**) and $\text{Rh}_4(\text{CO})_8(\text{dpm})_2$ (**3.9**) are structural derivatives of $\text{Rh}_4(\text{CO})_{12}$.

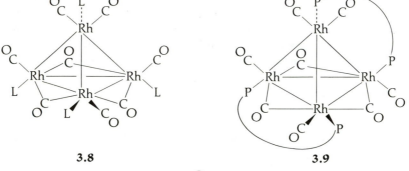

**3.8**                                          **3.9**

$\text{L} = \text{P}(\text{OPh}_3)_3$            $\widehat{\text{P P}} = \text{dpm} = bis(\text{diphenylphosphino})\text{methane}$

Each complex has three symmetrical $\mu_2$-CO ligands located in the basal plane and one phosphine ligand on each Rh atom. The dinuclear complex $\text{Rh}_2(\text{CO})_2(\text{PPh}_3)_4$,

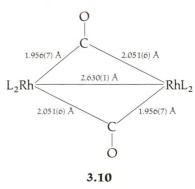

**3.10**

$\text{L} = \text{PPh}_3$

prepared by treating a benzene solution of $RhH(CO)(PPh_3)_3$ with CO at 25°C, contains two asymmetric $\mu_2$-CO ligands, as shown in **3.10**.

Whereas $Ir_4(CO)_{12}$ has only terminal CO ligands, the mono-, di- and trisubstituted complexes $Ir_4(CO)_{12-x}(PPh_3)_x$, where $x$ is 1 (**3.11**), 2 (**3.12**), or 3 (**3.13**), are structural analogues to $Co_4(CO)_{12}$; each complex has three symmetrical $\mu_2$-CO ligands located in the basal plane.

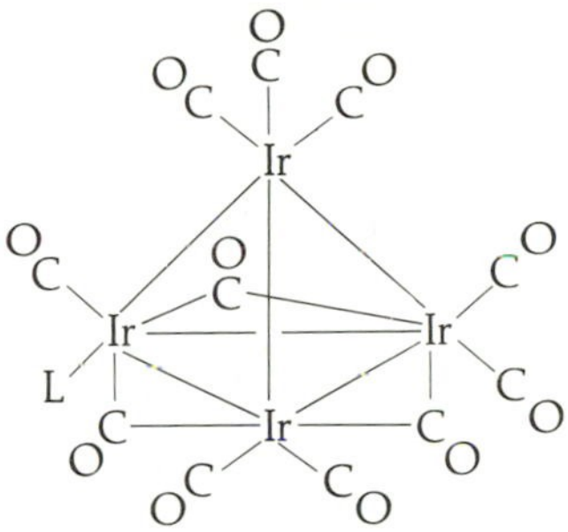

**3.11**

$L = Ph_3P$

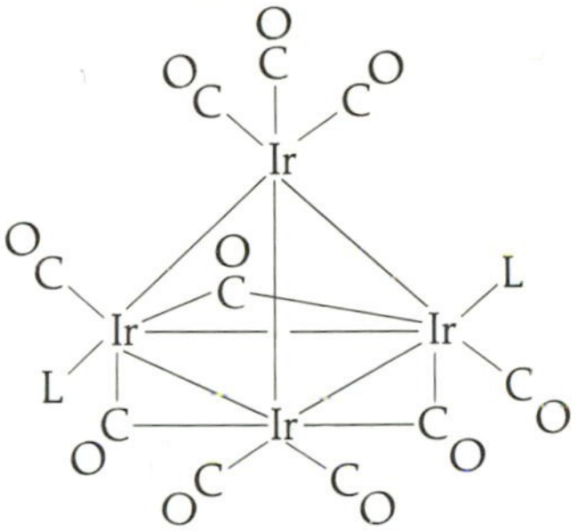

**3.12**

$L = Ph_3P$

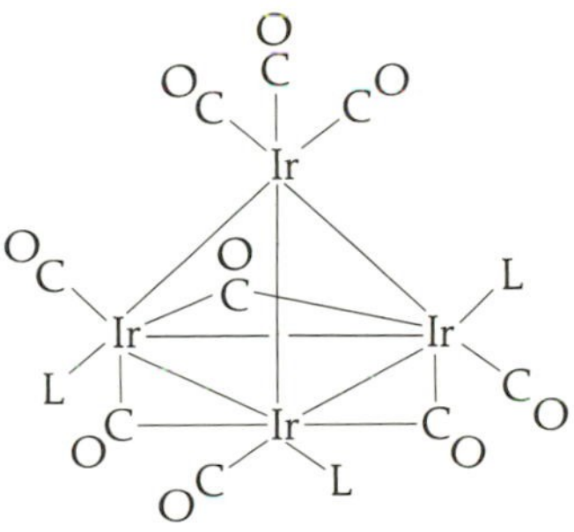

**3.13**

$L = Ph_3P$

This structural change is attributed to the increase in size of the "carbonyl polyhedron" that results from the incorporation of the bulky $PPh_3$ ligands. An enlarged substituted "carbonyl" polyhedron presumably best accommodates the $Ir_4$ core by shifting to bridging CO ligands.

Complex **3.14**, $Ir_4(CO)_{10}(diars)$, also adopts a structure like $Co_4(CO)_{12}$ except that two of the $\mu_2$-CO ligands form *asymmetrical* bridges while the other CO is bridging symmetrically as shown in **3.15**. Note that $Ir^1$ would have a 20-electron count if the $\mu_2$-CO ligands were terminal. The electron-rich $Ir^1$ atom donates this excess electron density to the electron-deficient $Ir^2$ and $Ir^3$ atoms through the $\mu_2$-CO ligands.

The nickel complex $Ni(CO)_4$ has an extensive ligand substitution chemistry. When $Ni(CO)_4$ is treated with o-phenanthroline, $Ni(CO)_2(phen)$ forms. Mixed-ligand substitution occurs with isocyanides and ligands having P, As or Sb donor

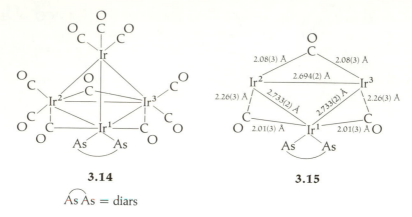

**3.14**                                    **3.15**

As As = diars

atoms, but intraligand steric interactions influence the stability and degree of substitution for each specific ligand.

Ligands that are good $\sigma$-donors and poor $\pi$-acceptors, such as those having N or O donor atoms, can react with metal carbonyls to effect disproportionation at the metal atom (eqs. 3.33–3.36).

$$3V(CO)_6 + 6L \longrightarrow [VL_6]^{2+}[V(CO)_6]_2^- + 6CO \qquad (3.33)$$

L = py or $R_3N$

$$3Mn_2(CO)_{10} + 12L \longrightarrow 2[MnL_6]^{2+}[Mn(CO)_5]_2^- + 10CO \qquad (3.34)$$

L = py or $R_3N$

$$5Fe(CO)_5 + 6L \longrightarrow [FeL_6]^{2+}[Fe_4(CO)_{13}]^{2-} + 12CO \qquad (3.35)$$

L = amine or some O-donor ligands

$$3Co_2(CO)_8 + 2nL \longrightarrow 2[CoL_n]^{2+}[Co(CO)_4]_2^- + 8CO \qquad (3.36)$$

n = 6 for monodentate amine
n = 3 for bidentate amine

Under appropriate reaction conditions, phosphorus ligands can induce disproportionation also. At elevated temperatures $Mn_2(CO)_{10}$ and diphos form an ionic product, $[Mn(CO)_2(diphos)_2]^+[Mn(CO)_5]^-$, and in polar solvents $Co_2(CO)_8$ and $PPh_3$ react to afford $[Co(CO)_3(PPh_3)_2]^+[Co(CO)_4]^-$. Notice that disproportionation occurs when a metal atom has both a stable +1 or +2 oxidation state and a stable anionic metal carbonyl complex. Strongly basic ligands stabilize highly positive metal oxidation states, and CO ligands stabilize reduced metal oxidation states.

Isocyanide ligands are sufficiently strong $\sigma$-donors that they form stable organometallic complexes with metals in high oxidation states, e.g., $Mn(CNCH_3)_6^{2+}$, $Fe(CNCH_3)_6^{2+}$, $Co(CNCH_3)_5^{2+}$ and $Ni(CNCH_3)_4^{2+}$. Equation 3.37 illustrates the strong $\sigma$-donating ability of isocyanides in an unusual *oxidative-ligation reaction*.

$$Mo(CO)_6 + I_2 + RNC(xs) \xrightarrow[\text{MeOH}]{\text{reflux}} [Mo(CNR)_7](I)_2 + 6CO \qquad (3.37)$$

$$\text{yellow solids, } 30\text{--}40\%$$

R = Me, t-butyl

Reaction conditions that are expected to produce ligand substitution can also afford complexes of higher nuclearity. When $Fe_2(CO)_9$ is treated with diars (**3.3**) in THF solution at $25°C$, $Fe_2(CO)_8$(diars) is isolated as well as a trinuclear complex, $Fe_3(CO)_{10}$(diars). This latter complex has a chelating, bidentate diars ligand and two semi-bridging CO ligands to equalize the electron density donated by the arsenic atoms. Similarly, when an ether solution of $Fe(CNEt)_5$ is photolyzed, a dinuclear complex, $Fe_2(CNEt)_9$, forms. This conversion is analogous to the preparation of $Fe_2(CO)_9$ from $Fe(CO)_5$.

Occasionally, CO displaces other ligands even under mild reaction conditions:

$$NiI_2(PMe_3)_3 + CO(g) \xrightarrow[\text{CH}_2\text{Cl}_2/\text{EtOH}]{20°} NiI_2(CO)(PMe_3)_2 + PMe_3 \qquad (3.38)$$

$$\text{brown solid, } 75\%$$

Presumably, the electron-rich Ni atom of the *tris*-trimethylphosphine complex prefers to substitute one of the $PMe_3$ ligands for CO, which is a much stronger $\pi$-acceptor. Intraligand steric repulsion is probably reduced also. The two phosphine ligands of the trigonal-bipyramidal product complex occupy axial sites.

## MECHANISMS OF LIGAND SUBSTITUTION REACTIONS AND STEREOSPECIFIC SYNTHESES

A vast amount of kinetic data has been accumulated for ligand substitution reactions. For brevity, we shall focus on reactions of Group VI metal carbonyls. The simplest example is thermally induced CO exchange (eq. 3.39).

$$M(CO)_6 + \overset{*}{C}O \underset{}{\overset{k_1}{\rightleftharpoons}} M(CO)_5(\overset{*}{C}O) + CO \qquad (3.39)$$

M = Cr, Mo, W

For Group VI hexacarbonyls, the rates of CO exchange in the gas phase are first order in complex concentration. These reaction rates and activation parameters are as follows:

| Complex | $T(°C)$ | $k_1 \times 10^4$ (sec$^{-1}$) | $\Delta H^*$ (kcal/mole) | Estimated $\Delta S^*$ (e.u.) |
|---|---|---|---|---|
| $Cr(CO)_6(g)$ | 117 | 0.20 | 38.7 | 18.5 |
| $Mo(CO)_6(g)$ | 116 | 0.75 | 30.2 | −0.4 |
| $W(CO)_6(g)$ | 142 | 0.026 | 39.8 | 11 |

Assuming that CO exchange proceeds by a dissociative mechanism involving an $M(CO)_5$ intermediate, we would expect the $\Delta H^*$ values should parallel the corresponding M—C bond enthalpies of 25.8, 36.3 and 42.5 kcal/mole, respectively, as given in Table 2.5. However, these $\Delta H^*$ and bond enthalpy values differ in both magnitude and relative order.

There are two explanations for this discrepancy. First, M—C bond enthalpies are average values for dissociation of all six CO ligands whereas the process of equation 3.39 involves dissociation of only the first CO ligand; and second, M—C bond enthalpies are relative to M and CO in their ground states whereas the presumed $M(CO)_5$ intermediate in this process clearly has M in an excited valence state. By adjusting the *mean* M—C bond enthalpies of several first-row metal carbonyls relative to metals in *excited* valence states (as we did earlier for $Cr(CO)_6$, eq. 2.22, one can rationalize the experimental trend in $\Delta H^*$ values for M—CO dissociation. However, in many instances these calculated $\Delta H^*$ values from adjusted bond enthalpy data are in considerable error because thermodynamic data represent only *average* bond dissociation enthalpies.

Ligand substitution reactions of Group VI metal carbonyls in solution with incoming ligands other than CO (eq. 3.40) proceed by a two term rate law:

$$M(CO)_6 + L \longrightarrow M(CO)_5L + CO \tag{3.40}$$

$$\text{rate} = k_1[M(CO)_6] + k_2[M(CO)_6][L]$$

where the $k_1$ and $k_2$ terms refer to dissociative and associative processes, respectively. Kinetic studies at low temperature sometimes permit the spectroscopic detection of intermediate species. Low-temperature photochemical studies indicate that the coordinatively unsaturated $M(CO)_5$ intermediate in a dissociative process is initially square pyramidal. This species thermally rearranges to the more preferred trigonal-bipyramidal geometry.

The contribution of the $k_2$ term to the overall rate depends on M and L; it increases as $Cr \ll Mo \approx W$ and $(PhO)_3P < Ph_3P < (n\text{-}C_4H_9)_3P$. Apparently, an associative mechanism is preferred between large metal atoms and more basic phosphine ligands. "Hard" Lewis bases—those that have relatively small and only slightly polarizable donor atoms, such as the nitrogen in amines—substitute CO ligands via a dissociative mechanism. "Soft" Lewis bases have relatively large and highly polarizable donor atoms, such as phosphorus.

Substitution of a second CO ligand affords disubstituted complexes, which exist as *cis* or *trans* isomers (eq. 3.41).

$$M(CO)_5L + L' \xrightarrow{-CO} \qquad \text{or} \tag{3.41}$$

cis                              trans

The relative thermodynamic stability of these geometrical isomers depends on electronic and steric factors.

The rate of dissociation of a second CO is influenced by L. If this rate is greater than the rate of CO dissociation in the binary metal carbonyl, then L is said to be a *labilizing ligand.* If the rates of CO dissociation in equations *3.40* and *3.41* are nearly equal, then L is a *nonlabilizing ligand.* A general observation is that labilizing ligands have N, O or halogen donor atoms, and nonlabilizing ligands have P, As, Sb or S donor atoms. Among the halogens, labilizing ability decreases in the order Cl > Br > I. Labilizing ligands are usually "hard" bases that have little or no $\pi$ bonding to the metal atom. This ability apparently results from a significant increase in the amount of negative charge placed on M by the labilizing ligand. This electronic effect reduces M—CO $\sigma$ *bonding* to the remaining CO ligands although M—CO $d\pi$-$p\pi^*$ back bonding should be enhanced (particularly to the CO *trans* to L because this CO ligand and L form $\pi$ bonds to the same metal $d$ orbital). Alternatively, labilizing ligands may better stabilize the coordinately unsaturated transition state species in the reaction. This stabilization would be expected of ligands that have additional available electron density on the donor atom, such as a lone pair of electrons.

One practical consequence of the labilizing ability of certain ligands is the stereo-specific preparation of substituted metal carbonyl complexes. Equation *3.42* shows how the *cis*-labilizing ability of piperidine can be used to generate a *cis*-diamine complex stereospecifically.

$$M(CO)_6 + C_5H_{10}NH(xs) \xrightarrow{-CO} \{M(CO)_5(NHC_5H_{10})\} \xrightarrow{-CO}$$
$$\text{piperidine} \qquad\qquad cis\text{-}M(CO)_4(NHC_5H_{10})_2 \qquad (3.42)$$

M = Cr, Mo, W

This disubstituted complex has been used in the preparation of specific isotopically labeled complexes (eqs. *3.43* and *3.44* and *cis*-M(CO)$_4$L$_2$ complexes by ligand substitution (eq. *3.45*).

$$>95\% \qquad (3.43)$$

$$cis\text{-}W(CO)_4(NHC_5H_{10})_2 + Ph_3P \longrightarrow cis\text{-}W(CO)_4(PPh_3)(NHC_5H_{10}) \xrightarrow{^{13}CO}$$
$$cis\text{-}W(CO)_4(^{13}CO)(PPh_3) \quad (3.44)$$

$$cis\text{-}Mo(CO)_4(NHC_5H_{10})_2 + 2L \longrightarrow cis\text{-}Mo(CO)_4L_2 \qquad (3.45)$$

L = PR$_3$, P(OR)$_3$, SbPh$_3$

The incoming CO or phosphines displace piperidine ligands because these ligands form stronger bonds to the metal atom. Such stereospecific synthetic pathways eliminate the need for difficult separations of the isomeric mixtures that would be obtained from direct CO substitution. Specific isomer synthesis is very useful for mechanistic and spectroscopic studies.

Steric factors are important in ligand substitution reactions as well. In reactions that represent the *reverse* of equation 3.41, the rates of dissociative loss of one L ligand from a series of phosphine complexes, *cis*-$Mo(CO)_4L_2$, indicate that steric factors are more important than electronic considerations. Bulkier L ligands dissociate at a faster rate than do less bulky ones. Conversely, rates of dissociative loss of one L ligand from a series of complexes, *trans*-$Cr(CO)_4L_2$, where L is a phosphorus or arsenic donor ligand, are more dependent on the $\pi$-acidity of the L ligands than on steric interactions. Strong $\pi$-acceptors dissociate at lower rates. Presumably, the larger steric interactions in the *cis* isomers predominate over any underlying electronic effect as revealed in the *trans* isomers.

Photochemically induced reactions of metal carbonyls are under active investigation. In the presence of $^{13}CO$ photochemical dissociation of a ligand from $M(CO)_5L$, where L is py or piperidine, gives $M(CO)_5(^{13}CO)$ or *cis*- and *trans*-$M(CO)_4(^{13}CO)L$, depending on which ligand is lost:

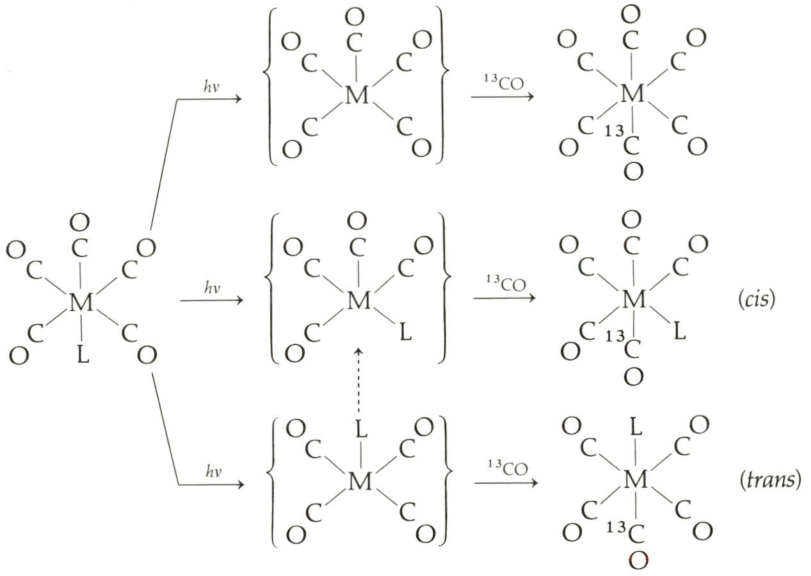

$$(3.46)$$

M = Cr, Mo, W; L = py or $NHC_5H_{10}$

Product analysis reveals that loss of L is the preferred photochemical process. No *trans* isomer is observed; so an axial $M(CO)_4L$ intermediate, if formed, probably isomerizes to the basal isomer as shown by the dashed arrow.

Conversely, when $Mo(CO)_5(PPh_3)$ is photolyzed at 366 nm in the presence of $^{13}CO$ or $PPh_3$ both *cis*- and *trans*-$Mo(CO)_4(PPh_3)L$ complexes form with only slight evidence of photochemical loss of $Ph_3P$. Apparently, an amine ligand is more photo-labile than a phosphine ligand.

We can now ask *how* photochemical activation accelerates ligand dissociation. A general answer is obtained by considering how an electronic transition at the metal atom changes the nature of metal–ligand bonding. Consider, for example, chromium hexacarbonyl. Figure 2.1 shows the bonding scheme of $Cr(CO)_6$ in its ground electronic state. Absorption of a UV photon of the proper energy can excite an electron from a $t_{2g}(\pi^*)$ orbital to one of the $e_g(\sigma^*)$ orbitals. In going to this excited electronic state, the metal–ligand bonding changes considerably. Metal-to-ligand $\pi$ back bonding decreases because one of the predominantly metal-localized $\pi$ electrons is now removed from that set of $\pi$ orbitals. In addition, this electron now occupies a $\sigma^*$ orbital, which is antibonding with respect to metal–ligand $\sigma$ bonding. (Recall that the $e_g^*$ orbitals are composed mainly of the metal $d_{z^2}$ and $d_{x^2-y^2}$ orbitals, so that an electron in these orbitals essentially repels the $\sigma$ electron density being donated by the ligands.) Both effects weaken the metal–ligand bonding and thereby enhance ligand dissociation. A strong $\sigma$-antibonding effect would favor the loss of the more strongly $\sigma$-donating ligand such as an amine.

## STERIC INFLUENCE IN SUBSTITUTED METAL CARBONYL CHEMISTRY

Steric effects (or steric strain energies) are very important in organometallic chemistry. It is now well established that the steric bulk of a particular ligand can influence the chemical behavior, bonding, and spectroscopic properties of a complex. The steric size of a ligand is expressed by the *cone angle*, $\theta$, (**3.16**).

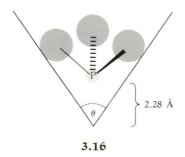

2.28 Å

**3.16**

For symmetrical ligands such as $PR_3$, the value of $\theta$ is equal to the apex angle of a cylindrical conical surface centered 2.28 Å from the phosphorus atom (representing, for example, the position of a nickel atom) and just touching the van der Waals radii of the outermost atoms of the R groups. When several rotational conformations of the R groups are possible, the one giving the minimum value of $\theta$ is used provided significant steric interaction between the R groups is avoided. Although phosphine ligands have been the most intensively studied, the cone angles for other ligand types have been determined also. Table 3.1 lists cone angles for several ligands.

We have already encountered steric effects in ligand substitution reactions. The degree of CO substitution by L in $M(CO)_6$ complexes depends on the size of L. When L is bulky, such as $P(NMe_2)_3$ or $P(i\text{-}Pr)_3$, only *trans*-$M(CO)_4L_2$ complexes form. However, less bulky phosphine ligands, such as $PMe_3$, form *cis*-$M(CO)_4L_2$, *fac*-$M(CO)_3L_3$ and *cis*-$M(CO)_2L_4$ complexes.

**TABLE 3.1** Values of the cone angle, $\theta$, for various ligands.

| Ligand | $\theta$ (°) | Ligand | $\theta$ (°) |
|---|---|---|---|
| H | 75 | diphos | 125 |
| $PH_3$ | 87 | t-butyl | 126 |
| $CH_3$ | 90 | $P(OPh)_3$ | 128 |
| F | 92 | $PEt_3$, $PPr_3$, $P(CH_2CH_2CN)_3$ | 132 |
| CO | 95 | $\eta^5$-$C_5H_5$ | 136 |
| Cl, Et | 102 | $PPh_3$ | 145 |
| $PF_3$ | 104 | $P(NMe_2)_3$ | 157 |
| Br, Ph | 105 | $P(i\text{-}Pr)_3$ | 160 |
| I, $P(OMe)_3$ | 107 | $P(t\text{-}butyl)_3$ | 182 |
| $P(OEt)_3$ | 109 | $P(C_6F_5)_3$ | 184 |
| $PMe_3$ | 118 | | |

Source: C. A. Tolman, *Chem. Rev.*, **77**, 313 (1977).

Frequently, the photochemical kinetically controlled products from CO dissociation in $M(CO)_5L$ complexes involve loss of a CO ligand *cis* to the bulkier L ligand.
Loss of a *cis*-CO ligand affords a five-coordinate intermediate having reduced
$L \cdots CO$ steric repulsion. Subsequent coordination of an incoming ligand L' would
give a *cis*-$M(CO)_4LL'$ complex that could then isomerize to the *trans* isomer, thus
reducing steric repulsion between L and L'.

Mechanistically different results are observed in thermally initiated ligand displacement reactions. When *cis*-$W(CO)_4(Me_2NCH_2CH_2CH_2NMe_2)$ (**3.17**) is treated
with a variety of phosphines, represented by L, the chelating amine ligand is displaced by L to afford *cis*- and *trans*-$W(CO)_4L_2$ complexes.

$$W(CO)_4(\text{tmpa}) + L \xrightarrow[40°C]{-\text{tmpa}}$$

**3.17**

tmpa $= Me_2N(CH_2)_3NMe_2$

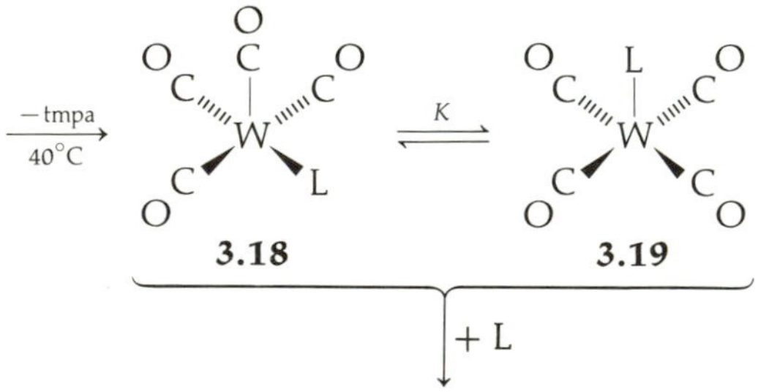

**3.18**    **3.19**

$+ L$

*cis*-$W(CO)_4L_2$    or    *trans*-$W(CO)_4L_2$

At 40°C *cis*–*trans* isomerization of the products does not occur. Therefore, the observed *cis/trans* ratio of the product complexes is presumably a reflection of steric
interactions attributable to the second incoming phosphine ligand as it attacks the
coordinately unsaturated intermediate **3.18** or **3.19**. Significant $L \cdots L$ steric
interactions favor formation of a *trans* isomer independent of the value of K.

As shown below for two $PPh_2R$ and two $P(aryl)_3$ ligands, the *trans* isomer is
favored as the steric bulk of L increases. A similar overall trend is observed among
all four L ligands, but the two different classes of phosphine ligands are slightly
different electronically and should not be compared.

| L in W(CO)$_4$L$_2$ | $\theta$ | % cis isomer | % trans isomer |
|---|---|---|---|
| PPh$_2$Et | 140 | 38 | 62 |
| PPh$_2$(t-butyl) | 157 | 21 | 79 |
| P(p-tolyl)$_3$ | 145 | 21 | 79 |
| P(o-tolyl)$_3$ | 194 | 0 | 100 |

For phosphine ligands, PR$_3$, increasing the steric bulk of R tends to (1) increase R—P—R angles; (2) increase the R$_3$P—M—L angles between a PR$_3$ ligand and the ligands *cis* to it; (3) increase M—P bond distances; (4) increase rates of ligand dissociation reactions but decrease rates of associative ligand substitution reactions; (5) favor lower coordination numbers; and (6) favor less crowded isomers. Clearly, it is important to include steric as well as electronic factors in any structural, spectroscopic, or chemical analysis of organometallic compounds.

## STRUCTURES OF SUBSTITUTED METAL CARBONYL COMPLEXES

To a first approximation the structures of substituted metal carbonyl complexes are analogous to those of the parent metal carbonyls (tables 2.2–2.4). Structural studies establish molecular geometry and provide precise bond distance and angle data. Chemical bonding is analyzed by comparing structural data for similar bonds within the same molecule or analogous complexes.

The nickel phosphine complex, (OC)$_3$NiPPh$_2$PPh$_2$Ni(CO)$_3$ (**3.20**) has a structure like that of Ni(CO)$_4$ with one of the CO ligands replaced by a diphenylphosphino ligand.

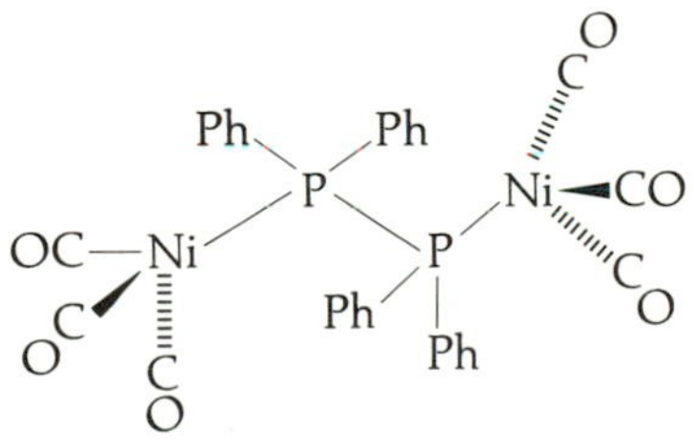

**3.20**

| Ni—P | 2.225(3) Å |
|---|---|
| Ni—C(av) | 1.803(3) Å |

The Ni atoms have nearly tetrahedral coordination geometry, and the diphenylphosphino ligands are not very strained sterically as the P—Ni—C angles in **3.20** have an average value of 107.4°C.

Monosubstituted complexes of the type M(CO)$_4$L, where M is a metal of the iron group and L has a Group V donor atom, have structures based on a trigonal bipyramid with L usually occupying an axial site. An example is Fe(CO)$_4$py (**3.21**).

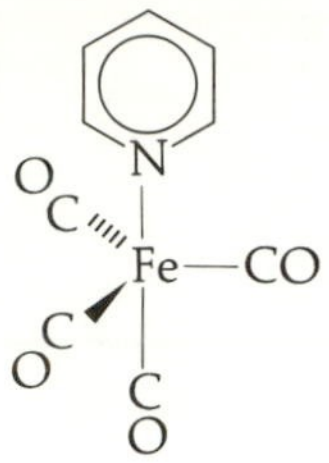

**3.21**

The average values of the OC(eq)—Fe—CO(eq) and OC(eq)—Fe—CO(ax) angles are 120° and 89.8°, respectively, and the N—Fe—C(ax) angle is 176.2°. The Fe—C(ax) distance 1.772(7) Å is slightly shorter than the average of the Fe—C(eq) distances, 1.805(7) Å.

Triphenylstibine complexes, $M(CO)_4(SbPh_3)$, of iron and ruthenium are not iso-structural as shown in **3.22** and **3.23**. Complex **3.22** is nearly trigonal bipyramidal and exists as the axial isomer. As in **3.21**, the Fe—C(ax) distance is slightly shorter than the Fe—C(eq) distance.

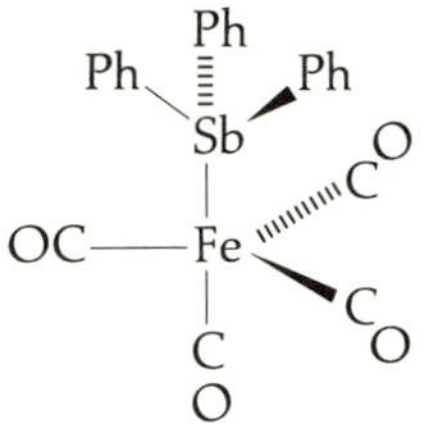

**3.22** 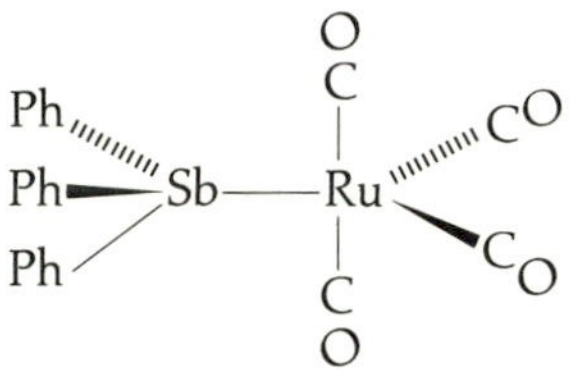                                              **3.23**

Complex **3.23**, however, is the first complex of its type having confirmed equatorial substitution. The Ru coordination geometry is slightly distorted toward a square-pyramidal structure as evidenced by essentially equal Ru—CO distances and a large OC(eq)—Ru—CO(eq) angle of 136.6°. Since $Ph_3PRu(CO)_4$ exists as the axial isomer, the reason for the unexpected geometry of **3.23** is not known. How-ever, the energy difference between trigonal-bipyramidal and square-pyramidal geometries is assumed to be quite small. Steric forces may greatly influence the stability of an observed structure. Intermolecular packing forces within the solid state can also significantly affect molecular structure.

Monosubstituted octahedral complexes of the type $M(CO)_5L$ can exist as only one isomer. Structural data for these complexes are relevant to our later discussion of M—L bonding. Variations in M—L bond distances reflect differences in M—L bonding and steric interactions. Table 3.2 provides some useful data for such an analysis. Several structural features are worth noting:

1. The *differences* in Mo—P and Cr—P distances for the $PPh_3$ and the $P(CH_2CH_2CN)_3$ complexes are 0.138 Å and 0.142 Å, respectively. This constant difference reflects the inherently larger covalent radius of molybdenum.

**TABLE 3.2** Structural parameters for the complexes $M(CO)_5PR_3$, including ligand cone angles.

| Complex | M—P (Å) | M—CO(ax) (Å) | Average M—CO(eq) (Å) | Cone angle $\theta$ of L (°) |
|---|---|---|---|---|
| $Cr(CO)_5PPh_3$ | 2.422(1) | 1.844(4) | 1.880(4) | 145 |
| $Cr(CO)_5P(OPh)_3$ | 2.309(1) | 1.861(4) | 1.896(4) | 128 |
| $Cr(CO)_5P(CH_2CH_2CN)_3$ | 2.364(1) | 1.876(4) | 1.891(4) | 132 |
| $Mo(CO)_5PPh_3$ | 2.560(1) | 1.995(3) | 2.046(4) | 145 |
| $Mo(CO)_5P(CH_2CH_2CN)_3$ | 2.506(1) | 2.008(4) | 2.044(5) | 132 |
| $Mo(CO)_5PF_3{}^a$ | 2.37(1) | 2.063(6) | 2.063(6) | 104 |
| $Mo(CO)_5(PTA)^a$ | 2.479(5) | 2.034(5) | 2.01(2) | 102 |
| $W(CO)_5P(t\text{-butyl})_3$ | 2.686(4) | 1.98(2) | 2.01(2) | 182 |
| $W(CO)_5PMe_3$ | 2.518 | | | 118 |

[a] Structure determined in gas phase by electron diffraction.
[b] PTA is phosphatriazaadamantane, a cyclic ligand of the type $P(CH_2N\diagup)_3$ (see abbreviation list).
Principal data source: F. A. Cotton, D. J. Darensbourg, and W. H. Ilsley, *Inorg. Chem.*, **20**, 578 (1981).

2. A difference of 0.113 Å in the Cr—P distances of the $PPh_3$ and $P(OPh)_3$ complexes is difficult to interpret since the phosphine ligands differ both electronically and sterically.

3. The Mo—P distance in the $PF_3$ complex is about 0.11 Å shorter than the Mo—P distance to the PTA ligand. Since $PF_3$ and PTA are similar sterically, this difference may indicate a greater Mo—P bond order in the $PF_3$ complex.

4. The larger W—P distance in the $P(t\text{-butyl})_3$ complex compared to that of the $PMe_3$ complex may reflect a larger steric interaction between the $P(t\text{-butyl})_3$ ligand and the $W(CO)_5$ moiety. These trialkylphosphine ligands are probably very similar electronically.

For di- and trisubstituted complexes of the types $M(CO)_4L_2$ and $M(CO)_3L_3$, the ancillary ligands usually prefer to be *trans to CO ligands*, as shown for the $PH_3$ complexes **3.24** and **3.25**.

**3.24**          **3.25**

| Cr—P | 2.349(2) Å | Cr—P | 2.346(4) Å |
|---|---|---|---|
| Cr—C(a) | 1.847(4) Å | Cr—C | 1.84(1) Å |
| Cr—C(b) | 1.914(7) Å | | |

Steric interaction between $PH_3$ ligands is not very significant because of similar cone angles for $PH_3$ and CO ligands, 87° and 95°, respectively. When L is $P(OPh)_3$, for which $\theta = 128°$, phosphine ligands in disubstituted complexes prefer to be *trans to each other*, thus minimizing L $\cdots$ L steric interactions as shown in **3.26**.

**3.26**                                    **3.27**

$$\text{Cr—P} \quad 2.252(1) \text{ Å} \qquad \text{Cr—P} \quad 2.429(8) \text{ Å}$$
$$\text{Cr—C} \quad 1.878(4) \text{ Å} \qquad \text{Cr—C} \quad 1.829(8) \text{ Å}$$

Complex **3.27** is interesting because it exists as the all *cis* or *facial* isomer although we might have expected the *meridional* isomer (with two *trans* phosphine ligands) to be preferred sterically. However, steric interactions in *facial* and *meridional* isomers may be rather similar because *meridional* isomers still have two L $\cdots$ L *cis* interactions. As we shall discuss shortly, a *facial* isomer is preferred electronically. Complex **3.27** is also unusual in that it has the longest $Cr^0$—P bond distances yet observed. Presumably, the expected large steric interactions between the $PEt_3$ ligands, $\theta = 132°$, cause elongation of the Cr—P distances.

Sterically bulky ligands, such as $PMe_2Ph$ ($\theta = 122°$), $PMePh_2$ ($\theta = 136°$) and $PEt_3$ ($\theta = 132°$) prefer to occupy axial positions in substituted dimanganese complexes $Mn_2(CO)_9L$ and $Mn_2(CO)_8L_2$, as shown in structures **3.28–3.30**.

**3.28**                          **3.29**                          **3.30**

| Mn—P | 2.24 Å | L = $PMePh_2$ | | L = $PEt_3$ | |
|---|---|---|---|---|---|
| Mn—Mn | 2.90 Å | Mn—P | 2.23 Å | Mn—P | 2.25 Å |
| | | Mn—Mn | 2.90 Å | Mn—Mn | 2.91 Å |

Phosphine substitution of $Mn_2(CO)_{10}$ does not perturb the molecular structure or the Mn—Mn bond distance significantly.

When $Fe_2(CO)_9$ is treated with 2,2′-bipyridyl (bipy) in THF solution, the dinuclear complex $Fe_2(CO)_7$(bipy) forms (**3.31**).

With the bipy ligand acting as a bidentate chelating ligand, complex **3.31** can be viewed as a stoichiometric derivative of $Fe_2(CO)_9$ (with the substitution of two CO ligands); however, it is not a structural derivative of $Fe_2(CO)_9$. There are only two bridging CO ligands, and one of these is semi-bridging. The observed structure can be derived from **3.32** (where both Fe atoms obey the EAN rule) by donating excess *d*-electron density on $Fe^a$ to the $\pi^*$ MO of one of the terminal CO ligands on $Fe^b$, thus affording a semi-bridging ligand. The $Fe^a$ atom is relatively electron rich because of the strong Lewis basicity of the two nitrogen donor atoms.

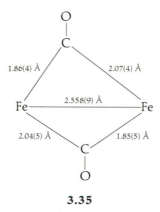

**3.31**                              **3.32**

The iron complexes $Fe_3(CO)_{11}(PPh_3)$ (**3.33**) and $Fe_3(CO)_9(PMe_2Ph)_3$ (**3.34**) are nearly isostructural to $Fe_3(CO)_{12}$.

**3.33**                              **3.34**

Each complex has two $\mu_2$-CO ligands. In complex **3.34** the $\mu_2$-CO ligands are essentially symmetrical; in **3.33** these ligands form a compensating set of asymmetrically bridging ligands, as shown in **3.35**.

**3.35**

The cause of such minor distortions is unknown, but it may reflect subtle energy differences between various intermediate stages of terminal or bridging CO ligands or subtle perturbations of the CO polyhedron when ligand substitution has occurred.

The structures of a Vaska-type complex, trans-$Ir(Cl)(CO)[P(o\text{-}CH_3C_6H_4)_3]_2$, and of the rhodium carbonyl chloride dimer $[Rh(CO)_2Cl]_2$ are shown in **3.36** and **3.37**. Both complexes have essentially square coordination about the metal atoms. In the rhodium complex the two coordination planes share an edge defined by the bridging chlorine atoms. Recent MO calculations on **3.37** indicate a weak Rh—Rh bonding

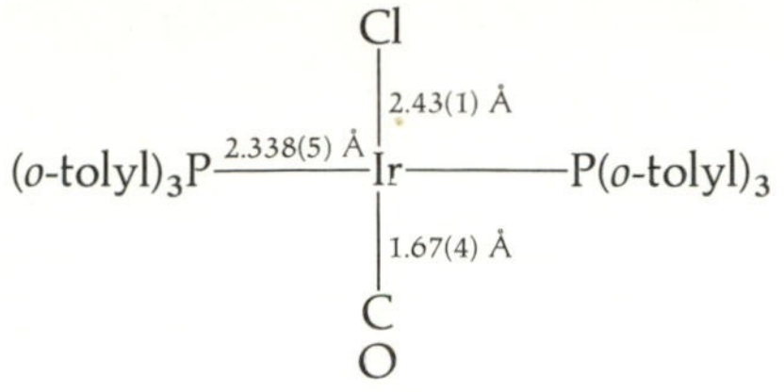

**3.36**

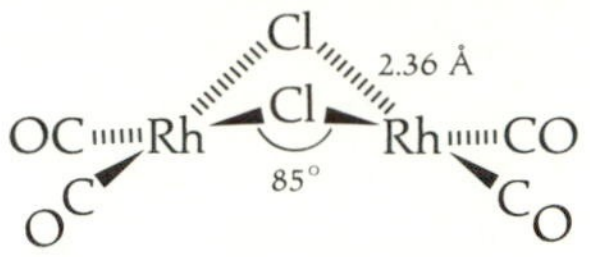

**3.37**
Rh—C = 1.81 Å

interaction. Metal carbonyl complexes that contain terminal or bridging halide ligands are quite common.

Isocyanides usually function as terminal or doubly bridging ligands. The complex
$Fe_2(CNEt)_9$ (**3.38**) is structurally similar to $Fe_2(CO)_9$. The terminal EtNC ligands
are essentially linear (mean CNC angle is 171°), and the bridging EtNC ligands are
bent (mean CNC angle is 123°). The Fe—Fe distance of 2.462(3) Å is 0.06 Å
shorter than the Fe—Fe distance in $Fe_2(CO)_9$.

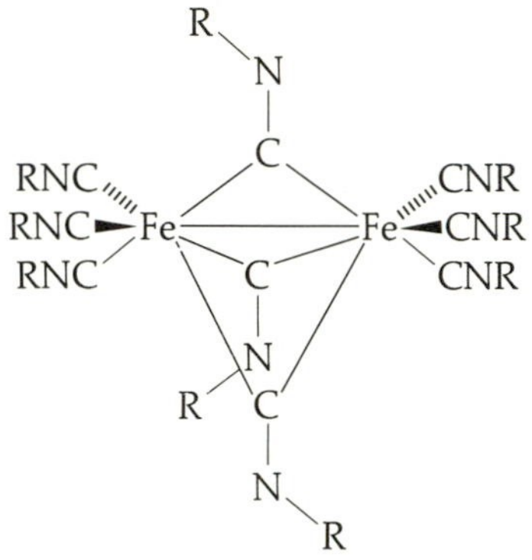

**3.38**

R = Et

The average Fe—CNEt(terminal) distance 1.84 Å is not significantly different from
the average Fe—CO(terminal) distance in $Fe_2(CO)_9$. The C—N bond order in
bridging RNC ligands is markedly less than that of terminal ligands.

An x-ray structure of $Co_2(t$-butyl $NC)_8$ reveals a structure similar to that of
$Co_2(CO)_8$. The cobalt atoms are connected by a single bond and two $\mu_2$-RNC
ligands. Each Co also has three terminal isocyanide ligands. Solution IR data reveal
only a bridged structure in contrast to the structure of $Co_2(CO)_8$ in which the all-
terminal isomer is also observed in solution.

## METAL-LIGAND BONDING IN SUBSTITUTED METAL CARBONYL COMPLEXES

Ligands having nitrogen or oxygen donor atoms, such as amines, alcohols or ethers,
are usually strong σ-donors. Coordination to a metal atom involves donation of an
electron pair on the donor atom to an empty valence orbital of the metal. Such co-

ordination places considerable electron density on the metal. When O or N donors are located in an unsaturated molecule, filled $\pi$ molecular orbitals of the ligand may also donate electron density to the metal. Pi-electron donation from $p\pi$ atomic orbitals, such as found on halogen ligands, also occurs. Unoccupied molecular orbitals of unsaturated ligands can receive electron density from the metal atom as $d\pi$-$p\pi^*$ back bonding. Such M—L $\pi$-bonding interactions explain the enhanced stability of complexes having $o$-phen rather than ethylenediamine as N-donor ligands.

In Werner coordination complexes containing metal ions in high oxidation states, amines afford more stable complexes than do phosphines. This trend is expected since N-donor ligands are stronger $\sigma$-donors than P-donor ligands. However, organometallic complexes containing phosphines tend to be more stable thermally than complexes containing amines, and, as discussed before, amine ligands appear to be more photo-labile than phosphine ligands. These observations can be rationalized by recalling that metal atoms in organometallic complexes are in reduced oxidation states. Only moderately strong $\sigma$-donors are preferred to very strong $\sigma$-donors to minimize the negative charge build-up on the metal atom. As we shall now see, there is an additional feature in M—$PR_3$ bonding.

Table 3.3 shows $v_{CO}$ data for a series of substituted metal carbonyl complexes. For $facial$-$L_3M(CO)_3$ complexes, two $v_{CO}$ bands of approximate relative intensities 1:2 (high-energy band to low-energy band) are observed. The calculated weighted-average frequencies allow a simple comparison of the relative amounts of electron density on the metal atoms. Consistent with the above predictions, $v_{CO}$ values for amine complexes are very low compared to those of phosphine complexes, reflecting the stronger $\sigma$-donation of amine ligands.

There has been considerable controversy over the interpretation of $v_{CO}$ data for phosphine complexes. One explanation is based solely on a $\sigma$-inductive argument.

**TABLE 3.3   C—O stretching frequencies of several phosphine and amine complexes of the type $fac$-$L_3Mo(CO)_3$.**

| L | $v_{CO}(cm^{-1})$ | Weighted average $(1:2)v_{CO}(cm^{-1})$ |
|---|---|---|
| py | 1888, 1746 | 1793 |
| dien | 1898, 1758 | 1805 |
| $CH_3CN$ | 1915, 1783 | 1827 |
| $PPh_3$ | 1934, 1835 | 1868 |
| $PMe_3$ | 1945, 1854 | 1884 |
| $PClPh_2$ | 1977, 1885 | 1916 |
| $P(OMe)_2Me$ | 1970, 1892 | 1918 |
| $P(OMe)_3$ | 1977, 1888 | 1918 |
| $P(OPh)_3$ | 1994, 1922 | 1946 |
| $PCl_2Ph$ | 2016, 1943 | 1967 |
| $PCl_2OEt$ | 2027, 1969 | 1988 |
| $PCl_3$ | 2040, 1991 | 2007 |
| $PF_3$ | 2090, 2055 | 2067 |

Principal data source: F. A. Cotton, *Inorg. Chem.*, **3**, 702 (1964).

For a series of phosphine ligands, $PR_3$, as the substituents R become more electronegative, the lone electron pair on P becomes more tightly held; and, therefore, the $PR_3$ molecule becomes a poorer ligand ($\sigma$-donation is decreased). In going from $Ph_3P$ to $F_3P$, electron donation to the metal atom is decreased substantially, and $v_{CO}$ should increase in energy—as it does.

A second and more popular explanation of these $v_{CO}$ data is based on a $\pi$-resonance back bonding argument. Elements in the sodium series, such as Si, P and S or heavier elements, like As, Sb, Se and Te, have empty $d$ orbitals in their *valence* shells in contrast to the lighter elements. These $d$ orbitals are of low enough energy that they should overlap with filled $d$ orbitals on a metal atom, thus affording $d\pi$-$d\pi$ back bonding (similar to the $d\pi$-$p\pi$ back bonding to CO ligands) as shown in figure 3.2.

If we assume that a series of $PR_3$ ligands have a *constant* $\sigma$-donation to a metal atom, then as R becomes more electronegative the empty $d$ orbitals on the phosphorus atom become more electronegative (i.e., lower in energy) and $d\pi$-$d\pi$ back bonding is enhanced. Since the back bonding removes electron density from the metal atom, the stretching frequencies of CO ligands on the metal atom shift to higher energy as R becomes more electronegative. This same rationale also explains the data in Table 3.3.

Although $d\pi$-$d\pi$ back bonding to ancillary ligands is now generally accepted, the *relative contribution* of $\sigma$-donation and $d\pi$-$d\pi$ back bonding to a given M—L chemical bond is very difficult to determine. However, based on $v_{CO}$ data for a large variety of complexes, the following ranking of common ligands according to *decreasing* $\pi$-acidity has been suggested:

$$CO \approx PF_3 \approx CNR > PCl_3 \approx AsCl_3 \approx SbCl_3 > PCl_2(OR) > PCl_2R >$$
$$PCl(OR)_2 > PClR_2 \approx P(OR)_3 > PR_3 \approx AsR_3 \approx SbR_3 \approx SR_2 > RCN >$$
$$NR_3 \approx OR_2 \approx ROH > H_2NC(O)R$$

This *qualitative* ranking of ligand $\pi$-acidity is useful to the synthetic chemist in adjusting the electron density on a metal atom in a given complex. However, it may

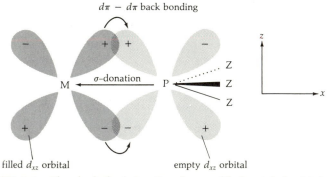

**FIGURE 3.2**    The $d\pi$-$d\pi$ back bonding from a filled metal $d$ orbital to an empty $3d$ orbital of a phosphorus atom in a phosphine ligand, $PR_3$, as shown in the $xz$ plane. A similar back bonding interation involving the $d_{xy}$ orbitals occurs in the $xy$ plane.

not properly account for differences in ligand $\sigma$-donation ability or steric bulk. The above trend follows the expected order based on the type of empty $\pi$ ligand orbitals and the relative electronegativity of the substituents on the donor atoms.

We can now explain why ancillary ligands usually prefer to be *trans* to a CO ligand. Because CO is one of the strongest $\pi$-acid ligands, M—CO $d\pi$-$p\pi^*$ back bonding is maximized when the other ligand attempting to back bond electron density from the *same* metal $d\pi$ orbital is a weaker $\pi$-acid. This competing ligand would be diametrically opposite to the CO ligand; therefore, weaker $\pi$-acid ligands prefer to be *trans* to stronger $\pi$-acid ligands. Of course, strong steric interactions can override these electronic structural preferences in highly substituted complexes.

Isocyanide coordination to a transition metal is represented by Lewis structures **3.39–3.41**.

$$:C{\equiv}\overset{+}{N}{-}R \qquad M{-}\overset{-}{C}{\equiv}\overset{+}{N}{-}R \longleftrightarrow M{=}C{=}\overset{..}{N}{-}R$$

**3.39**                                   **3.40**

free ligand                          terminal ligand

**3.41**

$\mu_2$-ligand

Metal-to-carbon back donation is represented by the Lewis structure, M=C=$\overset{..}{N}$—R. The bent structure and the long C—N distance of $\mu_2$-isocyanide ligands (**3.41**), result from formal $sp^2$ hybridization at the C and N atoms.

Since CNR and CO are isoelectronic, their molecular orbital descriptions are very similar. The M—CNR bonding is composed of M $\leftarrow$ C $\sigma$-donation from a $5\sigma$ MO and of M $\rightarrow$ C back donation into a $2\pi^*$ MO. Because the electronegativity of an NR group is less than that of an oxygen atom, the donor and acceptor molecular orbitals of a CNR ligand are higher in energy than those of CO. This energy difference makes isocyanide ligands much better $\sigma$-donors and slightly poorer $\pi$-acceptors than CO. The strong $\sigma$-donation by isocyanides is reflected in their ability to form stable complexes with metals in high formal oxidation states (see eq. 3.37).

As might be expected, C—N stretching frequencies of isocyanide ligands provide useful information about M—CNR bonding. In free isocyanides, $\nu_{CN}$ values are about 2130 cm$^{-1}$. Complexation to positively charged metals or binary metal carbonyl moieties shifts these frequencies by as much as 150 cm$^{-1}$ to *higher* energy. Such shifts are consistent with strong M $\leftarrow$ CNR $\sigma$-donation from the slightly antibonding $5\sigma$ MO and essentially no M $\rightarrow$ CNR back bonding. Complexation to low-valent metals may shift $\nu_{CN}$ to *lower* energy by 60–340 cm$^{-1}$, although the average shift is about 115 cm$^{-1}$. These shifts are explained by extensive $d\pi$-$p\pi^*$ M $\rightarrow$ CNR back bonding, which reduces the C—N bond order considerably. Bridging isocyanide ligands have $v_{CN}$ in the range $1870-1580$ cm$^{-1}$, a value consistent with the much longer C—N bond distances for these ligands.

## $v_{CO}$ STRETCHING BANDS AND THE STRUCTURES OF SUBSTITUTED METAL CARBONYL COMPLEXES

We have already used $v_{CO}$ stretching frequencies as an indirect measure of the relative electron densities on metal atoms within a related series of complexes. As less π-acidic ligands replace CO ligands, the electron density at a metal increases thereby enhancing $d\pi$-$p\pi^*$ back bonding to the remaining CO ligands and lowering the CO stretching frequencies. However, more chemical information can be obtained from these CO stretching bands. Independent of the $v_{CO}$ frequencies, the *number* of observed CO bands and their *relative intensities* give evidence of the geometry of the M(CO)$_x$ moiety. These spectral patterns exhibited by $v_{CO}$ bands reflect the three dimensional relative orientation of the CO ligands.

A substituted metal carbonyl complex of the type L$_5$M(CO) has only one $v_{CO}$ band, and the question of relative CO–ligand geometry does not exist. However, if two terminal CO ligands are present, as in L$_n$M(CO)$_2$, then either one or two $v_{CO}$ bands are observed. One band is predicted if the angle between the CO ligands is 180°; two bands are predicted if the OC—M—CO angle is between 180° and 0°. (Of course, angles near zero are chemically unreasonable.) The unique prediction for a linear geometry results from the selection rule for IR-active vibrations. IR-active vibrations must occur with a change in the molecular dipole moment. As shown in **3.42** and **3.43**, two CO ligands can vibrate either in phase or out of phase:

$$\overleftarrow{O}-\overleftarrow{C}-M-\overleftarrow{C}-\overleftarrow{O} \qquad \overleftarrow{O}-\overleftarrow{C}-M-\overrightarrow{C}-\overrightarrow{O}$$

**3.42**                                    **3.43**

in-phase vibration          out-of-phase vibration

Clearly, an in-phase, or *symmetric*, vibration does not alter the molecular dipole moment and therefore does not absorb IR radiation. In an out-of-phase, or *antisymmetric*, vibration, one CO bond is stretching while the other CO bond contracts. This vibration does produce a net change in the molecular dipole moment and so gives rise to a single IR absorption band.

For a nonlinear M(CO)$_2$ geometry, both symmetric (**3.44**) and antisymmetric (**3.45**) stretching vibrations are IR-active because the vector sum of the individual bond dipole displacements is nonzero.

**3.44**                    **3.45**

Furthermore, the relative intensity of the two IR bands is a simple function of the angle $2\phi$.

$$I_{sym}/I_{asym} = \cotan^2 \phi \qquad (3.47)$$

The relationship of equation 3.47 is derived from the fact that an IR band intensity is directly proportional to the square of the dipole moment change that occurs during the vibration. For CO stretching vibrations, a symmetric vibrational mode appears at higher frequency than its antisymmetric counterpart. More energy is required to stretch two CO bonds simultaneously than to execute one CO stretch and one CO contraction.

By recording the IR spectrum of an $L_nM(CO)_2$ complex and applying equation 3.47, we can obtain geometrical information about the complex very conveniently. For example, in a complex of the type $L_3M(CO)_2$, knowing the relative orientation of the CO ligands helps in deducing the molecular structure. If we assume that the metal coordination geometry is approximately trigonal bipyramidal, then the CO ligands could occupy (1) both axial sites, (2) one axial and one equatorial site, or (3) two equatorial sites.

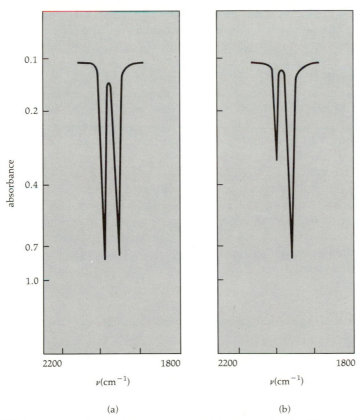

(a)                                            (b)

**FIGURE 3.3** Predicted relative intensities of the two $\nu_{CO}$ bands of the complex $L_3M(CO)_2$ when $2\phi$ is $90°$ (a) and $120°$ (b).

TABLE 3.4   Predicted $\nu_{CO}$ patterns for substituted octahedral complexes of the type, $L_nM(CO)_{6-n}$.

| Complex type | Number of IR active bands | Relative band intensities[a] by decreasing frequency |
|---|---|---|
| $M(CO)_6$ | 1 | — |
| $LM(CO)_5{}^b$ | 3 | w, vs, s |
| $LM(CO)_5{}^c$ | 4 | m, s(sh), vs, s(sh) |
| cis-$L_2M(CO)_4$ | 4 | m, vs, s(sh), s |
| trans-$L_2M(CO)_4$ | 1 | — |
| fac-$L_3M(CO)_3{}^b$ | 2 | s, vs |
| fac-$L_3M(CO)_3{}^c$ | 3 | s, s, s(sh) |
| mer-$L_3M(CO)_3$ | 3 | w, m, m |
| cis-$L_4M(CO)_2$ | 2 | see text |
| trans-$L_4M(CO)_2$ | 1 | — |
| $L_5M(CO)$ | 1 | — |

[a] w = weak, m = medium, s = strong, vs = very strong, and sh = shoulder.
[b] L is a very symmetrical ligand, such as a halogen atom.
[c] L is a relatively unsymmetrical ligand, such as py, $NHR_2$ or $PRR'_2$.

In the first case the two C—O bonds are colinear so only one IR $\nu_{CO}$ band would be observed. For the second and third cases two $\nu_{CO}$ bands would be expected. However, the relative intensities of those peaks would differ. In case 2, $2\phi$ would be nearly $90°$, therefore, $\phi$ would be about $45°$ and the $I_{sym}/I_{asym}$ ratio would be approximately 1.0. These two bands would have nearly equal intensity, as shown in figure 3.3a. In case 3, $2\phi$ would be nearly $120°$ and $\phi \approx 60°$; $I_{sym}/I_{asym} \approx 0.33$, and these two bands would have unequal relative intensities, as shown in figure 3.3b. It is important to stress the convenience and simplicity of using $\nu_{CO}$ spectra to provide both electronic and structural information for a given complex.

For other types of substituted metal carbonyl complexes, different numbers of $\nu_{CO}$ bands are IR-active. Relative intensity patterns provide structural information in a similar way although more complicated expressions than equation 3.47 must be used. Table 3.4 summarizes the expected IR information for octahedral substituted metal carbonyl complexes. Representative spectra for several of these complex types are given in figure 3.4. Familiarity with the general appearance of various $\nu_{CO}$ patterns is useful for identifying reaction products. Organometallic reactions are frequently monitored by IR. Observing changes in $\nu_{CO}$ *frequencies* and *patterns* often makes it possible to deduce what chemical change has taken place.

## STUDY QUESTIONS

1. Writing complete chemical equations, propose a stereospecific preparation of cis-$W(CO)_4(^{13}CO)(PPh_3)$ from $W(CO)_6$.
2. Explain when a disproportionation reaction may occur in ligand substitution reactions, and give one example.
3. Explain why RNC ligands form stable binary complexes to $M^{2+}$ ions although CO does not.

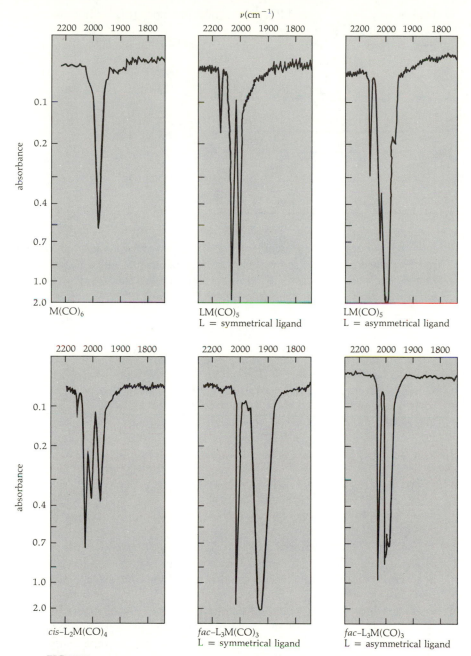

**FIGURE 3.4** Representative IR $\nu_{CO}$ spectra for octahedral complexes of the type $L_nM(CO)_{6-n}$.

4. Explain why CO, RNC and $PF_3$ form similar organometallic complexes.
5. Define the term *labilizing ligand*.
6. Explain the observed change in equilibrium constants for the following reaction as the ligand L is varied:

$$CoCl_2L_2(CO) \xrightleftharpoons{K} CO + CoCl_2L_2$$

| L | Cone angle | K × $10^4(M)$ |
|---|---|---|
| $PEt_3$ | 132° | 8.1 |
| $PEt_2Ph$ | 136° | 500 |
| $PEtPh_2$ | 140° | 1470 |
| $PPh_3$ | 145° | too large to measure |

7. Interpret and explain the experimental results obtained from a kinetic study of the ring closure reaction shown below. The experimental rate constant $k$ for ring closure varies with R in the following order: $k$ when R is Me ($\theta = 107°$) $< k$ when R is Ph ($\theta = 125°$) $< k$ when R is $C_6H_{11}$ ($\theta = 142°$).

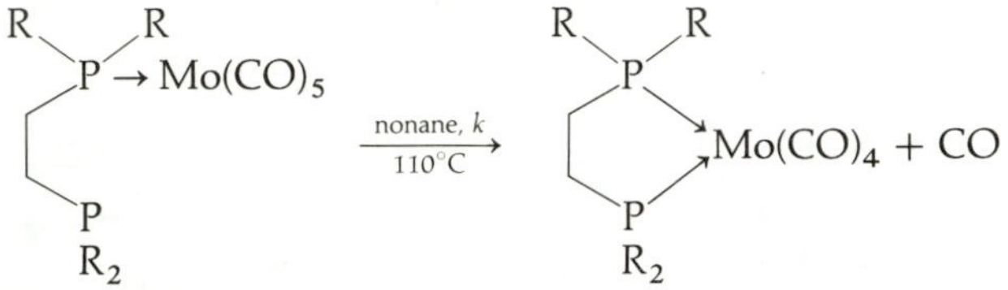

8. Match each $Mo(CO)_5L$ complex listed below with the correct C—O stretching frequency.

| L in $Mo(CO)_5L$ | $\nu_{CO}$ of most intense band ($cm^{-1}$) |
|---|---|
| 1. $PPh_3$ | a. 1990 |
| 2. py | b. 1952 |
| 3. $PF_3$ | c. 1944 |
| 4. DMF (O-donor) | d. 1924 |

9. When $Mo(CO)_6$ was heated at 130°C in diglyme for 30 minutes in the presence of excess en a yellow solid formed. The IR spectrum of this product is shown as spectrum A of figure 3.5. Sketch the structure of the complex, and rationalize your answer.

10. A student heated $W(CO)_6$ in acetonitrile in the presence of excess pyridine. The IR spectra of three yellow products are shown as spectra B, C and D of figure 3.5. Sketch a structure or structures for each product consistent with its IR spectrum, and show how proton NMR spectroscopy could be used to determine or confirm the correct structure of each product. Two of the products can be identified solely from the IR spectra.

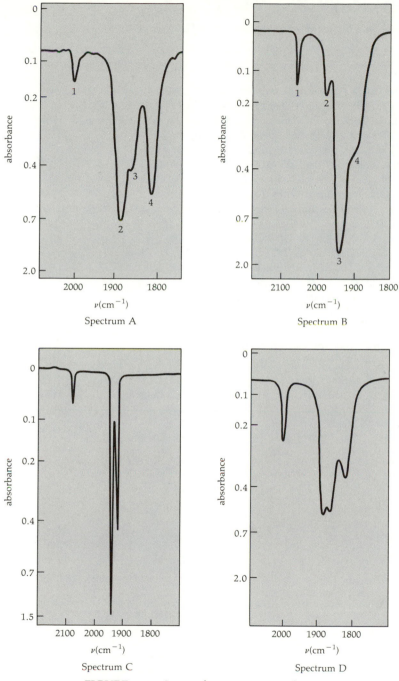

**FIGURE 3.5** Spectra for questions 9 and 10.

## SUGGESTED READING

### Synthesis and Structure of Substituted Metal Carbonyls

Angelici, R. J. *Organometal. Chem. Rev.* **3**, 173 (1968).

Basolo, F., and Pearson, R. G. *Mechanisms of Inorganic Reactions.* 2nd ed. New York: John Wiley, 1967.

Churchill, M. R. *Perspectives Struct. Chem.* **3**, 123 (1970).

Cotton, F. A. *Progr. Inorg. Chem.* **21**, 1 (1976).

Dobson, G. R., Stolz, I. W., and Sheline, R. K. *Adv. Inorg. Radiochem.* **8**, 1 (1966).

Hieber, W. *Adv. Organometal. Chem.* **8**, 1 (1970).

King, R. B. *Acc. Chem. Res.* **13**, 243 (1980).

Klabunde, K. J. *Acc. Chem. Res.* **8**, 393 (1975).

Kruck, Th. *Angew. Chem. Int. Ed. Engl.* **6**, 53 (1967).

Malatesta, L. *Progr. Inorg. Chem.* **1**, 283 (1959).

Malatesta, L., and Bonati, F. *Isonitrile Complexes of Metals.* New York: John Wiley & Sons, 1969.

Manuel, T. A. *Adv. Organometal. Chem.* **3**, 181 (1965).

Ozin, G. A., and Voet, A. V. *Acc. Chem. Res.* **6**, 313 (1973).

Strohmeier, W. *Angew. Chem. Int. Ed. Engl.* **3**, 730 (1964).

Timms, P. L., and Turney, T. W. *Adv. Organometal. Chem.* **15**, 53 (1977).

Tolman, C. A. *Chem. Revs.* **77**, 313 (1977).

Treichel, P. M. *Adv. Organometal. Chem.* **11**, 21 (1973).

Yamamoto, F. Y. *Coord. Chem. Rev.* **32**, 193 (1980).

### Infrared Spectroscopy and Photochemical Studies

Braterman, P. S. *Metal Carbonyl Spectra.* New York: Academic Press, 1975.

Braterman, P. S. *Struct. Bonding.* **26**, 1 (1976).

Geoffroy, G. L., and Wrighton, M. S. *Organometallic Photochemistry.* New York: Academic Press, 1979.

Kettle, S. F. A. *Topics Current Chem.* **71**, 111 (1977).

Kettle, S. F. A., and Paul, I. *Adv. Organometal. Chem.* **10**, 199 (1972).

Wrighton, M. *Chem. Rev.* **74**, 401 (1974).

# Cyclopentadienyl Metal Complexes

## CHAPTER 4

One of the most common organometallic ligands other than CO is the *cyclopentadienyl ligand*, $C_5H_5$. For electron-counting purposes, a $C_5H_5$ ligand is formally considered either an aromatic anion, $C_5H_5^-$, or a neutral radical, $C_5H_5$. Both conventions are in use, and the formal oxidation state of the metal atom must be adjusted appropriately (see chap. 1).

A cyclopentadienyl ligand usually coordinates to a metal atom in one of two principal structural arrangements (**4.1** and **4.2**).

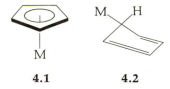

|  |  |
|:---:|:---:|
| **4.1** | **4.2** |

In **4.1** the $C_5H_5$ ligand coordinates to M as a symmetrical $\pi$-ligand; the five carbon atoms of the ring are essentially equidistant from the metal atom, and all five C—C distances within the $C_5H_5$ ring are nearly equal. This bonding geometry is referred to as a $\pi$-$C_5H_5$ or $\eta^5$-$C_5H_5$ since the $C_5H_5$ group acts as a *pentahapto* ligand. In **4.2**, the $C_5H_5$ group is a *monohapto* ligand having one M—C $\sigma$ bond and localized cyclopentadiene bonding within the ring. This type of $C_5H_5$ coordination is denoted as $\sigma$-$C_5H_5$ or $\eta^1$-$C_5H_5$. A common abbreviation for a $C_5H_5$ ligand of any structural type is *Cp*.

There are a few examples of M—$C_5H_5$ bonding different from both **4.1** and **4.2**. These structures are less symmetrical derivatives of **4.1**, as we shall discuss later in this chapter.

Substituted derivatives of the $C_5H_5$ ligand, such as $C_5H_4Me$, $C_5Me_5$, and even $C_5Cl_5$, form complexes analogous to **4.1** and **4.2**. Use of the $C_5Me_5$ ligand is growing rapidly because the larger size of this ligand compared to $C_5H_5$ imparts greater kinetic stability to its organometallic complexes. The methyl groups also increase the solubility of $C_5Me_5$ complexes in organic solvents and facilitate their crystallization from solution.

Cyclopentadienyl ligands are obtained directly from cyclopentadiene, $C_5H_6$, or from metallic salts or covalent compounds that contain the cyclopentadienide group. The most commonly used reagent is sodium cyclopentadienide, $NaC_5H_5$. This ionic compound is prepared as shown in equation 4.1. Dicyclopentadiene, $C_{10}H_{12}$, which is the Diels–Alder dimer of $C_5H_6$, is cracked thermally, and $C_5H_6$ is distilled at about 42°C.

$$\tfrac{1}{2}C_{10}H_{12} \xrightarrow{\Delta} C_5H_6 \xrightarrow[\text{THF}]{Na^0 \text{(sand)}} NaC_5H_5 + \tfrac{1}{2}H_2\uparrow \qquad (4.1)$$

Sodium sand is finely dispersed sodium metal, which is prepared by rapidly stirring molten sodium metal in hot toluene. Because of the high acidity of cyclopentadiene ($pK_a \approx 20$), the reaction of $C_5H_6$ with sodium is quantitative. Solutions of $NaC_5H_5$ are very air sensitive and are used immediately after preparation. Sodium cyclopentadienide is a white pyrophoric solid.

Occasionally, $NaC_5H_5$ not only provides $C_5H_5$ ligands but also reduces the metal ion. A less reducing $C_5H_5$ reagent is $TlC_5H_5$, which can be prepared in aqueous solution (eq. 4.2) but is also commercially available.

$$Tl_2SO_4 + 2KOH + 2C_5H_6 \xrightarrow{H_2O} \underset{\text{white solid, 95\%}}{2TlC_5H_5\downarrow} + K_2SO_4 \qquad (4.2)$$

Although $TlC_5H_5$ is much more stable to air than $NaC_5H_5$, its low solubility in most solvents makes it less attractive as a general reagent.

Pentamethylcyclopentadienyl ligands, $C_5Me_5$, are obtained from the corresponding Grignard reagent which is prepared by metalation of 1,2,3,4,5-pentamethylcyclopentadiene:

$$C_5Me_5H + i\text{-PrMgCl} \xrightarrow[\text{reflux}]{\text{toluene}} C_5Me_5MgCl + C_3H_8 \qquad (4.3)$$

1,2,3,4,5-Pentamethylcyclopentadiene is available commercially; however, it can be prepared in large quantities by the procedure shown in the following equation:

$$C_5Me_5H, 75\% \qquad (4.4)$$
$$\text{bp } 55\text{–}60°/13 \text{ torr}$$

The types of compounds discussed in this chapter include (1) binary cyclopentadienyl metal complexes, (2) cyclopentadienyl–metal carbonyl complexes, (3) selected derivatives of cyclopentadienyl–metal carbonyl complexes, and (4) complexes containing cyclopentadienyl like ligands. We shall discuss them in an order that nearly parallels the chronological discovery of cyclopentadienyl metal complexes.

## BINARY CYCLOPENTADIENYL METAL COMPLEXES

Binary cyclopentadienyl metal complexes, $M_x(C_5H_5)_y$, are either metallocene or nonmetallocene molecules. Metallocene complexes have the molecular formula $(C_5H_5)_2M$, where both cyclopentadienyl groups are *pentahapto* ligands.

### Metallocene Complexes

The discovery of ferrocene, $(\eta^5\text{-}C_5H_5)_2Fe$ (**1.2**) in 1951 marks the date of the "rebirth" of transition metal organometallic chemistry. Ferrocene has exceptional stability to exposure to air and heat compared to other known transition metal alkyl complexes. This unusual stability implied an unusual structure, which we now recognize as the $\eta^5\text{-}C_5H_5$ ligand.

Both neutral and ionic metallocene complexes have been prepared as shown in table 4.1. Neutral metallocenes are prepared conveniently by treating anhydrous metal salts, usually halide complexes, with $NaC_5H_5$ in THF or $MeOCH_2CH_2OMe$ (1,2-dimethoxyethane, DME) solution as shown in equation 4.5. Coordination complexes such as $[M(NH_3)_6]Cl_2$, where M is Co or Ni, are also useful reagents because these complexes are more soluble than the anhydrous chlorides.

$$MX_n + nNaC_5H_5 \xrightarrow[\text{or DME}]{\text{THF}} (\eta^5\text{-}C_5H_5)_2M + nNaX + (n-2)\{C_5H_5\cdot\} \quad (4.5)$$

| $MX_n$ | $(\eta^5\text{-}C_5H_5)_2M$ (% yield) | $MX_n$ | $(\eta^5\text{-}C_5H_5)_2M$ (% yield) |
|---|---|---|---|
| $VCl_3$ | 55 | $FeCl_2$ | 73 |
| $VCl_4$ | 47 | $RuCl_3$ | 52 |
| $CrCl_2$ | 82 | $OsCl_4$ | 22 |
| $CrCl_3$ | 82 | $CoCl_2$ | 90 |
| $MnBr_2$ | 70 | $NiCl_2$ | 37 |

Metals in high oxidation states are frequently reduced to the M(II) oxidation state by excess $NaC_5H_5$. Several metallocenes can be prepared directly from cyclopentadiene when an external base is present, as shown in equation 4.6. Ferrocene can be steam distilled from these reaction residues.

$$MCl_2 + 2C_5H_6 + 2Et_2NH \xrightarrow{\text{THF}} (\eta^5\text{-}C_5H_5)_2M + 2[Et_2NH_2]Cl \quad (4.6)$$
$$\sim 80\%$$

M = Fe, Co, Ni

Titanocene is a very reactive molecule and cannot be isolated as $(\eta^5\text{-}C_5H_5)_2Ti$. Instead of adopting this 14-electron structure, it reacts with another molecule to give a 16-electron $\eta^5\!:\!\eta^5$-fulvalene-di-$\mu_2$-hydrido species shown in **4.3**.

TABLE 4.1 Selected known metallocene complexes.

| Metal | Complex | Metal | Complex | Metal | Complex |
|---|---|---|---|---|---|
| Ti | $(C_{10}H_{10})Ti$[a] dark green solid dec. 130°C; $(C_5Me_5)_2Ti$ | Cr | $(C_5H_5)_2Cr$ scarlet solid mp 172°C; $(C_5H_5)_2Cr^+$; $(C_5H_5)_2Cr^-$ | Mn | $(C_5H_5)_2Mn$ purple solid mp 160°C; $(C_5H_5)_2Mn^+$; $(C_5H_5)_2Mn^-$ |
| V | $(C_5H_5)_2V$ purple solid mp 167°C; $(C_5H_5)_2V^+$; $(C_5H_5)_2V^-$ | Co | $(C_5H_5)_2Co$ purple-black solid, mp 173°C; $(C_5H_5)_2Co^+$ yellow solids; $(C_5H_5)_2Co^-$ | | |
| Nb | $(C_{10}H_{10})Nb$[a] | Ni | $(C_5H_5)_2Ni$ green solid dec. 173°C; $(C_5H_5)_2Ni^+$, $(C_5H_5)_2Ni^{2+}$ yellow solids; $(C_5H_5)_2Ni^-$ | | |
| Ta | $(C_{10}H_{10})Ta$[a] | Pd | $(C_5H_5)_2Pd$(?) red solid dec. ~ 40°C | | |
| Fe | $(C_5H_5)_2Fe$ amber solid mp 173°C; $(C_5H_5)_2Fe^+$ purple solids | Rh | $(C_5H_5)_2Rh^+$ yellow solids | | |
| Ru | $(C_5H_5)_2Ru$ pale yellow solid, mp 199°C; $(C_5H_5)_2Ru^+$ pale yellow solids | Ir | $(C_5H_5)_2Ir^+$ yellow solids | | |
| Os | $(C_5H_5)_2Os$ white solid mp 230°C | | | | |

[a] Not a true metallocene; see text.

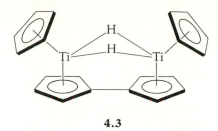

**4.3**

This structure apparently arises from a ring-to-metal hydrogen transfer. Interestingly, the permethyl derivative, $(C_5Me_5)_2Ti$, which lacks any ring protons, forms a dimer that does dissociate into monomers. Niobocene reacts similarly (**4.4**).

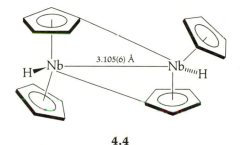

**4.4**

$$[(\eta^5\text{-}C_5H_5)(\eta^1\text{:}\eta^5\text{-}C_5H_4)NbH]_2$$

In this dimer one C—H bond of two of the $\eta^5$-$C_5H_5$ rings has undergone a formal oxidative-addition (cf. chap. 1) to the adjacent niobium atom forming the Nb—H and Nb—C($\sigma$) bonds. A single bond between the Nb atoms allows both metals to adopt an 18-electron structure. The tantalum analog has the same structure.

Some cationic metallocenes can be prepared from M(III) salts and $NaC_5H_5$ if reduction of the metal is prevented. A more general synthetic method involves chemical oxidation of the corresponding neutral metallocenes as shown in equations 4.7 and 4.8.

$$(\eta^5\text{-}C_5H_5)_2Fe + FeCl_3 \xrightarrow{\text{Et}_2O} [(\eta^5\text{-}C_5H_5)_2Fe]Cl + FeCl_2 \qquad (4.7)$$

$$\xrightarrow[\substack{p\text{-benzoquinone}}]{BF_3,\ C_6H_6} [(\eta^5\text{-}C_5H_5)_2Fe]BF_4 \qquad (4.8)$$

The $[(\eta^5\text{-}C_5H_5)_2M]^+$ ions, where M is Fe or Co, are referred to as ferricenium or cobalticenium complexes, respectively. Ferricenium ion is a relatively strong one-electron oxidant (its reduction product being ferrocene).

Anionic metallocenes have been prepared electrochemically. These complexes have not been examined in much detail.

Metallocenes would appear to have rather simple structures. Interligand repulsion between the two $\eta^5$-$C_5H_5$ ligands, if significant, would be minimized when the rings are diametrically opposed as two parallel planes. In this structure, the two rings can adopt one of two idealized relative orientations—an eclipsed conformation, as in **4.5**, where the carbon atoms of one ring lie directly over those of the other ring, or a staggered conformation, structure **4.6**, where the carbon atoms of one ring lie between two adjacent carbons of the other ring.

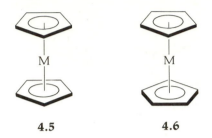

**4.5**                                **4.6**

eclipsed structure        staggered structure

The lower-energy structure of ferrocene is eclipsed and the small deviation from this structure in the solid state is due to crystal-packing forces. The $C_5H_5$ hydrogen atoms are displaced slightly *toward* the Fe atom, presumably because this shift permits better overlap of the Fe orbitals with the $p\pi$ orbitals of the rings.

Decamethylferrocene, $(\eta^5\text{-}C_5Me_5)_2Fe$, has a staggered structure with the methyl groups tilted *away* from the Fe atom to minimize steric repulsion. The eclipsed structure is about 2 kcal/mole less stable than the staggered conformation. The Fe—C and C—C distances in ferrocene and decamethylferrocene are the same.

Other than the relative orientation of the two $C_5H_5$ rings, the most interesting structural parameter in metallocenes is the M—C distance. The M—C distances for the $3d^n$ metallocenes vanadium through nickel are as shown in the table.

| $(\eta^5\text{-}C_5H_5)_2M$ | M—C (Å) |
|---|---|
| V | 2.280(5) |
| Cr | 2.169(4) |
| Mn, $(\eta^5\text{-}C_5Me_5)_2Mn$ | 2.380(6), 2.112(2) |
| Fe | 2.064(3) |
| Co | 2.119(3) |
| Ni | 2.196(4) |

Since atomic radii should decrease from left to right across a series of elements, it is surprising to observe a *decrease* in M—C distances from V to Fe followed by an *increase* in M—C distances from Fe to Ni. (Manganocene is an unusual high-spin organometallic complex with five unpaired $3d$ electrons. Decamethylmanganocene is a low-spin complex, and its M—C distance is used in the above comparison.)

Another interesting structural observation is that the M—C distances of cobalticenium ion are about 0.06 Å *shorter* than those of cobaltocene, while the M—C distances of ferricenium ion are about 0.06 Å *longer* than those of ferrocene. These data imply that M—$C_5H_5$ bonding in $(\eta^5\text{-}C_5H_5)_2Co^+$ is similar to that in ferrocene and M—$C_5H_5$ bonding in $(\eta^5\text{-}C_5H_5)_2Fe^+$ is similar to that in $(\eta^5\text{-}C_5Me_5)_2Mn$.

We now need to examine M—$C_5H_5$ bonding in metallocenes to explain these structural trends and their observed stabilities.

A simple bonding analysis begins with application of the EAN rule. In a neutral metallocene (**4.5** or **4.6**), a metal atom is coordinated to two $\eta^5$-$C_5H_5$ radicals where each $C_5H_5$ ligand acts as a five-electron donor (one $p\pi$ electron in each of the five $2p_z$ atomic orbitals). The total number of electrons in the valence shell of the metal atom is then ten plus the number of valence electrons on the neutral metal atom. For the $3d$ metallocenes, the valence electron counts on M are

| $(\eta^5\text{-}C_5H_5)_2M$ | V | Cr | Mn | Fe | Co | Ni |
|---|---|---|---|---|---|---|
| valence electrons | 15 | 16 | 17 | 18 | 19 | 20 |

Obviously, ferrocene is the only neutral metallocene that obeys the EAN rule. Since metallocenes on both sides of Fe should have less ideal electronic configurations than ferrocene, the M—$C_5H_5$ bonding may be weaker in these complexes as reflected in the variation of M—C distances shown above in going from vanadocene to nickelocene.

Similarly, cobalticenium ion is isoelectronic to ferrocene, and ferricenium ion is isoelectronic to decamethylmanganocene. This correspondence explains the high stability and short M—C distances of cobalticenium ion as well as the relatively high chemical reactivity and long M—C distances of ferricenium ion.

However, since it is relatively rare for organometallic complexes to disobey the EAN rule, why are 15- and 20-electron metallocenes isolable? This question is answered by considering an orbital view of metallocene bonding. Figure 4.1 shows a qualitative bonding scheme for ferrocene. The Fe—$C_5H_5$ bonding results from $\pi$ interactions between the 10 $p\pi$ atomic orbitals of the two $C_5H_5$ ligands and the $3d$, $4s$ and $4p$ atomic orbitals of the iron atom. As shown in figure 2.1, only orbitals that have the same spatial overlap and nearly the same energy interact strongly. The different possible combinations of the ten $2p_z$ atomic orbitals of the $C_5H_5$ ligands are denoted as the $1\pi \to 6\pi$ orbitals. For example, the $1\pi$ and $2\pi$ combinations are shown in **4.7** and **4.8**, respectively.

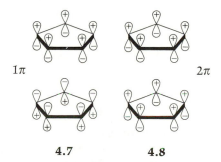

|  |  |
|:---:|:---:|
| $1\pi$ | $2\pi$ |
| **4.7** | **4.8** |

Each Fe valence atomic orbital is also classified as having one of these spatial orientations. When Fe and ligand orbitals interact, they yield the molecular orbitals of ferrocene shown in the center of figure 4.1.

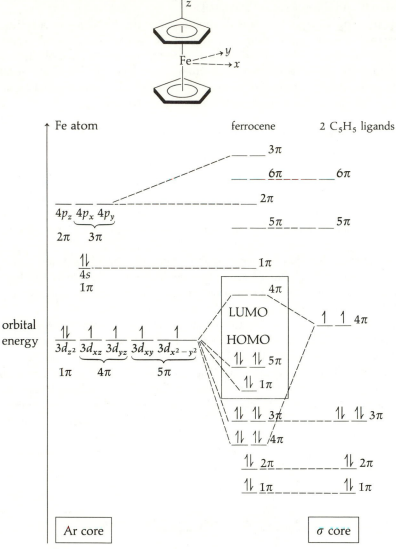

**FIGURE 4.1**   A qualitative bonding scheme for $(\eta^5\text{-}C_5H_5)_2Fe$.

We can understand $Fe$—$C_5H_5$ bonding qualitatively without calculating an exact orbital description of each MO or its orbital energy. Major $Fe$—$C_5H_5$ bonding occurs between the $4\pi$ orbitals of $Fe$, the $3d_{xz}$ and $3d_{yz}$ orbitals, and the $4\pi$ orbitals of the $C_5H_5$ ligands. Less significant $Fe$—$C_5H_5$ bonding arises from the interaction of the iron $4p$ atomic orbitals with the $2\pi$ and $3\pi$ molecular orbitals of the ligands.

However, notice that the $5\pi$ HOMO and the next lower $1\pi$ MO have primarily metal character and that these orbitals are only slightly bonding with respect to

Fe—$C_5H_5$ bonding. Therefore, removing electrons from the $5\pi$ and $1\pi$ orbitals, as is done when going from ferrocene to vanadocene, does not *drastically* affect the net amount of M—$C_5H_5$ bonding although M—C distances do increase slightly. Similarly, population of the $4\pi$ LUMO orbital, which is slightly antibonding with respect to M—$C_5H_5$ bonding, in cobaltocene and nickelocene decreases the net amount of M—$C_5H_5$ bonding and yields longer M—C distances than observed in ferrocene. Ferrocene, or ferrocenelike metallocenes, such as $(\eta^5\text{-}C_5H_5)_2Co^+$, have a completely filled set of bonding molecular orbitals.

Because the molecular orbitals of ferrocene are composed of filled and empty metal and $C_5H_5$ atomic orbitals, synergistic M—$C_5H_5$ bonding is possible. Electron density can be donated from the $C_5H_5$ ligands to the metal atom and back bonded from the metal atom to the $C_5H_5$ ligands. Because of the relatively weak $\pi$-acidity of $C_5H_5$ ligands, the net effect is transfer of electron density from the cyclopentadienyl ligands to the iron atom. This $C_5H_5 \rightarrow Fe$ donation should decrease the $\pi$-bonding electron density *within* the $C_5H_5$ ligands and thereby reduce the C—C bond order relative to that of the free ligand. In crystalline ferrocene, C—C distances range from 1.427(2) Å to 1.438(2) Å with an average distance of 1.433(6) Å. A recent structure of $[Na(Me_2NCH_2CH_2NMe_2)][C_5H_5]$ reveals an essentially ionic solid having discrete $C_5H_5^-$ ions. The average C—C distance in the $C_5H_5^-$ anion is 1.38(1) Å. As anticipated, the average C—C distance in $\eta^5\text{-}C_5H_5$ ligands is slightly longer than that of "free" cyclopentadienide anion.

As depicted inside the square in figure 4.1, the valence electronic configurations of $3d$ metallocenes are represented by an appropriate filling of the $1\pi$, $5\pi$ and $4\pi$ molecular orbitals. These electron configurations are shown in table 4.2 along with the corresponding number of unpaired electron spins and the calculated average dissociation energy, $D(M^{2+}\text{—}C_5H_5^-)$, for removing one $C_5H_5^-$ ligand from the $M^{2+}$ ion.

For vanadocene and chromocene there is only one electron in the $1\pi$ orbital in the most stable ground state structure. Notice that all of the metallocenes are paramagnetic except ferrocene. As might be expected from the trend in M—C distances, it is most difficult to disrupt the $Fe^{2+}\text{—}C_5H_5^-$ bond, and the value of $D(M^{2+}\text{—}C_5H_5^-)$ decreases across the series in either direction from ferrocene.

**TABLE 4.2    Valence electronic configuration, number of unpaired electrons, and average $M^{2+}\text{—}C_5H_5^-$ bond disruption enthalpies for selected metallocenes.**

| Complex | Valence electronic configuration | Number of unpaired electrons | $D(M^{2+}\text{—}C_5H_5^-)$ (kcal/mole) |
|---|---|---|---|
| $(C_5H_5)_2V$ | $(1\pi)^1(5\pi)^2$ | 3 | 292 |
| $(C_5H_5)_2Cr$ | $(1\pi)^1(5\pi)^3$ | 2 | 318 |
| $(C_5Me_5)_2Mn$ | $(1\pi)^2(5\pi)^3$ | 1 | — |
| $(C_5H_5)_2Fe$ | $(1\pi)^2(5\pi)^4$ | 0 | 352 |
| $(C_5H_5)_2Co$ | $(1\pi)^2(5\pi)^4(4\pi)^1$ | 1 | 334 |
| $(C_5H_5)_2Ni$ | $(1\pi)^2(5\pi)^4(4\pi)^2$ | 2 | 315 |

From heat of combustion data, the heat of reaction for the process can be determined.

$$(\eta^5\text{-}C_5H_5)_2Fe \longrightarrow Fe(g)^0 + 2C_5H_5(g) \qquad \Delta H_{rx} \qquad (4.9)$$

The calculated average $Fe$—$C_5H_5$ bond-disruption enthalpy is 73 kcal/mole. If a neutral $C_5H_5$ ligand is taken as a five-electron donor, then each donated electron contributes a bond energy of 14.6 kcal/mole.

A large body of experimental data attests to the high stability of the $Fe$—$C_5H_5$ bonds in ferrocene. Ferrocene is stable to air, moisture, and temperatures up to 470°C. It does not decompose in boiling concentrated HCl or 10% NaOH. In addition, very few examples of ligand substitution in ferrocene are known.

In contrast, other $3d$ metallocenes should exhibit more facile $\eta^5\text{-}C_5H_5$ displacement or transfer reactions. These metallocenes have lower $M$—$C_5H_5$ bond energies, and the metals should undergo reactions which afford 18-electron products. This is particularly true for nickelocene which is a 20-electron complex with two electrons in the antibonding $4\pi$ MO. Some examples of these $C_5H_5$ transfer or displacement reactions are shown in equations 4.10–4.13.

$$(\eta^5\text{-}C_5H_5)_2Cr + FeCl_2 \xrightarrow[\text{65°C, 1 hr}]{\text{THF}} (\eta^5\text{-}C_5H_5)_2Fe + \{CrCl_2\} \qquad (4.10)$$
$$72\%$$

$$(\eta^5\text{-}C_5H_5)_2Cr + CO(g) \xrightarrow[\text{250°C}]{\text{100 atm}} Cr(CO)_6 \qquad (4.11)$$

$$(\eta^5\text{-}C_5H_5)_2Ni + PF_3(g) \xrightarrow[\text{60°C}]{\text{press}} Ni(PF_3)_4 + \text{organic products} \qquad (4.12)$$
$$95\%$$
$$+ 4L$$
$$\xrightarrow[\text{L = Ph}_3\text{P, PhNC}]{\text{C}_6\text{H}_{12}, \text{ 120°C}} NiL_4 + \text{organic products} \qquad (4.12)$$
$$100\%$$

$$(\eta^5\text{-}C_5H_5)_2Fe + RuCl_3 \xrightarrow{\text{250°C}} (\eta^5\text{-}C_5H_5)_2Ru + \{FeCl_3\} \qquad (4.13)$$
$$42\%$$

## Nonmetallocene Complexes

When nickelocene is treated with a variety of electrophilic reagents, such as a protic acid, a unique "triple-decker sandwich" cationic complex, $[(\eta^5\text{-}C_5H_5)_3Ni_2]^+$, forms. As shown in equation 4.14, the proton adds to one of the $\eta^5\text{-}C_5H_5$ ligands converting it into a $\eta^4$-cyclopentadiene ligand. Loss of $C_5H_6$ affords a 14-electron $(C_5H_5)Ni^+$ complex (4.9), which can be isolated as a $BF_4^-$ or $SbF_6^-$ salt. Intermediate 4.9 reacts with nickelocene to give the triple-decker sandwich (4.10). The structure of the $BF_4^-$ salt of 4.10 has been established by x-ray diffraction. Such triple-decker sandwich complexes are predicted by MO theory to be stable if they have either 30 or 34 total valence electrons. Complex 4.10 has 34 valence electrons.

Most of the remaining nonmetallocene binary metal cyclopentadienyl complexes are monomeric complexes, $(C_5H_5)_xM$, where M is a Sc, Ti, V or Cr group metal or

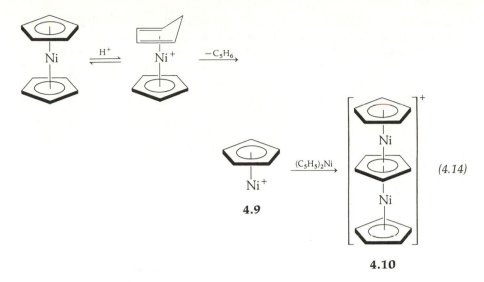

a lanthanide or actinide metal and where $x$ is 3 or 4. These complexes are usually prepared from anhydrous metal halide complexes and $NaC_5H_5$ or $KC_5H_5$ as shown in equations 4.15–4.21. Reduction of the metal ion occurs in some instances.

$$(C_5H_5)_2TiCl + NaC_5H_5 \xrightarrow[\text{2 hrs}]{\text{THF, }\Delta} (C_5H_5)_3Ti + NaCl \qquad (4.15)$$
$$\text{green solid, 70\%}$$

$$(C_5H_5)_2MCl_2 + 2NaC_5H_5 \xrightarrow{\text{THF}} (C_5H_5)_4M + 2NaCl \qquad (4.16)$$
$$\text{Ti: dark green solid, 80\%}$$
$$\text{Zr: yellow solid, 54\%}$$

$$HfCl_4 + 4NaC_5H_5 \xrightarrow[\text{40°C}]{C_6H_6} (C_5H_5)_4Hf + 4NaCl \qquad (4.17)$$
$$\text{yellow solid, 44\%}$$
$$\text{dec. 207°C}$$

$$MoCl_5 + NaC_5H_5(xs) \xrightarrow{Et_2O} (C_5H_5)_4Mo \qquad (4.18)$$
$$\text{red solid, 39\%}$$
$$\text{dec. 220°C}$$

$$MCl_3 + 3NaC_5H_5 \xrightarrow{\text{THF}} (C_5H_5)_3M + 3NaCl \qquad (4.19)$$

M = Sc, Y, La, Ce, Pr, Nd, Sm, Gd, Dy, Er or Yb

$$2MCl_3 + 3(C_5H_5)_2Be \xrightarrow[\text{melt}]{65°C} 2(C_5H_5)_3M + 3BeCl_2 \qquad (4.20)$$

M = Pu, Am, Bk or Cf

$$MCl_4 + 4KC_5H_5 \xrightarrow{C_6H_6} (C_5H_5)_4M + 4KCl \qquad (4.21)$$

M = Th, U or Np

Since U(IV) is isovalent to the M(IV) ions of the titanium group, a structural comparison of these $(C_5H_5)_4M$ complexes is possible. Tetracyclopentadienyl titanium, $(C_5H_5)_4Ti$, is actually $(\eta^1\text{-}C_5H_5)_2(\eta^5\text{-}C_5H_5)_2Ti$ as shown in **4.11**. The Ti atom has essentially tetrahedral coordination geometry as defined by the two Ti—C($\sigma$) bonds and the two Ti—$C_5H_5$ (centroid) axes. Localized $\pi$-bonding within the $\eta^1$-$C_5H_5$ ligands is reflected in the intraring C—C distances shown in **4.12**.

**4.11**                                                    **4.12**

Unexpectedly, all five carbon atoms of these $\eta^1$-$C_5H_5$ ligands are essentially coplanar; an envelope bend does not appear to be characteristic of $\eta^1$-$C_5H_5$ ligand coordination. Although complex **4.11** is a 16-electron Ti complex, the isostructural Mo complex, $(\eta^1\text{-}C_5H_5)_2(\eta^5\text{-}C_5H_5)_2Mo$ obeys the EAN rule.

Tetracyclopentadienyl zirconium, $(C_5H_5)_4Zr$, has one $\eta^1$-$C_5H_5$ and three $\eta^5$-$C_5H_5$ ligands, but the Zr—C($\sigma$) bond and Zr—$C_5H_5$ (centroid) distances are significantly longer than normal Zr(IV)—C distances to these types of ligands. Presumably, there is a balance between steric crowding of the ligands and the tendency of the metal atom to attain an 18-electron configuration. The long Zr—$C_5H_5$ (centroid) distances may indicate that these ligands are effectively acting as something less than five-electron donors.

Unlike $(C_5H_5)_4Zr$, $(C_5H_5)_4Hf$ has two $\eta^5$-$C_5H_5$ and two $\eta^1$-$C_5H_5$ ligands. The Hf—C($\sigma$) distances are short relative to the Zr—C($\sigma$) distance in $(C_5H_5)_4Zr$, a contraction that may reflect the greater electron deficiency of the hafnium ion in this 16-electron complex. Although zirconium and hafnium organometallic complexes are usually isostructural, the structures of $(C_5H_5)_4Zr$ and $(C_5H_5)_4Hf$ are obviously different.

Tetracyclopentadienyl uranium, $(C_5H_5)_4U$, has four $\eta^5$-$C_5H_5$ ligands, but the slightly long U—C distances of 2.81(2) Å indicate some steric crowding of the ligands. Less crowded U(IV)—$C_5H_5$ complexes have U—C distances of approximately 2.74 Å. The centroids of the $\eta^5$-$C_5H_5$ ligands in $(C_5H_5)_4U$ define a tetrahedral coordination geometry. The analogous thorium and neptunium complexes are isostructural.

Clearly, $(C_5H_5)_4M$ complexes, where M is titanium or uranium, represent two structural extremes with the zirconium complex having an intermediate geometry. Since M(IV) ionic radii increase in the order $Ti^{4+}$ (0.61 Å), $Zr^{4+} \approx Hf^{4+}$ (0.72 Å), $U^{4+}$ (0.97 Å), the larger metal atoms can presumably accommodate more $\pi$-ligands because of weaker interligand steric interaction.

# CYCLOPENTADIENYL METAL CARBONYL COMPLEXES

## Preparation of Cyclopentadienyl Metal Carbonyl Complexes

Table 4.3 lists several known neutral homonuclear cyclopentadienyl metal carbonyl complexes. Such complexes have the general formula $(C_5H_5)_xM_y(CO)_z$ and are usually prepared (1) from a metallocene and CO or a binary metal carbonyl, (2) from cyclopentadiene or dicyclopentadiene and a binary metal carbonyl, (3) by a reductive-ligation reaction, or (4) by photolysis or thermolysis of another cyclopentadienyl metal carbonyl complex.

Examples of syntheses using a metallocene and a CO source are shown in equations 4.22–4.27.

$$\text{"}(C_5H_5)_2Ti\text{"} + 2CO \xrightarrow[\text{1 atm}]{C_6H_6,\ 25°C} (\eta^5\text{-}C_5H_5)_2Ti(CO)_2 \qquad (4.22)$$
$$\text{dec. } 90°C$$

$$(\eta^5\text{-}C_5H_5)_2V + CO \xrightarrow[\text{10 min}]{\text{heptane, } 25°C} (\eta^5\text{-}C_5H_5)_2V(CO) \qquad (4.23)$$
$$81\%$$

$$\xrightarrow[\text{7 hrs}]{CO\ (60\ atm)\ 117°C} (\eta^5\text{-}C_5H_5)V(CO)_4 \qquad (4.24)$$
$$23\%,\ \text{dec. } 138°C$$

$$(\eta^5\text{-}C_5H_5)_2Cr + CO \xrightarrow[\text{100 atm}]{150-160°C} [(\eta^5\text{-}C_5H_5)_2Cr(CO)_3]_2 \qquad (4.25)$$
$$\text{dec. } 163-168°C$$

$$(\eta^5\text{-}C_5H_5)_2Co + CO \xrightarrow[\text{THF, 10 hrs}]{100\ atm,\ 130°C} (\eta^5\text{-}C_5H_5)Co(CO)_2 \qquad (4.26)$$
$$25\%$$

$$(\eta^5\text{-}C_5H_5)_2Ni + Ni(CO)_4 \xrightarrow[\text{12 hrs}]{C_6H_6,\ 70°C} [(\eta^5\text{-}C_5H_5)Ni(CO)]_2 + 2CO \qquad (4.27)$$
$$50\%,\ \text{dec. } 146°C$$

Usually either CO is used at or above one atmosphere of pressure, or else a labile binary metal carbonyl such as $Ni(CO)_4$ (eq. 4.27) is used as a CO source. Disadvantages of this synthetic method are the inconvenience of handling the more reactive metallocenes and the formation of small amounts of polymeric organic material that the formal loss of a $C_5H_5^{\cdot}$ radical usually affords.

Direct reactions between a binary metal carbonyl and a cyclopentadiene source are shown in equations 4.28–4.31.

$$Co_2(CO)_8 + 2C_5H_6 \xrightarrow[\text{reflux}]{CH_2Cl_2} 2(\eta^5\text{-}C_5H_5)Co(CO)_2 + 4CO + H_2 \qquad (4.28)$$
$$95\%$$

$$2Mo(CO)_6 + C_{10}H_{12} \xrightarrow[\text{135-145°C}]{\text{neat liquid}} [(\eta^5\text{-}C_5H_5)Mo(CO)_3]_2 + 6CO + H_2 \qquad (4.29)$$
$$80\%,\ \text{dec. } 180°C$$

$$2Fe(CO)_5 + C_{10}H_{12} \xrightarrow[\text{reflux, 16 hrs}]{\text{neat liquid}} [(\eta^5\text{-}C_5H_5)Fe(CO)_2]_2 + 6CO + H_2 \qquad (4.30)$$
$$91\%,\ \text{dec. } 194°C$$

**TABLE 4.3  Selected known neutral homonuclear cyclopentadienyl metal carbonyl complexes.**

| Metal | Complex | Metal | Complex | Metal | Complex |
|---|---|---|---|---|---|
| Ti | $(\eta^5\text{-}C_5H_5)_2Ti(CO)_2$ red-brown solid | V | $(\eta^5\text{-}C_5H_5)V(CO)_4$ orange solid<br><br>$(\eta^5\text{-}C_5H_5)_2V(CO)$ dark brown solid<br><br>$(\eta^5\text{-}C_5H_5)_2V_2(CO)_3(\mu_2\text{-}CO)_2$ dark green solid | Cr | $(\eta^5\text{-}C_5H_5)_2Cr_2(CO)_6$ dark green solid<br><br>$(\eta^5\text{-}C_5H_5)_2Cr_2(\mu_2\text{-}CO)_4$ green solid |
| Zr | $(\eta^5\text{-}C_5H_5)_2Zr(CO)_2$ violet solid | Nb | $(\eta^5\text{-}C_5H_5)Nb(CO)_4$ red solid<br><br>$(\eta^5\text{-}C_5H_5)_3Nb_3(CO)_6(\mu_3\text{-}CO)$ black solid | Mo | $(\eta^5\text{-}C_5H_5)_2Mo_2(CO)_6$ purple-red solid<br><br>$(\eta^5\text{-}C_5H_5)_2Mo_2(\mu_2\text{-}CO)_4$ dark brown solid<br><br>$(\eta^5\text{-}C_5H_5)_2Mo(CO)$ bright green solid |
| Hf | $(\eta^5\text{-}C_5H_5)_2Hf(CO)_2$ purple solid | Ta | $(\eta^5\text{-}C_5H_5)_2Ta(CO)_4$ orange solid | W | $(\eta^5\text{-}C_5H_5)_2W_2(CO)_6$ red-violet solid<br><br>$(\eta^5\text{-}C_5H_5)(\eta^3\text{-}C_5H_5)W(CO)_2$ black solid<br><br>$(\eta^5\text{-}C_5H_5)_2W(CO)$ green solid |
| Mn | $(\eta^5\text{-}C_5H_5)Mn(CO)_3$ yellow solid | Fe | $(\eta^5\text{-}C_5H_5)_2Fe_2(CO)_2(\mu_2\text{-}CO)_2$ cis and trans isomers purple-red solid<br><br>$(\eta^5\text{-}C_5H_5)_4Fe_4(\mu_3\text{-}CO)_4$ dark green solid | Co | $(\eta^5\text{-}C_5H_5)Co(CO)_2$ red liquid<br><br>$(\eta^5\text{-}C_5H_5)_2Co_2(\mu_2\text{-}CO)_2$ green solid<br><br>$(\eta^5\text{-}C_5H_5)_3Co_3(\mu_2\text{-}CO)_2(\mu_3\text{-}CO)$ black solid |

| Metal | Complex | Metal | Complex | Metal | Complex |
|---|---|---|---|---|---|
| Tc | $(\eta^5\text{-}C_5H_5)Tc(CO)_3$ white solid | Ru | $(\eta^5\text{-}C_5H_5)_2Ru_2(CO)_2(\mu_2\text{-}CO)_2$ orange-red solid | Rh | $(\eta^5\text{-}C_5H_5)Rh(CO)_2$ orange liquid <br> $(\eta^5\text{-}C_5H_5)_2Rh_2(CO)_2(\mu_2\text{-}CO)$ dark red solid <br> $(\eta^5\text{-}C_5H_5)_3Rh_3(\mu_2\text{-}CO)_3$ black solid <br> $(\eta^5\text{-}C_5H_5)_3Rh_3(CO)(\mu_2\text{-}CO)_2$ dark red solid |
| Re | $(\eta^5\text{-}C_5H_5)Re(CO)_3$ white solid <br> $(\eta^5\text{-}C_5H_5)_2Re_2(CO)_4(\mu_2\text{-}CO)$ yellow solid | Os | $(\eta^5\text{-}C_5H_5)_2Os_2(CO)_4$ yellow solid | Ir | $(\eta^5\text{-}C_5H_5)Ir(CO)_2$ yellow liquid |
| Ni | $(\eta^5\text{-}C_5H_5)_2Ni_2(\mu_2\text{-}CO)_2$ dark red solid <br> $(\eta^5\text{-}C_5H_5)_3Ni_3(\mu_3\text{-}CO)_2$ dark green solid | | | | |
| Pd | None | | | | |
| Pt | $(\eta^5\text{-}C_5H_5)_2Pt_2(CO)_2$ red solid | | | | |

$$[Na(diglyme)_2][V(CO)_6] + C_5H_5HgCl \xrightarrow[THF]{25°C}$$

$$(\eta^5\text{-}C_5H_5)V(CO)_4 + 2CO + Hg^0 + 2 \text{ diglyme} + NaCl \quad (4.31)$$
$$50\text{-}80\%$$

The thermal reactions of equations 4.29 and 4.30 exhibit variable reproducibility, and the formal loss of hydrogen may actually occur via reduction of an olefinic bond in $C_5H_6$ to give cyclopentene. Equation 4.31 is another example of a formal *oxidative-ligation* reaction.

Examples of reduction-ligation syntheses are shown in equations 4.32–4.34. Metal halide complexes and electropositive metals are the usual reagents. In equation 4.34, both CO and $C_5H_5$ ligands are introduced in a single step.

$$(\eta^5\text{-}C_5H_5)_2Zr(BH_4)_2 + CO \xrightarrow[Et_3N, \ 2 \ days]{100 \ atm, \ 50°C} (\eta^5\text{-}C_5H_5)_2Zr(CO)_2 \quad (4.32)$$
$$15\%$$

$$(\eta^5\text{-}C_5H_5)_2MCl_2 + CO \xrightarrow[THF, \ 3 \ hrs]{Na/Hg, \ 1 \ atm} (\eta^5\text{-}C_5H_5)_2M(CO) \quad (4.33)$$
$$35\%$$

M = Mo or W

$$2MnCl_2(py)_2 + Mg^0 + 2C_5H_6 + 6CO \xrightarrow[300 \ atm, \ 180°C]{DMF, \ CO/H_2}$$

$$2(\eta^5\text{-}C_5H_5)Mn(CO)_3 + 2pyH^+Cl^- + 2py + MgCl_2 \quad (4.34)$$
$$70\%$$

Photolytic or thermal activation of many cyclopentadienyl metal carbonyl complexes affords new complexes as shown in equations 4.35–4.40.

$$(\eta^5\text{-}C_5H_5)V(CO)_4 \xrightarrow[hv, \ N_2 \ purge]{THF, \ 25°C} (\eta^5\text{-}C_5H_5)_2V_2(CO)_5 + 3CO \quad (4.35)$$
$$75\%$$

$$(\eta^5\text{-}C_5H_5)Nb(CO)_4 \xrightarrow[hv, \ 20 \ min]{hexane, \ 25°C} (\eta^5\text{-}C_5H_5)_3Nb_3(CO)_7 + 5CO \quad (4.36)$$
$$70\%$$

$$[(\eta^5\text{-}C_5H_5)M(CO)_3]_2 \xrightarrow[reflux]{toluene} [(\eta^5\text{-}C_5H_5)M(CO)_2]_2 + 2CO \quad (4.37)$$
$$\text{up to } 90\%$$

M = Cr or Mo

$$(\eta^5\text{-}C_5H_5)Re(CO)_3 \xrightarrow[hv, \ 2.5 \ hrs]{C_6H_{12}, \ 25°C} (\eta^5\text{-}C_5H_5)_2Re_2(CO)_5 + CO \quad (4.38)$$
$$20\%$$

$$2[(\eta^5\text{-}C_5H_5)Fe(CO)_2]_2 + PPh_3(xs) \xrightarrow[7 \ hrs]{xylene, \ reflux} [(\eta^5\text{-}C_5H_5)Fe(CO)]_4 + 4CO \quad (4.39)$$
$$54\%$$

$$3[(\eta^5\text{-}C_5H_5)Ni(CO)]_2 \xrightarrow{130°C} 2(\eta^5\text{-}C_5H_5)_3Ni_3(CO)_2 + 2CO \quad (4.40)$$
$$82\%$$

This method usually affords dinuclear or cluster complexes with loss of CO. Yields of the molybdenum tetracarbonyl dimer (eq. 4.37) range from 0 to 90%, and there is some speculation that the dimeric starting material undergoes Mo—Mo bond cleavage to give radical intermediates. The postulated intermediate in equation *4.39* is a $Ph_3P$ complex, $(\eta^5\text{-}C_5H_5)_2Fe_2(CO)_3(PPh_3)$, which decomposes thermally to give the tetranuclear complex.

## Structures of Cyclopentadienyl Metal Carbonyl Complexes

The Group IV metal complexes have structures based on idealized tetrahedral coordination geometry about the metal atoms as shown in **4.13**. Values of the angles $\theta$ and $\theta'$ are approximately 89° and 140°, respectively.

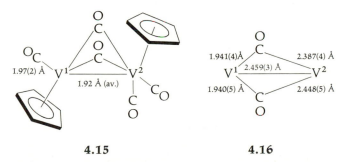

**4.13**                    **4.14**

M = Ti, Zr, Hf

Structural data for $(\eta^5\text{-}C_5H_5)V(CO)_4$ and the niobium analogue indicate that these Group V metal complexes have a "four-legged piano-stool" structure as shown in **4.14** for the vanadium complex. Presumably, the tantalum complex has a similar structure.

Considerable discussion persists over the interpretation of the structural data of $(\eta^5\text{-}C_5H_5)_2V_2(CO)_5$, **4.15**. The argument centers around the magnitude of the V—V bond order and the type of $\mu_2$-CO ligands present in the complex (see **4.16**).

**4.15**                    **4.16**

From a structural viewpoint, the two $\mu_2$-CO ligands of **4.15** form a noncompensating set of asymmetrical bridging ligands.

The $\mu_3$-CO ligand in $(\eta^5\text{-}C_5H_5)_3Nb_3(CO)_7$ is *not* bonded symmetrically to the three niobium atoms, as shown in **4.17**.

The carbon and oxygen atoms of the $\mu_3$-CO ligand are within bonding distances of three and two Nb atoms, respectively. This type of asymmetric $\mu_3$-CO coordination can be designated as $\mu_3\text{-}\eta^3\text{-}, \eta^2\text{-CO}$. Since the Nb—C—O angle within

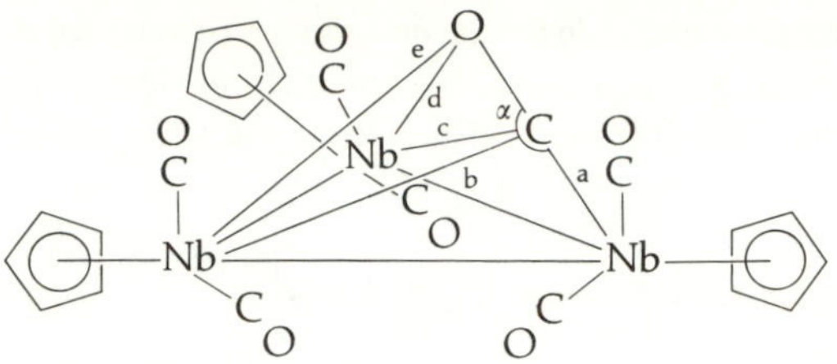

$$a = 1.97\ \text{Å},\ b = 2.32\ \text{Å}$$
$$c = 2.28\ \text{Å},\ d = 2.24\ \text{Å}$$
$$e = 2.21\ \text{Å},\ \alpha = 170°$$
$$\text{Nb—Nb} = 3.004,\ 3.181,$$
$$3.320\ \text{Å}$$

**4.17**

this ligand is about 170°, this bonding can be classified as a "linear" $\mu_3$-CO ligand where *both* filled $\pi$ orbitals of the CO ligand are donating electron density to two different Nb atoms as shown in **4.18**.

**4.18**

The exceptionally long C—O distance of 1.30(1) Å and the very low $\nu_{CO}$ stretching frequency of 1330 cm$^{-1}$ are consistent with this type of bonding. This CO ligand acts as a formal *six-electron donor*.

The Group VI metal hexacarbonyl dimers, $[(\eta^5\text{-}C_5H_5)M(CO)_3]_2$, crystallize with a *trans* orientation of the $C_5H_5$ ligands as shown in **4.19**.

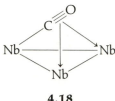

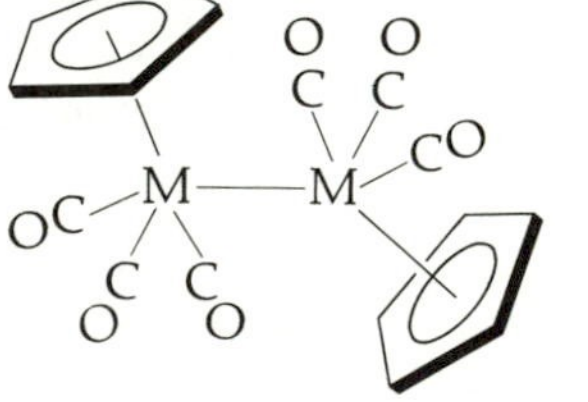

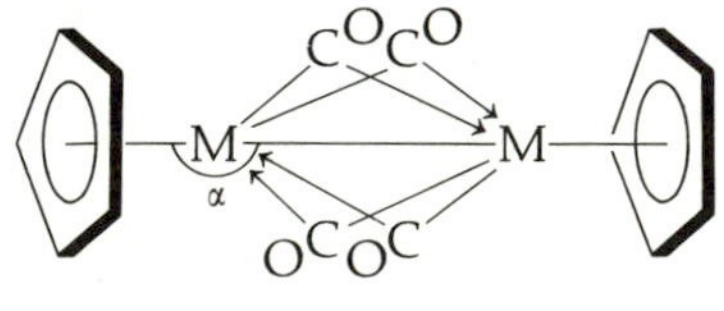

**4.19**                                                                    **4.20**

M = Cr, Mo, W                                                      M = Cr, Mo

The Cr—Cr distance in the chromium dimer is significantly *greater* than the M—M distance of the molybdenum or tungsten analogues. The weak Cr—Cr bond that is reflected by this length also gives rise to the known high reactivity of the chromium dimer. Apparently, strong intramolecular $C_5H_5 \cdots CO$ steric interactions between the two Cr moieties results in a very weak Cr—Cr bond.

A different structure is observed for the chromium and molybdenum tetracarbonyl dimers, $[(\eta^5\text{-}C_5H_5)M(CO)_2]_2$ (**4.20**). Since the M—M distances in these complexes are about 1 Å shorter than in the hexacarbonyl dimers, the M—M bonds are described as triple bonds, causing both metals to obey the EAN rule. However, all four CO ligands are presumably weakly bonded to the adjacent metal atom as shown in **4.21**.

| M | ā (Å) | b̄ (Å) | ᾱ (°) | β̄ (°) |
|---|---|---|---|---|
| Cr | 1.93 | 2.49 | 74 | 170 |
| Mo | 2.13 | 2.56 | 67 | 176 |

**4.21**

The large values of the $\beta$ angles imply that these groups act as "linear" $\mu_2$-CO ligands where there is some $\pi$-donation of electron density from each CO to the second metal atom. The $\nu_{CO}$ frequencies (1900 and 1850 cm$^{-1}$ for the Mo complex) are slightly lower than those expected for normal terminal CO ligands. Perhaps a better bonding description for these complexes is to have an M—M bond order between 2.0 and 3.0 with some degree of $\pi$-electron donation from each terminal CO to the other metal atom. This combination of bonding interactions would satisfy the EAN rule and saturate the coordination sphere of each metal atom. The generation of free coordination sites on the metal atoms by displacing these linear $\mu_2$-CO ligands is being examined for potential stoichiometric and catalytic applications. (See chap. 13 for a discussion of this principle.)

In order for the complex $(C_5H_5)_2W(CO)_2$ to satisfy the EAN rule, one of the $C_5H_5$ ligands must act as a three-electron donor. The x-ray structure confirms this unique form of $C_5H_5$ coordination (**4.22**).

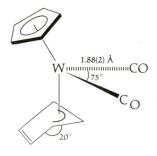

**4.22**

The carbon atoms of the $\eta^5$-$C_5H_5$ ring are coplanar to within $\pm 0.003$ Å and have an average W—C distance of 2.37 Å. However, the other $C_5H_5$ ligand has a folded conformation with only three carbon atoms bonded to the tungsten atom: W—C distances are 2.40(2) Å to the two outer carbon atoms and 2.82(2) Å to the central, or inner, carbon atom. The remaining two carbon atoms of the ring are at a non-bonding distance of 2.98(3) Å from the W atom, and the C—C distance of 1.35(3) Å between these atoms is consistent with a localized C—C double bond. Complex **4.22** is the first structurally characterized example of $\eta^3$-$C_5H_5$ coordination.

The Group VII metal tricarbonyl complexes have a three-legged piano-stool structure as shown in **4.23** for the manganese complex.

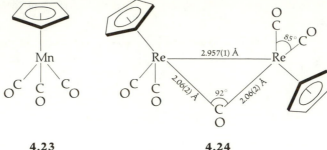

**4.23**　　　　　　　　　　　　　　**4.24**

The dinuclear complex, $(\eta^5\text{-}C_5H_5)_2Re_2(CO)_5$, has a single symmetrical $\mu_2$-CO ligand, and it is the only example of a $\mu_2$-CO ligand between two rhenium atoms (**4.24**).

The structure of trans-$[(\eta^5\text{-}C_5H_5)Fe(CO)_2]_2$ is shown in **4.25**. The $Fe_2(\mu_2\text{-}CO)_2$ atoms are coplanar with an average Fe—C distance of 1.914(5) Å to the symmetrical $\mu_2$-CO ligands. The structure of the cis analogue to **4.25** has been determined also.

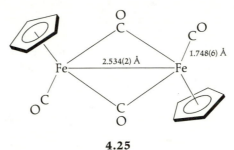

**4.25**

The tetrameric complex, $[(\eta^5\text{-}C_5H_5)Fe(CO)]_4$, has a tetrahedral $Fe_4$ core with a symmetrical $\mu_3$-CO ligand centered above each triangular face of the $Fe_4$ tetrahedron as shown in **4.26**. The $\nu_{CO}$ stretching frequency is 1620 cm$^{-1}$.

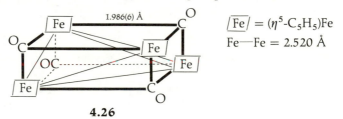

$$\boxed{Fe} = (\eta^5\text{-}C_5H_5)Fe$$
$$Fe\text{—}Fe = 2.520 \text{ Å}$$

**4.26**

Metals of the cobalt group form monomeric complexes of the type $(\eta^5\text{-}C_5H_5)M(CO)_2$, where M is cobalt, rhodium, or iridium. These complexes are liquids at room temperature, and they are believed to have the structure shown in **4.27**. Using equation 3.47 and solution $\nu_{CO}$ data for the cobalt complex, we can calculate the value of $2\phi$ to be 95°.

A dimeric cobalt complex, $[(\eta^5\text{-}C_5H_5)Co(CO)]_2$, is known, but its high reactivity has prevented a structure determination. However, the structure of the $C_5Me_5$ analogue has been reported (**4.28**).

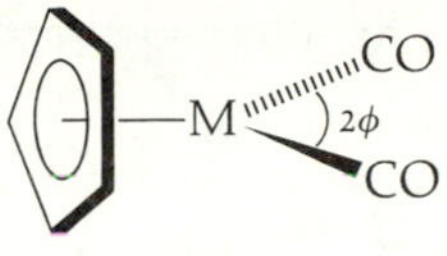

**4.27**

M = Co, Rh, Ir

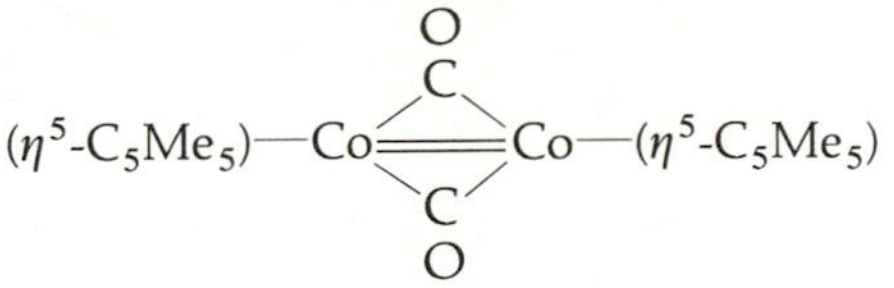

$(\eta^5\text{-}C_5Me_5)$—Co$=\!\!=$Co—$(\eta^5\text{-}C_5Me_5)$   Co—Co = 2.327 Å

**4.28**

The relatively short Co—Co distance is interpreted as a double bond, which satisfies the EAN rule for each Co atom. Each $\mu_2$-CO ligand is bonded symmetrically. The analogous $C_5H_5$ complex presumably has a similar structure since IR data show only a single $\mu_2$-CO stretching frequency at 1798 cm$^{-1}$.

The trimeric complex $[(\eta^5\text{-}C_5H_5)Co(CO)]_3$ has in the solid state a triangular $Co_3$ core with two $\mu_2$-CO ligands and one $\mu_3$-CO ligand as shown in **4.29**.

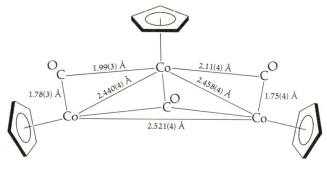

**4.29**

The $\mu_3$-CO ligand coordinates symmetrically, and its $\nu_{CO}$ frequency is 1673 cm$^{-1}$. Structural classification of the $\mu_2$-CO ligands ranges from two semi-bridging ligands to one asymmetric and one semi-bridging ligand type. A semi-bridging formulation based on Co—Co bond polarity is not easily derived because of the presence of a $\mu_3$-CO ligand. Each Co atom would obey the EAN rule with one terminal CO ligand. The C—O stretching frequencies of the $\mu_2$-CO ligands are 1833 and 1775 cm$^{-1}$.

Detailed structural analysis of **4.29** reveals that the three CO ligands are nearly equivalent in occupying the free space between the $C_5H_5$ ligands. Structure **4.29** is presumably the lowest-energy arrangement in the crystal; the one that minimizes intra- and intermolecular steric interactions.

The dirhodium complex, $(\eta^5\text{-}C_5H_5)_2Rh_2(CO)_3$, has a single, symmetrical $\mu_2$-CO ligand bridging the two Rh atoms as shown in **4.30**. Each Rh has a $C_5H_5$ ligand

and one terminal CO ligand. A Rh—Rh single bond satisfies the EAN requirement for each rhodium atom.

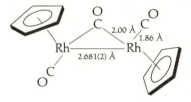

**4.30**

The trimeric rhodium complex [($\eta^5$-C$_5$H$_5$)Rh(CO)]$_3$ exists in the solid state as two isomers, **4.31** and **4.32**.

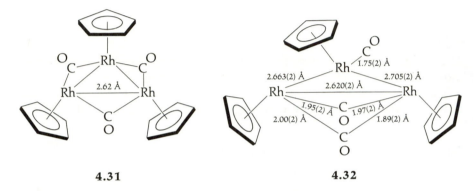

**4.31**                                             **4.32**

Complex **4.31** has three symmetrical $\mu_2$-CO ligands and three Rh—Rh single bonds. Complex **4.32** is less symmetrical in that one Rh has a terminal CO ligand while the other two Rh atoms are bridged by one symmetric and one asymmetric $\mu_2$-CO ligands. The Rh atoms of each isomer obey the EAN rule.

Structural data for several Rh—CO complexes reveal Rh—C distances of 1.75 Å, 2.00 Å and 2.17 Å for Rh—CO (terminal), Rh-($\mu_2$-CO) and Rh-($\mu_3$-CO) coordination geometries, respectively. This trend reflects the expected Rh—C bond lengthening as a CO coordinates to one to three metal atoms, as discussed in chap. 2.

The two cyclopentadienyl nickel carbonyls, [($\eta^5$-C$_5$H$_5$)Ni(CO)]$_2$ and ($\eta^5$-C$_5$H$_5$)$_3$-Ni$_3$(CO)$_2$, have structures **4.33** and **4.34**, respectively.

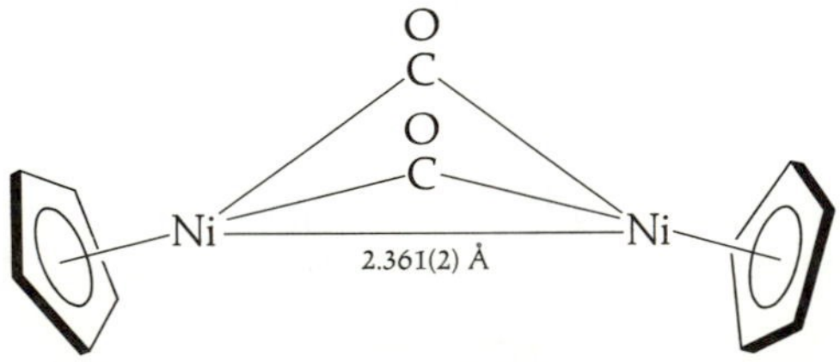

$\overline{\text{Ni—CO}}$ = 1.88(1) Å
Ni—C$_5$H$_5$ (centroid) = 1.74(1) Å

**4.33**

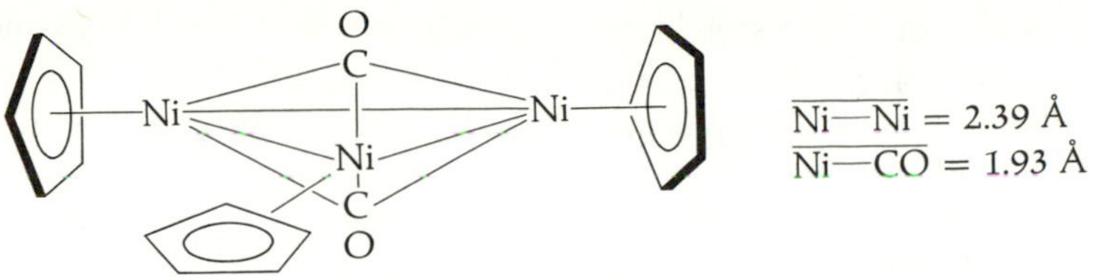

$$\overline{\text{Ni}\text{—}\text{Ni}} = 2.39 \text{ Å}$$
$$\overline{\text{Ni}\text{—}\text{CO}} = 1.93 \text{ Å}$$

**4.34**

Complex **4.33** has two symmetrical $\mu_2$-CO ligands and an Ni—Ni single bond; complex **4.34** has a symmetrical $\mu_3$-CO ligand centered over each side of a triangular Ni$_3$ core. The trinuclear complex has one unpaired electron and so is paramagnetic.

## Heteronuclear Cyclopentadienyl Metal Carbonyl Complexes

Several heteronuclear cyclopentadienyl metal carbonyl complexes have been prepared. Instead of providing a comprehensive discussion of these complexes, we shall look at selected complexes that demonstrate the structural variety within this class of molecules.

Two examples of dinuclear complexes having only terminal CO ligands are shown in **4.35** and **4.36**. Both complexes have an Mn—M single bond.

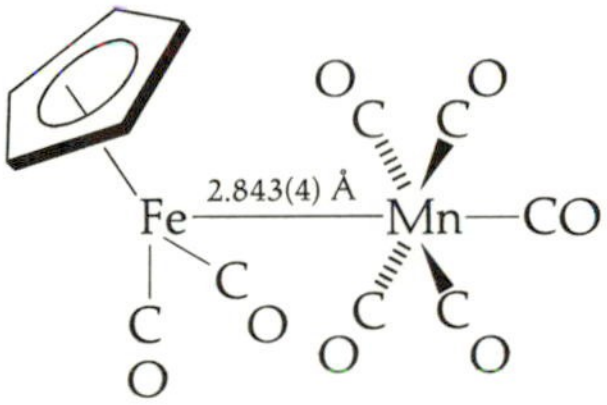

**4.35** 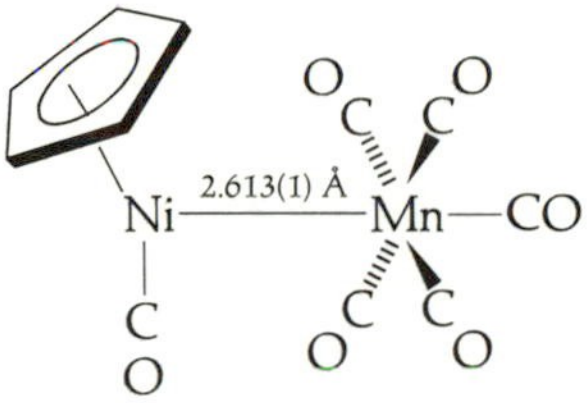                                        **4.36**

Each metal obeys the EAN rule, and each complex is composed of an Mn(CO)$_5$ fragment that is derived formally from cleaving the Mn—Mn bond in Mn$_2$(CO)$_{10}$ and a metal fragment derived from formal cleavage of the M—M bonds in **4.25** and **4.33**.

The Mo—Re complex, **4.37**, has two semi-bridging $\mu_2$-CO ligands ($\nu_{CO} = 1650$ cm$^{-1}$).

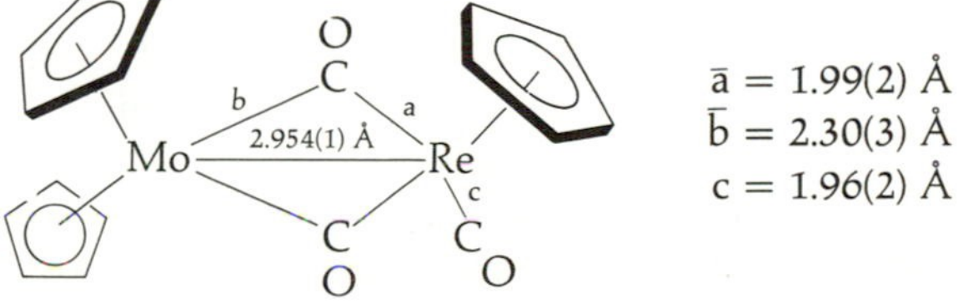

$$\overline{a} = 1.99(2) \text{ Å}$$
$$\overline{b} = 2.30(3) \text{ Å}$$
$$c = 1.96(2) \text{ Å}$$

**4.37**

This is a classic example of semi-bridging ligands since both metals obey the EAN rule in the all-terminal isomer, $(\eta^5\text{-}C_5H_5)_2\overset{-}{Mo}\text{---}\overset{+}{Re}(CO)_3(\eta^5\text{-}C_5H_5)$. This complex has the correct M—M′ bond polarity to produce the observed semi-bridging $\mu_2$-CO coordination.

Complex **4.38** is interesting in that no symmetrical $\mu_2$-CO structure affords 18 electrons at both metal atoms.

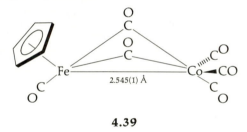

$$a = 2.272(8) \text{ Å}$$
$$b = 2.431(5) \text{ Å}$$

**4.38**

Consequently, the complex adopts a structure having an asymmetric $\mu_2$-CO, which formally donates one electron to each metal, and a linear $\mu_2$-CO ligand, which donates two electrons to each metal atom.

Conversely, complex **4.39** has two symmetrical $\mu_2$-CO ligands, and each metal obeys the EAN rule. This complex represents the formal coupling of the mononuclear halves of $Co_2(CO)_8$ and **4.25**.

The two fragments that couple to give **4.39** are

$$(\eta^5\text{-}C_5H_5)(OC)_2Fe \text{ and } (OC)_4Co$$

These metal moieties are 17-electron species in which each metal atom has a valence orbital occupied by a single electron. Because of the similarity of the total electron count and the electron population of the valence orbitals between these metal atoms, the metals are said to be "isolobal." When these fragments bond to each other, both metals satisfy the EAN rule, and their valence orbitals become completely filled. An analogy from organic chemistry would be the formal coupling of the radicals $\cdot CH_3$ and $\cdot CF_3$ to give $H_3C\text{---}CF_3$. The formation of many polynuclear or cluster complexes can be regarded formally as a coupling of several isolobal metal fragments.

A heterotrinuclear cyclopentadienyl metal carbonyl complex is shown in **4.40**. Complex **4.40** is prepared by irradiating a THF solution of $Fe_2(CO)_9$ and $[(\eta^5\text{-}C_5Me_5)Co(CO)]_2$ (**4.28**) for two days. It is structurally similar to $[(\eta^5\text{-}C_5H_5)Co(CO)]_3$ (**4.29**) since it has two $\mu_2$-CO ligands bridging different M—M

**4.40** **4.41**

bonds and a $\mu_3$-CO ligand. Complex **4.40** represents the formal insertion of an $Fe(CO)_4$ fragment across the Co—Co double bond in **4.28** with a concomitant formation of the bridging CO ligand arrangement.

Complex **4.41** is an example of a heterotetranuclear cyclopentadienyl metal carbonyl complex. Although the $\mu_2$-CO ligand between the rhodium atoms is symmetrical, the two $\mu_2$-CO ligands that bridge two of the Rh—Fe bonds are asymmetric.

## SELECTED DERIVATIVES OF CYCLOPENTADIENYL METAL CARBONYL COMPLEXES

One general reaction type that affords a cyclopentadienyl metal complex is treatment of a metal halide with a $C_5H_5$ reagent, which yields a complex having both $C_5H_5$ and non-$C_5H_5$ ligands (other than CO) on the metal atom. Equations 4.41 and 4.42 show examples of syntheses using binary metal halides as reagents.

$$TiCl_4 + 2NaC_5H_5 \xrightarrow{\text{THF}} (\eta^5\text{-}C_5H_5)_2TiCl_2 + 2NaCl \quad (4.41)$$
$$\text{red-orange solid, 72–90\%}$$
$$\text{mp 290°C}$$

$$NbBr_5 + 2NaC_5H_5 \xrightarrow{\text{THF}} (\eta^5\text{-}C_5H_5)_2NbBr_3 + 2NaBr \quad (4.42)$$
$$\text{red-brown solid, 70\%}$$
$$\text{dec. 260°C}$$

However, halide displacement by $C_5H_5$ occurs in many instances where other ligands are also present:

$$\tfrac{1}{2}[RhL_2Cl]_2 + NaC_5H_5 \xrightarrow{\text{xylene}} (\eta^5\text{-}C_5H_5)RhL_2 + NaCl \quad (4.43)$$

$L = CO$ or $\eta^2\text{-}C_2H_4$

Two very important reactions of cyclopentadienyl metal carbonyls, and of binary metal carbonyls as well, are the formal *oxidative cleavage* of M—M single bonds by halogens affording monomeric metal halide complexes and the *reductive cleavage* of M—M single bonds by alkali metals or other reducing agents affording mononuclear anionic complexes. Examples are shown in equations 4.44–4.46.

$$\left.\begin{array}{l}[(\eta^5\text{-}C_5H_5)Mo(CO)_3]_2 \\ [(\eta^5\text{-}C_5H_5)Fe(CO)_2]_2 \\ [(\eta^5\text{-}C_5H_5)Ni(CO)]_2\end{array}\right\} \xrightarrow[\text{CHCl}_3]{\text{I}_2} \left\{\begin{array}{l}2(\eta^5\text{-}C_5H_5)Mo(CO)_3I \\ 2(\eta^5\text{-}C_5H_5)Fe(CO)_2I \\ 2(\eta^5\text{-}C_5H_5)Ni(CO)I\end{array}\right. \qquad (4.44)$$

$$M_2(CO)_{10} + X_2 \xrightarrow{\text{CHCl}_3} 2M(CO)_5X \qquad (4.45)$$

$M = Mn, Tc, Re$

$$\left.\begin{array}{l}[(\eta^5\text{-}C_5H_5)Mo(CO)_3]_2 \\ [(\eta^5\text{-}C_5H_5)Fe(CO)_2]_2 \\ [(\eta^5\text{-}C_5H_5)Ni(CO)]_2 \\ Co_2(CO)_8 \\ M_2(CO)_{10}\end{array}\right\} \xrightarrow[\text{THF}]{Na^0 \text{ or } K^0/Hg} \left\{\begin{array}{l}2[(\eta^5\text{-}C_5H_5)Mo(CO)_3]^- \\ 2[(\eta^5\text{-}C_5H_5)Fe(CO)_2]^- \\ 2[(\eta^5\text{-}C_5H_5)Ni(CO)]^- \\ 2[Co(CO)_4]^- \\ 2[M(CO)_5]^-\end{array}\right. \qquad (4.46)$$

$M = Mn, Tc, Re$

Such metal halide complexes have great synthetic utility because of the highly reactive metal-halogen bond. Anionic complexes like those shown above have an electron pair in the valence shell of the metal atom that makes these complexes very versatile reagents. These anions act as Brønsted bases or as nucleophiles. Protonation affords neutral metal hydride complexes; alkylation or acylation affords neutral alkyl or acyl complexes with the metal. Although we shall mention other methods for preparing anionic organometallic complexes in chap. 8, the frequent use of these simple anionic complexes as chemical reagents justifies their inclusion at this point.

As equations 4.44–4.46 show, cyclopentadienyl metal carbonyls and binary metal carbonyls have considerable chemistry in common. Inspection of tables 2.1 and 4.3 reveals that for the chromium to nickel groups, those metals that exist as mononuclear binary metal carbonyls form dinuclear cyclopentadienyl metal carbonyls, and those metals that exist in their simplest form as dinuclear binary metal carbonyls form mononuclear cyclopentadienyl metal carbonyl complexes. This trend results from a $\eta^5\text{-}C_5H_5$ ligand being a five-electron donor. One $\eta^5\text{-}C_5H_5$ ligand and one M—M single bond are electronically equivalent to three terminal CO ligands according to electron counting formalism. It is not surprising then, that most M—M chemical reactivity is common to both classes of organometallic compounds.

Cyclopentadienyl metal carbonyls are such common reagents that many can be purchased commercially. At 1984 prices, 25 g quantities of $[C_5H_5Fe(CO)_2]_2$, $[C_5H_5Mo(CO)_3]_2$, $C_5H_5Co(CO)_2$, $C_5H_5Mn(CO)_3$, $[C_5H_5Ni(CO)]_2$, $C_5H_5V(CO)_4$ and $C_5H_5Nb(CO)_4$ cost approximately $25, $100, $100, $144, $200, $375, and $600, respectively. These prices are higher than those of the corresponding binary metal carbonyls in most cases.

Two classes of cyclopentadienyl metal complexes that have particularly interesting structural features are substituted cyclopentadienyl metal carbonyl complexes and those complexes with significantly asymmetrical coordination of a $C_5H_5$ ligand. Substituted cyclopentadienyl metal carbonyl complexes are prepared by methods discussed in chap. 3. Two-electron donor ligands that are used frequently to replace one or more CO ligands in these complexes are P-donor ligands, $PR_3$, and isocyanides, RNC. As with substituted metal carbonyls, the structures of these complexes are analogous to the structures of the corresponding unsubstituted cyclopentadienyl metal carbonyls.

Complexes **4.42**–**4.44** demonstrate this similarity. Complex **4.42** has a $Ph_3P$ ligand in place of the unique terminal CO of $(C_5H_5)_2V_2(CO)_5$; **4.43** and **4.44** have a $CH_3NC$ or $(PhO)_3P$ ligand, respectively, substituted for one of the terminal CO ligands in the corresponding parent complex. Intramolecular bond distances and angles within these complexes do not differ greatly from those of the unsubstituted complexes.

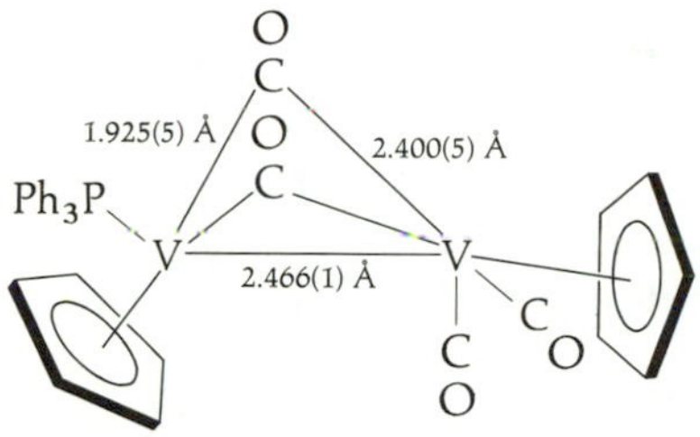

**4.42** 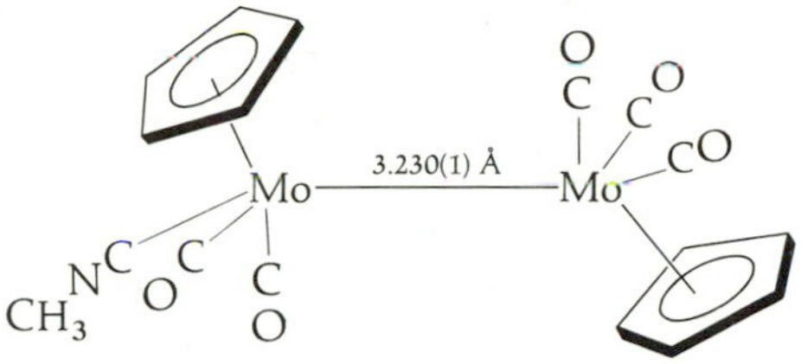                                                                  **4.43**

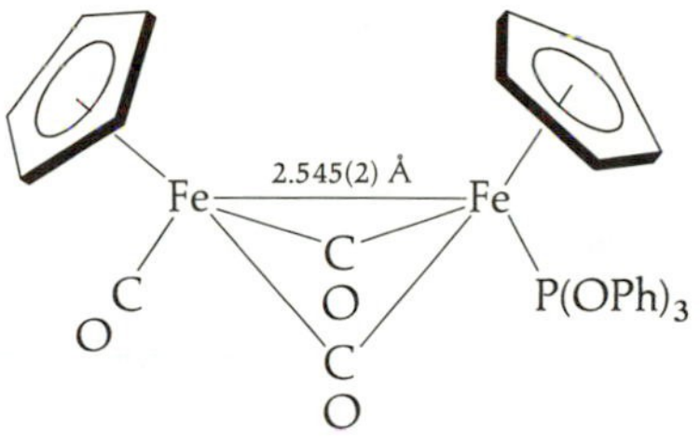

**4.44**

An interesting series of complexes that demonstrates the isovalent equivalence of CO and isocyanides is shown by the substituted $[(C_5H_5)Fe(CO)_2]_2$ derivatives **4.45**–**4.48**.

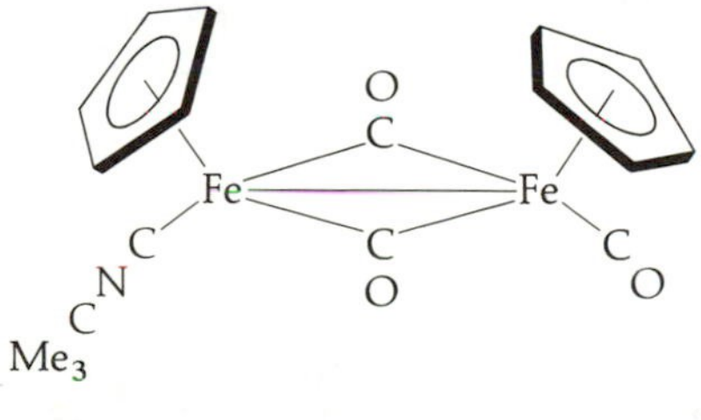

**4.45** 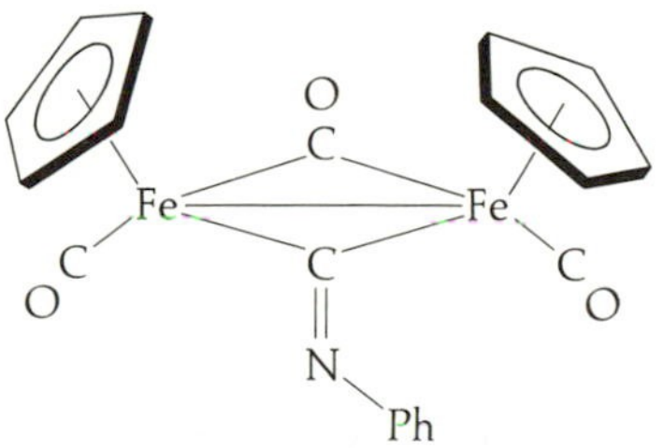                                                                  **4.46**

Isocyanide ligands can replace a terminal CO, a $\mu_2$-CO, both $\mu_2$-CO or all four CO ligands of the parent dinuclear complex.

*Pentahapto*-cyclopentadienyl ligands usually coordinate to metal atoms in a very symmetrical fashion. Inter-ring C—C distances are usually in the range of 1.39 Å to 1.42 Å with an average value of 1.40 Å. The five M—C (ring) distances are

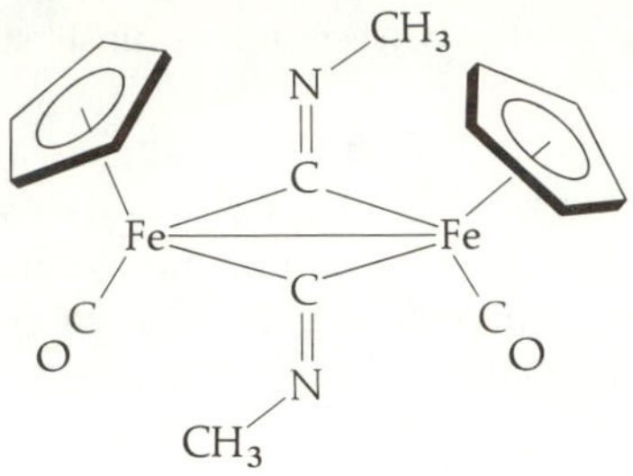

**4.47** 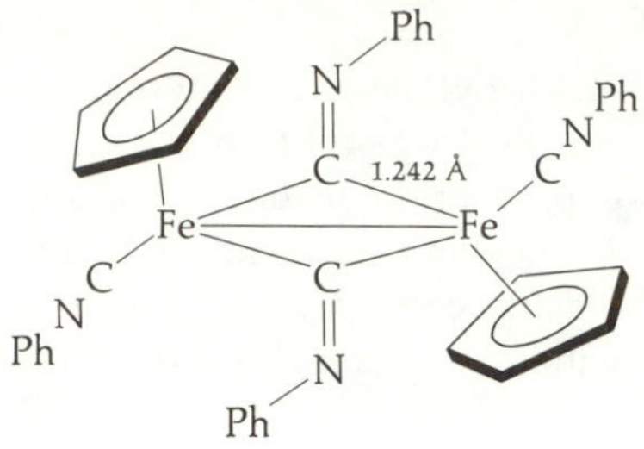                                   **4.48**

usually equal to within about 0.10 Å or less. This symmetrical coordination is observed even in complexes of relatively low symmetry.

However, several molecules possess $\eta^5$-$C_5H_5$ ligands that coordinate less symmetrically—although in most instances the structural asymmetry in C—C and M—C bond distances are barely significant statistically. An example of a distorted mode of $\eta^5$-$C_5H_5$ coordination is shown in **4.49**.

**4.49**

Complexes showing this allylic distortion include $(\eta^5$-$C_5Me_5)Rh(PPh_3)(\eta^2$-$C_2H_4)$ and $(\eta^5$-$C_5H_5)Rh(\eta^2$-$C_2H_4)(\eta^2$-$C_2F_4)$. An obvious extension of a distorted $\eta^5$-$C_5H_5$ coordination geometry to actual $\eta^3$-$C_5H_5$ coordination has been observed in the tungsten complex **4.22**.

Cyclopentadienyl ligands can also bridge two metal atoms, as shown in **4.50**. (The other bridging group is an $\eta^3$-allyl ligand; see chap. 5.) Both palladium atoms are bonded to the same $\pi$-$C_5H_5$ ligand. Several complexes of this type are known for the late transition metals.

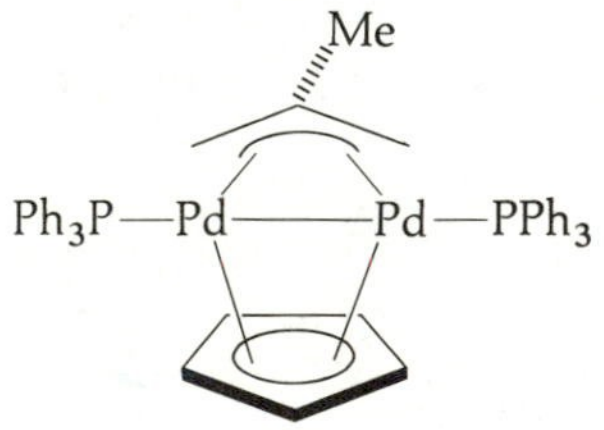

**4.50**

## COMPLEXES CONTAINING CYCLOPENTADIENYL-LIKE LIGANDS

There are several *pentahapto* ligands that act as six $\pi$-electron donors and form complexes isovalent and isostructural to those of cyclopentadienide. The more important ligands of this type include indenyl (**4.51**), fluorenyl (**4.52**), pyrrolyl (**4.53**), azulene (**4.54**), and 7,8-$B_9C_2H_{11}{}^{2-}$, the dicarbollyl ion (**4.55**).

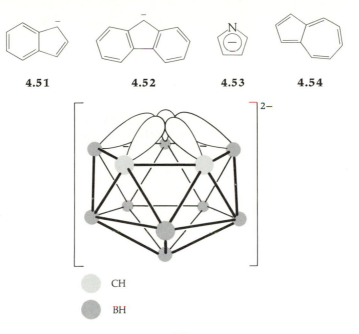

**4.51**          **4.52**          **4.53**          **4.54**

○ CH

● BH

**4.55**

Indenyl and fluorenyl ligands behave essentially as benzo cyclopentadienyls. A convenient source of these ligands is the corresponding sodium salts, which are prepared from indene or fluorene and sodium metal as shown in equation *4.47*. Syntheses of known complexes that contain these ligands are shown in equations *4.48–4.52*.

$$\left.\begin{array}{c} \text{indene} + \text{Na}^0 \\[2em] \text{fluorene} + \text{Na}^0 \end{array}\right\} \xrightarrow[\text{3–12 hrs}]{\text{THF, reflux}} \left\{\begin{array}{c} \text{NaC}_9\text{H}_7 \quad + \tfrac{1}{2}\text{H}_2{\uparrow} \\ \text{sodium indenide} \\[1em] \text{NaC}_{13}\text{H}_9 \quad + \tfrac{1}{2}\text{H}_2{\uparrow} \\ \text{sodium fluorenide} \end{array}\right. \qquad (4.47)$$

$$(\eta^5\text{-}C_5H_5)Fe(CO)_2I + NaC_9H_7 \xrightarrow[\text{5 hrs}]{\text{THF, 25°C}}$$
$$(\eta^5\text{-}C_5H_5)Fe(CO)_2(\eta^1\text{-}C_9H_7) + NaI \quad (4.48)$$
$$\text{red solid, 20\%}$$
$$\text{mp 63°C}$$

$$Ni(acac)_2 + 2LiC_9H_7 \xrightarrow{\text{THF, 25°C}} (\eta^3\text{-}C_9H_7)_2Ni \quad + 2Li[acac] \quad (4.49)$$
$$\text{deep brown solid, 33\%}$$
$$\text{mp 150°C}$$

$$W(CO)_6 + indene(xs) \xrightarrow[\text{diglyme}]{175°C}$$
$$(\eta^3\text{-}C_9H_7)(\eta^5\text{-}C_9H_7)W(CO)_2 + [(\eta^5\text{-}C_9H_7)W(CO)_3]_2 \quad (4.50)$$

$$Mn(CO)_5Br + NaC_9H_7 \xrightarrow[\Delta]{THF} (\eta^5\text{-}C_9H_7)Mn(CO)_3 + NaBr + 2CO \quad (4.51)$$
$$\text{yellow-orange solid, } 56\%$$
$$\text{mp } 50°C$$

$$ZrCl_4 + 2Na \text{ fluorenide} \xrightarrow{DME} (\eta^3\text{-}C_{13}H_9)(\eta^5\text{-}C_{13}H_9)ZrCl_2 + 2NaCl \quad (4.52)$$

Indenyl radical coordinates to metals via the five-membered ring as a *mono-*, *tri-*, or *pentahapto* ligand. *Trihapto* coordination of indenyl is more common than with $C_5H_5$ because of the aromatic stabilization effect of the six-membered ring of indene.

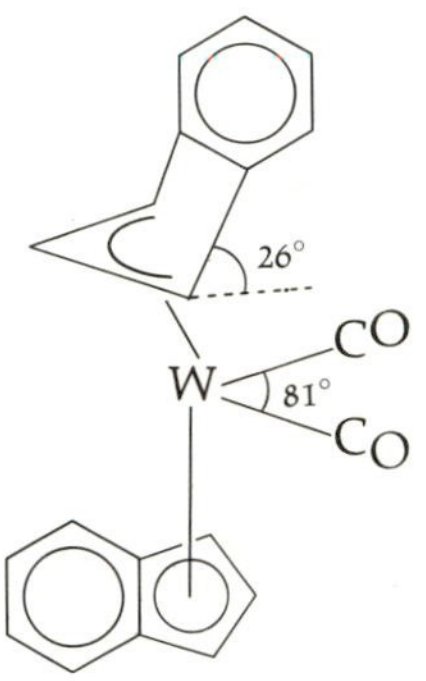

**4.56** 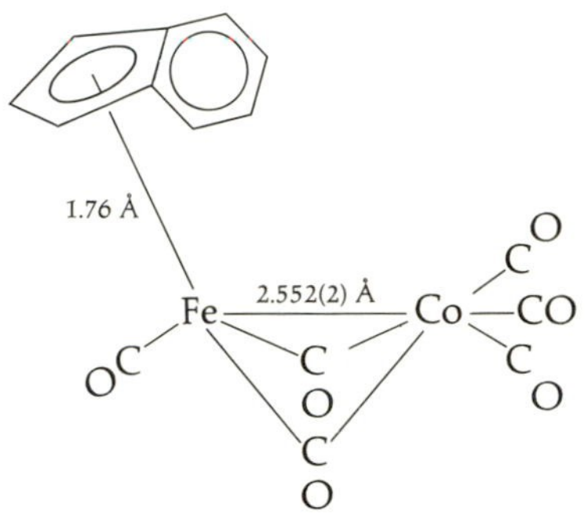                     **4.57**

The structures of **4.56** and **4.57** show the basic $\eta^3$- and $\eta^5$-$C_9H_7$ coordination geometry; the metals in both complexes obey the EAN rule. The hypothetical facile conversion of an $\eta^5$-indenyl ligand into an $\eta^3$-indenyl ligand might be relevant to the design of organometallic catalysts. An $\eta^5$-$\eta^3$ conversion would free an available coordination site on the metal atom for possible bonding to an incoming substrate molecule (see chap. 14).

Fluorenyl radical usually coordinates to metals as a *tri-* or *pentahapto* ligand as shown in **4.58**.

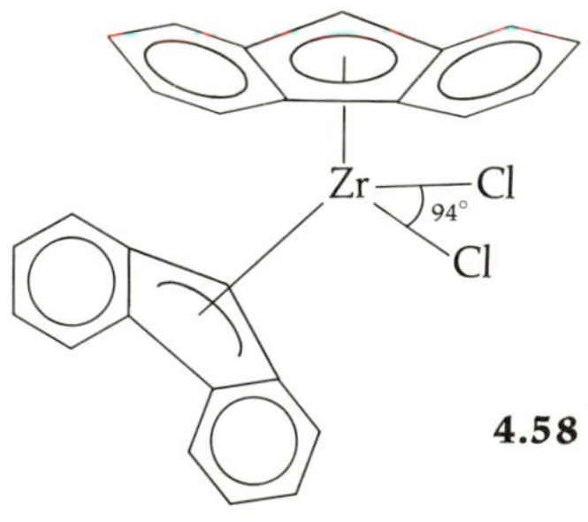

**4.58**

When it coordinates as an $\eta^3$-$C_{13}H_9$ ligand, the two carbon atoms of the five-membered ring that are not coordinated to the metal are usually 0.2–0.4 Å farther from the metal atom than are the three carbon atoms that are coordinated.

Pyrrolyl complexes are prepared from pyrrole or its conjugate base as shown in equations 4.53–4.55. Pyrrolyl ligands form *pentahapto* organometallic complexes that are generally less stable than the corresponding $\eta^5$-$C_5H_5$ compounds. The nitrogen atom of a pyrrolyl ligand is an isoelectronic substitute for a CH group in a $C_5H_5^-$ ligand.

$$Mn_2(CO)_{10} + 2 \underset{}{\boxed{\phantom{xx}N-H}} \xrightarrow[\text{6 hrs}]{\text{diglyme, 130°C}}$$

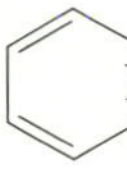

$$2(\eta^5\text{-}C_4H_4N)Mn(CO)_3 + H_2 + 4CO$$
$$66\%, \text{ mp } 41°C \qquad\qquad (4.53)$$

$$FeCl_2 + NaC_4H_4N + NaC_5H_5 \xrightarrow{\text{THF}} (\eta^5\text{-}C_5H_5)(\eta^5\text{-}C_4H_4N)Fe + 2NaCl \quad (4.54)$$
$$\text{red solid, } 1\%$$
$$\text{mp } 114°C$$

$$(\eta^5\text{-}C_5H_5)Fe(CO)_2I + KC_4H_4N \xrightarrow[\text{reflux}]{C_6H_6} (\eta^5\text{-}C_5H_5)(\eta^5\text{-}C_4H_4N)Fe + KI + 2CO$$
$$22\%$$

$$(4.55)$$

Azulene is unique among this series of $C_5H_5$-like ligands, since it is neutral. In several azulene complexes the five-membered ring is coordinated to a metal atom as a *pentahapto* five-electron donor. Two such complexes are **4.59** and **4.60**.

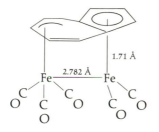

**4.59**

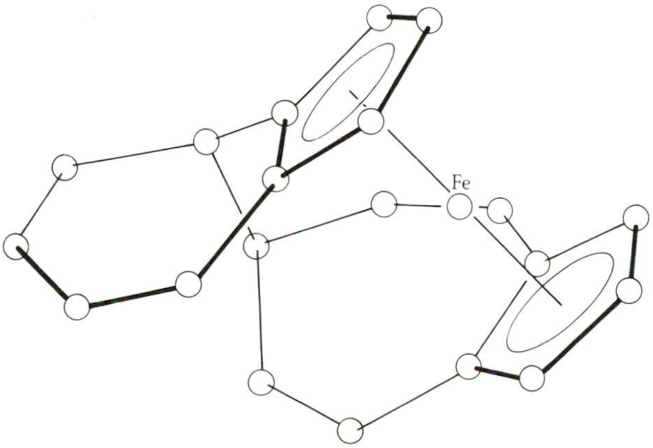

**4.60**

These reveal two common features of azulene coordination: (1) Carbon atoms within the seven-membered ring can coordinate to a second metal atom, as revealed in the $\eta^3$-coordination in **4.59**; or (2) two azulene ligands may undergo an *ortho-para* C—C bond coupling to afford a chelating ligand as shown in **4.60**.

A large series of *pentahapto* complexes of the dicarbollyl dianion have been prepared. This carborane anion has an open pentagonal face with suitable valence orbitals on the three boron and two carbon atoms to permit coordination to a metal located above the center of this open face. The structures of these complexes are consistent with a dicarbollyl ligand being isoelectronic to a $\eta^5$-$C_5H_5^-$ ligand. The preparation and structure of one of these complexes is shown in equation 4.56. Many analogous cobalt complexes and complexes of other metal ions have also been prepared.

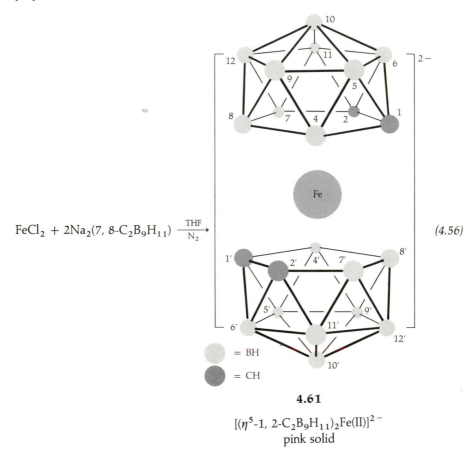

$$FeCl_2 + 2Na_2(7,\ 8\text{-}C_2B_9H_{11}) \xrightarrow[N_2]{THF}$$  (4.56)

○ = BH

● = CH

**4.61**

$$[(\eta^5\text{-}1,\ 2\text{-}C_2B_9H_{11})_2Fe(II)]^{2-}$$
pink solid

Dicarbollyl ligands stabilize high metal oxidation states better than $C_5H_5^-$ does. Thus, the iron(II) complex, **4.61**, which is a ferrocene analogue, is readily oxidized to a more stable iron(III) complex. This result is not unexpected due to the high negative charge of the dicarbollyl ligand. (Note that the numbering scheme of **4.61** indicates the positions of the two carbon atoms within the dicarbollyl framework.)

Many dicarbollylmetal carbonyl complexes are also known. Complexes **4.62** and **4.63** are representative examples. Complex **4.62** is prepared from the dicarbollyl dianion and $BrRe(CO)_5$.

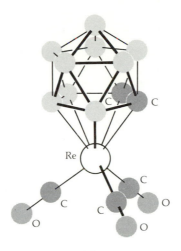

**4.62**

$[(\eta^5\text{-}1, 2\text{-}C_2B_9H_{11})Re(CO)_3]^-$

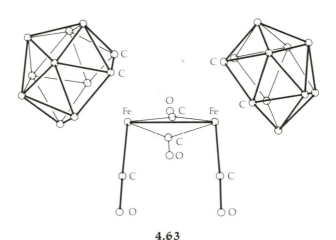

**4.63**

$[(\eta^5\text{-}1, 2\text{-}C_2B_9H_{11})Fe(CO)(\mu_2\text{-}CO)]_2^{2-}$

One other type of $C_5H_5$-like ligand, the *tris*(pyrazoyl)borate ligands (**4.64**) should be mentioned here.

This ligand is monoanionic, and acts as a six-electron donor, like $C_5H_5^-$, although it does not coordinate to a metal atom as a *pentahapto* ligand. Even though these ligands form only M—N bonds, they comprise a large series of complexes that

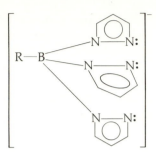

**4.64**

$RB(pz)_3^-$, R = H, alkyl, aryl or pz

are structurally analogous to the metallocenes or other $C_5H_5$ metal complexes as shown in equations 4.57–4.59.

$$2K^+[HB(pz)_3]^- + MCl_2 \longrightarrow [HB(pz)_3]_2M + 2KCl \qquad (4.57)$$

M = transition metal ion

$$K^+[HB(pz)_3]^- + Mo(CO)_6 \longrightarrow [HB(pz)_3Mo(CO)_3]^- + 3CO \qquad (4.58)$$

$$\xrightarrow[-KBr]{BrMn(CO)_5} HB(pz)_3Mn(CO)_3 + 2CO \qquad (4.59)$$

The ferrocene analogue is a diamagnetic solid at room temperature, but it is paramagnetic at temperatures above 27°C. Presumably, an $HB(pz)_3^-$ ligand acts more like a moderately strong-field, classical N-donor ligand than like a $\eta^5$-$C_5H_5^-$ ligand. A series of related pyrazolyl borate anions can be prepared as shown in equation 4.60. Complexes of these anions are known also.

$$KBH_4 + 2\ \text{pyrazole} \xrightarrow[\text{melt}]{105°C} K[H_2B(pz)_2] \xrightarrow[\text{pyrazole}]{180°C}$$

**4.65**

$$K[HB(pz)_3] \xrightarrow[\text{pyrazole}]{220°C} K[B(pz)_4] \qquad (4.60)$$

**4.66**                                              **4.67**

## STUDY QUESTIONS

1. For each first transition metal series complex shown in Table 4.3, determine all possible compounds of the type $(\eta^5$-$C_5H_5)_aM_b(CO)_c(NO)_d$ that obey the EAN rule. Assume that NO is a three-electron ligand and that $a$ and $b$ have the same value as listed in Table 4.3. For polynuclear complexes, determine whether any mononuclear complexes of this type obey the EAN rule.

2. The $\eta^5$-$C_5H_5$ ligand is susceptible to both electrophilic and nucleophilic attack. Predict the relative reactivity of ferrocene and cobalticenium complexes in these types of reactions.

3. Predict other reactions that a metal–metal single bond may undergo in metal carbonyl or cyclopentadienyl metal carbonyl complexes.

4. Propose five specific reactions of metal carbonyl anions or cyclopentadienyl metal carbonyl anions in which the organometallic anion acts as a nucleophile. Choose a wide variety of electrophilic reagents.

5. Summarize the structural types of $C_5H_5$ coordination geometries.

6. If one CO ligand in **4.39** were replaced by an RNC ligand, then how many structural isomers would be possible?

7. Give a rational description of the structure **4.15** by specifying the magnitude of the V—V bond order and the type of $\mu_2$-CO ligands.

8. Predict the appearance of the proton NMR spectral resonances for $\eta^5$- and $\eta^1$-$C_5H_5$ ligands.

9. Predict the products of the following reactions:

   (a) $(\eta^5\text{-}C_5H_5)Fe(CO)_2Cl + NaC_5H_5 \xrightarrow{\text{THF}}$

   (b) $(\eta^5\text{-}C_5H_5)Fe(CO)_2I + AgBF_4 \xrightarrow{\text{THF}}$

   (c) $(\eta^5\text{-}C_5H_5)Mn(CO)_3 \xrightarrow[\text{THF}]{hv}$

10. For each reaction below, identify the organometallic products and explain how the IR spectral data in figure 4.2 are consistent with your answer. For products **B** and **C**, use equation 3.47 to calculate the angle between the CO ligands.

    (a) $[(\eta^5\text{-}C_5H_5)Mo(CO)_3]_2 + I_2 \xrightarrow[25°C]{CHCl_3} \mathbf{A} \xrightarrow[C_6H_6, \Delta]{Ph_3P} \mathbf{B}$

    $(\bar{\nu}_{CO} = 1958 \text{ cm}^{-1})$

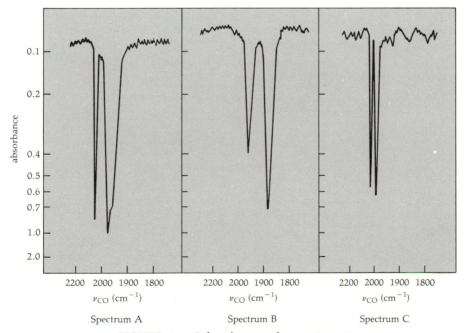

**FIGURE 4.2**   Infrared spectra for question 10.

**(b)** $[(\eta^5\text{-}C_5H_5)Fe(CO)_2]_2$ $+ Cl_2$ $\xrightarrow{CCl_4}$ **C**

$(\bar{v}_{CO} = 1975 \text{ cm}^{-1}$ for
terminal CO ligands)

## SUGGESTED READING

### General References

Baker, E. C., Halstead, G. W., and Raymond, K.N. *Structure and Bonding.* **25,** 23 (1976).
Birmingham, J. M. *Adv. Organometal. Chem.* **2,** 365 (1964).
Callahan, K. P., and Hawthorne, M. F. *Adv. Organometal. Chem.* **14,** 145 (1976).
Fischer, E. O., and Fritz, H. P. *Adv. Inorg. Chem. Radiochem.* **1,** 55 (1959).
Haaland, A. *Acc. Chem. Res.* **12,** 415 (1979).
Rubezhov, A. Z., and Gubin, S. P. *Adv. Organometal. Chem.* **10,** 347 (1972).
Siebert, W. *Adv. Organometal. Chem.* **18,** 301 (1980).
Trofimenko, S. *Acc. Chem. Res.* **4,** 17 (1971).
Werner, H. *J. Organometal. Chem.* **200,** 335 (1980).
Wilkinson, G., and Cotton, F. A. *Progr. Inorg. Chem.* **1,** 1 (1959).
Wilkinson, G. *J. Organometal. Chem.* **100,** 273 (1975).

### References to Isolobal Concepts

Albright, T. A. *Tetrahedron* **38,** 1339 (1982).
Hoffmann, R. *Angew. Chem. Int. Ed. Engl.* **21,** 711 (1982).
Stone, F. G. A. *Angew. Chem. Int. Ed. Engl.* **23,** 89 (1984).

# π-Complexes of Other Carbocyclic Ligands and Complexes Containing Alkenes, Alkynes, and π-Enyl Ligands

## CHAPTER 5

Organometallic complexes that have a primarily π-bonding interaction between an unsaturated organic molecule or radical and a metal atom are fundamentally important in applications of organometallic chemistry. Ligands of this type include (1) *carbocyclic ligands*, $C_nH_n$, such as $C_5H_5$ (chap. 4), (2) conjugated, nonconjugated, or cumulated *olefins*, (3) *alkynes*, and (4) *allylic* or *dienyl radicals*.

### COMPLEXES CONTAINING $\eta^n$-CARBOCYCLIC LIGANDS

Like the *pentahapto*-cyclopentadienyl metal complexes, several other carbocyclic molecules or radicals, $C_nH_n$ or $C_nR_n$ species, coordinate to metal atoms as π ligands. Each carbon atom of the carbocyclic ring is bonded to the metal affording the maximum $\eta^n$-hapticity for the ligand.

### $\eta^3$-$C_3R_3$ Complexes

The simplest carbocyclic ligand would have a three-membered ring and would afford a *trihapto* complex as shown in **5.1**.

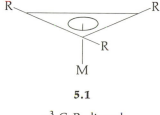

**5.1**

$\eta^3$-$C_3R_3$ ligand

Such molecules are referred to as *cyclopropenyl complexes*. Neutral cyclopropenyl ligands act as formal three-electron donors; an aromatic cyclopropenium cation, $\eta^3$-$C_3R_n^+$, would be a formal two-electron donor. Triphenylcyclopropenium salts, such as $[Ph_3C_3]X$, where X is halide, $BF_4^-$ or $ClO_4^-$, can be isolated and are frequently used as reagents for synthesizing cyclopropenyl complexes. An x-ray structure of $[Ph_3C_3]ClO_4$ reveals three equal C—C bond distances averaging 1.373(5) Å.

Dicobalt octacarbonyl reacts with $[Ph_3C_3]BF_4$ to afford $(\eta^3\text{-}C_3Ph_3)Co(CO)_3$ (eq. *5.1*). The cobalt atoms undergo a formal disproportionation reaction.

$$4[Ph_3C_3]BF_4 + 3Co_2(CO)_8 \xrightarrow[25°C,\ 2\ days]{CH_2Cl_2}$$

$$4(\eta^3\text{-}C_3Ph_3)Co(CO)_3 + 2Co(BF_4)_2 + 12\ CO \quad (5.1)$$
$$\text{golden solid, 34\%}$$
$$\text{mp 104°C}$$

Reaction of $Ni(CO)_4$ with $[Ph_3C_3]X$, where X is chloride or bromide, affords an amorphous nickel carbonyl complex (eq. *5.2*). Substitution of the CO ligands by pyridine affords crystalline complexes.

$$Ni(CO)_4 + [Ph_3C_3]X \xrightarrow[\Delta]{MeOH} (C_3Ph_3)Ni(CO)_2X$$

X = Cl or Br

$$-2CO \Big| py$$

$$(\eta^3\text{-}C_3Ph_3)Ni(\eta^5\text{-}C_5H_5) \xleftarrow[\substack{C_6H_6 \\ X\ =\ Br}]{TlC_5H_5} (\eta^3\text{-}C_3Ph_3)Ni(py)_2X \cdot py \qquad (5.2)$$
$$\text{red-orange solid, 78\%}$$
$$\text{mp 137−138°C}$$

X = Br

The structure of the chloride compound reveals an $\eta^3\text{-}C_3Ph_3$ ligand. Thallium cyclopentadienide displaces the pyridine and halide ligands to give a mixed-sandwich complex, $(\eta^3\text{-}C_3Ph_3)Ni(\eta^5\text{-}C_5H_5)$.

The structures of two triphenylcyclopropenyl complexes are shown below as **5.2** and **5.3**.

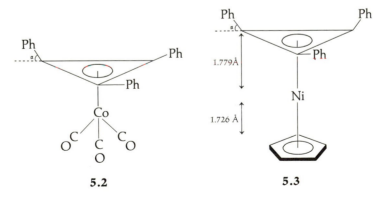

**5.2**                                                                 **5.3**

In each, the $\eta^3\text{-}C_3Ph_3$ ligand coordinates symmetrically to the metal atom with the phenyl substituents tilted about 20° out of the $C_3$ plane away from the metal atom. This tilt angle, $\alpha$, may result from intramolecular steric repulsions or from a change in hybridization of the ring carbon atoms. The average C—C distance of 1.42 Å within an $\eta^3$-$C_3Ph_3$ ligand is 0.05 Å *longer* than the average C—C bond length in $[Ph_3C_3]ClO_4$. A similar trend was discussed earlier for $\eta^5$-$C_5H_5$ ligands. As revealed in structure **5.3**, the line from the Ni atom normal to the $C_5H_5$ plane is shorter than the corresponding line to the $Ph_3C_3$ plane. In general, as the size of the carbocyclic ligand increases, the minimum distance from the metal to the ligand plane decreases.

## $\eta^4$-$C_4R_4$ Complexes

Because cyclobutadiene does not conform to the Hückel $4n + 2$ π-electron rule for aromatic molecules, its electronic and geometrical structure has been investigated intensely both theoretically and experimentally. Experimental isolation of cyclobutadiene can be accomplished only at low temperature, 8–20°K, using matrix isolation techniques.

In 1959 two research groups isolated from reactions of diphenylacetylene with iron carbonyls a complex that was later confirmed by x-ray analysis to be ($\eta^4$-tetraphenylcyclobutadiene)Fe(CO)$_3$. Pettit's report in 1965 of the isolation of the first authenticated cyclobutadiene metal complex represents one of the first applications of transition metal organometallic chemistry to a problem of great interest to organic chemists. Many cyclobutadiene metal complexes are stable crystalline solids that can be handled conveniently. As we shall discuss later, reaction chemistry can be performed on the cyclobutadiene ligand with subsequent cleavage of the metal moiety (an iron complex usually) to liberate the organic ligand. This sequence constitutes a convenient source for a class of otherwise unstable organic molecules.

Some of the principal synthetic routes to cyclobutadiene complexes are (1) from halocyclobutenes, (2) from α-pyrone, (3) from alkynes and metal carbonyls, and (4) by ligand transfer reactions. In the halocyclobutene reactions, a metal carbonyl or an external reducing agent, such as Na/Hg, acts as a halide acceptor (eqs. 5.3 and 5.4).

$$\xrightarrow[-\,FeCl_2]{Fe_2(CO)_9} (\eta^4\text{-}C_4H_4)Fe(CO)_3$$
yellow oil, 45%
bp 47°C (3 mm)

$$\xrightarrow[Na/Hg]{M(CO)_6,\ M\,=\,Mo,\ W} (\eta^4\text{-}C_4H_4)M(CO)_4$$
orange-red solids, 35%
(M = Mo)
mp 17°C (M = Mo)

$$\xrightarrow[Et_2O,\ h\nu]{Cr(CO)_6} (\eta^4\text{-}C_4H_4)Cr(CO)_4 \qquad\qquad (5.3)$$
yellow solid, 1%
mp 83°C

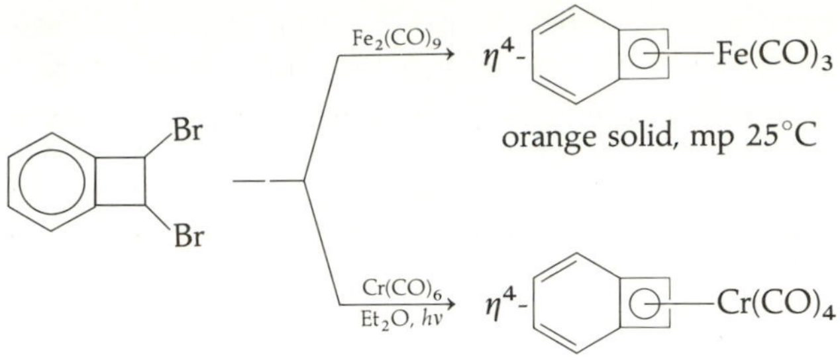

$$\eta^4\text{-}\left[\begin{array}{c}\text{C}_6\text{H}_4\end{array}\right]\text{—Fe(CO)}_3$$

orange solid, mp 25°C

$$\xrightarrow[\text{Et}_2\text{O, }hv]{\text{Cr(CO)}_6} \quad \eta^4\text{-}\left[\begin{array}{c}\text{C}_6\text{H}_4\end{array}\right]\text{—Cr(CO)}_4 \qquad (5.4)$$

Photolytic decarboxylation of photo-α-pyrone in the presence of a metal carbonyl complex affords cyclobutadiene complexes via ligand substitution (eq. 5.5).

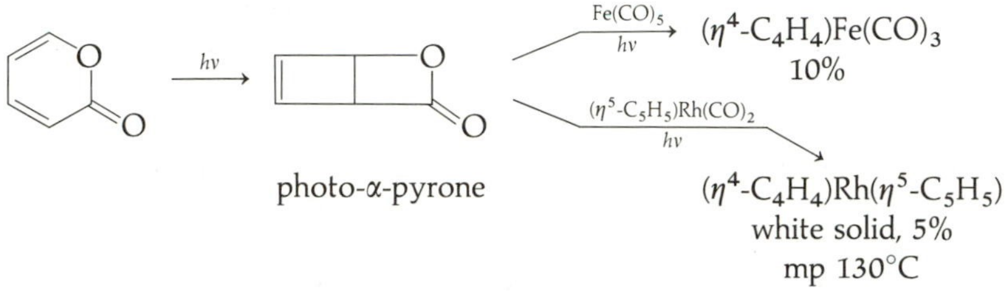

$$\xrightarrow[hv]{\text{Fe(CO)}_5} (\eta^4\text{-C}_4\text{H}_4)\text{Fe(CO)}_3$$
10%

$$\xrightarrow[hv]{(\eta^5\text{-C}_5\text{H}_5)\text{Rh(CO)}_2} (\eta^4\text{-C}_4\text{H}_4)\text{Rh}(\eta^5\text{-C}_5\text{H}_5) \quad (5.5)$$
white solid, 5%
mp 130°C

photo-α-pyrone

Thermal reactions between alkynes and organometallic compounds lead to a variety of products in most instances. As shown in equations 5.6 and 5.7, cyclobutadiene complexes are among these products. As might be expected, cobaltocene eliminates a $C_5H_5$ ligand to form $(\eta^4\text{-C}_4\text{R}_4)\text{Co}(\eta^5\text{-C}_5\text{H}_5)$ complexes that obey the EAN rule.

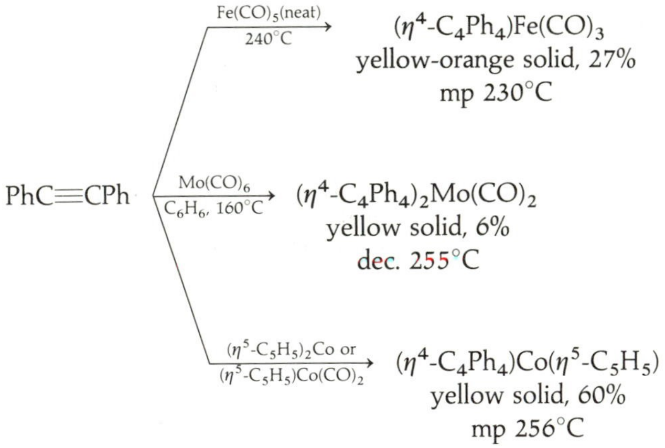

$$\xrightarrow[240°\text{C}]{\text{Fe(CO)}_5\text{(neat)}} (\eta^4\text{-C}_4\text{Ph}_4)\text{Fe(CO)}_3$$
yellow-orange solid, 27%
mp 230°C

$$\text{PhC}\equiv\text{CPh} \xrightarrow[\text{C}_6\text{H}_6,\ 160°\text{C}]{\text{Mo(CO)}_6} (\eta^4\text{-C}_4\text{Ph}_4)_2\text{Mo(CO)}_2$$
yellow solid, 6%
dec. 255°C

$$\xrightarrow[(\eta^5\text{-C}_5\text{H}_5)\text{Co(CO)}_2]{(\eta^5\text{-C}_5\text{H}_5)_2\text{Co or}} (\eta^4\text{-C}_4\text{Ph}_4)\text{Co}(\eta^5\text{-C}_5\text{H}_5) \qquad (5.6)$$
yellow solid, 60%
mp 256°C

$$\xrightarrow{(\eta^5\text{-C}_5\text{H}_5)\text{Co(CO)}_2} [\eta^4\text{-(CH}_2)_m \quad (CH_2)_5] \quad (5.7)$$

$m = 4$ or 5

$$\text{Co}(\eta^5\text{-C}_5\text{H}_5)$$
yellow solids

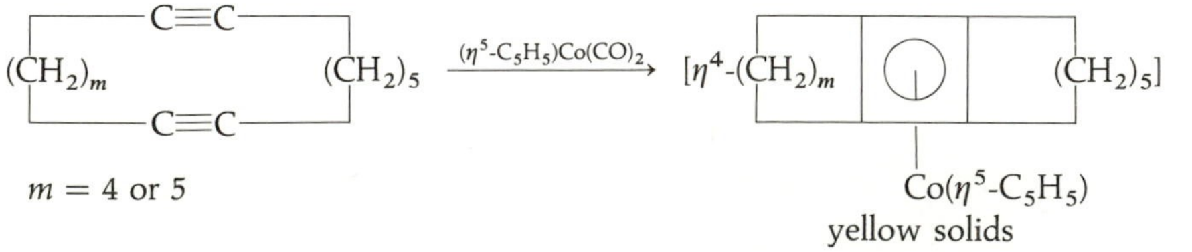

Clearly two acetylene molecules couple together during these reactions to afford an $\eta^4$-$C_4R_4$ ligand. An interesting application of this coupling reaction to effect intramolecular transannular coupling is shown in equation 5.7.

Intermolecular transfer of $\eta^4$-$C_4R_4$ ligands between different metals (eq. 5.8) is a powerful synthetic route to $\eta^4$-$C_4R_4$ complexes that cannot be prepared by other methods.

$$[(\eta^4\text{-}C_4Ph_4)PdBr_2]_2$$

$$\xrightarrow[\substack{\text{xylene, reflux} \\ \text{20 min}}]{Fe(CO)_5} (\eta^4\text{-}C_4Ph_4)Fe(CO)_3 \atop 88\%$$

$$\xrightarrow[\substack{C_6H_6,\ \text{reflux} \\ \text{2.5 hrs}}]{Ni(CO)_4} [(\eta^4\text{-}C_4Ph_4)NiBr_2]_2 \atop 47\%$$

$$\xrightarrow[\substack{\text{xylene, reflux} \\ \text{2.5 hrs}}]{(\eta^5\text{-}C_5H_5)_2Co} (\eta^4\text{-}C_4Ph_4)Co(\eta^5\text{-}C_5H_5) \atop 12\%$$

$$\xrightarrow[\substack{\text{toluene, reflux} \\ \text{10 min}}]{(\eta^5\text{-}C_5H_5)V(CO)_4} (\eta^4\text{-}C_4Ph_4)V(\eta^5\text{-}C_5H_5)(CO)_2 \atop \substack{\text{orange solid, 15\%} \\ \text{dec. 235°C}}$$

(5.8)

The structure of cyclobutadiene complexes was of considerable interest because of the initial lack of structural information for free cyclobutadienes. The structures of four substituted cyclobutadienes have now been determined. Two structures that have alkyl substituents and only minimal intramolecular steric strain are shown in **5.4.** The four carbon atoms of the cyclobutadiene ring define a rectangle of dimensions 1.341(2) Å by 1.598(2) Å. This structure is consistent with localized diene bonding as shown in **5.4.** Theoretical calculations of cyclobutadiene reveal a similar rectangular structure, **5.5,** for the ground-state molecule.

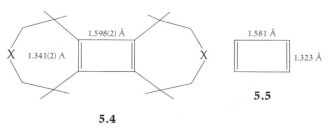

**5.4**

X = S or CH$_2$

At 10−15 kcal/mole higher energy, cyclobutadiene adopts a square structure having a C—C distance of 1.428 Å. The molecule is still diamagnetic in this conformation.

Cyclobutadienes coordinate to metal atoms as four-electron π-carbocyclic ligands, as shown in **5.6−5.9.**

The carbon atoms of the C$_4$ ring are coplanar and define a nearly square geometry with C—C distances of about 1.44 Å. The average C—C distance in $\eta^4$-C$_4$H$_4$ ligands is nearly equivalent to the calculated C—C distance in the diamagnetic, square geometry for uncomplexed C$_4$H$_4$. The metal atom is usually equidistant from all four carbon atoms of the C$_4$ ring. Substituents on the C$_4$ ring carbon

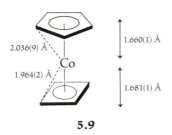

**5.6**                          **5.7**                          **5.8**

atoms are displaced slightly from the $C_4$ plane *away* from the metal atom. This displacement is quite large with phenyl substituents, as it is in $\eta^3$-$C_3Ph_3$ complexes. Even the hydrogen atoms of the cyclobutadiene ligand in **5.9** are displaced about 0.07 Å from the $C_4$ plane away from the cobalt atom. Hydrogen atoms of $\eta^5$-$C_5H_5$ ligands, on the other hand, either lie in the $C_5$ plane or are displaced slightly *toward* the metal atom.

**5.9**

## $\eta^6$-$C_6R_6$ Complexes

The first $\pi$-arene complexes were prepared by Hein as early as 1919 using chromium, but their correct structure was not recognized until 1954. These complexes were isolated from the reaction of $CrCl_3$ with $PhMgBr$. More direct synthetic methods for preparing $\eta^6$-arene complexes include (1) the Fischer–Hafner procedure, (2) reductive-ligation, (3) metal vapor-arene cocondensation, and (4) ligand substitution.

The Fischer–Hafner method involves reacting an anhydrous metal halide with an arene in the presence of an aluminum halide (eq. 5.9).

$$CoCl_2 + 2C_6Me_6 + 2AlCl_3 \xrightarrow{90-120°C} [(\eta^6\text{-}C_6Me_6)_2Co][AlCl_4]_2$$

$$\downarrow 2NH_4PF_6$$

$$[(\eta^6\text{-}C_6Me_6)_2Co][PF_6]_2 \qquad (5.9)$$
yellow-brown solid, 73%
dec. 140°C

Reduction of the metal occurs when a reducing agent such as Al powder is added (eq. 5.10). This reaction is formally a reductive-ligation.

$$3CoCl_2 + 6C_6Me_6 + 3AlCl_3 + Al^0 \longrightarrow 3[(\eta^6\text{-}C_6Me_6)_2Co][AlCl_4]$$

$$\Big\downarrow NH_4PF_6$$

$$[(\eta^6\text{-}C_6Me_6)_2Co]PF_6 \qquad (5.10)$$
$$\text{yellow solid, } 95+\%$$
$$\text{dec. } 170°C$$

In both cases, the $AlCl_3$ acts as a halide acceptor. Dibenzene chromium is conveniently prepared by forming the monocationic complex using the Fischer-Hafner procedure and then reducing the cation to the neutral complex with aqueous dithionite:

$$3CrCl_3 + C_6H_6(xs) + AlCl_3 + 2Al^0 \longrightarrow 3[(\eta^6\text{-}C_6H_6)_2Cr][AlCl_4]$$

$$H_2O \Big\downarrow Na_2S_2O_4$$

$$(\eta^6\text{-}C_6H_6)_2Cr \qquad (5.11)$$
$$\text{black solid, } 95\%$$
$$\text{mp } 284°C$$

Direct reductive-ligation can occur, as shown in equations 5.12 and 5.13, where a Grignard reagent or lithium metal is the reducing agent.

$$MnCl_2 + C_5H_5MgBr + C_6H_6 \longrightarrow (\eta^6\text{-}C_6H_6)(\eta^5\text{-}C_5H_5)Mn \qquad (5.12)$$
$$\text{red solid, } 1\text{--}2\%$$
$$\text{dec. } 205°C$$

$$CrCl_3 + 2\,\text{biphenyl} + 3Li^0 \xrightarrow[(2)KI]{(1)THF} [(\eta^6\text{-}C_6H_5\text{---}C_6H_5)_2Cr]I \qquad (5.13)$$
$$\text{orange solid, } 8\%$$

The manganese complex is an example of a "mixed-sandwich" complex.

Metal vapor cocondensation (figure 3.1) with a potential arene ligand affords a variety of $(\eta^6\text{-arene})_2M$ complexes, as shown in equation 5.14 for chromium compounds.

$$Cr^0(g) + \text{arene}(g) \xrightarrow[(2)\text{ warm to RT}]{(1)\text{ cold probe}} (\eta^6\text{-arene})_2Cr \qquad (5.14)$$

| Arene | % yield | Arene | % yield |
|---|---|---|---|
| $C_6H_6$ | 13 | $1,2,4,5\text{-}C_6H_2Me_4$ | 21 |
| toluene | 51 | $C_6H_5Cl$ | 24 |
| o-xylene | 25 | $C_6H_5F$ | 32 |
| $1,3,5\text{-}C_6H_3Me_3$ | 13 | $C_6H_5NMe_2$ | 6 |

This method can be used to prepare complexes that are not accessible using the Fischer–Hafner method, such as $(\eta^6\text{-}C_6H_6)_2Ti$ and $(\eta^6\text{-}C_6F_6)(\eta^6\text{-}C_6H_6)Cr$.

Substituted $\eta^6$-arene complexes can be prepared by ligand substitution. A very large series of $(\eta^6$-arene)$M(CO)_3$ complexes have been prepared by this method (eq. *5.15*).

$$M(CO)_6 + \text{arene} \xrightarrow{\Delta} (\eta^6\text{-arene})M(CO)_3 + 3CO \qquad (5.15)$$

M = Cr, Mo, W

This procedure is adequate for substituted or fused-ring arenes, but the benzene complex is obtained only in low yield after prolonged heating. Higher yields are obtained under milder reaction conditions if trisubstituted complexes, such as $(NH_3)_3Cr(CO)_3$ or $(CH_3CN)_3W(CO)_3$, are used instead. Chromium tricarbonyl complexes of bromo- or iodobenzene are best prepared by this procedure. Heating $Cr(CO)_6$ and benzene in the presence of 2-picoline gives a nearly quantitative yield of $(\eta^6$-$C_6H_6)Cr(CO)_3$. Presumably, *in situ* formation of (2-picoline)$_nCr(CO)_{6-n}$ complexes facilitates CO substitution by benzene. Photolytic dissociation of CO ligands in the presence of arenes can also afford $\eta^6$-arene complexation.

The structure of dibenzene chromium (**5.10**) reveals an eclipsed orientation of the two $\eta^6$-$C_6H_6$ ligands.

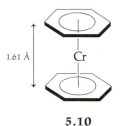

1.61 Å        Cr

**5.10**

Intra-ring C—C distances are not significantly different and average 1.420 Å. In the gas phase, the average C—C distance is 1.423(2) Å. Structure **5.10** is obviously analogous to the sandwich structure of the metallocenes.

*Hexahapto*-arene chromium tricarbonyl complexes have either a staggered (**5.11**) or eclipsed (**5.12**) structure as defined by the relative orientation of the Cr—CO bond vectors and the ring carbon atoms.

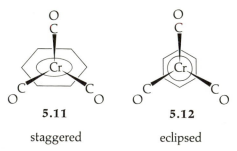

**5.11**              **5.12**

staggered            eclipsed

Low temperature structural determinations of $(\eta^6$-$C_6H_6)Cr(CO)_3$ reveal a staggered structure where the C—C bonds located over each Cr—CO bond are slightly longer than the other three intra-ring C—C bonds. The six arene hydrogen atoms are displaced 0.021(3) to 0.038(3) Å out of the $C_6$ plane *toward* the Cr atom. A similar displacement is observed in ferrocene. The average Cr—CO distance of 1.842 Å is 0.07 Å less than the average Cr—CO distance of 1.913 Å in $Cr(CO)_6$; this difference indicates that an $\eta^6$-$C_6H_6$ ligand is a stronger donor and weaker π-acid than is a CO ligand.

Other ($\eta^6$-arene)$Cr(CO)_3$ complexes that have staggered structures include complexes of $C_6Me_6$, acetylbenzene, phenanthrene (outside ring), naphthalene, and anthracene (outside ring). Eclipsed structures are observed when the arene ligand is anisole, toluidine, or methylbenzoate, among others. In solution, PMR spectra indicate that both structures can be present. Presumably, the observed structural preferences are attributed to a combination of intramolecular steric and electronic effects.

Arene displacement of three terminal carbonyl ligands from a metal atom in a *cluster* is known also. A particularly interesting example is the *carbido-cluster* (**5.13**).

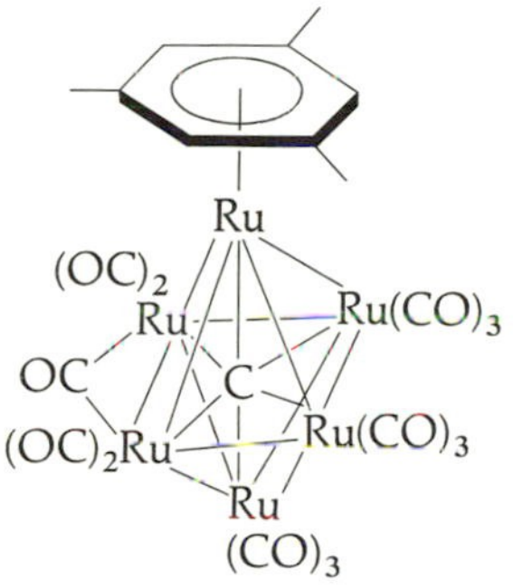

**5.13**

$$Ru_6C(CO)_{14}(\eta^6\text{-}1,3,5\text{-}C_6H_3Me_3)$$

This compound is isolated, along with the parent carbonyl cluster, $Ru_6C(CO)_{17}$, when $Ru_3(CO)_{12}$ is heated at reflux in mesitylene. Complex **5.13** is isoelectronic to $Rh_6(CO)_{16}$, and the $\eta^6$-toluene and xylene analogues can be prepared similarly. These complexes are examples of clusters that have *encapsulated* atoms within the metal core. In this carbido-cluster, a carbon atom is located very near the center of the slightly distorted $Ru_6$ octahedron.

Many phosphine and arsine ligands have aryl substituents. Two examples are $Ph_3P$ and $Ph_2AsCH_2AsPh_2$, which can act as $\eta^6$-arene ligands, as shown in **5.14–5.16**.

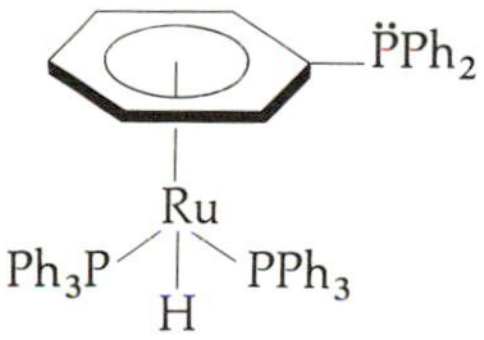

**5.14** 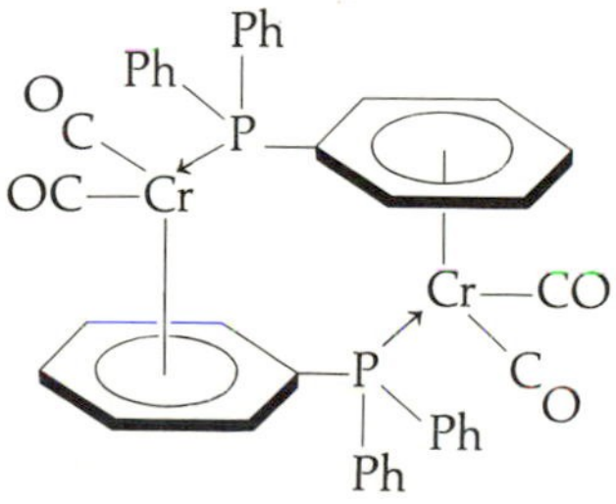          **5.15** 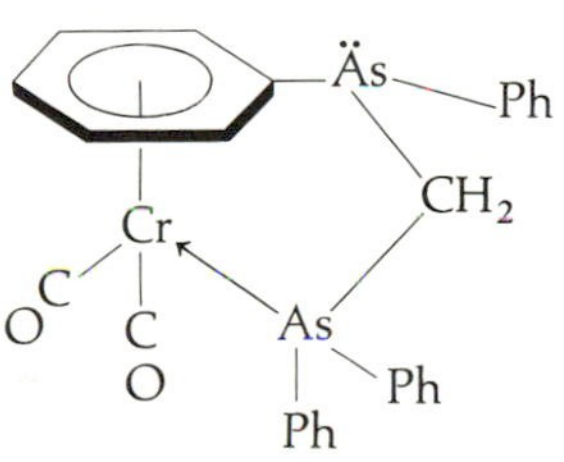          **5.16**

Such molecules can coordinate solely through an aryl ring (**5.14**), or through both an aryl substituent and a heteroatomic donor atom (**5.15**). Dibasic ligands can exhibit both coordination modes, as shown in **5.16**. Usually, such $\eta^6$-arene coordination occurs only under forcing conditions. For example, complexes **5.15** and **5.16** are formed when $Cr(CO)_6$ is heated at reflux in decalin with $Ph_3P$ or $Cr(CO)_5(Ph_2AsCH_2AsPh_2)$, respectively.

Even counterions possessing aryl substituents, such as $BPh_4^-$, are known to act as $\eta^6$-arene ligands. Coordination of the aryl group occurs when more weakly bound ligands, such as $\pi$-olefins, are displaced. The zwitterionic complex **5.17** is an example of $[(\eta^6\text{-}C_6H_5)B(C_6H_5)_3]^-$ coordination.

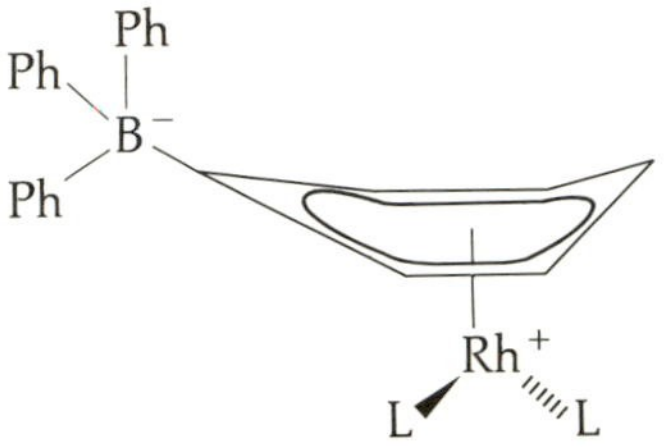

Rh—C range = 2.30–2.41 Å
C—C range = 1.39–1.44 Å
$\overline{\text{Rh—P}}$ = 2.18 Å

**5.17**

L = P(OCH$_3$)$_3$

A slight distortion of the $\eta^6$-arene ring toward a boat conformation is observed. However, this distortion may not be statistically significant. Coordination of these aryl groups is usually evident from the proton NMR spectrum by a 1–2 ppm upfield shift of the aromatic proton resonances.

In some complexes, an arene ligand is severely distorted because it acts as a four-electron donor, $\eta^4$-C$_6$R$_6$ ligand, as shown in **5.18**.

**5.18**

$(\eta^4\text{-}C_6Me_6)(\eta^6\text{-}C_6Me_6)Ru$

The structure permits the ruthenium atom to obey the EAN rule with this unusual distortion of the aromatic $\pi$-electron delocalization. The structure of the isoelectronic complex, $[(\eta^6\text{-}C_6Me_6)_2Co]^+$, is an alternate possibility where the arene ligands remain *hexahapto*, but the metal–ring distance is unusually long. Complex **5.18** is a known catalyst for arene hydrogenation presumably due to activation of an arene ligand via $\eta^4$-coordination. Considerable effort is being expended to understand the mechanism of these catalytic hydrogenations. Even $\eta^2$-arene coordination, which is known for $d^{10}$-metal complexes, may be involved in the complete hydrogenation of arenes.

## $\eta^7$-C$_7$R$_7$ Complexes

Tropylium salts, such as $C_7H_7{}^+Br^-$, are isolable compounds because of the aromatic stabilization of the $C_7H_7{}^+$ ion. In contrast to cyclopropenium salts, tropylium salts react with organometallic anions to give redox reactions (eq. 5.16)

$$Na[Mn(CO)_5] + [C_7H_7]Br \longrightarrow Mn_2(CO)_{10} + NaBr + \text{organic products} \quad (5.16)$$

rather than $\eta^7$-*cycloheptatrienyl complexes*. For this reason, all direct syntheses of $\eta^7$-C$_7$H$_7$ complexes use 1,3,5-cycloheptatriene, $C_7H_8$, as the source of the $C_7$ ring via a formal loss of $H^-$ or a hydrogen atom.

Group VI metal hexacarbonyls react with $C_7H_8$ under heating to form ($\eta^6$-C$_7$H$_8$)-M(CO)$_3$ complexes, where M is Cr, Mo or W. Each of the three olefinic bonds in cycloheptatriene displaces a CO ligand from the M(CO)$_6$ molecule. Hydride abstraction from the coordinated $C_7H_8$ ligand by [Ph$_3$C]BF$_4$ generates a cationic $\eta^7$-C$_7$H$_7$ complex, as shown for Cr and Mo in equation 5.17. For the Mo complex,

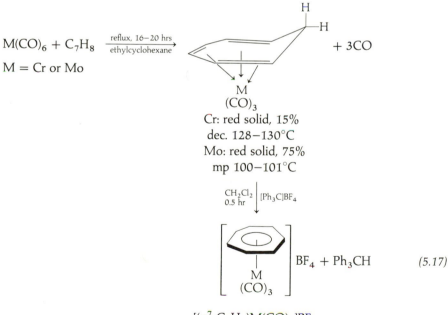

$$M(CO)_6 + C_7H_8 \xrightarrow[\text{ethylcyclohexane}]{\text{reflux, 16–20 hrs}} \qquad + 3CO$$

M = Cr or Mo

M
(CO)$_3$

Cr: red solid, 15%
dec. 128–130°C
Mo: red solid, 75%
mp 100–101°C

$$\begin{array}{c} \text{CH}_2\text{Cl}_2 \\ 0.5\ \text{hr} \end{array} \Big| [\text{Ph}_3\text{C}]\text{BF}_4$$

M
(CO)$_3$

$\text{BF}_4 + \text{Ph}_3\text{CH}$ $\qquad$ (5.17)

[($\eta^7$-C$_7$H$_7$)M(CO)$_3$]BF$_4$
Cr: orange solid, 99%
dec. 230°C
Mo: orange solid, 90%
mp >270°C

conversion of the neutral $C_7H_8$ complex into the cationic cycloheptatrienyl complex causes a shift of 50 cm$^{-1}$ to higher energy for $\nu_{CO}$ (average). This shift is consistent with the formation of a cation.

Several mixed-sandwich compounds having $\eta^7$-$C_7H_7$ ligands are known, but $(\eta^7$-$C_7H_7)_2M$ complexes are unknown. Reductive-ligation and oxidative-ligation reactions are used to prepare mixed-sandwich complexes, as shown in equations 5.18–5.20.

$$(\eta^5\text{-}C_5H_5)\text{TiCl}_3 + C_7H_8 + i\text{-}C_3H_7\text{MgBr} \xrightarrow{\text{Et}_2O} (\eta^5\text{-}C_5H_5)\text{Ti}(\eta^7\text{-}C_7H_7) \quad (5.18)$$
$$\text{blue solid, 33\%}$$
$$\text{subl. 125°C/1 mm}$$

$$2(\eta^5\text{-}C_5H_5)\text{V(CO)}_4 \xrightarrow[\text{reflux, 8 hrs}]{C_7H_8} 2(\eta^5\text{-}C_5H_5)\text{V}(\eta^7\text{-}C_7H_7) + 8\text{CO} + H_2 \quad (5.19)$$
$$\text{purple solid, 40–60\%}$$
$$\text{subl. 80°C/0.1 mm}$$

$$(\eta^5\text{-}C_5H_5)\text{Cr}(\eta^6\text{-}C_6H_6) + C_7H_8 \xrightarrow[\substack{(2)\ H^+/H_2O, \\ PF_6}]{(1)\ \text{AlCl}_3} [(\eta^5\text{-}C_5H_5)\text{Cr}(\eta^7\text{-}C_7H_7)]\text{PF}_6$$
$$\text{yellow-green solid, 75\%}$$

$$\Big\downarrow \text{Na}_2\text{S}_2\text{O}_4$$

$$(\eta^5\text{-}C_5H_5)\text{Cr}(\eta^7\text{-}C_7H_7) \quad (5.20)$$
$$\text{blue-green solid}$$
$$\text{mp 230°C}$$

The vanadium and cationic chromium compounds have one unpaired electron, while the neutral chromium sandwich complex obeys the EAN rule and is isoelectronic to $(\eta^6$-$C_6H_6)_2$Cr.

The structures of $\eta^7$-$C_7H_7$ complexes reveal a planar, symmetrical ring with no significant differences in intra-ring C—C bond distances. This symmetrical coordination is reflected in the proton NMR spectrum, which shows a single sharp resonance for the $\eta^7$-$C_7H_7$ ligand.

Complexes **5.19** and **5.20** have a three-legged piano-stool structure. The average value of the intra-ring C—C distances is 1.40 Å in each complex.

**5.19**        **5.20**

Unfortunately, disorder of the $C_7H_7$ ring in tropylium salts has precluded a precise determination of the intra-ring C—C distance in uncomplexed $C_7H_7^+$ ions. However, intra-ring C—C distances of complexed $\eta^7$-$C_7H_7$ ligands are similar to those observed for $\eta^5$-$C_5H_5$ and $\eta^6$-$C_6H_6$ ligands. Intra-ring C—C—C angles in **5.19** and **5.20** average 128.4° and 128.5°, respectively—not significantly different from the theoretical value of 128.57° for a regular heptagon.

The structures of two mixed-sandwich complexes, ($\eta^5$-$C_5H_5$)M($\eta^7$-$C_7H_7$), where M is titanium or vanadium, are shown in **5.21** and **5.22**. In the vanadium compound the average V—C distances to *each* ring are essentially equal, but distances from V to the rings are very different, the $C_7$ plane being significantly closer to the V atom. The titanium compound, **5.21**, has a similar structure.

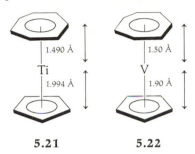

**5.21**            **5.22**

A common coordination mode for $C_7H_7$ ligands is as a *trihapto* ligand. We shall discuss examples of this type of bonding later in this chapter.

## $\eta^8$-$C_8R_8$ Complexes

Cyclooctatetraene and its substituted analogues form a large series of π-olefin complexes as expected for a molecule containing four olefinic bonds. Only $\eta^8$-$C_8H_8$ coordination is discussed here; many others are described later in this book. The usual source of $\eta^8$-$C_8H_8$ ligands is the aromatic dianion of cyclooctatetraene, $C_8H_8^{2-}$ ($COT^{2-}$), which contains 10 $p\pi$ electrons. This dianion is prepared by alkali metal reduction of COT or through use of an electron transfer agent, such as potassium naphthalenide, KNp, as shown in equation 5.21.

$$2K + COT \xrightarrow[-30°C]{THF}$$
$$\searrow K_2[COT] \qquad (5.21)$$
$$\nearrow COT, -70°C$$
$$2K + 2\ naphthalene \xrightarrow[25°C,\ 6\ hrs]{THF} 2KNp$$

Dianions of substituted COT molecules are prepared similarly. An x-ray structure of [K(diglyme)]$_2$[1,3,5,7-tetramethylcyclooctatetraenide] reveals a planar $C_8$ ring of aromatic structure in which all intra-ring C—C distances are essentially equal [C—C = 1.407(6) Å].

Because of the large number of $p\pi$ electrons and the large size of an $\eta^8$-$C_8H_8^{2-}$ ring, only large and highly electron deficient metal ions form complexes with this ligand. Typically these metals are scandium and titanium group metals including the lanthanides and actinides. Complexes of the chromium group metals are known, but they do not have $\eta^8$-$C_8H_8$ coordination. The synthesis of a titanium complex is shown in equation 5.22. This reaction is a reductive-ligation.

$$(\eta^5\text{-}C_5H_5)TiCl_3 + 1.5K_2[COT] \xrightarrow[\text{15 min, 42\%}]{\text{THF, reflux}}$$

$$(\eta^5\text{-}C_5H_5)Ti(\eta^8\text{-}C_8H_8) \qquad (5.22)$$
$$\text{dark green solid}$$
$$\text{dec. } 160°C$$

$$TiCl_4 + NaC_5H_5 + K_2[COT] \xrightarrow[\text{1 hr, 60\%}]{\text{toluene, reflux}}$$

Lanthanide-COT complexes are prepared similarly, as shown in equations 5.23 and 5.24. These complexes exhibit a variety of colors depending on the metal ion present.

$$LnCl_3 + 1.5K_2[COT] + 0.5COT \xrightarrow{\text{THF}} K[(COT)_2Ln]^- \qquad (5.23)$$
$$\text{50--80\%}$$
Ln = Ce, Pr, Nd, Sm, Tb
$$\text{dec. } \sim 160°C$$

$(\eta^5\text{-}C_5H_5)LnCl_2 \cdot 3THF + K_2[COT] \xrightarrow{\text{THF}}$

Ln = Y, Sm, Ho, Er

$$\qquad\qquad\qquad\qquad\qquad (\eta^5\text{-}C_5H_5)Ln(\eta^8\text{-}C_8H_8)$$

$[(\eta^8\text{-}C_8H_8)Ln(Cl)(THF)_2]_2 + NaC_5H_5 \xrightarrow{\text{THF}}$
$$\qquad\qquad\qquad\qquad\qquad\qquad\qquad (5.24)$$

Ln = Y, Nd

The synthesis of "uranocene," $(\eta^8\text{-}C_8H_8)_2U$, in 1968 led to intensive development of organometallic lanthanide and actinide chemistry. Uranocene complexes are prepared from $UCl_4$ via formal reductive-ligation (eq. 5.25). Isostructural actinide complexes, $(\eta^8\text{-}C_8H_8)_2M$, where M is Th, Pa, Np and Pu, are prepared similarly.

$$UCl_4 + 2K_2[C_8H_7R] \xrightarrow[\text{0°C--25°C}]{\text{THF}} (\eta^8\text{-}C_8H_7R)_2U + 4KCl \qquad (5.25)$$
$$\text{green solids, 58--80\%}$$

R = H, alkyl, OR, $NR_2$

Although $(\eta^5\text{-}C_5H_5)Ti(\eta^8\text{-}C_8H_8)$ has a typical mixed-sandwich structure (**5.23**), $(COT)_2Ti$ has one $\eta^8\text{-}C_8H_8$ and one $\eta^4\text{-}C_8H_8$ ligand, as shown in **5.24**.

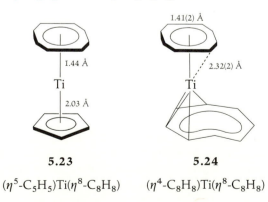

**5.23**                                        **5.24**

$(\eta^5\text{-}C_5H_5)Ti(\eta^8\text{-}C_8H_8)$         $(\eta^4\text{-}C_8H_8)Ti(\eta^8\text{-}C_8H_8)$

In complex **5.23**, the average intra-ring C—C and Ti—C distances for the two rings are not very different even though the minimum distance from the Ti atom to the plane of the $\eta^8$-$C_8H_8$ ligand is much shorter than the corresponding distance of the $\eta^5$-$C_5H_5$ ligand. *Tetrahapto* coordination of $C_8H_8$, as in **5.24**, is fairly common.

Lanthanide and actinide complexes of the type $(COT)_2M^n$ have the "uranocene" structure shown in **5.25**.

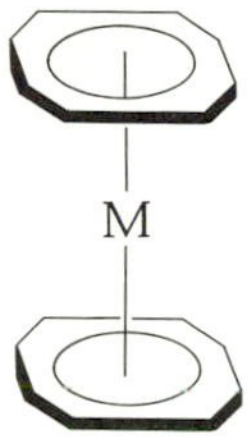

| M | n |
|---|---|
| U | 0 |
| Th | 0 |
| Ce | −1 |

M

**5.25**

$(\eta^8$-$C_8H_8)_2M^n$

The two rings can be staggered, eclipsed, or partially staggered relative to each other. In uranocene and thoracene, the rings are eclipsed. Detailed analysis of M—C distances indicates some degree of covalent M—C bonding in the lanthanide complexes.

## Bonding to $\eta^n$-Carbocyclic Ligands

As with the metallocenes, a detailed description of M—R bonding to carbocyclic ligands is best obtained from detailed MO calculations. However, it is advantageous to develop a simplified bonding scheme that rationalizes most of the structural data and that may help in predicting reaction chemistry.

In general, the predominant M—L bonding interactions between a metal atom and a π-carbocyclic ligand involve the metal valence orbitals and the $p\pi$ molecular orbitals of the carbocyclic ligand. Such bonding interactions are analogous to those presented in figure 4.1 for metallocene complexes. The strength of M-ring bonding would depend on the relative energies of the metal and ring orbitals, that is, on *the orbital energy match,* and on the degree of orbital overlap. Strong bonding occurs between orbitals having similar energy and good overlap.

Figure 5.1 shows a comparison of the energies of the $p\pi$ molecular orbitals for $C_3$ through $C_8$ rings to the $3d$ atomic orbital energies for gaseous metal atoms of the first transition metal series. Since the metal atoms are neutral, the electronic configuration of the $C_nH_n$ carbocycles are shown for the neutral molecule or radical. Metal $4s$ and $4p$ atomic orbitals, which are not shown, lie slightly above the $3d$ orbitals.

Although these orbital energies are only approximate, the figure does reveal a generally favorable energy match between valence orbitals of the rings and the metal atoms. The predominant metal-$C_nH_n$ π-bonding occurs between the metal $3d$, $4s$ and $4p$ orbitals and the $2\pi$ and $3\pi$ molecular orbitals of the rings. According to

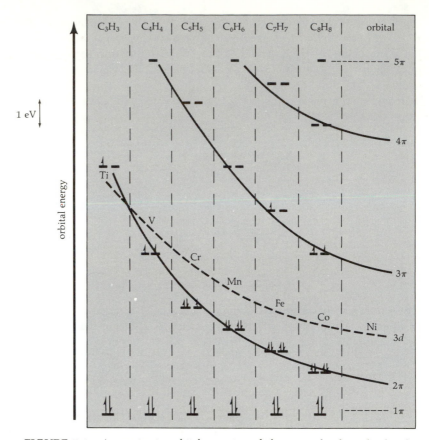

**FIGURE 5.1** Approximate orbital energies of the $p\pi$-molecular orbitals of $C_3H_3$ to $C_8H_8$ carbocyclic molecules or radicals and the $3d$ atomic orbitals of the first series transition metals.

figure 5.1, orbital energy depends on both ring size and the position of a metal within the series. As the ring size increases, the energies of the $2\pi$ and $3\pi$ molecular orbitals decrease, while the $3d$ orbital energies decrease across the series with increasing atomic number of M. For a particular metal, the exact nature of $M$—$C_nH_n$ bonding varies with ring size. In titanium complexes, the Ti($3d$)-$C_nH_n$($2\pi$) energy match decreases with ring size, and the Ti($3d$)-$C_nH_n$($3\pi$) energy match increases with ring size, in going from $C_4H_4$ to $C_6H_6$, for example.

Orbital overlap between metal and $p\pi$ ligand orbitals is a function of the size of the orbitals, their relative orientation, and the distance between the orbitals. These geometrical parameters obviously depend on the size of the ring and metal atom. For some $M$—$C_nH_n$ combinations, effective overlap may occur without any distortion of the ring $p\pi$ orbitals as shown in **5.26.**

For smaller rings, overlap with the same metal may require a hybridization like that shown in **5.27** where the $p\pi$ orbital is tilted away from the metal as evidenced

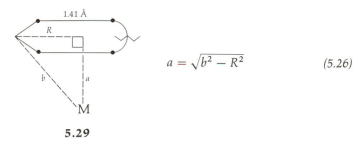

**5.26**          **5.27**          **5.28**

by displacement of the hydrogen atoms out of the $C_n$ plane *away* from M. For larger rings, effective overlap may require the opposite distortion as shown in **5.28**. Here the hydrogen atoms are displaced *toward* M. Interestingly, hydrogen atom positions in $(\eta^4\text{-}C_4H_4)Fe(CO)_3$ and ferrocene follow patterns **5.27** and **5.28**, respectively. Calculations have confirmed the tilting of $p\pi$ orbitals for $C_3H_3$ through $C_8H_8$ rings. The crossover from **5.27** to **5.28** occurs at $C_5H_5$.

Three important distances that characterize $\pi$-carbocyclic ligand coordination are the M—C distances, the M—$C_n$ (ring) minimum distances, and the intra-ring C—C distances. Since meaningful structural comparisons are best made between ligands within the *same* molecule, a comparison between $\eta^5\text{-}C_5H_5$ and $\eta^n\text{-}C_nH_n$ ligands in mixed-sandwich complexes seems appropriate. These data are shown in table 5.1. Notice that the intra-ring C—C distances are nearly equal and average about 1.41 Å. The slightly longer C—C distances in $\eta^4\text{-}C_4H_4$ rings may indicate that $\eta^4\text{-}C_4H_4$ ligands are slightly better donors than $\eta^5\text{-}C_5H_5$ ligands. Within each mixed-sandwich complex, the two average M—C distances are remarkably similar (to within 0.10 Å when excluding the $\eta^3\text{-}C_3Ph_3$ ligand because of the unknown electronic and steric influence of the Ph substituents). Of course, the actual *value* of the average M—C distances depends on the covalent radius of the metal. For example, in $(\eta^5\text{-}C_5H_5)Ti(\eta^8\text{-}C_8H_8)$, $\overline{M\text{—}C}$ is 2.33 Å, but in $(\eta^4\text{-}C_4H_4)Co(\eta^5\text{-}C_5H_5)$, $\overline{M\text{—}C}$ is only 2.00 Å. However, notice that the M—$C_nH_n$ (ring) distances change dramatically relative to the M—$C_5H_5$ (ring) distance. Rings smaller than $C_5$ become progressively farther from M, and rings greater than $C_5$, progressively closer.

This dependence of minimum M—$C_nH_n$ (ring) distance on the value of $n$ is due solely to geometrical considerations. If two carbocyclic rings of unequal size coordinate to the same metal atom with approximately the same M—C distances, then, by necessity, the plane of the larger ring must be closer to the metal atom. This relationship can be demonstrated easily, as shown in **5.29** and equation 5.26.

$$a = \sqrt{b^2 - R^2} \qquad (5.26)$$

**5.29**

A carbocyclic ligand defines a regular polygon in which each side has a length of 1.41 Å. The distance from the center of the polygon to a vertex (or C atom) is $R$,

**TABLE 5.1** Relative differences in metal–ring coordination for the series of complexes $(\eta^5\text{-}C_5H_5)M(\eta^n\text{-}C_nH_n)$, where $n$ is 3 to 8.

| Complex | Distance (Å) | | |
|---|---|---|---|
| | $C-C(C_5) - [C-C(C_n)]$ | $M-C(C_5) - [M-C(C_n)]$ | $M-C_5(\text{ring}) - [M-C_n(\text{ring})]$ |
| $(C_5H_5)Ni(C_3Ph_3)$ | −0.023 | 0.139 | −0.053 |
| $(C_5H_5)Co(C_4H_4)$ | −0.047 | 0.072 | −0.021 |
| $(C_5H_5)_2M$ | 0 | 0 | 0 |
| $(C_5H_5)Cr(C_6H_6)^a$ | — | 0.100 | — |
| $(C_5H_5)V(C_7H_7)$ | 0.02 | −0.02 | +0.40 |
| $(C_5H_5)Ti(C_8H_8)$ | 0.001 | +0.03 | +0.59 |

[a] Structure calculated by energy minimization methods. See D. W. Clack and K. D. Warren, *J. Organometal. Chem.*, **157**, 421 (1978).

**TABLE 5.2    Calculated values of the center-to-vertex distance, $R$, for various regular polygons in which the length of each side is   or 1.41 Å.**

| Regular polygon | Number of sides | $R$ (in units of $s$) | $R$ (for $s = 1.41$ Å) |
|---|---|---|---|
| Triangle | 3 | 0.57735 $s$ | 0.814 Å |
| Square | 4 | 0.70711 $s$ | 0.997 Å |
| Pentagon | 5 | 0.85065 $s$ | 1.199 Å |
| Hexagon | 6 | 1.00000 $s$ | 1.410 Å |
| Heptagon | 7 | 1.1524 $s$ | 1.625 Å |
| Octagon | 8 | 1.3066 $s$ | 1.842 Å |

**TABLE 5.3    Calculated and observed perpendicular metal-to-ring distances, $a$, for $(\eta^3\text{-}C_3Ph_3)Ni(\eta^5\text{-}C_5H_5)$ and $(\eta^5\text{-}C_5H_5)Ti(\eta^8\text{-}C_8H_8)$.**

| Complex | $b^a$, $C_5$ ring | $b^a$, $C_n$ ring | Observed $a$ ($C_5$) | Observed $a$ ($C_n$) | Calculated $a$ ($C_5$) | Calculated $a$ ($C_n$) |
|---|---|---|---|---|---|---|
| $(\eta^3\text{-}C_3Ph_3)Ni(\eta^5\text{-}C_5H_5)$ | 2.10 Å | 1.96 Å | 1.73 Å | 1.78 Å | 1.72 | 1.78 |
| $(\eta^5\text{-}C_5H_5)Ti(\eta^8\text{-}C_8H_8)$ | 2.35 Å | 2.32 Å | 2.03 Å | 1.44 Å | 2.02 | 1.41 |

$^a$ $b$ is the M—C distance to the designated carbocyclic ligand.

$a$ is the minimum distance between the plane of the ring and the metal atom, and $b$ is the M—C distance. From equation 5.26, if $b$ remains essentially constant for two rings of different size, then the value of $a$ to the larger ring must become smaller as the ring size increases because $R$ must be greater for the larger ring. The value of $R$ depends on the number of vertices in the regular polygon and the length of each side $s$ as shown in table 5.2. Values of $R$ calculated for $s = 1.41$ Å are provided.

Table 5.3 shows the calculated and observed perpendicular metal-ring distances, $a$, for $(\eta^3\text{-}C_3Ph_3)Ni(\eta^5\text{-}C_5H_5)$ (**5.3**) and $(\eta^5\text{-}C_5H_5)Ti(\eta^8\text{-}C_8H_8)$ (**5.23**). The calculated M-ring perpendicular separation is obtained by using equation 5.26 and the appropriate $R$ values from table 5.2. *Thus, minimum metal-ring distances are clearly predictable from geometrical considerations.*

Not many thermochemical data are available for carbocyclic complexes. For the series of compounds, $(\eta^5\text{-}C_5H_5)Cr(\eta^n\text{-}C_nH_n)$, where $n$ is 5, 6 and 7, bond dissociation energies for the $C_nH_n$ ligands have been measured by electron impact. Bond energies *per Cr—C bond* for these ligands are 14.77 kcal/mole for $C_5H_5$, 11.15 kcal/mole for $C_6H_6$, and 11.13 kcal/mole for $C_7H_7$. Slightly stronger bonding to the carbon atoms of the $\eta^5\text{-}C_5H_5$ ligand may reflect a better energy match (see figure 5.1) or better overlap of the Cr orbitals with the orbitals of the smaller ring.

Extensive MO calculations have been performed on a number of mixed-sandwich complexes to provide a detailed understanding of metal-ring bonding in these compounds and to ascertain relative donor/acceptor abilities of the carbocyclic ligands. These studies lead to some general bonding conclusions:

1. Although the dominant M—$C_nH_n$ bonding occurs as $M(3d)$—$C_nH_n$ ($2\pi$, $3\pi$) *pi bonding*, significant $M(4s, 4p)$—$C_nH_n$ ($\sigma$-skeleton MO) *sigma-bonding* interactions are observed.

2. The $M(3d)$—$C_nH_n$ $(1\pi)$ interactions are essentially nonbonding because of poor overlap.
3. Some degree of M—$C_nH_n$ synergistic bonding is present in these complexes.

**TABLE 5.4** Calculated charge distributions in several mixed-sandwich complexes.*

| Complex | Charge on $C_nH_n$ | Charge on M | Charge on $C_5H_5$ |
|---|---|---|---|
| $(\eta^3$-$C_3H_3)Ni(\eta^5$-$C_5H_5)$ | $+0.27$ | $-0.83$ | $+0.56$ |
| $(\eta^4$-$C_4H_4)Co(\eta^5$-$C_5H_5)$ | $+0.12$ | $-0.67$ | $+0.55$ |
| $(\eta^6$-$C_6H_6)Mn(\eta^5$-$C_5H_5)$ | $+0.42$ | $-0.58$ | $+0.16$ |
| $(\eta^7$-$C_7H_7)M(\eta^5$-$C_5H_5)$ | $-1.05$ | $+1.75$ (Ti) | $-0.70$ |
|  | $-0.96$ | $+1.71$ (V) | $-0.75$ |
|  | $-0.87$ | $+1.67$ (Cr) | $-0.80$ |
| $(\eta^8$-$C_8H_8)Ti(\eta^5$-$C_5H_5)$ | $-0.28$ | $+0.28$ | $0.00$ |

* See *Suggested Reading* for references.

More specific analysis would be appropriate only for individual complexes. One property of interest to the preparative chemist is the distribution of charge within a molecule. Table 5.4 shows these calculated charge distributions for several mixed-sandwich systems. For the $C_3H_3$ to $C_6H_6$ complexes, the carbocyclic ligands have a net positive charge while M is negative. From figure 5.1, the $3d$ orbitals of Mn, Co and Ni appear to have a better energy match with the filled $\pi$ molecular orbitals of the ligands than with the empty ligand orbitals, thus affording a net ring-to-metal donation of electron density. Conversely, the lighter metals (e.g., Ti, V and Cr) have a better energy match with the empty ligand $\pi$ molecular orbitals of the larger rings. This causes a net metal-to-ring transfer of electron density, affording negatively charged carbocyclic ligands and positively charged metals.

Although calculated atomic charges are only qualitative measures of charge distribution, such calculations indicate a positive charge on arene ligands in $(\eta^6$-$C_6R_6)Cr(CO)_3$ complexes, also. As we will discuss in chapters 11 and 12, arene ligands in these complexes are very susceptible to nucleophilic attack, but they undergo electrophilic attack only under forcing conditions.

In $(\eta^5$-$C_5H_5)Ti(\eta^8$-$C_8H_8)$, the $C_5H_5$ ring is more positively charged than the $C_8H_8$ ring. Deprotonation by $n$-butyllithium occurs on the $C_5H_5$ ring as expected from these results.

Clearly, the above bonding analysis, based primarily upon orbital energy match, is too simplistic. Unfortunately the results of MO calculations, which include all orbital interactions, are usually not readily available.

## COMPLEXES CONTAINING ALKENES AS LIGANDS

Since the preparation of Zeise's salt, $K[(\eta^2$-$C_2H_4)PtCl_3]\cdot H_2O$, in 1827, a very large number of $\eta^2$-alkene complexes have been prepared. Essentially every type of olefin has been coordinated to metal atoms. These ligands include ethylene; alkyl- or aryl-substituted olefins; heteroatomic-substituted olefins; mono-olefins, di- or polyolefins that are either conjugated or nonconjugated; and cumulated olefins, such as allene.

Application of transition metal organometallic chemistry in organic chemistry and industrial catalysis frequently depends on formation of alkene complexes. Coordination of an olefinic bond may either activate or protect that bond from subsequent chemical attack, and the presence of the metal moiety frequently affects regio- or stereospecificity or permits resolution of optical or geometrical isomers. Such applications are discussed in chapters 12 and 13.

## Preparation of Complexes Containing Alkene Ligands

Important synthetic routes to alkene complexes include (1) reductive-ligation, (2) ligand substitution, (3) $\beta$-anion abstraction from an alkyl ligand, (4) $\gamma$-protonation of an unsaturated alkyl ligand, and (5) direct coordination to unsaturated metal atoms. Several examples of reductive-ligation are shown in equations 5.27–5.29.

$$\text{Ni(acac)}_2 + 2\text{Ph}_3\text{P} + \text{C}_2\text{H}_4 + \text{AlEt}_3 \xrightarrow[0°C]{\text{Et}_2\text{O}} (\eta^2\text{-C}_2\text{H}_4)\text{Ni(PPh}_3)_2 \quad (5.27)$$
$$\text{yellow solid, 85\%}$$

$$2\text{RhCl}_3 \cdot 3\text{H}_2\text{O} + 2 \text{ 1,5-cyclooctadiene} \xrightarrow[80°C, 3 \text{ hrs}]{\text{EtOH}}$$
$$\text{(1,5-COD)}$$

$$[(\eta^4\text{-1,5-COD})\text{RhCl}]_2 + 2\text{CH}_3\text{C(O)H} + 4\text{HCl} + 6\text{H}_2\text{O} \quad (5.28)$$
$$\text{yellow solid, 60\%}$$
$$\text{dec. 250°C}$$

$$\text{PtCl}_4 + \text{C}_2\text{H}_4(\text{xs}) \xrightarrow[\Delta]{\text{C}_6\text{H}_6} [(\eta^2\text{-C}_2\text{H}_4)\text{PtCl}_2]_2 + 2\text{PtCl}_6 \text{ (reduced to Pt by alkene)}$$
$$\text{orange solid,}$$
$$\text{dec. 160°C}$$

$$\text{H}_2\text{O} \Big| 2\text{KCl}$$

$$2\text{K}[(\eta^2\text{-C}_2\text{H}_4)\text{PtCl}_3] \cdot \text{H}_2\text{O} \quad (5.29)$$
$$\text{yellow solid}$$

In equation 5.28, ethanol is oxidized to acetaldehyde as Rh(III) is reduced to Rh(I). Similar solvent-induced reductions were encountered previously in equations 3.7 and 3.9.

Ligand substitution is the most general route to alkene complexes. Both thermal and photochemical activation are used although heat and light can induce polymerization and oligomerization of many olefinic reactants and cause decomposition of alkene complexes. In these instances, ligand substitution must be performed under mild thermal conditions.

As shown in equations 5.30–5.34, thermal substitution of CO ligands affords $\eta^2$-alkene complexes directly.

$$\text{Mo(CO)}_6 + \left[\bigcirc\!\!\!\!\!\triangleright\right](\text{xs}) \xrightarrow[\text{reflux, 16 hrs}]{\text{methylcyclohexane}}$$

$$\left(\eta^2\!-\!\left[\bigcirc\!\!\!\!\!\triangleright\right]\right)_2 \text{Mo(CO)}_4 \qquad + 2\text{CO} \quad (5.30)$$
$$\text{yellow solid}$$
$$30\text{–}50\%$$
$$\text{dec. 76°C}$$

$$\text{Fe(CO)}_5 + \underset{}{\text{(butadiene)}} \xrightarrow[\text{pressure}]{140°\text{C, 12 hrs}} (\eta^4\text{-C}_4\text{H}_6)\text{Fe(CO)}_3 + 2\text{CO} \qquad (5.31)$$

$$\text{orange liquid, 42\%}$$
$$\text{mp 19°C}$$

$$\text{Fe(CO)}_5 + \underset{}{\text{(1,3-cyclohexadiene)}} \xrightarrow[\text{pressure}]{150°\text{C, 20 hrs}} (\eta^4\text{-1,3-C}_6\text{H}_8)\text{Fe(CO)}_3 + 2\text{CO} \qquad (5.32)$$

$$\text{yellow liquid, 19\%}$$
$$\text{mp 8°C}$$

$$(\eta^5\text{-C}_5\text{H}_5)\text{Co(CO)}_2 + \underset{}{\text{(cyclooctadiene)}} \xrightarrow[\text{reflux, 16 hrs}]{\text{ethylcyclohexane}}$$

$$(\eta^5\text{-C}_5\text{H}_5)\text{Co}(\eta^4\text{-1,5-COD}) + 2\text{CO} \quad (5.33)$$
$$\text{orange solid, 48\%}$$
$$\text{mp 104°C}$$

$$\text{Fe}_2(\text{CO})_9 + \underset{}{\text{(maleic anhydride)}} \xrightarrow[\text{45°C, 4 hrs}]{\text{C}_6\text{H}_6} \left( \eta^2\text{-}\underset{}{\text{(maleic anhydride)}} \right) \text{Fe(CO)}_4 + \text{Fe(CO)}_5 \quad (5.34)$$

$$\text{yellow solid, 78\%}$$
$$\text{dec. 148°C}$$

Usually the olefin is present in excess; the maleic anhydride complex in equation 5.34, however, is formed from an equimolar mixture of the two reactants. All of the binary iron carbonyls have been used to prepare olefin complexes. As shown in equation 5.32, the $\text{Fe(CO)}_3$ group so strongly prefers to bond to 1,3-dienes that 1,4-dienes are isomerized. Even 1,5-COD is isomerized to 1,3-COD by $\text{Fe(CO)}_5$. Both mono- and polyolefin complexes are prepared by CO substitution. Equation 5.17 shows how $\eta^6$-cycloheptatriene complexes are prepared by this method.

When mild reaction conditions are necessary, thermal substitution of more labile ligands is employed. Reagents such as $(\text{MeCN})_3\text{M(CO)}_3$, where M is Cr, Mo and W, give $\eta^6$-triene complexes, and the reaction of $\text{Fe}_2(\text{CO})_9$ with donor solvents, such as acetone or THF, affords reactive $\text{Fe(CO)}_x(\text{solvent})_y$ complexes (eq. 3.30), which are useful in preparing $\eta^4$-diene complexes. Two examples of THF complex formation and subsequent displacement of the THF ligand by an alkene are shown in equations 5.35 and 5.36. Yields for equation 5.35 can be increased if gaseous $\text{BF}_3$ is used to remove the THF ligand.

Examples of alkene substitution of halide ligands and other alkene ligands are shown in equations 5.37 and 5.38. Facile halide displacement frequently occurs in the presence of a strong Lewis acid, which acts as a halide acceptor. Alkene exchange using the Fe-isobutylene complex shown in equation 5.38 is a common route to cationic Fe-alkene complexes.

Photochemically-induced ligand substitution affording $\eta^2$-alkene complexes is also a general synthetic method. Olefin substitution of CO ligands in $(\eta^5\text{-C}_5\text{H}_5)$

$$(\eta^5\text{-}C_5H_5)Fe(CO)_2I + AgBF_4 \xrightarrow[\text{3 hrs}]{\text{THF, 25°C}} [(\eta^5\text{-}C_5H_5)Fe(CO)_2(THF)]^+BF_4^- + AgI\downarrow$$

red solid, 97%<br>
dec. 103°C

$$\Big\downarrow \begin{array}{c} \text{olefin(xs)} \\ \hline \end{array}$$ CH_2Cl_2, 25–40°C

$$[(\eta^5\text{-}C_5H_5)Fe(CO)_2(\eta^2\text{-olefin})]^+BF_4^- \quad (5.35)$$

orange solids, 92%<br>
(for C_2H_4)

$$(\eta^5\text{-}C_5H_5)Mn(CO)_3 \xrightarrow[hv]{\text{THF}} (\eta^5\text{-}C_5H_5)Mn(CO)_2(THF) + CO$$

red solution

$$\Big\downarrow \text{COT} \quad 25°$$

$$(\eta^5\text{-}C_5H_5)Mn(CO)_2(\eta^2\text{-COT}) \quad (5.36)$$

orange solid, 20%<br>
mp 68°C

$$(\eta^5\text{-}C_5H_5)M(CO)_xCl + AlCl_3 \xrightarrow[\text{alkene}]{\Delta} [(\eta^5\text{-}C_5H_5)M(CO)_x(\eta^2\text{-alkene})]^+AlCl_4^- \quad (5.37)$$

M = Mo, x = 3
M = Fe, x = 2

$$[(\eta^5\text{-}C_5H_5)Fe(CO)_2(\eta^2\text{-}CH_2{=}CMe_2)]^+BF_4^- + \text{olefin} \xrightarrow[\text{60°C, 10 min}]{1,2\text{-}C_2H_4Cl_2}$$

$$[(\eta^5\text{-}C_5H_5)Fe(CO)_2(\eta^2\text{-olefin})]^+BF_4^- + H_2C{=}CMe_2 \quad (5.38)$$

$Mn(CO)_3$ cannot be accomplished by thermal activation, but it can be effected via photolysis (eq. 5.39).

$$(\eta^5\text{-}C_5H_5)Mn(CO)_3 + C_2H_4 \xrightarrow{hv} (\eta^5\text{-}C_5H_5)Mn(CO)_2(\eta^2\text{-}C_2H_4) + CO$$

orange-red solid, dec. 116°C

$$\Big\downarrow \xrightarrow[\text{norbornadiene (NOR)}]{\text{hexane, } hv, \text{ 5 hrs}} (\eta^5\text{-}C_5H_5)Mn(CO)_2(\eta^2\text{-NOR}) + CO \quad (5.39)$$

$$(\eta^6\text{-}C_6H_3Me_3)Mo(CO)_3 + C_2H_4 \xrightarrow[\text{hexane}]{hv} (\eta^6\text{-}C_6H_3Me_3)Mo(CO)_2(\eta^2\text{-}C_2H_4) + CO$$

yellow solid, dec. 85°C

$$\Big\downarrow \xrightarrow{hv} (\eta^4\text{-}1,3\text{-}C_4H_6)_2Mo(CO)_2 + C_6Me_3H_3 + CO \quad (5.40)$$

yellow solid, 15%<br>
dec. 110°C

In photolysis of $(\eta^6\text{-arene})Mo(CO)_3$ complexes, mono-olefins displace a CO ligand, whereas dienes displace the arene ligand as shown in equation 5.40. Normal CO substitution by conjugated or nonconjugated olefins is common.

Ethylene and propylene complexes are prepared frequently by β-anion abstraction (eqs. 5.41–5.44).

$$(\eta^5\text{-}C_5H_5)Fe(CO)_2(\overset{\alpha}{C}H_2\overset{\beta}{C}H_3) + [Ph_3C]BF_4 \xrightarrow{CH_2Cl_2}$$

$$[(\eta^5\text{-}C_5H_5)Fe(CO)_2(\eta^2\text{-}C_2H_4)]^+ BF_4^- + Ph_3CH \quad (5.41)$$
$$\text{yellow solid}$$

$$(\eta^5\text{-}C_5H_5)Mo(CO)_3(CH_2CH_3) + [Ph_3C]BF_4 \xrightarrow{CHCl_3}$$

$$[(\eta^5\text{-}C_5H_5)Mo(CO)_3(\eta^2\text{-}C_2H_4)]^+ BF_4^- + Ph_3CH \quad (5.42)$$
$$\text{golden solid, 30\%}$$

$$CH_3CH_2Mn(CO)_5 + [Ph_3C]BF_4 \xrightarrow{\text{neat}}$$

$$[(\eta^2\text{-}C_2H_4)Mn(CO)_5]^+ BF_4^- + Ph_3CH \quad (5.43)$$
$$\text{white solid, 10\%}$$

$$Na[(\eta^5\text{-}C_5H_5)Fe(CO)_2] + \overset{O}{\triangle} \xrightarrow[\text{30 min}]{THF, \, 0° - 25°C} Na[(\eta^5\text{-}C_5H_5)(OC)_2FeCH_2CH_2O^-]$$

$$\Big\downarrow Et_2O \Big| 2HBF_4 \quad (5.44)$$

$$[(\eta^5\text{-}C_5H_5)Fe(CO)_2(\eta^2\text{-}C_2H_4)]BF_4\downarrow$$
$$90\%$$

$\beta$-Hydride abstraction from alkyl ligands by $Ph_3C^+$ cation affords cationic alkene complexes (eqs. 5.41−5.43). As shown in equation 5.44, the $C_5H_5Fe(CO)_2^-$ anion attacks ethylene oxide to give an anionic alkyl complex in which the alkyl ligand is a $\beta$-alkoxide anion. Elimination of water upon acidification generates the cationic olefin complex.

*Monohapto*-allyl ligands can be protonated at the $\gamma$-carbon atom to give $\eta^2$-alkene complexes, as shown in equation 5.45. Yields can be greater than 90%. This method cannot generate $\eta^2$-$C_2H_4$ complexes.

$$(\eta^5\text{-}C_5H_5)(OC)_xM\text{—}\overset{\alpha}{C}H_2\overset{\beta}{C}(R)\text{=}\overset{\gamma}{C}HR \xrightarrow{HX}$$

$$[(\eta^5\text{-}C_5H_5)(OC)_xM(\eta^2\text{-}CH_2\text{=}C(R)CH_2R)]X \quad (5.45)$$

M = Fe, x = 2; M = Mo or W, x = 3; R = H or alkyl

In many reactions, a coordinatively unsaturated metal atom reacts with an olefin to form an $\eta^2$-alkene complex. Such electron-deficient metallic species are probably generated *in situ* in thermal and photochemical ligand substitution reactions. Moderately stable electron-deficient metal species, such as metal vapors or Vaska-type 16-electron complexes, coordinate with alkenes as shown in equations 5.46 and 5.47.

$$Pd^0(\text{vapor}) + \text{alkene(xs)} \longrightarrow Pd(\text{alkene})_x \quad (5.46)$$

| Olefin | x | Yield (%) |
|---|---|---|
| 1,5-COD | 2 | 20 |
| bicyclo[2.2.1]heptene | 3 | 40 |

$$\text{trans-Ir(CO)(PPh}_3)_2\text{Cl} + \text{C}_2(\text{CN})_4 \xrightarrow{\text{CHCl}_3} \text{Ir(CO)[}\eta^2\text{-C}_2(\text{CN})_4\text{](PPh}_3)_2\text{Cl} \quad (5.47)$$
$$\text{white solid, 19\%}$$

In solution, $\text{Pt(PPh}_3)_4$ undergoes ligand dissociation to form $\text{Pt(PPh}_3)_3$ and $\text{Pt(PPh}_3)_2$. The latter compound reacts with alkenes to form $\eta^2$-alkene complexes (eq. 5.48).

$$\text{PtL}_4 \underset{-L, K_1}{\rightleftharpoons} \text{PtL}_3 \underset{-L, K_2}{\rightleftharpoons} \text{PtL}_2 \xrightarrow{\text{alkene}} \text{PtL}_2(\eta^2\text{-alkene}) \quad (5.48)$$

$L = \text{Ph}_3\text{P}, K_1 = \text{large}, K_2 = 1.6 \times 10^{-4}$

Thermally induced 1,2-addition of the Ta—H bond across the alkene C—C double bond in a Ta(hydride)(alkene) complex frees a coordination site for complexation to COT, as shown in equation 5.49. An $n$-propyl ligand is formed from the 1,2-addition.

$$(\eta^5\text{-C}_5\text{H}_5)_2\text{Ta(H)}(\eta^2\text{-CH}_2{=}\text{CHMe}) + \text{C}_8\text{H}_8 \xrightarrow[\text{60°C, 32 hrs}]{\text{toluene}}$$
$$(\eta^5\text{-C}_5\text{H}_5)_2\text{Ta(C}_3\text{H}_7)(\eta^2\text{-C}_8\text{H}_8) \quad (5.49)$$
$$\text{red solid, 67\%}$$

Direct formation of cationic Fe-alkene complexes from $[(\eta^5\text{-C}_5\text{H}_5)\text{Fe(CO)}_2]_2$ is accomplished by *oxidative-ligation* using *p*-benzoquinone as shown in equation 5.50.

$$[(\eta^5\text{-C}_5\text{H}_5)\text{Fe(CO)}_2]_2 + 2\,\text{O}{=}\!\!\left\langle\!\!\bigcirc\!\!\right\rangle\!\!{=}\text{O} + 2\text{HBF}_4 + \text{alkene(xs)} \xrightarrow[\text{25°C}]{\text{CH}_2\text{Cl}_2}$$
$$[(\eta^5\text{-C}_5\text{H}_5)\text{Fe(CO)}_2(\eta^2\text{-alkene})]^+\text{BF}_4^- \quad (5.50)$$
$$\text{yellow solids, 7–39\%}$$

This reaction represents the oxidation of each Fe atom forming the $(\eta^5\text{-}\text{C}_5\text{H}_5)\text{Fe(CO)}_2^+$ cation which then coordinates alkenes. A mechanistically different oxidative method is shown in equation 5.51. Trimethylamine-N-oxide oxidizes CO ligands to $\text{CO}_2$, thus freeing the coordination sites necessary to bind dienes.

$$\text{Fe(CO)}_5 + \text{1,3-diene} + 2\text{Me}_3\text{NO} \xrightarrow[\text{0°C}]{\text{C}_6\text{H}_6}$$
$$(\eta^4\text{-1,3-diene})\text{Fe(CO)}_3 + 2\text{CO}_2 + 2\text{NMe}_3 \quad (5.51)$$
$$\text{62–80\%}$$

This method is very useful for removing CO ligands under very mild reaction conditions. The mechanism of this CO oxidation to $\text{CO}_2$ is discussed in chap. 11.

## Structures of Metal Alkene Complexes

In mono-olefin or nonconjugated polyolefin complexes, the major structural features of interest are the orientation of the olefinic C—C bond vector relative to the principal molecular coordination plane, the M—C and C—C (alkene) distances, and the nonplanarity of the bound alkene as measured by angles $\alpha$ and $\beta$ in **5.30**. The value of $\beta$ is defined as the angle between the C—C bond vector and a line normal to the $\text{CR}_2$ plane, while $\alpha$ is the angle between these two normal lines, as shown.

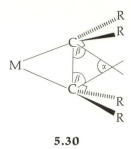

**5.30**

As the R substituents bend away from M and a planar olefin geometry, the value of $\alpha$ increases from $0°$ and the values of $\beta$ decrease from $90°$.

For $d^{10}$-ML$_3$ complexes in which at least one ligand is an olefin, the olefinic C—C bond vector lies *in* the coordination plane of M, as shown in **5.31**. The olefinic C—C distances of 1.44 Å are approximately 0.1 Å *longer* than corresponding distances in the free alkene. When R is CN or F, $\alpha$ is $64°$ or $80°$, respectively.

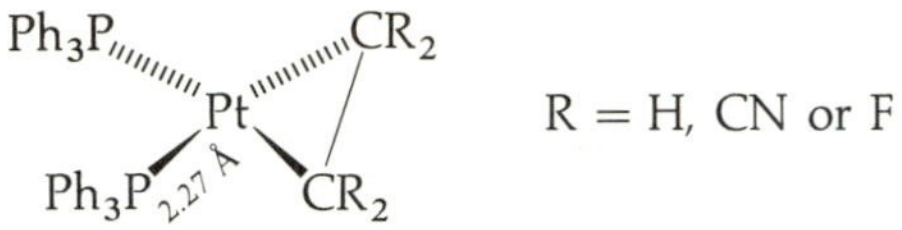

$$Ph_3P\quad\substack{\quad\quad\quad\quad\\\diagdown}\quad\substack{CR_2\\ \diagup}$$

Ph$_3$P$\cdots$ Pt $\cdots$ CR$_2$    R = H, CN or F

Ph$_3$P $^{2.27\ Å}$ CR$_2$

**5.31**

In $d^8$-ML$_4$ complexes containing alkene ligands, the C—C bond vector is usually *perpendicular* to the molecular coordination plane, as shown in **5.32** for Zeise's salt.

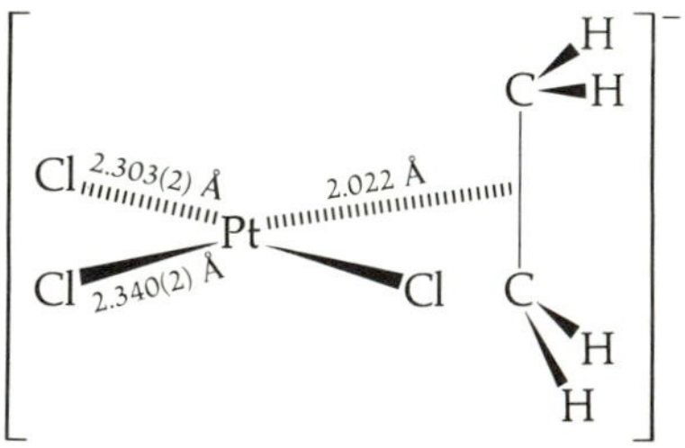

$$\left[ Cl\substack{2.303(2)\ Å}\ \overset{2.022\ Å}{Pt}\ Cl \quad \begin{array}{c} H \\ C{\blacktriangleleft}H \\ \\ C \\ {\diagdown}H \\ H \end{array} \right]^-$$

Cl $^{2.340(2)\ Å}$

**5.32**

$$K[(\eta^2\text{-}C_2H_4)PtCl_3] \cdot H_2O$$

The hydrogen atoms were located by neutron diffraction. The C atoms are 0.164 Å out of the H$_4$ plane with an $\alpha$ angle of $32.5°$. The C—C bond length is 1.375(4) Å.

Theoretical calculations indicate that this perpendicular coordination mode is preferred primarily because of unfavorable steric interactions in the "in-plane" geometry. By preparing a complex of a nonconjugated, cyclic diene in which the two olefinic bonds are fixed perpendicular to each other, it is possible to override these steric effects and form a complex having one perpendicular and one in-plane olefinic orientation as shown in **5.33**.

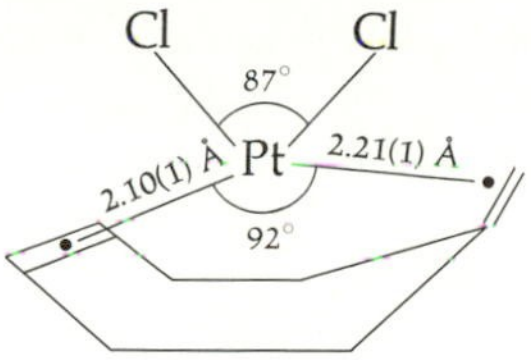

**5.33**

$(\eta^4\text{-}5\text{-methylenecycloheptene})PtCl_2$

The exocyclic olefin has adopted an "in-plane" orientation; however, the Cl ligand *cis* to this olefin bends away slightly, presumably relieving steric strain.

As a result of favorable M-alkene orbital interactions, $(\eta^2\text{-alkene})ML_4$ complexes, such as $(\eta^2\text{-}C_2H_4)Fe(CO)_4$ or $(\eta^2\text{-}C_2F_4)Fe(CO)_4$, which are trigonal bipyramidal, have the C—C bond vectors lying in the equatorial coordination plane. Such structures are *pseudo*-octahedral if each alkene carbon atom is regarded as a donor atom.

An interesting result of the complexation of alkenes to a metal moiety is that the plane of symmetry in the free olefin may not exist in the complex. Substituted olefins of types *other than* $X_2C{=}CX_2$, *cis*-$XYC{=}CXY$, or $X_2C{=}CY_2$ afford chiral complexes upon coordination. This feature is used for isomer separation and for the stereospecific control of subsequent reaction chemistry as is discussed in chap. 12.

Conjugated or nonconjugated polyenes usually form $\eta^4$- or $\eta^6$-complexes, as shown in equation 5.17 for cycloheptatriene and in **5.34** for 1,5-cyclooctadiene. However, structures **5.35** and **5.36** reveal that $\eta^2$-complexes of such ligands are also possible.

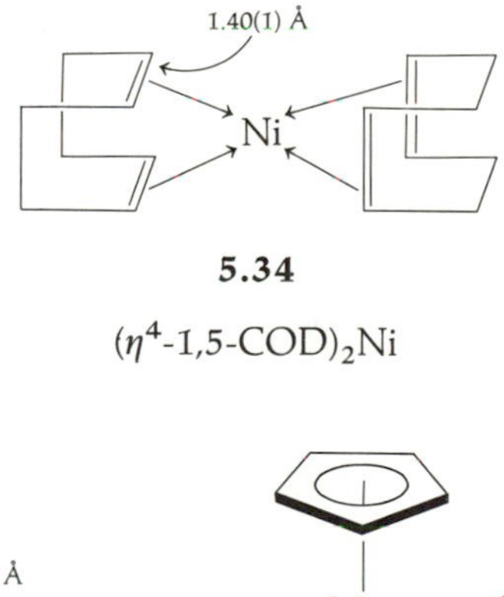

**5.34**

$(\eta^4\text{-}1,5\text{-COD})_2Ni$

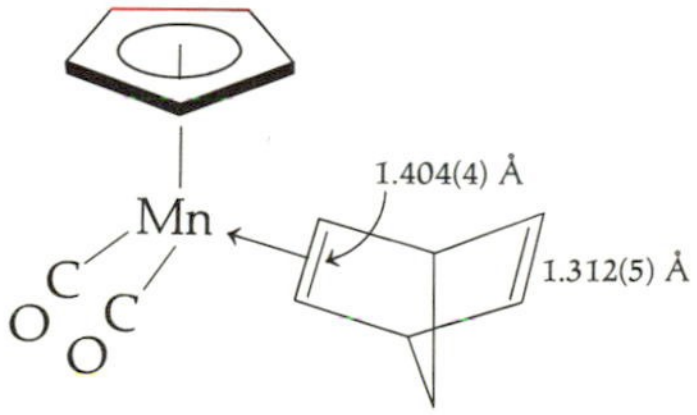

**5.35**

$(\eta^5\text{-}C_5H_5)Mn(CO)_2(\eta^2\text{-NOR})$

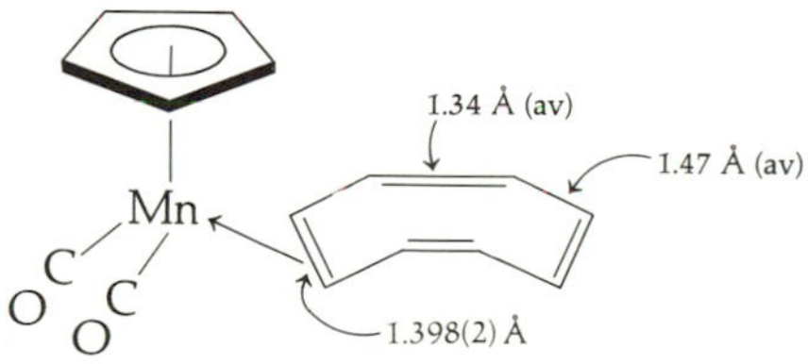

**5.36**

$(\eta^5\text{-}C_5H_5)Mn(CO)_2(\eta^2\text{-COT})$

Complexation of conjugated olefins is exemplified best in $\eta^4$-1,3-diene complexes. Free 1,3-butadiene has structure **5.37**, and the two observed *limiting geometries* for complexed $\eta^4$-1,3-dienes are shown in **5.38** and **5.39**.

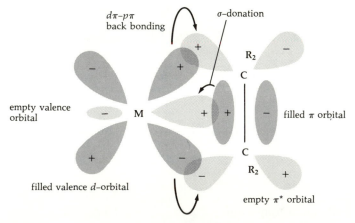

As shown by **5.38** and **5.39**, coordination of a diene elongates the diene olefinic bonds. The resulting C—C distances of coordinated dienes are essentially equal when coordinated to metals having strong $\pi$-acid ligands, such as the Fe(CO)$_3$ group in **5.38**, or exhibit a long-short-long pattern when coordinated to fairly electron-rich metals as in **5.39**. A short-long-short pattern as in **5.37** has not been observed for any mononuclear diene complex. Clearly, a bonding model for $\eta^4$-1,3-diene coordination must provide at least a qualitative rationalization of these limiting structural differences.

## Bonding in Metal Alkene Complexes

Metal-ligand bonding to mono-olefins or nonconjugated alkenes is represented generally by the M-$\eta^2$-C$_2$H$_4$ structure, as observed in Zeise's salt (**5.32**). A qualitative description of metal-alkene bonding as proposed originally by Dewar (and adopted by Chatt and Duncanson) is shown in figure 5.2. In the *Dewar–Chatt–*

**FIGURE 5.2**    A qualitative orbital description of metal-alkene bonding.

*Duncanson model,* the filled π-MO of the olefin donates σ-electron density to an empty valence orbital of the metal. This empty orbital could be a pure metal *d* orbital or a hybrid orbital. Concomitant *d*π-*p*π back donation of electron density from a filled valence *d* orbital on M to the empty π* MO on the alkene completes a synergistic bonding interaction.

For Zeise's salt, **5.40**, olefin → Pt σ-donation utilizes the platinum $5d_{z^2}$ orbital with some participation by the *6s* and *6p* orbitals of Pt.

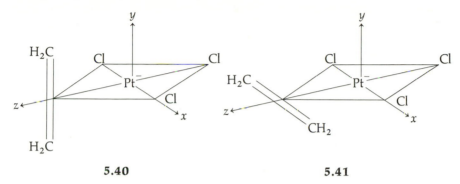

**5.40**                                                **5.41**

Platinum → olefin back bonding utilizes the Pt $5d_{yz}$ orbital. Detailed calculations of Zeise's anionic complex indicate a net σ-donation of 0.24 electrons and a net *d*π-*p*π back bonding of 0.22 electrons. The calculated bond disruption enthalpy of the ethylene ligand is +28.5 kcal/mole. In this complex, the Pt atom has a calculated atomic charge of +1.40.

MO calculations repeatedly confirm the qualitative validity of the Dewar–Chatt–Duncanson model. Since both depopulation of the alkene π MO and population of the alkene π* MO decrease the C=C bond order, this model correctly explains the observed increase in the olefinic C=C distance upon coordination. Electronegative substituents on the alkene, such as F, Cl or CN groups, increase the electronegativity of the olefin π* MO and thereby enhance M → alkene back bonding. Such complexes usually have greater olefinic C=C distances and shorter M—C distances than ethylene or alkyl-substituted alkene ligands.

Similarly, the oxidation state of M affects metal-alkene bonding. The metals in platinum (0) $d^{10}$-complexes back bond more electron density to the alkene ligand than do the more electronegative metals in Pt(II) $d^8$-complexes. The former usually have shorter Pt-C(olefin) distances and longer C=C bonds than do analogous Pt(II) complexes. Ancillary ligands that are strong donors also enhance M → alkene back bonding. The relative importance of σ-donation and *d*π-*p*π* back bonding in any particular complex depends on all of these factors.

Calculations indicate that the degree to which the olefin substituents are bent away from the metal atom depends upon the formal oxidation state of M and is a direct result of synergistic bonding. Large substituents also prefer the bent conformation to minimize steric interactions with M or other ligands.

Complexation appears to alter the electronic charge on coordinated olefins as well. Coordination to low-valent metals possessing strong donor ancillary ligands transfers negative charge to the olefin and makes it susceptible to *electrophilic attack.*

Conversely, coordination to high-valent metals having poor donors as ancillary ligands leads to a net electron loss from the olefin making it susceptible to *nucleophilic attack*. Metal stabilization of transition state species that form during such electrophilic or nucleophilic attack probably also occurs (see chap. 11).

The orientation of the alkene C—C bond vector relative to the molecular coordination plane is of some interest. Calculations reveal that all $L_2M(\eta^2$-alkene) complexes prefer an in-plane geometry because an M $\rightarrow \pi^*$ back-bonding overlap is lost in going to a perpendicular geometry. Energy barriers for rotation about the M-alkene bond are greater than 20 kcal/mole. For $L_3M(\eta^2$-alkene) complexes, such as Zeise's salt, a perpendicular orientation of the olefin is preferred primarily to minimize steric repulsions with adjacent ligands. Rotational energy barriers are usually in the range 10–20 kcal/mole. For Zeise's salt, about 70% of the energy difference between structures **5.40** and **5.41** is attributable to *cis*-Cl steric interactions in the in-plane structure (**5.41**). Only 10% of the rotational barrier in going from **5.40** to **5.41** results from electronic differences in Pt-alkene bonding.

A common difficulty in interpreting the electronic structure of molecules that possess synergistic bonding is the formal description of the chemical bonding in terms of localized valence-bond structures. Limiting structures representing M-($\eta^2$-alkene) bonding are shown in **5.42** and **5.43**.

$$L_nM \leftarrow \| \begin{matrix} CR_2 \\ CR_2 \end{matrix} \qquad L_nM \begin{matrix} {}^{\text{\tiny\textbackslash}}C^{\text{\tiny\textbackslash\textbackslash\textbackslash}}R \\ \searrow R \\ \nearrow R \\ {}_{\text{\tiny/}}C_{\text{\tiny\textbackslash\textbackslash\textbackslash}}R \end{matrix}$$

**5.42**                          **5.43**

Structure **5.42** shows donation of the alkene π-bonding electron pair to M; structure **5.43** represents an insertion or oxidative-addition of $L_nM$ into the π bond of the olefin affording a metallacyclopropane. The important observation is that in each representation, the alkene acts as a formal two-electron donor. Each formulation represents a limiting structure of the Dewar–Chatt–Duncanson model depending on the relative importance of M $\leftarrow$ olefin σ-donation and M $\rightarrow$ olefin $d\pi$-$p\pi^*$ back bonding.

Bonding in conjugated alkene complexes can be represented by ($\eta^4$-*cis*-1,3-butadiene)$ML_n$ complexes. Butadiene ligands act as formal four-electron donors since both olefinic bonds are coordinated to the metal atom. However, conjugated dienes are electronically different from nonconjugated dienes because of extensive $p\pi$-$p\pi$ overlap of the $2p_z$ atomic orbitals in the extended π electron system. For this reason, metal-butadiene bonding is usually analyzed as interactions between the valence orbitals of M and the π molecular orbitals of the butadiene molecule.

Figure 5.3 shows the π molecular orbitals of *cis*-1,3-butadiene. Free butadiene has four π electrons, which fill the 1π and 2π orbitals affording a short-long-short pattern for the three C—C distances (**5.37**).

Upon complexation, metal valence orbitals, particularly the $d_{xz}$ and $d_{yz}$ orbitals, interact *primarily* with the HOMO and LUMO butadiene orbitals, 2π and 3π, re-

| orbital combination | molecular orbital | occupancy |
|---|---|---|

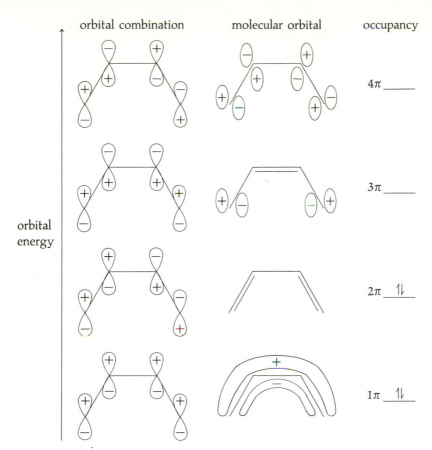

orbital energy

$4\pi$ _____

$3\pi$ _____

$2\pi$ _⥮_

$1\pi$ _⥮_

**FIGURE 5.3** π-Molecular orbitals of *cis*-1,3-butadiene.

spectively, because of the favorable orbital energy match. Synergistic bonding be-tween M and the butadiene ligand would involve donation of electron density from $2\pi$ (and $1\pi$) to the metal and back bonding of electron density from M to $3\pi$ (and $4\pi$). The net effect might be a complexed butadiene having a structure similar to that of the first excited electronic state of butadiene shown in **5.44**. In this electronic configuration, population of the $3\pi$ MO increases the double-bond character of the central C—C bond so that the pattern of C—C distances is now long-short-long.

1.39 Å

1.45 Å  1.45 Å          $1\pi^2\ 2\pi^1\ 3\pi^1$

**5.44**                    electronic
                           configuration

This pattern of C—C distances is observed in complexes such as **5.39** where appreciable M → butadiene back bonding is expected. Less back bonding may gen-erate a structure between the ground- and excited-state structures (**5.37** and **5.44**) which has nearly equal C—C distances of approximately 1.40 Å, as in **5.38**. [More precisely, $(\eta^4$-1,3-butadiene)Fe(CO)$_3$ complexes have outer C—C distances of ca. 1.42 Å with a central C—C distance nearly 0.02 Å shorter.] By using the following equation

$$\text{C}{=\!=\!=}\text{C distance (in Å)} = 1.517 - 0.18\,p \qquad (5.52)$$

and assuming a C—C butadiene distance of 1.410 Å, we can calculate the $\pi$-*bond order*, $p$, of C—C bonds in $(\eta^4$-1,3-butadiene)Fe(CO)$_3$ complexes to be 0.59. The total C—C bond order is then 1.59.

Detailed calculations on $(\eta^4$-1,3-butadiene)Fe(CO)$_3$ indicate a total charge of $-0.88$ electrons on the butadiene ligand. Even with strongly $\pi$-acidic CO ligands, there appears to be a net back bonding of electron density from Fe to the butadiene ligand. Therefore, butadiene ligands may be generally susceptible to electrophilic attacks such as protonation—a prediction that has been observed for several complexes.

## Complexes Having Cumulated Alkenes as Ligands

Allenes and cumulenes usually coordinate to a metal atom as a formal two-electron donor like an alkene. Coordination of only one of the olefinic bonds to a metal atom is expected since adjacent $\pi$-bonds in a cumulene are perpendicular to each other. Because of the facile oligomerization of allene (**5.45**), stable allene complexes are known only with Rh, Ir, Pd and Pt. Complexes of substituted allenes and cumulenes are more common.

$$\text{H}_2\text{C}{=\!=}\text{C}{=\!=}\text{CH}_2 \text{ (allene)}$$

**5.45**

Allene and cumulene complexes are prepared most frequently via ligand-substitution reactions, as shown in equations 5.53–5.56.

$$\text{Fe}_2(\text{CO})_9 + \text{Me}_2\text{C}{=\!=}\text{C}{=\!=}\text{CMe}_2 \longrightarrow (\eta^2\text{-C}_3\text{Me}_4)\text{Fe(CO)}_4 \qquad (5.53)$$

$$\text{Fe}_2(\text{CO})_9 + \text{Ph}_2\text{C}{=\!=}\text{C}{=\!=}\text{C}{=\!=}\text{CPh}_2 \xrightarrow[25°C]{\text{C}_6\text{H}_6} (\eta^2\text{-C}_4\text{Ph}_4)\text{Fe(CO)}_4 \qquad (5.54)$$
$$\text{golden solid, 75\%}$$
$$\text{dec. }145°\text{C}$$

$$(\eta^5\text{-C}_5\text{H}_5)\text{Mn(CO)}_3 + \text{Ph}_2\text{C}{=\!=}\text{C}{=\!=}\text{CPh}_2 \xrightarrow{h\nu}$$
$$(\eta^5\text{-C}_5\text{H}_5)\text{Mn(CO)}_2(\eta^2\text{-C}_3\text{Ph}_4) + \text{CO} \quad (5.55)$$

$$[\text{PtCl}_2\text{L}]_2 + \text{C}_3\text{H}_4 \xrightarrow[25°C]{\text{CH}_2\text{Cl}_2} (\eta^2\text{-C}_3\text{H}_4)\text{PtL}_2\text{Cl} + \text{L} \qquad (5.56)$$
$$\text{colorless solids, 52–80\%}$$
$$\text{L} = \text{PPr}_3, \text{PMe}_2\text{Et}, \text{PMe}_2\text{Ph} \qquad\qquad \text{mp }112–140°\text{C}$$

Thermal or photochemical substitution of CO, phosphine, $\eta^2$-alkene or other ligands affords $\eta^2$-allene or cumulene complexes. Protonation of propargyl ligands also generates $\eta^2$-allene complexes (eq. 5.57).

$$(\eta^5\text{-}C_5H_5)Fe(CO)_2CH_2C\equiv CR \xrightarrow[C_6H_6]{HClO_4}$$

$$[(\eta^5\text{-}C_5H_5)Fe(CO)_2(\eta^2\text{-}CH_2CCHR)]^+ClO_4^- \quad (5.57)$$

As with $\eta^2$-alkene complexes, coordination of one of the olefinic bonds of an allene or cumulene leads to C—C bond lengthening and a bending back of the substituents located on these carbon atoms. For butadiene and hexapentaene ligands only the central olefinic bond coordinates to M. Two structural types are observed in mononuclear complexes: one has the coordinated C—C bond vector perpendicular to the molecular coordination plane; the other has the coordinated C—C bond vector essentially in the molecular coordination plane.

Structural parameters of four allene complexes are listed in table 5.5. The terminal $CH_2$ groups tilt away from the metal affording C—C—C angles of 150–160°. Phi, $\phi$, is the angle between the coordinated C—C bond vector and the molecular coordination plane. Metal-allene or cumulene bonding can be rationalized using the Dewar–Chatt–Duncanson model. In-plane and perpendicular geometries are as expected based on electronic and steric effects as discussed for $\eta^2$-alkene complexes.

TABLE 5.5   Structural parameters for several allene complexes.

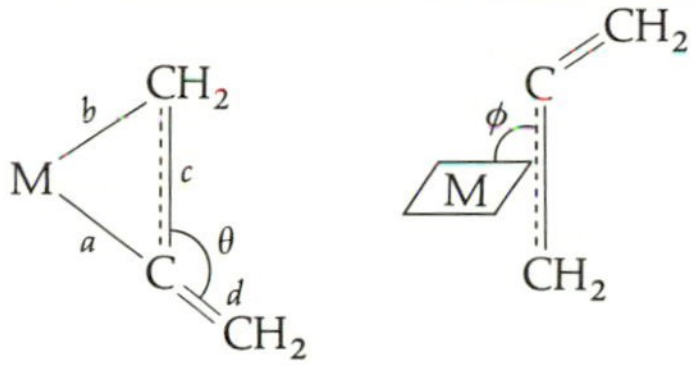

| Complex | a (Å) | b (Å) | c (Å) | d (Å) | $\theta$(°) | $\phi$ (°) |
|---|---|---|---|---|---|---|
| 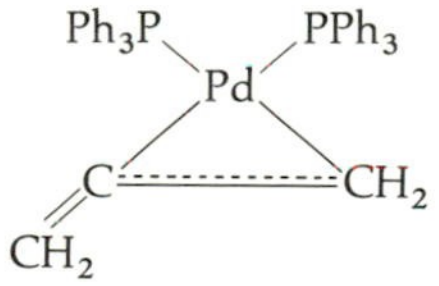 5.46 | 2.07 | 2.12 | 1.40 | 1.38 | 148 | 9 |
| 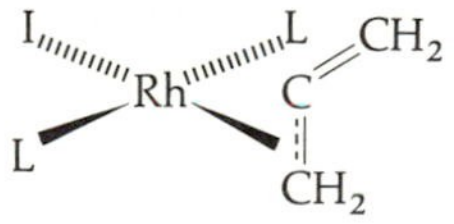 5.47, L = PPh$_3$ | 2.04 | 2.17 | 1.35 | 1.34 | 158 | ~90 |
| 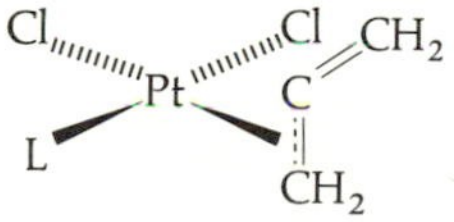 5.48, L = PPh$_3$ | 2.10 | 2.15 | 1.42 | 1.29 | 158 | ~90 |
| 5.49, L = PMe$_2$Ph | 2.09 | 2.15 | 1.40 | 1.31 | 158 | ~90 |

Allene C=C=C stretching frequencies decrease upon complexation by 180 to 260 cm$^{-1}$ from the free allene frequency of 1940 cm$^{-1}$. The greatest frequency shift is observed for the in-plane complex, **5.46**. These shifts are consistent with a decrease in the C—C bond order upon coordination to a metal.

## COMPLEXES CONTAINING ALKYNE LIGANDS

Alkynes, like alkenes, coordinate to metals via their $\pi$-electron pairs. Unlike alkenes, however, alkynes have *two* filled $\pi$ molecular orbitals. Alkynes can use one $\pi$ bond to coordinate to one metal atom or both $\pi$ bonds to coordinate to two, three, or four metal atoms. In such polynuclear coordination, alkynes act as bridging ligands. For some mononuclear complexes, where bonds to alkyne ligands are exceptionally strong, both $\pi$ bonds may be involved in M—C$_2$R$_2$ bonding. The high reactivity of alkynes leads to relatively facile oligomerization and polymerization reactions. Many of these reactions are metal catalyzed, so mild reaction conditions should be used if these coupling reactions are to be avoided.

### Preparation of Alkyne Complexes

Alkyne complexes are usually prepared by one of three methods: (1) reductive-ligation, (2) complexation to coordinatively unsaturated metal atoms, or (3) ligand substitution. Equations *5.58* and *5.59* illustrate reductive-ligation reactions using Na amalgam or LiAlH$_4$ as reducing agents.

$$(\eta^5\text{-}C_5H_5)_2MoCl_2 + C_2H_2 \xrightarrow[\text{1 atm, 18 hrs}]{\text{Na/Hg, toluene}} (\eta^5\text{-}C_5H_5)Mo(\eta^2\text{-}C_2H_2) + 2NaCl \quad (5.58)$$
$$\text{orange solid, 25\%}$$

$$Mo(TTP)Cl_2 + LiAlH_4 + C_2Ph_2(xs) \xrightarrow[\text{25°C, 1hr}]{\text{THF}} (\eta^2\text{-}C_2Ph_2)Mo(TTP) \quad (5.59)$$
$$\text{violet solid}$$

TTP = *meso*-tetra-*p*-tolylporphyrin

In each reaction, Mo(IV) is reduced to Mo(II) with concomitant complexation of the alkyne. (The TTP ligand is a planar porphyrin molecule, which coordinates to the metal through four nitrogen atoms in a nearly square-planar arrangement.)

Complexation of alkynes to coordinatively unsaturated metals occurs either with isolable reagents (eq. *5.60*) or with transient metal complexes (eqs. *5.61* and *5.62*).

$$[(\eta^5\text{-}C_5H_5)Mo(CO)_2]_2 + C_2H_2 \xrightarrow[\text{4 hrs}]{\text{toluene}} [(\eta^5\text{-}C_5H_5)Mo(CO)_2]_2(\mu_2\text{-}\eta^2\text{-}C_2H_2) \quad (5.60)$$
$$\text{red solid, 62\%}$$

$(\eta^5\text{-}C_5H_5)Fe(CO)_2I + AgBF_4 \xrightarrow[\text{C}_2\text{R}_2]{\text{CH}_2\text{Cl}_2,}$

R = CH$_3$, Et, Ph
$$[(\eta^5\text{-}C_5H_5)Fe(CO)_2(\eta^2\text{-}C_2R_2)]^+BF_4^- + AgI \quad (5.61)$$
$$\text{yellow to orange solids, 68–71\%}$$

$[(\eta^5\text{-}C_5H_5)Mo(CO)_3]_2 + 2AgBF_4 + C_2Me_2 \xrightarrow[\text{12 hrs}]{\text{CH}_2\text{Cl}_2,}$

$$[(\eta^5\text{-}C_5H_5)Mo(CO)(\eta^2\text{-}C_2Me_2)_2]^+BF_4^- \quad (5.62)$$
$$\text{yellow solid, 64\%}$$

Addition of acetylene to a Mo—Mo triple bond (eq. *5.60*) affords a bridging acetylene ligand that acts as a four-electron donor. The Mo—Mo bond order is reduced
to one. In equation *5.61*, $AgBF_4$ is used to remove $I^-$ from the reactant complex
forming a solvent stabilized $(\eta^5\text{-}C_5H_5)Fe(CO)_2^+$ cation, which then bonds to an
alkyne. In equation *5.62*, however, $Ag^+$ ion is used to oxidatively cleave the Mo
dimer to give $(\eta^5\text{-}C_5H_5)Mo(CO)_3^+$. This cation loses two CO ligands when the
dimethylacetylene complex is formed, because alkynes are better *σ*-donors than is
CO. The 14-electron species $Pt(PPh_3)_2$ coordinates alkynes as well as alkenes, as
shown in equation *5.48*.

Ligand substitution is the most general synthetic route to alkyne complexes.
Carbonyl substitution can be effected photochemically (eqs. *5.63* and *5.64*) or thermally (eqs. *5.65* and *5.66*).

$$(\eta^5\text{-}C_5H_5)Mn(CO)_3 + C_2Ph_2 \xrightarrow[h\nu]{\text{hexane}}$$

$$(\eta^5\text{-}C_5H_5)Mn(CO)_2(\eta^2\text{-}C_2Ph_2) + CO \quad (5.63)$$
$$\text{brown solid, 34\%}$$
$$\text{mp 104°C}$$

$$(\eta^5\text{-}C_5H_5)_2Mo(CO) + C_2H_2 \xrightarrow[h\nu]{\text{toluene}} (\eta^5\text{-}C_5H_5)_2Mo(\eta^2\text{-}C_2H_2) + CO \quad (5.64)$$
$$\text{orange solid, 95\%}$$

$$(\eta^5\text{-}C_5H_5)_2Ti(CO)_2 + C_2Ph_2(xs) \xrightarrow[25°C, \ 3\ \text{hrs}]{\text{heptane}}$$

$$(\eta^5\text{-}C_5H_5)_2Ti(CO)(\eta^2\text{-}C_2Ph_2) + CO \quad (5.65)$$
$$\text{yellow solid, 80\%}$$

$$Co_2(CO)_8 + C_2Ph_2 \xrightarrow[25°C, \ 16\ \text{hrs}]{\text{pentane}} [Co(CO)_3]_2(\mu_2\text{-}\eta^2\text{-}C_2Ph_2) + 2CO \quad (5.66)$$
$$\text{purple solid, 82\%}$$
$$\text{mp 110°C}$$

Olefinic ligands, such as ethylene or 1,5-COD, are also displaced by alkynes (eqs.
*5.67* and *5.68*).

$$Rh(acac)(\eta^2\text{-}C_2H_4)_2 + C_2(CF_3)_2 \xrightarrow[-78°C, \ 2\ \text{hrs}]{\text{Et}_2O}$$

$$Rh(acac)(\eta^2\text{-}C_2H_4)[\eta^2\text{-}C_2(CF_3)_2] + C_2H_4 \quad (5.67)$$
$$\text{yellow solid, 85\%}$$
$$\text{mp 58°C}$$

$$Pt(\eta^4\text{-}1,5\text{-COD})_2 + 2C_2Ph_2 \xrightarrow[25°C]{\text{pentane}} Pt(\eta^2\text{-}C_2Ph_2)_2 + 2\ 1,5\text{-COD} \quad (5.68)$$
$$\text{white solid}$$
$$\text{dec. 140°C}$$

Nickelocene reacts with alkynes to form an 18-electron dinuclear complex with loss
of one $C_5H_5$ ligand (eq. *5.69*). From these substitution reactions, it is clear that
alkynes form strong bonds to metal atoms.

$$2(\eta^5\text{-}C_5H_5)_2Ni + 2C_2R_2 \longrightarrow$$

$$[(\eta^5\text{-}C_5H_5)Ni]_2(\mu_2\text{-}\eta^2\text{-}C_2R_2) + 2\{C_5H_5\cdot\} \quad (5.69)$$

## Structures of Alkyne Complexes

When alkynes coordinate to metals, the acetylenic $C\equiv C$ distance increases from about 1.20 Å in a free alkyne to 1.24 to 1.40 Å. In addition, the linear $R-C\equiv C$ angle in free alkynes decreases to $168°-140°$ as the substituents bend away from the metal moiety. Specific structural types of M-alkyne coordination range in complexity from one-metal, one-alkyne interactions up to four-metal, one-alkyne interactions. In all complexes an alkyne ligand acts as a *dihapto* ligand.

One-metal, one-alkyne interactions are characterized by valence bond structures **5.50** to **5.52**.

|       |                   |       |
|-------|-------------------|-------|
| **5.50** | **5.51** | **5.52** |

Structures **5.50** and **5.51** are, respectively, $\pi$- and $\sigma$-*limiting formulations* of M-alkyne coordination when the alkyne acts as a formal two-electron donor. These formulations are analogous to the $\pi$- and $\sigma$-formulations of M-$\eta^2$-alkene bonding encompassed in the Dewar–Chatt–Duncanson bonding model. Metal-alkyne coordination is usually described in the literature as **5.51** because the observed $M-C$ and $C=C$ distances and the $R-C=C$ angles of coordinated alkynes best fit this *metallacyclopropene* structure.

Electron-deficient metals may interact strongly with *both* $\pi$ molecular orbitals of an alkyne, as shown in **5.52**. This "$\sigma,\pi$-structural type" is characterized by shorter $M-C$ distances and longer $C=C$ distances than in related complexes of type **5.51**. In such $\sigma,\pi$-complexes, the alkyne acts as a four-electron donor. Unfortunately, there is some confusion in the literature over the classification of alkyne ligands as either two- or four-electron donors. Some alkynes are classified as four-electron donors to satisfy the EAN rule rather than to rationalize structural features. It is possible that two-electron alkyne ligands may effectively occupy two coordination sites of the metal atom, thereby stabilizing 16- or 14-electron complexes through *coordination saturation*.

The major structural features of a one-metal, one-alkyne interaction are illustrated by complexes **5.53–5.57**.

These five complexes have acetylenic $C-C$ distances of 1.27–1.29 Å, a range that reflects $C\equiv C$ bond lengthening of about 0.08 Å upon coordination. In these complexes, the alkyne ligands act as two-electron donors as represented by the metallacyclopropene formalism of **5.51**. The coordination geometry of acetylene, $C_2H_2$, diphenylacetylene (sometimes referred to as *tolane*) and hexafluorobut-2-yne, $C_2(CF_3)_2$, is quite similar for a wide variety of complexes. Of complexes **5.53–5.57**, all except **5.55** and **5.57** are formal 16-electron species. Compound **5.55** is a 14-electron complex in which the nearly tetrahedral coordination geometry imposed by the two alkyne ligands apparently stabilizes the electron-deficient $Pt^0$ atom by

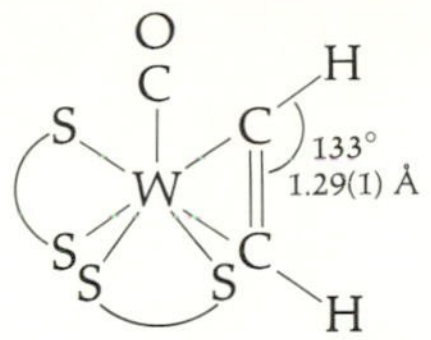

**5.53**

W(CO)($\eta^2$-C$_2$H$_2$)(S$_2$CNEt$_2$)$_2$

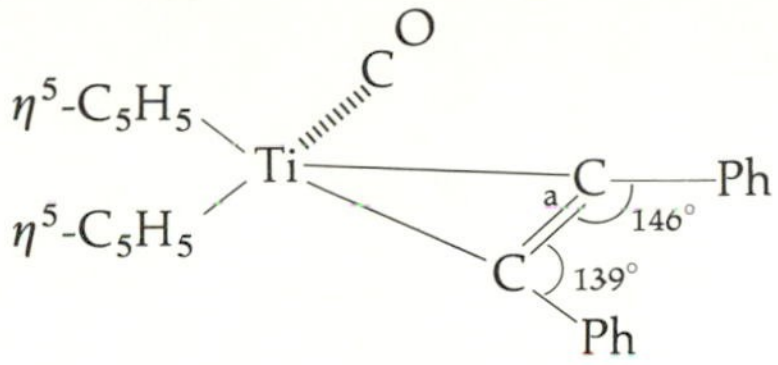

**5.54**

($\eta^5$-C$_5$H$_5$)$_2$Ti(CO)($\eta^2$-C$_2$Ph$_2$)
a = 1.29(1) Å

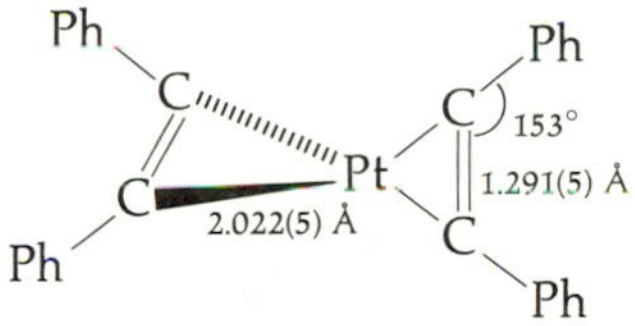

**5.55**

($\eta^2$-C$_2$Ph$_2$)$_2$Pt

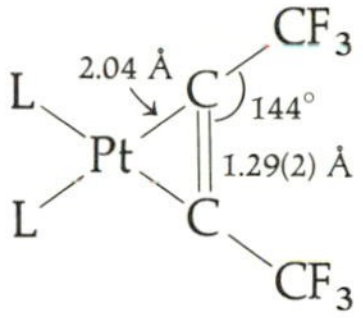

**5.56**

L = P(C$_6$H$_{11}$)Ph$_2$, P(C$_6$H$_{11}$)$_2$Ph

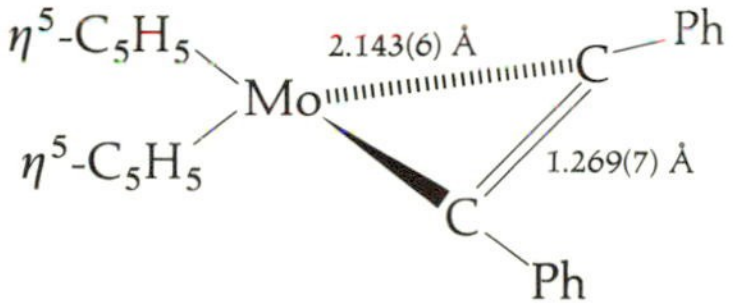

**5.57**

($\eta^5$-C$_5$H$_5$)$_2$Mo($\eta^2$-C$_2$Ph$_2$)

coordination saturation. Structural comparison of **5.55** to the planar, 16-electron Pt$^0$ complex in **5.56** reveals very similar alkyne C—C and Pt—C distances. Notice that **5.56** has an in-plane orientation of the alkyne ligand as observed in analogous alkene complexes, such as **5.31**. Compound **5.57** is an 18-electron complex.

Compounds **5.58**, **5.59** and **5.60** are all 16-electron complexes in which the alkyne ligands are acting as formal four-electron donors, as shown in valence bond structure **5.52**.

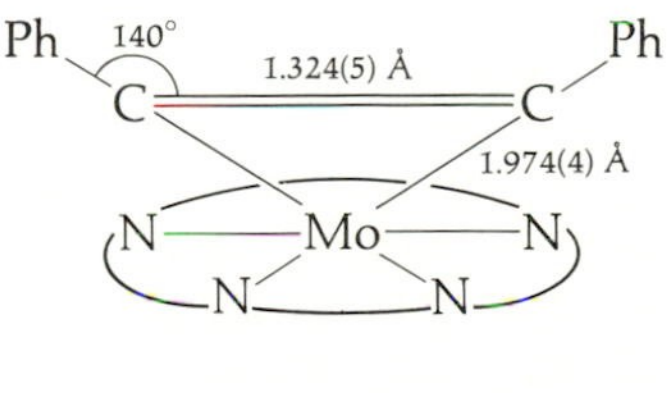

**5.58**

($\eta^2$-C$_2$Ph$_2$)Mo(TTP)

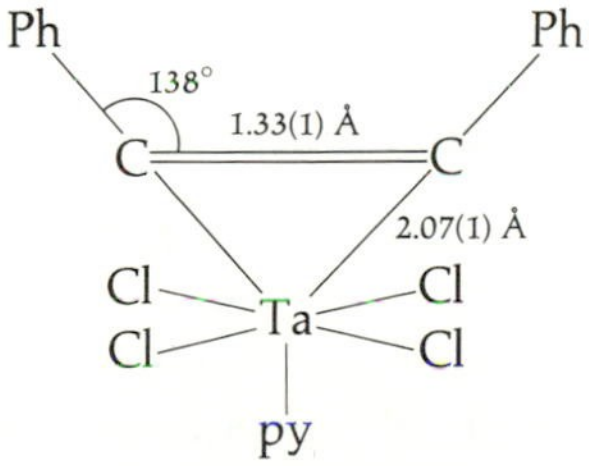

**5.59**

($\eta^2$-C$_2$Ph$_2$)TaCl$_4$py$^-$

$\eta^5$-C$_5$Me$_5$

Ta——— 2.071(6) Å
Cl                          C——Ph
    Cl                      | 1.337(8) Å
                            C   140°

                            Ph

**5.60**

$(\eta^5$-C$_5$Me$_5)$Ta$(\eta^2$-C$_2$Ph$_2)$Cl$_2$

Acetylenic C—C distances are in the range 1.32–1.34 Å, about 0.05 Å *longer* than those of two-electron donors. Comparison of the two Mo(II) complexes, **5.57** and **5.58**, also reveals a shorter Mo—C distance to the four-electron alkyne ligand of **5.58**. Structural features of the Ta-tolane bonding in the two Ta(III) complexes, **5.59** and **5.60**, are essentially identical, even though the ancillary ligands are quite different electronically.

Two-metal, one-acetylene interactions may be of structural types **5.61** or **5.62**. In both structures, the alkyne acts as a $\mu_2$-bridging ligand.

R              R
  C=C
M———————M              R           R                R           R
                        C=C    ≡ M----||----M ≡ M      C———C
**5.61**               R      C                          M———————M
                        M———M           C
                                        |
                                        R
                                 **5.62**

CF$_3$                   CF$_3$                                    139°    H
    C —1.27(1) Å— C                           O    H  1.337(5) Å  C
2.04(1) Å        110°                          C    C              |      $\eta^5$-C$_5$H$_5$
Rh···CO         Rh         OC—Mo              Mo
       2.682(1) Å                  $\eta^5$-C$_5$H$_5$  2.980(1) Å    C   C
$\eta^5$-C$_5$H$_5$    C  $\eta^5$-C$_5$H$_5$                          O   O
                    O

**5.63**                                              **5.64**

*trans*-[$(\eta^5$-C$_5$H$_5)$Rh(CO)]$_2[\mu_2$-$\eta^2$-C$_2$(CF$_3)_2$]        [$(\eta^5$-C$_5$H$_5)$Mo(CO)$_2]_2(\mu_2$-$\eta^2$-C$_2$H$_2$)

Complex **5.63** exemplifies structure **5.61**. The acetylenic C—C bond vector is *parallel* to the Rh—Rh bond vector, and the C≡C distance of 1.27 Å indicates a formal two-electron donation (one electron to each Rh). Each Rh atom obeys the EAN rule. Structural type **5.61** is called a *σ-bridging alkyne ligand*.

The most common $\mu_2$-bridging alkyne structure is **5.62**. Here an alkyne ligand is formally a four-electron donor. Each of the two perpendicular π bonds donates two electrons to one of the metal atoms. The acetylenic C—C bond vector is approximately *perpendicular* to the M—M bond vector. Complex **5.64** illustrates this type of coordination. In this complex, the average value of the Mo—C(C$_2$H$_2$) distances is 2.17 Å. The C$_2$H$_2$ ligand reduces the Mo—Mo triple bond of [($\eta^5$-C$_5$H$_5$)Mo(CO)$_2$]$_2$ (4.18) to a single bond in 5.64 when it bridges the two Mo atoms. Acetylenic C—C distances for this type of alkyne coordination are usually 1.34 Å to 1.38 Å. Complexes like [Co(CO)$_3$]$_2$($\mu_2$-$\eta^2$-C$_2$R$_2$), [($\eta^5$-C$_5$H$_5$)Ni]$_2$ ($\mu_2$-$\eta^2$-C$_2$R$_2$), and [($\eta^4$-1,5-COD)Ni]$_2$($\mu_2$-$\eta^2$-C$_2$R$_2$) also have the $\mu_2$-bridging alkyne structure 5.62.

Three-metal, one-alkyne interactions usually belong to one of three "$\mu_3$"-structural types (5.65−5.67).

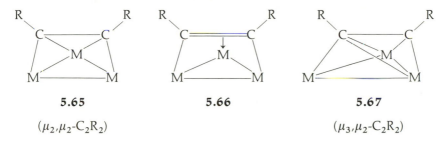

5.65             5.66             5.67

($\mu_2$,$\mu_2$-C$_2$R$_2$)             ($\mu_3$,$\mu_2$-C$_2$R$_2$)

Structures 5.65 and 5.66 are essentially equivalent representations (see 5.62 for related representations). When Fe$_2$(CO)$_9$ is treated with C$_2$Ph$_2$, complex 5.68 forms. This complex has a structure like 5.67. These structural types are distinguished by the relative magnitudes of the six M—C(alkyne) distances and the relative orientations of the C≡C and M—M bond vectors.

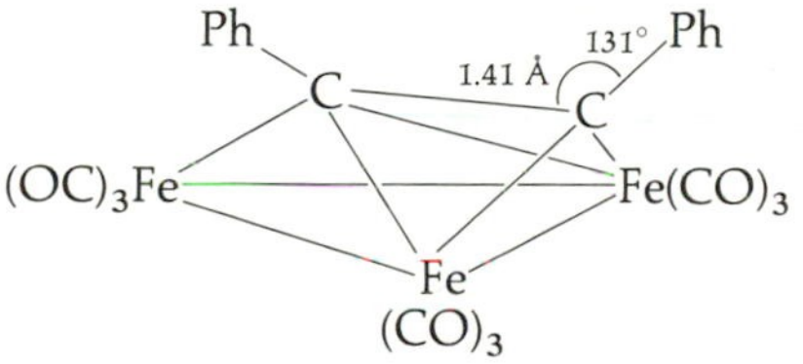

5.68

Fe$_3$(CO)$_9$($\mu_3$, $\mu_2$-C$_2$Ph$_2$)

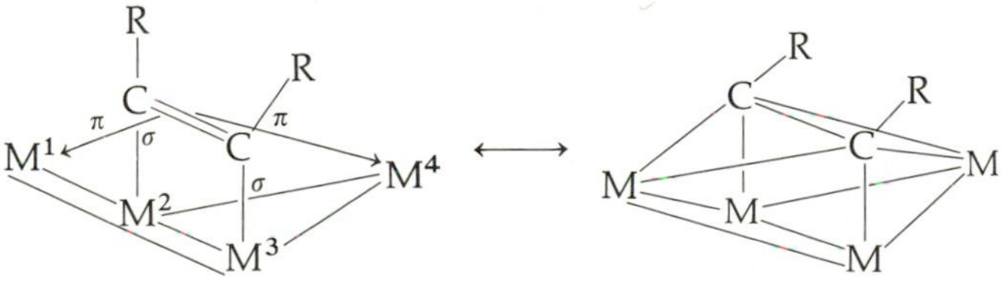

(2$\sigma$,2$\mu$-$\pi$)             ($\mu_3$,$\mu_3$)

5.69

Four-metal, one-alkyne interactions usually adopt the "$\mu_4$"-structure (5.69). The four metal atoms form a *butterfly-shaped cluster*. The acetylenic C—C bond vector is parallel to the M$^2$—M$^3$ bond vector, but perpendicular to the M$^1$ ⋯ M$^4$ vector. Such a structure can be represented either as having two M—C $\sigma$ bonds that involve one of the alkyne $\pi$ bonds and two M-alkyne $\pi$ bonds that involve the *other* acetylenic filled $\pi$ MO (thus affording a bent "$\mu$-$\pi$" bond) or as an all "$\sigma$"-bonding $\mu_3,\mu_3$-structure (**5.69**).

Two complexes, $Co_4(CO)_{10}(C_2Et_2)$ and $Ru_4(CO)_{11}C_8H_{10}$ (where $C_8H_{10}$ is cyclo-oct-1-ene-5-yne), have structures of the type **5.69**. The cobalt complex (**5.70**) is prepared from $Co_4(CO)_{12}$ and 3-hexyne. The acetylenic C—C distance is 1.44(2) Å.

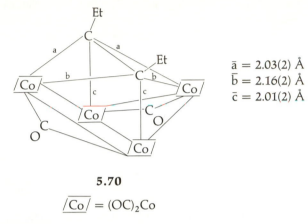

$\bar{a} = 2.03(2)$ Å
$\bar{b} = 2.16(2)$ Å
$\bar{c} = 2.01(2)$ Å

**5.70**

$\boxed{Co} = (OC)_2Co$

Bonding in alkyne clusters having more than two metal atoms is probably best analyzed using MO theory. Many of the all $\sigma$-bonding representations exceed the normal covalency of four for a carbon atom. In these complexes, the alkyne still acts as a four-electron donor. However, as the number of alkyne-metal interactions increases, the acetylenic C—C distance increases, too. Presumably, alkyne ligands in such clusters are either donating or back bonding more electron density than alkyne ligands that are interacting with only one or two metal atoms.

**Bonding in Alkyne Complexes**

When alkynes coordinate to metals, the acetylenic $v_{C\equiv C}$ stretching vibration shifts from around 2200 cm$^{-1}$ for free acetylenes to 1700–2000 cm$^{-1}$. This shift to lower energy is consistent with a decrease in the acetylenic C—C bond order upon complexation. Proton NMR resonances for acetylenic protons shift downfield toward the olefinic region, and $^{13}$C—NMR C—H coupling constants indicate more $p$-character in the alkyne C—R bond.

Comparison of metal-alkyne and metal-alkene structures reveals two differences: First, M—C distances to alkyne ligands are about 0.07 Å shorter than M—C distances to alkene ligands, and second, $\eta^2$-C$\cdots$C bond lengthening upon complexation is slightly greater for alkenes than for alkynes in comparable molecules. Both observations are consistent with metal-alkyne interactions being generally stronger than those of metal alkenes. *For equivalent M—C bonding,* M—C distances to alkyne ligands should be *only* 0.03 Å shorter than M—C distances to alkene ligands because of the difference in formal carbon hybridization in the M—C bonds ($sp$ to $sp^2$ for alkynes and $sp^2$ to $sp^3$ for alkenes).

The second observation noted above appears to support stronger M-alkene bonding than M-alkyne bonding. However, bond distance-bond order correlations are not linear, as demonstrated in figure 2.3 for CO. The relatively smaller amount of acetylenic C—C bond lengthening upon coordination may indicate a considerably greater lowering of the C—C bond order because an alkyne C—C bond order changes formally from three to two upon coordination, while olefinic C—C bond order changes formally from two to one.

A simplistic orbital description of metal-alkyne bonding is shown in figure 5.4. This description is based on the Dewar–Chatt–Duncanson model. The interaction of figure 5.4a parallels that shown in figure 5.2 for M-$\eta^2$-alkene bonding. Synergistic bonding between M and the $\pi$ and $\pi^*$ molecular orbitals of the $\pi$ bond that lies *in* the $MC_2$ plane is shown. Both interactions are strongly bonding and give rise to the metallacyclopropene valence-bond structure of **5.51**.

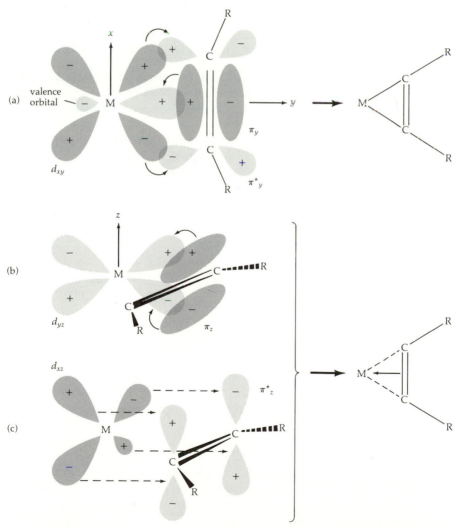

**FIGURE 5.4**    A qualitative orbital description of metal-alkyne bonding.

The interactions shown in parts b and c of figure 5.4 describe synergistic bonding between M and the *second* set of $\pi$ and $\pi^*$ molecular orbitals, which are *perpendicular* to the $MC_2$ plane. For most complexes, interaction c should be only weakly bonding because of the poor orbital overlap in this *δ-bonding interaction*, in which all four lobes of the $d$ orbitals participate in $\pi$ overlap with the alkyne orbital. Interaction b is strongly bonding if a $d_{yz}$ orbital on M is empty. For $d^{10}$-metals, such as $Pd^0$ or $Pt^0$, interaction b should be very weak. Bonding denoted by b and c represents a $\pi$-bonded valence-bond structure.

In complexes where interaction a predominates, the alkyne acts as a two-electron donor, but when b and c become significant, the alkyne acts as a four-electron donor. Since **5.55** and **5.56** are $Pt^0$-$d^{10}$ complexes, interaction a will dominate M-alkyne bonding. As reflected in the structures of these complexes, the alkyne ligands act as formal two-electron donors and stabilize the electron-deficient metal atoms via coordination saturation.

## Metal-Promoted Oligomerization of Alkyne Ligands

Several organometallic complexes catalyze the polymerization or oligomerization of alkynes. Nickel-catalyzed cyclotetramerization of acetylene to cyclooctatetraene and cobalt-catalyzed cyclotrimerization of alkynes to substituted benzenes are well-known examples.

Organometallic complexes are frequently isolated from reactions with alkynes where oligomerization or coupling of alkyne ligands has already occurred. These complexes have more complex structures and are usually formed at higher temperatures. An excellent example is shown in equation 5.70.

$$2(\eta^5\text{-}C_5H_5)_2Ti(CO)(\eta^2\text{-}C_2Ph_2) \xrightarrow[>30°C]{\text{heptane}}$$

$$\textbf{5.54}$$

$$(\eta^5\text{-}C_5H_5)_2Ti(CO)_2 + \eta^5C_5H_5 \quad (5.70)$$

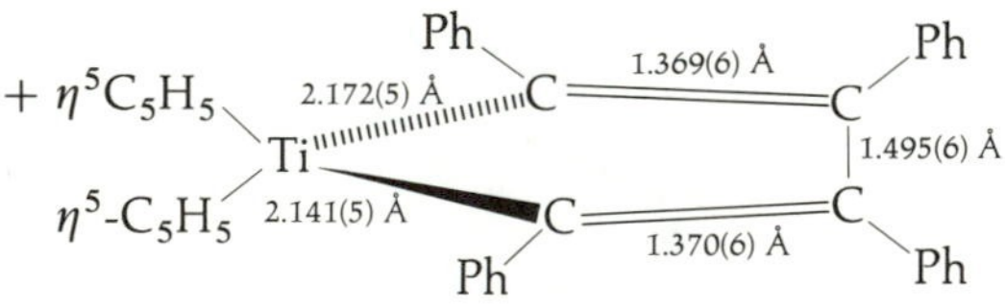

$$\textbf{5.71}$$

When the titanium-diphenylacetylene complex (**5.54**) is heated above 30°C, intermolecular ligand exchange occurs with the formation of a titanacyclopentadiene complex (**5.71**). Partial localization of the $\pi$-bonding electron density is observed within this metallacyclopentadiene ring.

Some of the organometallic products isolated from the thermal reaction of $Fe_3(CO)_{12}$ and $C_2Ph_2$ are shown below (**5.72–5.75**). Uncomplexed tetraphenyl-cyclopentadieneone, the ligand in **5.73,** (also known as tetracyclone) is isolated also. Note that metallacyclic complexes and $\pi$-complexes of cyclic organic molecules, such as tetracyclone and tetraphenylcyclobutadiene, are formed. From other reactions, $\pi$-complexes of substituted *p*-quinones, benzenes and cycloheptatrienones are isolated as well.

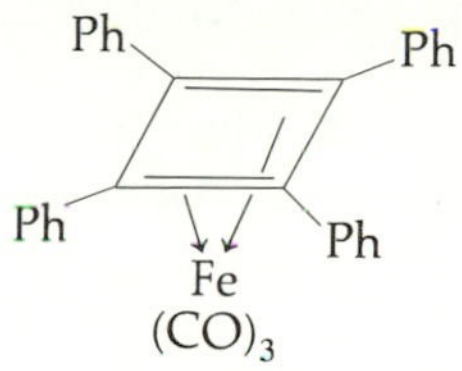

**5.72**

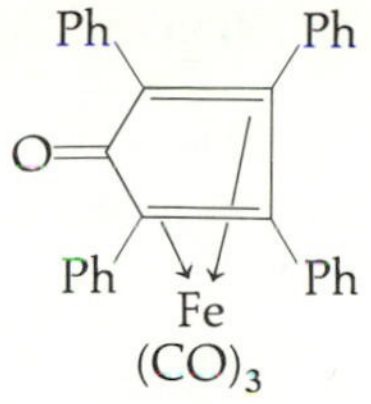

**5.73**

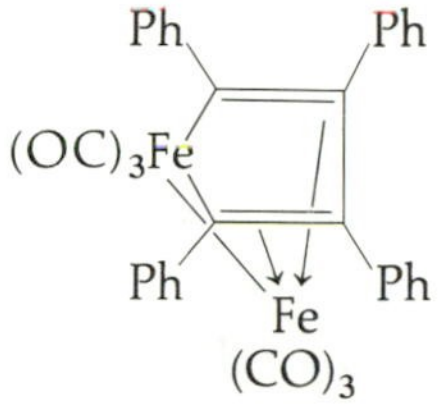

**5.74**

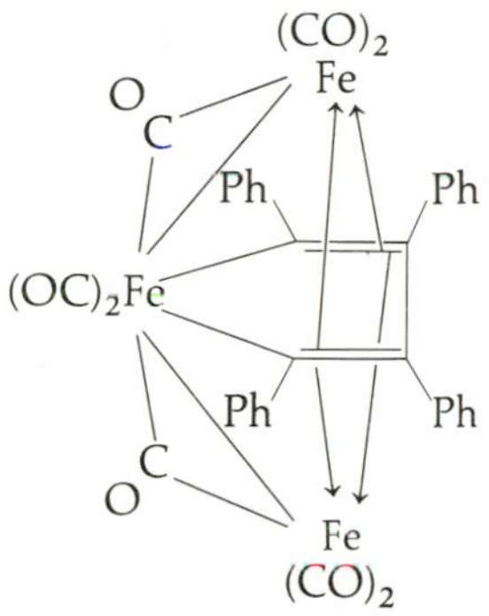

**5.75**

Mechanistic studies of these alkyne oligomerizations indicate that two alkyne ligands couple to give a metallacyclopentadiene complex such as **5.71**. Coordination of an additional alkyne molecule to the metal followed by insertion of this alkyne ligand into one of the M—C bonds of the metallacyclopentadiene ring gives a metallacycloheptatriene ring. Reductive-elimination of the metal fragment liberates the free unsaturated cyclic organic product, as shown in equation 5.71.

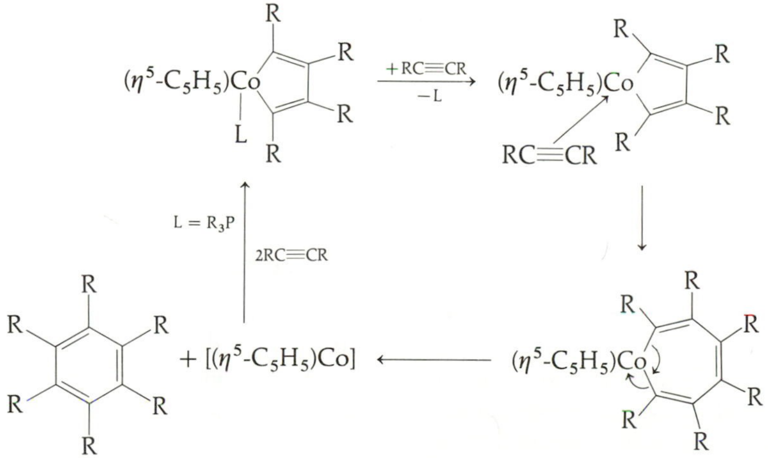

$$( \eta^5\text{-}C_5H_5)Co \xrightarrow[-L]{+RC\equiv CR} ( \eta^5\text{-}C_5H_5)Co$$

$$L = R_3P$$

$$2RC\equiv CR$$

$$+ [(\eta^5\text{-}C_5H_5)Co] \longleftarrow (\eta^5\text{-}C_5H_5)Co$$

(5.71)

Applications of such alkyne insertions to the synthesis of natural products are presented in chap. 12.

## COMPLEXES CONTAINING $\eta^3$-ALLYL AND $\eta^5$-DIENYL LIGANDS

Allylic reagents, such as $CH_2$=$C(H)$—$CH_2$—$Z$, where Z is usually halide or hydroxyl group, react with metals or metal complexes to give four types of coordination geometries (**5.76–5.79**).

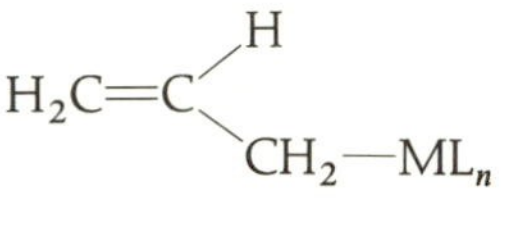

|            |            |            |
| :--------: | :--------: | :--------: |
|  **5.76**  |  **5.77**  |  **5.78**  |
| $\eta^1$-, or $\sigma$-allyl | $\eta^2$-olefin | $\eta^3$-, or $\pi$-allyl |
| one-electron donor | two-electron donor | three-electron donor |

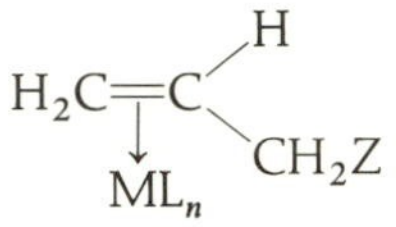

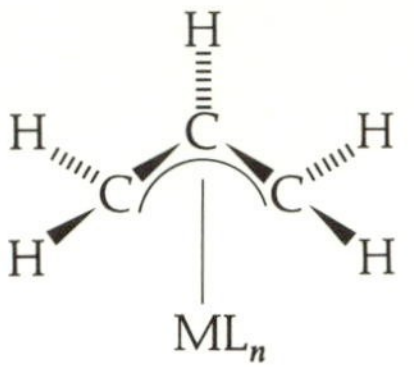

In **5.76**, the neutral allyl ligand acts as a one-electron donor, like an alkyl ligand. This coordination mode is frequently referred to as a $\sigma$-allyl. Examples of the $\eta^2$-alkene coordination (**5.77**) are relatively rare. Most complexes of this type occur with platinum and palladium. The $\eta^3$- or $\pi$-allyl coordination (**5.78**) is a very common structure in which the neutral allyl ligand acts as a three-electron donor. Substituted *trihapto*-allylic ligands are referred to as $\eta^3$-*enyl ligands*. A few complexes are known where $\eta^3$-allylic ligands bridge two metal atoms as shown in **5.79**.

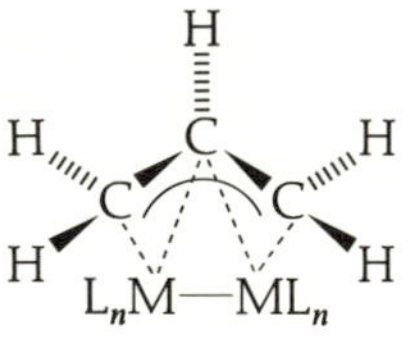

**5.79**

$\mu_2$-$\eta^3$-allyl
three-electron donor

### Preparation of Complexes Containing Allylic Ligands

Allyl complexes are usually prepared from (1) metals or metal complexes and allyl Grignard, allyl halide, or allyl alcohol reagents; (2) hydride or $\beta$-anion abstraction from alkene ligands; or (3) protonation of $\eta^4$-1,3-diene ligands. Allyl Grignards form allyl complexes via reductive-ligation, as shown in equation 5.72, or by displacing an anionic ligand, as in equation 5.73.

$$NiBr_2 + 2C_3H_5MgCl \xrightarrow[-10°C]{Et_2O} (\eta^3\text{-}C_3H_5)_2Ni + 2\{MgBrCl\} \qquad (5.72)$$
$$\text{yellow solid, 20\%}$$
$$mp \sim 1°C$$

$$(\eta^5\text{-}C_5H_5)_2Ni + C_3H_5MgCl \xrightarrow[0°C]{THF} (\eta^5\text{-}C_5H_5)Ni(\eta^3\text{-}C_3H_5) + C_5H_5MgCl \quad (5.73)$$
$$\text{deep red liquid, 50\%}$$
$$\text{bp 40°C (0.02 mm)}$$

Allyl halides react with organometallic anions to give $\eta^1$-, or $\eta^3$-allyl complexes, as shown in equations 5.74 and 5.75.

$$Na[(\eta^5\text{-}C_5H_5)Fe(CO)_2] + C_3H_5Cl(xs) \xrightarrow[25°C]{THF} (\eta^5\text{-}C_5H_5)Fe(CO)_2(\eta^1\text{-}C_3H_5) + NaCl$$
$$\text{amber oil, 34\%}$$
$$\text{dec. 65°C}$$

$$\text{neat oil} \left| \begin{array}{c} h\nu \\ 6\text{ hrs} \end{array} \right.$$

$$(\eta^5\text{-}C_5H_5)Fe(CO)(\eta^3\text{-}C_3H_5) + CO \quad (5.74)$$
$$\text{yellow solid, 80\%}$$
$$\text{dec. 65°C}$$

$$Li[Mn(CO)_5] + C_3H_5Cl \xrightarrow[25°C,\ 18\ hrs]{THF} (\eta^1\text{-}C_3H_5)Mn(CO)_5 + LiCl$$
$$\text{yellow liquid, 71\%}$$
$$\text{bp 45°C (15 mm)}$$

$$\left| \begin{array}{c} 86°C \end{array} \right.$$

$$(\eta^3\text{-}C_3H_5)Mn(CO)_4 + CO \quad (5.75)$$
$$\text{yellow solid, 88\%}$$
$$\text{mp 55°C}$$

*Monohapto*allyl complexes can be converted to $\eta^3$-allyl complexes by thermal or photochemical ligand substitution. This conversion shown in equation 5.74 occurs *only* with photochemical activation. Allyl halides also react with metals or low-valent metal complexes to afford allyl complexes by formal oxidative-addition (eqs. 5.76 and 5.77) or by substitution of an anionic ligand (eq. 5.78).

$$Pd^0 + C_3H_5Br \xrightarrow{\Delta} [(\eta^3\text{-}C_3H_5)PdBr]_2 \quad (5.76)$$
$$\text{yellow solid, 80\%}$$
$$\text{dec. 135°C}$$

$$Fe_2(CO)_9 + C_3H_5Br \xrightarrow[25°C]{hexane} (\eta^3\text{-}C_3H_5)Fe(CO)_3Br + 2CO \quad (5.77)$$
$$\text{brown solid, 42\%}$$
$$\text{mp 86°C}$$

$$Na_2[PdCl_4] + C_3H_5Cl \xrightarrow[CO\ atm]{MeOH} [(\eta^3\text{-}C_3H_5)PdCl]_2 + 2NaCl \quad (5.78)$$
$$\text{yellow solid, 84\%}$$
$$\text{dec. 130°C}$$

Allyl alcohols react with $PdCl_2$ to give dimeric, halide-bridged complexes. With some metals, allyl alcohols form $\eta^2$-alkene complexes, which are converted to $\eta^3$-allyl complexes via protonation followed by elimination of water (eq. 5.79).

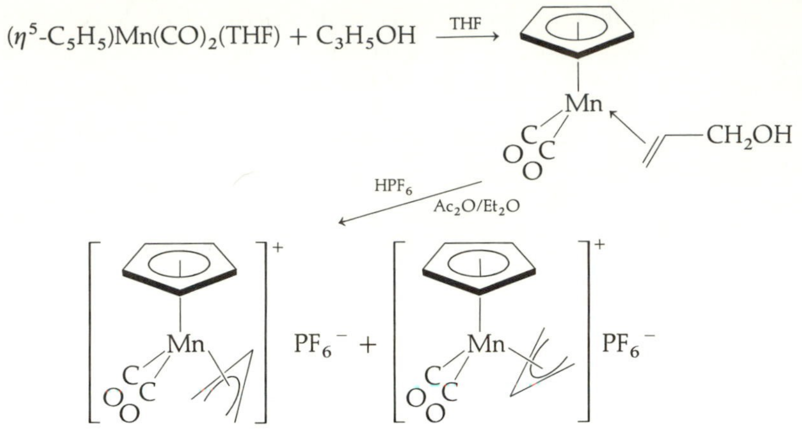

$$(\eta^5\text{-}C_5H_5)Mn(CO)_2(THF) + C_3H_5OH \xrightarrow{THF}$$

$$\xleftarrow[\text{Ac}_2\text{O/Et}_2\text{O}]{\text{HPF}_6}$$

$$(5.79)$$

*exo* isomer       *endo* isomer

yellow solids, $\sim 25\%$

The *exo* or *endo* conformations refer to the location of the central carbon atom of the allyl ligand relative to the $C_5H_5$ plane.

Highly substituted $\eta^3$-allyl complexes of palladium can be prepared directly from $PdCl_2$ and mono-alkenes, as shown in equation 5.80.

$$2PdCl_2 + 2CH_3CH{=}C(Et)CH_2CH_3 \longrightarrow$$

$$[(\eta^2\text{-}CH_3CH{=}C(Et)CH_2CH_3)PdCl_2]_2$$

$$\xrightarrow[H_2O]{50-100°C}$$

$$[(\eta^3\text{-}1,3\text{-}Me_2\text{-}2\text{-}Et\text{-}C_3H_2)PdCl]_2 + 2HCl \quad (5.80)$$
$$\text{yellow solid, } 98\%$$
$$\text{dec. } 130°C$$

Dimeric $\eta^2$-alkene complexes are intermediates in this reaction. Other metal complexes also convert $\eta^2$-substituted alkenes into $\eta^3$-allyl ligands by eliminating an allylic hydrogen atom as HCl or $H_2$. These conversions might proceed through oxidative-addition of an allylic C—H bond of the $\eta^2$-alkene ligand to the metal atom followed by reductive-elimination (see chapters 1 and 10) of HX as follows:

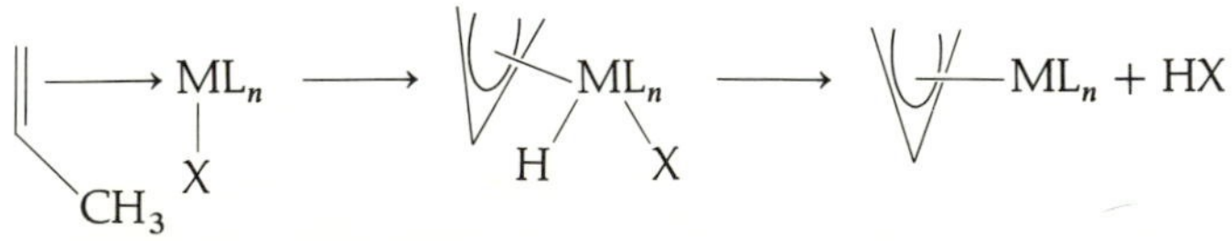

$$(5.81)$$

Reversible conversions of an $\eta^2$-propene complex to the corresponding $\eta^3$-allyl metal hydride complex has been observed.

Protonation of $\eta^4$-1,3-diene complexes affords $\eta^3$-allyl complexes (eq. 5.82). Incorporation of the entire protic acid molecule into the complex occurs when the counter ion is a good ligand, such as Cl$^-$.

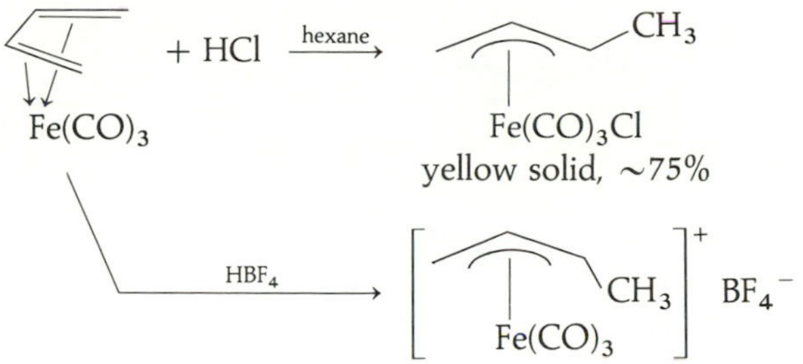

$$(5.82)$$

Such protonation reactions represent electrophilic attack on coordinated molecules, a topic that is discussed in chap. 11.

## Structures of $\eta^3$-Allyl Complexes

The structures of three $\eta^3$-C$_3$H$_5$ complexes are shown as **5.80–5.82.**

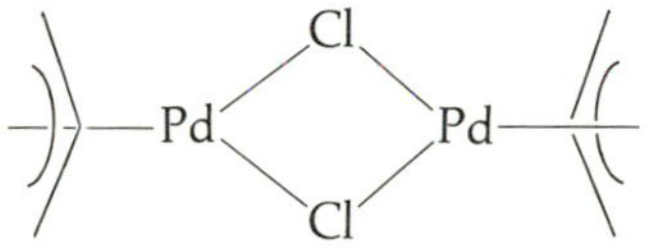

| **5.80** | **5.81** | **5.82** |
|---|---|---|
| [($\eta^3$-C$_3$H$_5$)PdCl]$_2$ | ($\eta^3$-C$_3$H$_5$)Co(CO)$_3$ | ($\eta^3$-C$_3$H$_5$)Fe(CO)$_3$Br |

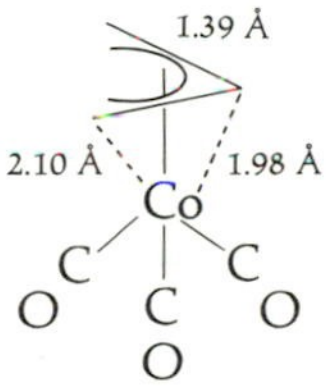

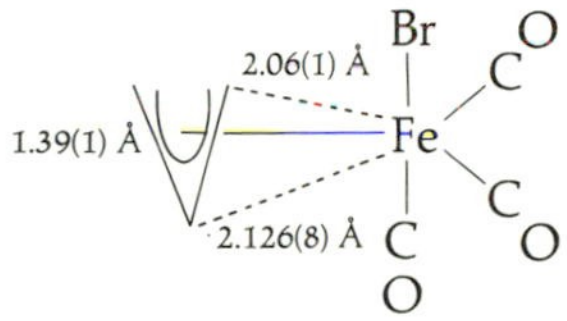

The principal structural features of $\eta^3$-C$_3$H$_5$ coordination are:

1. The allylic carbon atoms are essentially $sp^2$ hybridized giving an internal C—C—C angle near 120°.

2. Allylic C—C distances are approximately 1.40 Å, as observed in carbocyclic ligands.

3. The metal atom lies out of the C$_3$-allyl plane essentially below the centroid of the C$_3$ triangle defined by the allylic carbon atoms.

4. The M—C distances to the two terminal allylic carbon atoms are usually 0.05 to 0.15 Å longer than the M—C distance to the central allylic carbon atom. In complexes having very different ancillary ligands, some asymmetry in allylic M—C and C—C distances is observed.

5. When a principal coordination plane for the metal moiety can be defined, the C$_3$-allyl plane forms a dihedral angle of 95° to 120° with respect to this coordination plane.

Substituted $\eta^3$-allyl ligands are also well characterized. Structure **5.83** reveals that the 2-methyl substituents are displaced by about 0.5 Å *toward* the metal atoms.

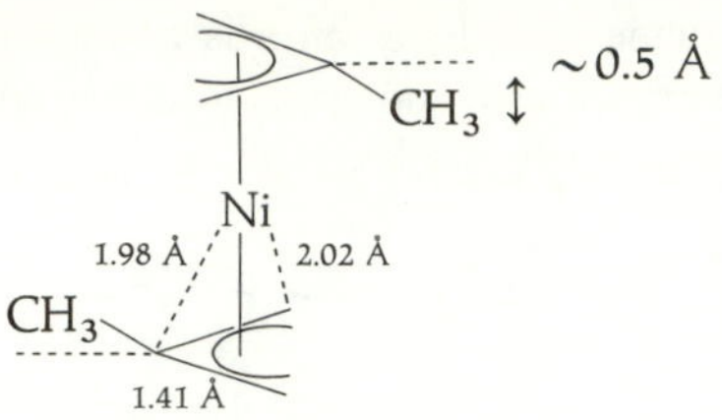

**5.83**

$(\eta^3\text{-2-methylallyl})_2\text{Ni}$

This displacement is opposite to that expected from steric effects and probably indicates a tilting of the $2p_z$ orbital of the central carbon atom toward the nickel atom for better orbital overlap.

Substituted $\eta^3$-allyl ligands exhibit geometrical isomerism depending on the degree of substitution and the number of nonequivalent substituents. The nomenclature used to distinguish these isomers is shown in **5.84** and **5.85** for an $\eta^3$-1-methylallyl ligand. When a substituent is *cis* or *trans* relative to the substituent on the central carbon atom, the isomer is denoted as *syn* or *anti*, respectively. This nomenclature is also used to label the two different types of terminal hydrogen atoms in an $\eta^3$-$C_3H_5$ ligand (**5.86**).

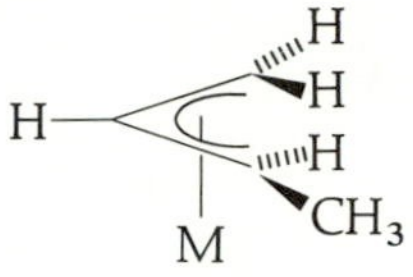

|  |  |  |
|---|---|---|
| *syn* isomer | *anti* isomer |  |
| **5.84** | **5.85** | **5.86** |

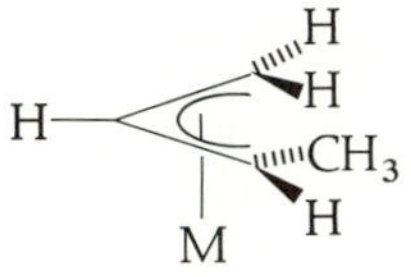

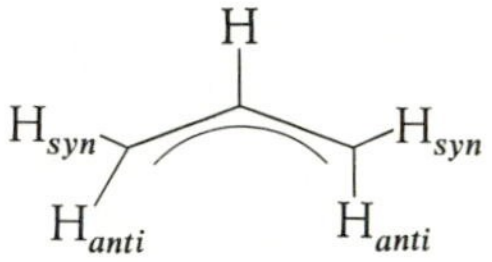

Two special types of $\eta^3$-allyl coordination are exhibited by $\eta^3$-benzyl and $\eta^3$-complexes of $C_7R_7$ or $C_8R_8$, as shown in **5.87** and **5.88**.

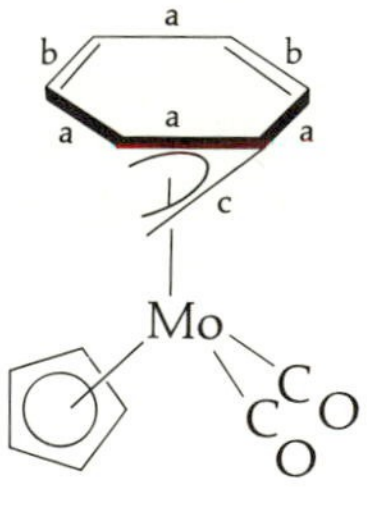

$\bar{a} = 1.42(1)$ Å
$\bar{b} = 1.34(1)$ Å
$c = 1.42(1)$ Å

**5.87**                                                  **5.88**

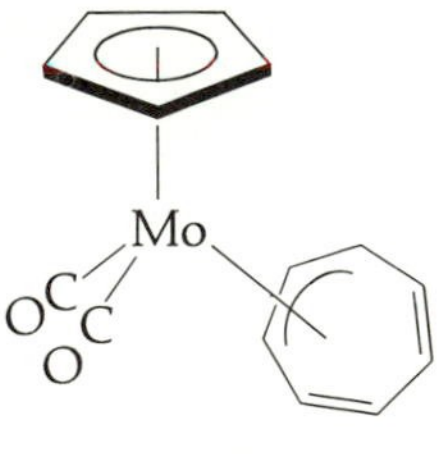

$(\eta^5\text{-}C_5H_5)\text{Mo(CO)}_2(\eta^3\text{-}C_7H_7)$

*Trihapto*benzyl complexes are prepared by methods similar to those used for preparing $\eta^3$-allyl complexes. Structure **5.87** reveals significant localized bonding within the phenyl ring. Complex **5.88** forms when $(\eta^7\text{-}C_7H_7)Mo(CO)_2I$ is treated with $NaC_5H_5$. Structures of other $\eta^3\text{-}C_7H_7$ complexes also reveal significant localized bonding within the seven-membered ring.

## Bonding in $\eta^3$-Allyl Complexes

An $\eta^3\text{-}C_3H_5$ unit constitutes a neutral, three-electron *π-enyl* ligand. Such ligands are formally radicals, and the valence bond structures for the allyl radical are shown in **5.89**. A delocalized description of the π molecular orbitals of an allyl radical are shown in figure 5.5.

**5.89**

Valence orbitals of a metal atom interact with the π molecular orbitals 1π, 2π and 3π when $\eta^3$-coordination occurs. (Although allyl radical is a three-electron donor, allyl cation or allyl anion are, respectively, formal two- or four-electron donors.) As with other π-ligands, $\eta^3$-allyl-M bonding is synergistic. Ligand → metal donation occurs from 1π and 2π; metal → ligand back bonding into 2π and 3π is possible also. Since 2π is partially filled in allyl radical, it can either donate or accept electron density. As usual, the final distribution of charge in a complex depends on the degree of orbital interaction between the allyl ligand and M and on how much electron density is available on M.

Because of the different spatial orientations of 2π and 3π, M-allyl orbital overlap is maximized when the $C_3$ plane of the allyl ligand is tilted slightly with respect

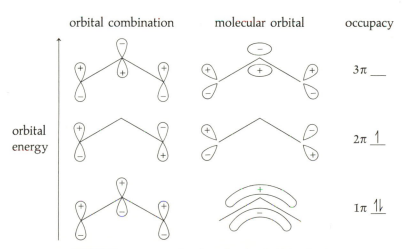

FIGURE 5.5    π-Molecular orbitals of allyl radical.

to the molecular (or $xz$) plane of the metal moiety, as observed in actual molecular structures. Efficient M-allyl overlap is also attained when the $p$ orbitals of the allyl ligand are directed toward M. This effect apparently causes the observed displacement of substituents on the central carbon atom *toward* the metal atom.

Molecular orbital calculations reveal that the strongest M-allyl orbital interaction is between $2\pi$ and the appropriate metal $d$ orbital. Calculations on $(\eta^3\text{-}C_3H_5)_2M$ and $[(\eta^3\text{-}C_3H_5)MCl]_2$ complexes where M is nickel, palladium, or platinum indicate donation of electron density from $1\pi$ to M in both series of complexes. In $(\eta^3\text{-}C_3H_5)_2Pd$, the Pd acceptor orbitals are predominantly $s$ and $p$ atomic orbitals. Interestingly, in the $(\eta^3\text{-}C_3H_5)_2M$ complexes, electron density drifts *from* M *into* $2\pi$, while in the Cl-bridged complexes, electron density drifts *from* $2\pi$ *into* $d$ orbitals on M. Presumably, the high electronegativity of the Cl ancillary ligands increases the Lewis acidity of M to the point where $2\pi$ becomes a net donor of electron density. The *bis*(allyl) complexes lack electronegative ligands, so the relatively electron-rich metal atoms back bond electron density into the $2\pi$ MO.

The donor/acceptor capability of $2\pi$ is important in the reaction chemistry of $\eta^3$-allyl ligands. Depending on the electronic structure of the entire complex, $2\pi$ could be either a HOMO or LUMO for electrophilic or nucleophilic reagents, so *trihapto*-allyl ligands should be susceptible to electrophilic or nucleophilic attack. Because the charge distribution of $2\pi$ resides only on the terminal carbon atoms, these incoming reagents should bond at only these terminal positions. Furthermore, since $2\pi$ is a nearly nonbonding ligand MO having a relatively diffuse charge distribution, only relatively "soft" nucleophiles with polarizable electron pairs should prefer to form a bond with the orbital. As we shall discuss in chap. 11, these predictions are observed experimentally.

### *Pentahapto*-dienyl Complexes

Acyclic analogues to the cyclic $C_5H_5\text{·}$ radical are $\eta^5$- or $\pi$-*dienyl ligands*. These complexes are usually prepared by (1) reductive-ligation (eq. 5.83), (2) abstraction of a hydride ion or heteroatomic substitutent from a carbon atom adjacent to a 1,3-diene (eqs. 5.84–5.86), or (3) hydride addition to cationic $\eta^6\text{-}C_6H_6$ complexes (eq. 5.87).

$$\text{FeCl}_2 + 2\text{Li}^+ \left[ \text{R—} \underset{R}{\overset{R}{\diagup}} \right]^- \xrightarrow[\substack{-78°C \\ (-2\text{ LiCl})}]{\text{THF}} \quad \text{R—} \underset{R}{\overset{R}{\diagup}} \text{Fe} \tag{5.83}$$

R = H and CH$_3$

red-orange solids

**5.90**

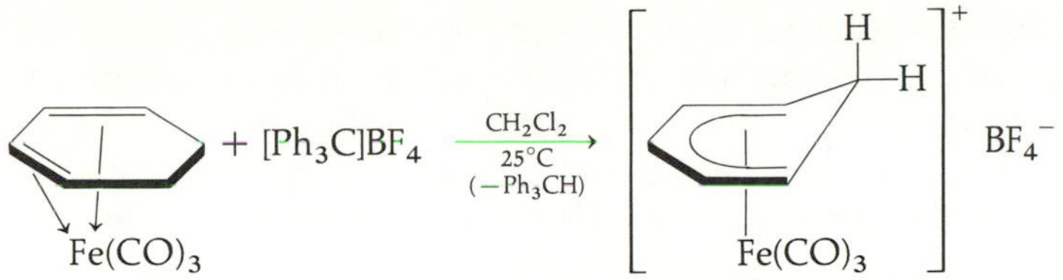

$$+ [Ph_3C]BF_4 \xrightarrow[\substack{25°C \\ (-Ph_3CH)}]{CH_2Cl_2} \quad BF_4^- \quad (5.84)$$

Fe(CO)₃

yellow solid
dec. 190°C

$$-CH_2OH \xrightarrow[\substack{Ac_2O \\ (-H_2O)}]{HClO_4} \quad ClO_4^- \quad (5.85)$$

Fe(CO)₃

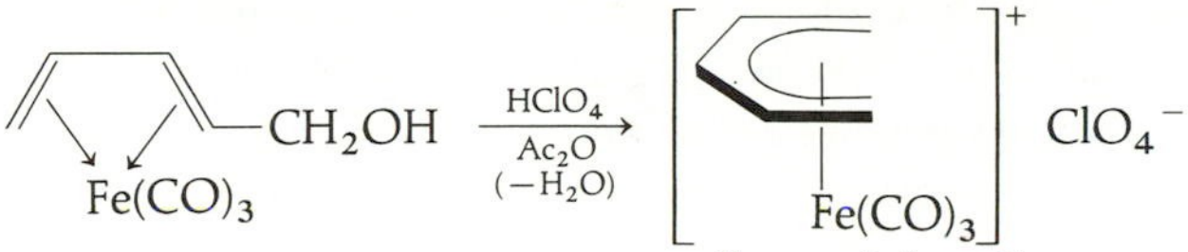

yellow solid, 84%
explodes on heating

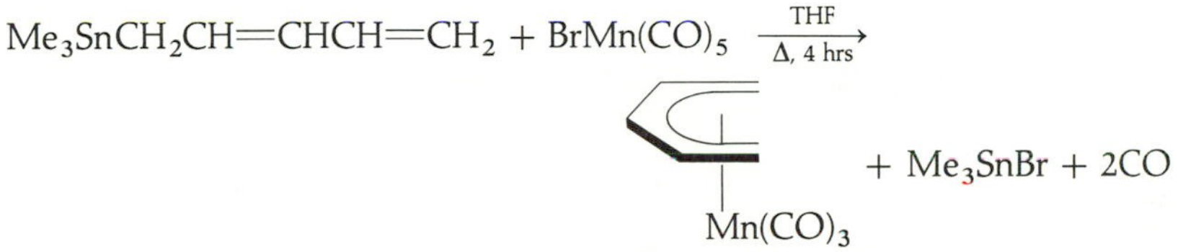

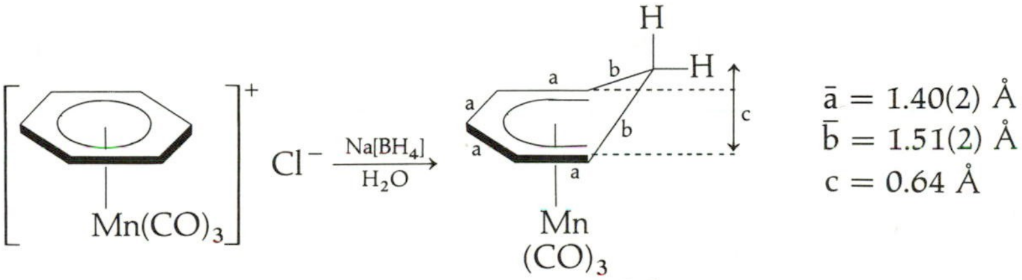

$$Me_3SnCH_2CH{=}CHCH{=}CH_2 + BrMn(CO)_5 \xrightarrow[\Delta, \ 4 \ hrs]{THF}$$

$$+ \ Me_3SnBr + 2CO \quad (5.86)$$

Mn(CO)₃
yellow solid, 52%
mp 101°C

$$Cl^- \xrightarrow[H_2O]{Na[BH_4]}$$

$$\begin{aligned} \bar{a} &= 1.40(2) \ \text{Å} \\ \bar{b} &= 1.51(2) \ \text{Å} \quad (5.87) \\ c &= 0.64 \ \text{Å} \end{aligned}$$

Mn(CO)₃

Mn(CO)₃
yellow solid, 18%
mp 78°C

**5.91**

Complex **5.90** is an "open"-analogue to ferrocene. *Pentahapto*-dienyl ligands are formal five-electron donors like the $\eta^5$-$C_5H_5$ ligand. The structure of **5.91** reveals an $\eta^5$-cyclohexadienyl ligand having dienyl C—C distances of 1.40 Å as expected for a delocalized π electron charge distribution.

## STUDY QUESTIONS

1. According to equation 5.26, as the carbocyclic ring size increases, the M-ring perpendicular distance $a$ decreases if $b$ is constant. At some ring size, $a$ may be small enough for M $\cdots$ ring *core-electron* repulsion to become significant. Such

core-electron repulsion for a Ti—$C_nH_n$ interaction may become noticeable at values of $a$ around 1.18 Å (Ti$^{4+}$ radius is 0.68 Å and a carbon core radius is taken to be 0.50 Å). Use equation 5.26 to calculate $a$ for a Ti—$C_9H_9$ interaction where $b$ and $R$ (for a nonagon) are 2.32 Å and 2.06 Å, respectively. Would core-electron repulsions be expected? How might such repulsions affect the observed structure?

2. The Ti—$C(C_7H_7)$ distance in **5.21** is anomalously short relative to the Ti—$C(C_5H_5)$ distance or relative to the two Ti—C(ring) distances in **5.23**. Rationalize this short Ti—C distance in **5.21** by using figure 5.1.

3. In the complexes $(\eta^7\text{-}C_7H_7)M(\eta^5\text{-}C_5H_5)$, where M is Ti, V or Cr, the charge on M becomes less positive as the atomic number of M increases (see table 5.4). Use figure 5.1 to explain this observation in terms of differences in M to ring back bonding.

4. The bis(fulvalene)diiron complex shown below can be prepared from 1,1'-diiodoferrocene. Provide a specific procedure for effecting this reaction.

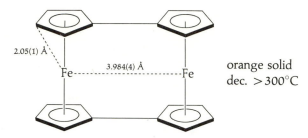

orange solid
dec. >300°C

5. Sketch the structures of the optical isomers of the complex $(\eta^2\text{-propene})$ $Fe(CO)_4$.

6. (a) Do complexes **5.74** and **5.75** obey the EAN rule for each Fe atom?

   (b) The complex shown below is analogous to **5.74**. The x-ray structure reveals an Fe—Fe single bond and *one* semi-bridging $\mu_2$-CO ligand. For this complex rationalize an electron count that satisfies the EAN rule.

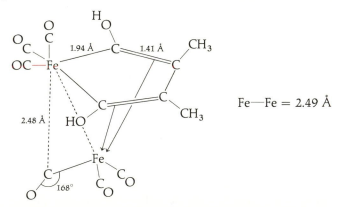

Fe—Fe = 2.49 Å

   (c) When $Fe(CO)_5$ is treated with 1,7-cyclododecadiyne, complexes **A** and **B** are isolated. Formulate an electron-counting description of complex **A**, and speculate as to the mechanism of formation of **A**.

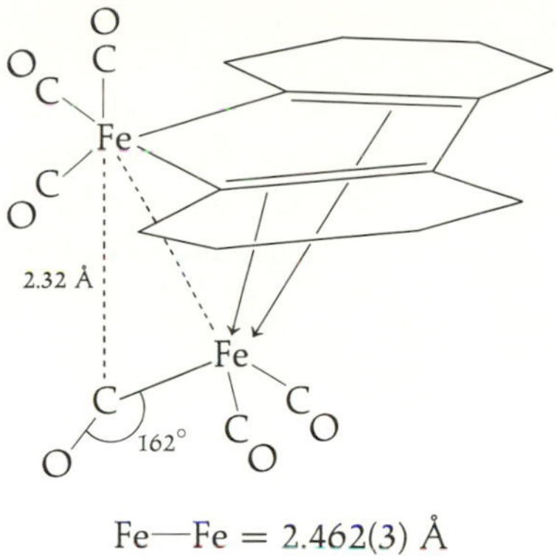

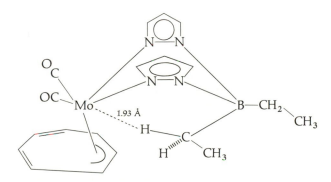

Fe—Fe = 2.462(3) Å

**A**

Fe(CO)₃

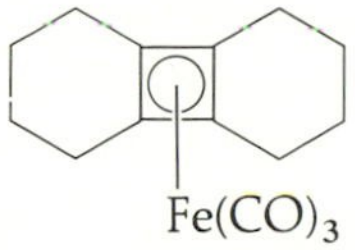

**B**

7. Sketch the orbital interactions involved in a $\mu_2$-acetylene interaction with two metal atoms as shown in **5.62.**
8. Consult the available literature, and compare proton NMR data for each pair of compounds shown below:
    (a) $\eta^1$-C₃H₅ and $\eta^3$-C₃H₅ ligands
    (b) free and complexed olefins having at least one olefinic proton
    (c) free and complexed alkynes having at least one acetylenic proton
    (d) C₆H₆ and an $\eta^6$-C₆H₆ ligand
9. The $\eta^3$-C₇H₇ complex shown below has a moderately strong bonding interaction between the molybdenum atom and the C—H bonding electron pair of one of the methylene C—H bonds. Rationalize this structure based on the EAN rule.

10. Consult the available literature for the $\pi$ molecular orbitals of the *cis,cis*-pentadienyl radical. Determine which metal $d$ orbitals could interact with each of these ligand molecular orbitals upon $\eta^5$-complexation. Identify the molecular orbitals that would be HOMO or LUMO orbitals if the ligand were to have a net +1 or −1 charge in the complex.
11. Explain why the $pK_a$ of the complex shown below is observed to be *higher* than the measured $pK_a$ of the free ligand.

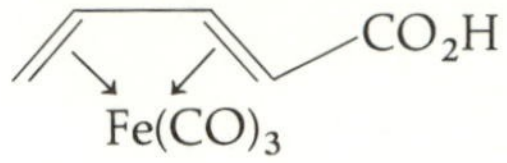

$$\text{(structure of a diene-Fe(CO)}_3\text{ complex with } CO_2H \text{ substituent)}$$

Fe(CO)$_3$

12. Other resonance forms for mononuclear and dinuclear metal-alkyne bonding are shown below. If these contribute strongly to the molecular electronic structure, then what geometrical structural features should be evident?

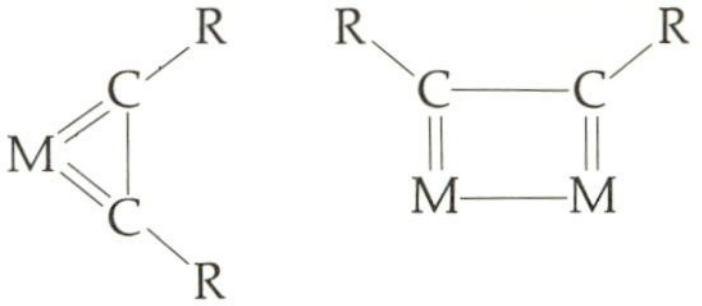

A                    B

## SUGGESTED READING

### Complexes Containing Carbocyclic Ligands

Baker, E. C., Halstead, G. W., and Raymond, K. N. *Structure and Bonding.* **25,** 23 (1976).
Bennett, M. A. *Adv. Organometal. Chem.* **4,** 353 (1966).
Efraty, A. *Chem Rev.* **77,** 691 (1977).
Maitlis, P. M. *Adv. Organometal. Chem.* **4,** 95 (1966).
Maitlis, P. M. *J. Organometal. Chem.* **200,** 161 (1980).
Pettit, R. *J. Organometal. Chem.* **100,** 205 (1975).
Silverthorn, W. E. *Adv. Organometal. Chem.* **13,** 47 (1975).
Wilkinson, G., and Cotton, F. A. *Progr. Inorg. Chem.* **1,** 1 (1959).

### Complexes Containing Olefins and Allenes as Ligands

Birch, A. J., and Jenkins, I. D. *Transition Metal Organometallics in Organic Synthesis.* Vol. 1. H. Alper, ed. Chapter 1. New York: Academic Press, 1976.
Bowden, F. L., and Giles, R. *Coord. Chem. Rev.* **20,** 81 (1976).
Churchill, M. R., and Mason, R. *Adv. Organometal. Chem.* **5,** 93 (1967). Discusses allyl and alkyne complexes also.
Fischer, E. O., and Werner, H. *Metal π-Complexes.* Vol. 1, *Complexes with Di- and Oligo-Olefinic Ligands.* New York: Elsevier, 1966. Discusses allyl complexes also.
Guy, R. G. and Shaw, B. L. *Adv. Inorg. Chem. Radiochem.* **4,** 77 (1962). Discusses alkyne and $\eta^3$-allyl complexes also.
Hartley, F. R. *Chem. Rev.* **73,** 163 (1973).
Herberhold, M. *Metal π-Complexes.* Vol. 2, *Complexes with Mono-Olefinic Ligands.* New York: Elsevier, 1972.
Ittel, S. D., and Ibers, J. A. *Adv. Organometal. Chem.* **14,** 33 (1976). Discusses alkyne complexes also.
Jones, R. *Chem. Rev.* **68,** 785 (1968).
King, R. B. *The Organic Chemistry of Iron.* Vol. 1. E. A. Koerner Von Gustorf, F.-W. Grevels and I. Fischler, eds. pp. 397−626. New York: Academic Press, 1978. Discusses allyl complexes also.
Kruger, C., Barnett, B. L., and Brauer, D. *ibid.* pp. 1−112.
Mingos, D. M. P. *Adv. Organometal. Chem.* **5,** 1 (1977).
Nelson, J. H., and Jonassen, H. B. *Coord. Chem. Rev.* **6,** 27 (1971).

Pettit, R., and Emerson, G. F. *Adv. Organometal. Chem.* **1**, 1 (1964).
Quinn, H. W., and Tsai, J. H. *Adv. Inorg. Chem. Radiochem.* **12**, 217 (1969).
Thayer, J. S. *J. Chem. Ed.* **46**, 442 (1969).

### Complexes Containing Alkyne Ligands

Dickson, R. S., and Fraser, P. J. *Adv. Organometal. Chem.* **12**, 323 (1974).
Maitlis, P. M. *J. Organometal. Chem.* **200**, 161 (1980).
Nicholas, K. M., Nestle, M. O., and Seyferth, D. *Transition Metal Organometallics in Organic Synthesis.* Vol. 2. H. Alper, ed. Chapter 1. New York: Academic Press, 1978.
Otsuka, S., and Nakamura, A. *Adv. Organometal. Chem.* **14**, 245 (1976).

### Complexes Containing Allylic Ligands

Chiusoli, G. P., and Cassar, L. *Organic Syntheses via Metal Carbonyls.* Vol. 2., I. Wender and P. Pino, eds. p. 297. New York: John Wiley & Sons, 1977.
Green, M. L. H., and Nagy, P. L. I. *Adv. Organometal. Chem.* **2**, 325 (1964).
Wilke, G., et al., *Angew. Chem. Internat. Edit.* **5**, 151 (1966).

### Theoretical Calculations on Carbocyclic Complexes

Clack, D. W., and Warren, K. D. *J. Organometal. Chem.* **157**, 421 (1978).
Clack, D. W., and Warren, K. D. *J. Organometal. Chem.* **149**, 401 (1978).
Clack, D. W., and Warren, K. D. *Inorg. Chim. Acta.* **27**, 105 (1978).
Clack, D. W., and Warren, K. D. *Inorg. Chim. Acta.* **24**, 35 (1977).
Zeinstra, J. D., and Nieuwpoort, W. C. *Inorg. Chim. Acta.* **30**, 103 (1978).

# Complexes Containing Nitric Oxide and Other Small-Molecule or Monatomic Ligands

## CHAPTER 6

Coordination of small molecules to transition metals has always been of interest to organometallic chemists. Not only do such complexes incorporate gaseous molecules in solid or liquid complexes, but complexation may alter the chemistry of a small molecule from that of the free species. For many molecules such as CS, CSe, or SO, isolable complexes are convenient reagents for exploring the chemistry of otherwise unstable or transient diatomic molecules.

Many cluster complexes contain monatomic ligands of the representative elements. These complexes are of interest because a single atom is stabilized solely via bonding to transition metal atoms.

Specific selected ligands to be discussed as examples of these types of organometallic compounds include (1) NO, NS, $N_2$, and related molecules; (2) CS, CSe, $CS_2$, and other heterocumulenes; (3) $S_2$, SO, $SO_2$, $SR_2$, and sulfonium ions; and (4) complexes containing monatomic ligands.

## COMPLEXES CONTAINING NO AND OTHER SMALL MOLECULES AS LIGANDS

### Complexes of NO, NS, $N_2$, and Related Molecules

Nitric oxide, NO, which is a gas at standard temperature and pressure, coordinates through the nitrogen atom. Compounds having NO ligands are referred to as *nitrosyl complexes.*

The electronic structures of NO and CO are very similar except that NO has one more electron, which occupies the partially filled $2\pi$ or $\pi^*$ MO (see figure 2.1).

CO     [$\sigma$ molecular orbitals]$^8$     $(1\pi)^4(5\sigma)^2(2\pi)^0$

NO     [$\sigma$ molecular orbitals]$^8$     $(1\pi)^4(5\sigma)^2(2\pi)^1$

This additional electron means that NO is a radical with a formal bond order of 2.5. The N—O distance of NO(g) is 1.151 Å, and the stretching frequency is 1904 cm$^{-1}$. Loss of the $\pi^*$ electron affords the diamagnetic *nitrosonium ion*, NO$^+$, which is isoelectronic to CO. As would be expected, when an electron is removed from an antibonding MO, in NO$^+$ the N—O distance decreases (by 0.088 Å), $\nu_{NO}$ increases (to ~2376 cm$^{-1}$), and the formal bond order increases to 3.0. Several nitrosonium salts are commercially available.

Because of the electronic similarity between CO and NO or NO$^+$, one might expect a similarity in their coordination chemistry, and some similarity is observed. However, a neutral NO ligand can act as either a formal one-electron or three-electron donor. As a three-electron donor, NO coordination parallels that of CO. The structural consequences arising from NO acting as a formal one-electron donor are discussed later in this chapter.

Organometallic nitrosyl complexes are usually prepared via one of six routes.

**1.** From NO gas (eqs. *6.1–6.6*),

$$V(CO)_6 + NO(g) \xrightarrow[0°C]{C_6H_{12}} V(CO)_5(NO) + CO \qquad (6.1)$$
$$\text{violet-red complex}$$

$$Cr(CO)_6 + NO(g) \xrightarrow[\text{pentane}]{h\nu} Cr(NO)_4 + 6CO \qquad (6.2)$$
$$\text{brown-black solid, 50\%}$$
$$\text{mp 38°C, } \nu_{NO} = 1721 \text{ cm}^{-1}$$

$$(\eta^5\text{-}C_5H_5)CrCl_2 + NO(g) \xrightarrow{THF} (\eta^5\text{-}C_5H_5)Cr(NO)_2Cl + \{ClNO\}$$
$$\text{olive-green solid, 33\%}$$
$$\text{dec. 140°C}$$

$$\Big\downarrow 2NaBH_4 \Big| C_6H_6/H_2O$$

$$[(\eta^5\text{-}C_5H_5)Cr(NO)_2]_2 \qquad (6.3)$$
$$\text{red-violet solid, 5\%}$$
$$\text{dec. 158°C}$$

$$MnBr_2 + NaC_5H_5 + NO(g) \xrightarrow{MeOCH_2CH_2OMe} (\eta^5\text{-}C_5H_5)_3Mn_3(NO)_4 \qquad (6.4)$$
$$\text{black solid, 2\%}$$

$$Co_2(CO)_8 + NO(g) \xrightarrow[\Delta]{\text{sealed tube}}$$

$$Co(CO)_3(NO) \xrightarrow[\Delta \text{ or } h\nu]{NO(g)} Co(NO)_3 \qquad (6.5)$$
$$\text{dark red liquid} \qquad\qquad \text{black solid}$$
$$\text{mp } -1°C \qquad\qquad \nu_{NO} = 1860, 1795 \text{ cm}^{-1}$$

$$[(\eta^5\text{-}C_5H_5)Fe(CO)_2]_2 + NO(g) \xrightarrow[120°C]{\text{octane}} [(\eta^5\text{-}C_5H_5)Fe(NO)]_2 + 4CO \qquad (6.6)$$
$$\text{dark green solid, 74\%}$$

**2.** From nitrosonium salts (eq. *6.7*),

$$Ir(CO)L_2Cl + [NO]BF_4 \xrightarrow[\Delta]{C_6H_6} \qquad [Ir(CO)(NO)L_2Cl]BF_4 \qquad (6.7)$$

L = PPh$_3$

$$\text{violet solid}$$
$$\nu_{NO} = 1680 \text{ cm}^{-1}, \ \nu_{CO} = 2050 \text{ cm}^{-1}$$

**3.** From nitrosyl halides (eq. *6.8*),

$$Et_4N[HB(3,5\text{-}Me_2\text{-}pz)_3Mo(CO)_3] + ClNO \xrightarrow{CH_2Cl_2}$$

$$HB(3,5\text{-}Me_2\text{-}pz)_3Mo(CO)_2(NO) + CO + [Et_4N]Cl \quad (6.8)$$
$$\text{orange solid, 85\%}$$
$$\text{dec. 270°C,}$$
$$\nu_{NO} = 1673 \text{ cm}^{-1}$$

**4.** From N-nitrosamides (eq. *6.9*),

$$HMn(CO)_5 + \qquad p\text{-tolylSO}_2N(Me)(NO) \xrightarrow{Et_2O}$$
$$\text{N-nitroso-N-methyl-}p\text{-toluenesulfonamide}$$
$$\text{"Diazald"}$$

$$Mn(CO)_4(NO) \quad + CO + p\text{-tolylSO}_2NHMe \quad (6.9)$$
$$\text{dark red liquid, 60\%}$$
$$\text{mp} \sim 0°C$$

**5.** From nitrite salts (eq. *6.10*),

$$NaCo(CO)_4 + NaNO_2 \xrightarrow[2HOAc]{H_2O} Co(CO)_3(NO) + CO + 2NaOAc \quad (6.10)$$
$$\text{dark red liquid, 50\%}$$

**6.** From nitronium (NO$_2{}^+$) salts (eq. *6.11*),

$$(\eta^5\text{-}C_5H_5)Re(CO)_3 + [NO_2]PF_6 \xrightarrow{CH_3CN}$$

$$[(\eta^5\text{-}C_5H_5)Re(CO)_2(NO)]PF_6 + CO_2 \quad (6.11)$$
$$\text{yellow solid, 90\%}$$

The most general source of NO is NO gas, which is available commercially.

In most of these reactions, the NO acts as a three-electron ligand by displacing three electrons from the valence shell of metals in 18-electron complexes or by pairing up its unpaired electron with the unpaired electron in 17-electron complexes. For example, in equation *6.1*, the NO displaces one CO ligand and pairs with the unpaired electron in the 17-electron V(CO)$_6$ complex affording V(CO)$_5$(NO), in which the vanadium atom satisfies the EAN rule. In the second step of equation *6.5*, six donor electrons from three CO ligands are displaced by two NO groups, and in equation *6.2* six CO ligands are displaced by four NO groups. In the first step of equation *6.5*, one NO displaces a CO ligand and a one-electron metal fragment. Similarly, in equation *6.9*, an NO molecule displaces CO and a neutral hydrogen atom.

In equations *6.3* and *6.4*, nitrosyl complexes are formed in reductive-ligations with NO and C$_5$H$_5{}^-$ anion, respectively, as reducing agents. Reagent sources of NO other than NO gas act as formal sources of nitrosonium ion, NO$^+$. For example, ClNO gives $NO^+$ and $Cl^-$; $NO_2^-$ plus two $H^+$ gives $NO^+$ and $H_2O$; and $NO_2^+$ plus CO gives $NO^+$ and $CO_2$. Salts of the *nitronium ion*, $NO_2^+$, are also available commercially. These reagents frequently convert anionic complexes into neutral nitrosyl complexes, as in equations *6.8* and *6.10*, or neutral complexes into cationic nitrosyl complexes, as in equation *6.11*.

Most of the structural types of M—NO coordination are analogous to those of M—CO coordination. The more familiar modes of coordination include linear terminal, $\mu_2$-bridging, and $\mu_3$-bridging.

$$M{\Longleftarrow}N{=\!=\!=}O \longleftrightarrow M{=\!=\!=}\overset{+}{\underset{}{N}}{=\!=\!=}O$$

**6.1**

linear terminal

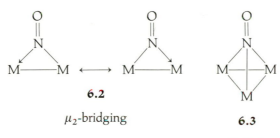

**6.2**

$\mu_2$-bridging

**6.3**

$\mu_3$-bridging

Structure **6.1** is like that of terminal CO ligands in which considerable M—NO synergistic bonding is present. Formally, a neutral NO ligand as in **6.1** is a three-electron donor, and an $NO^+$ ligand is a two-electron donor. The NO stretching frequency is usually in the range 1800–1900 cm$^{-1}$, but it can be as low as 1645 cm$^{-1}$ depending on the charge and type of ancillary ligands on the metal atom. As with CO ligands, the number and relative intensities of the observed $\nu_{NO}$ bands reflect the relative orientation of the NO ligands.

Structure **6.2** represents doubly bridging NO ligands. Both symmetrical and asymmetrical $\mu_2$-NO ligands are known. In valence bond notation, a $\mu_2$-NO ligand donates two electrons to one M and one electron to the other M, like a $\mu_2$-halide atom. When two $\mu_2$-NO ligands bridge the same two M atoms, the total number of electrons donated to each metal by these NO ligands is three. Therefore, as with CO, two $\mu_2$-NO ligands are electronically equivalent to one terminal NO ligand on *each* M atom. For $\mu_2$-NO ligands, $\nu_{NO}$ is usually in the range 1550–1400 cm$^{-1}$.

Triply bridging NO ligands (**6.3**) are regarded as one-electron donors to each M atom. This coordination type is quite rare, and $\nu_{NO}$ bands are observed around 1320 cm$^{-1}$.

One coordination mode of NO that is not observed with CO is the *bent terminal* structure (**6.4**). In it, the NO ligand is a one-electron donor having a formal $sp^2$-hybridized nitrogen atom, similar to the nitrosyl halides (**6.5**) or nitroso compounds (**6.6**). Values of the M—N—O angle are near 120°. An alternative orbital description is shown in **6.7**. Here, a metal atom formally donates a pair of valence electrons into an empty $\pi^*$ MO of an $NO^+$ ligand.

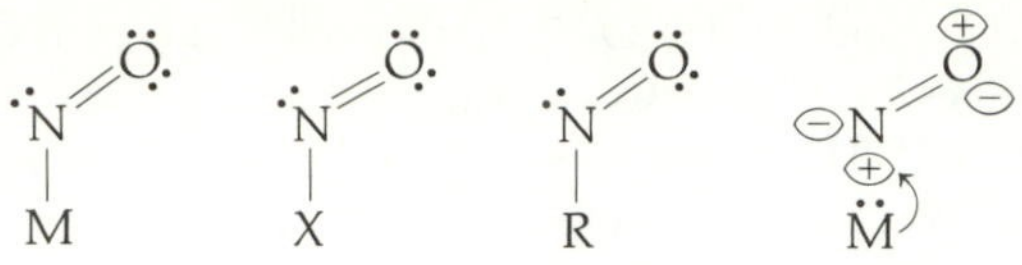

$$\text{6.4} \qquad \text{6.5} \qquad \text{6.6} \qquad \text{6.7}$$

Considerable research has been directed toward understanding the detailed bonding involved in bent terminal NO coordination and in determining how this bonding mode differs from linear terminal NO coordination. Apparently, whether a terminal NO ligand is bent or linear depends on the coordination number $CN$ of M, the coordination geometry about M, the number $n$ of $d$ electrons on M (when all ligands are assumed to be two-electron donors, i.e., NO as $NO^+$), and the atomic orbitals that participate in the HOMO of the complex.

In MO terminology, an M—NO geometry will be linear unless the NO $\pi^*$ orbital is occupied in the complex. From this theory, the geometry of an [M—NO] functional group can be predicted for a given type of complex, as shown below:

| CN of M | Geometry of [M—NO] in $L_{CN-1}M(NO)$ for $d^n$-metals |
|---------|-------------------------------------------------------|
| 6 | linear when $n \leq 6$ |
|   | bent when $n \geq 7$ |
| 5 | linear when $n \leq 6$ |
|   | (more complex when $n = 7$) |
|   | linear when $n = 8$ in trigonal bipyramidal complexes |
|   | bent when $n = 8$ in square pyramidal complexes |
| 4 | linear when $n = 10$ in tetrahedral complexes |
|   | bent when $n = 10$ in square-planar complexes |

Less rigorously, complexes of the type $L_mM(NO)$ will have bent terminal NO ligands when the $L_mM$ fragment gives a 17-electron metal atom. For these complexes, the NO ligand acts as a one-electron ligand, as in **6.4**, and thereby permits M to obey the EAN rule. In 16-electron square-planar complexes, the M atom frequently has a lone pair of electrons located in its valence shell (e.g., a filled $d_{z^2}$ orbital in $d^8$-metal ions), and this lone pair can be donated to an approaching $NO^+$ ion to form a bent terminal NO ligand as shown in **6.7.** The formal electron count on M remains the same although the formal atomic charge on M becomes more positive. Equation 6.7 is an example of this type of reaction.

Conversion of a linear terminal NO ligand to its bent terminal NO counterpart might be important to the development of potentially catalytically active complexes. This conversion would actually free a coordination site for an incoming two-electron donor *without requiring dissociation of a ligand.* For this reason, examples of linear–bent M—NO equilibration are quite important. One such example is shown in equation 6.12.

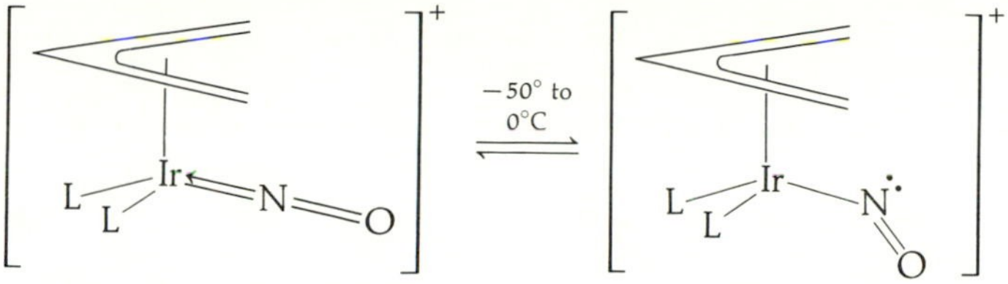

$$[(\eta^3\text{-}C_3H_5)IrL_2(NO)]^+$$
$$L = PPh_3$$
18-electron complex                 16-electron complex

$$(6.12)$$

In the temperature range shown, PMR spectra indicate an $\eta^3$-$C_3H_5$ ligand, and IR spectra revealed *two* $\nu_{NO}$ bands—one at 1763 cm$^{-1}$ and a second at 1631 cm$^{-1}$. Since the relative intensities of these $\nu_{NO}$ bands change with temperature, the equilibration of two species, as shown in equation 6.12, has been proposed. The structure of the 16-electron complex will be discussed shortly.

The structures of several metal nitrosyl complexes representing the various types of M—NO coordination geometries are shown in **6.8–6.12**. Complex **6.8** has a linear terminal NO ligand, which acts as a three-electron donor thereby affording an 18-electron complex. As with "linear" CO ligands, the M—N—O angles for "linear" NO ligands may deviate somewhat from 180°. The linear terminal and $\mu_2$-NO ligands in **6.9** have $\nu_{NO}$ bands at 1677 and 1518 cm$^{-1}$, respectively. The $\mu_2$-NO ligands are bonded symmetrically to each 18-electron chromium atom.

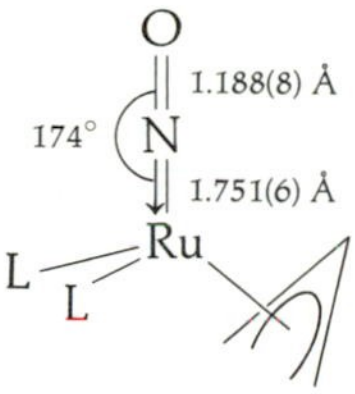

**6.8**

$$(\eta^3\text{-}C_3H_5)Ru(NO)(PPh_3)_2$$

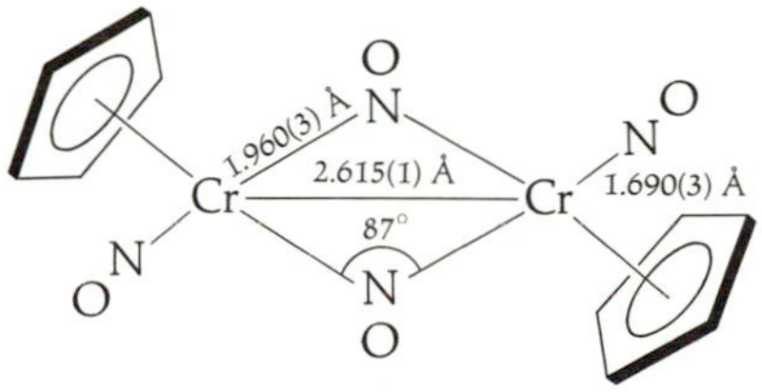

**6.9**

$$[(\eta^5\text{-}C_5H_5)Cr(NO)(\mu_2\text{-}NO)]_2$$

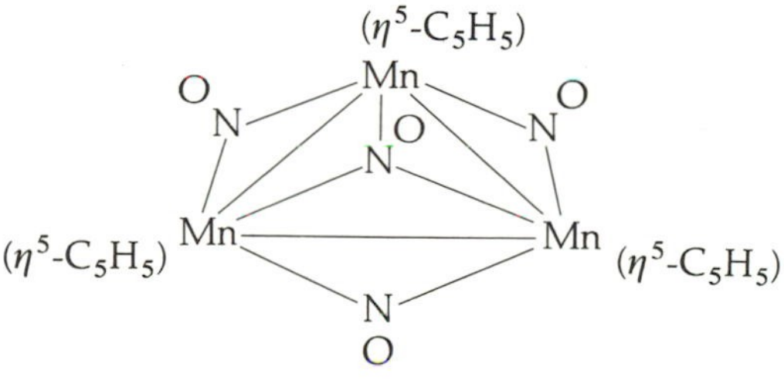

**6.10**

$$[(\eta^5\text{-}C_5H_5)Mn(\mu_2\text{-}NO)]_3(\mu_3\text{-}NO)$$

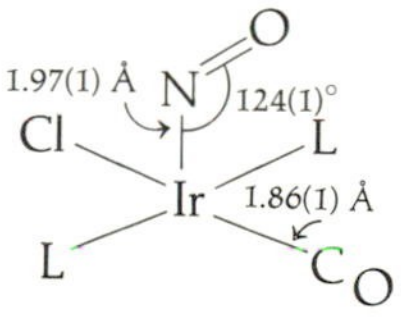

**6.11**

$$[Ir(CO)(NO)(PPh_3)_2Cl]^+$$

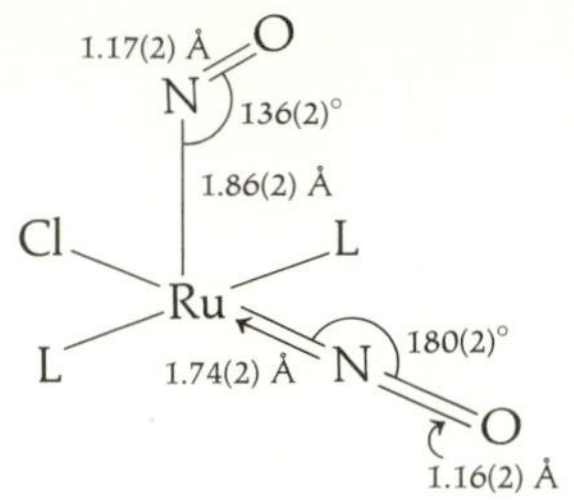

**6.12**

$$[Ru(NO)_2(PPh_3)_2(Cl)]^+$$

Complex **6.10** has three $\mu_2$-NO ligands and one $\mu_3$-NO ligand. The average Mn—Mn distance of 2.50 Å is consistent with a Mn—Mn single bond. The $\nu_{NO}$ value for the $\mu_3$-NO ligand is 1328 cm$^{-1}$.

The cationic complex **6.11** has a bent terminal NO ligand. This complex can be regarded as a Lewis salt adduct of a neutral, 16-electron $L_n\ddot{I}r$ Lewis base with the cationic Lewis acid NO$^+$, as shown in **6.7**. In fact, complex **6.11** is prepared from [NO]BF$_4$ and Vaska's complex as shown in equation 6.7. The Ir atoms in each neutral $L_n\ddot{I}r$ moiety can be considered formally $d^8$-metal ions. Complex **6.12** provides a direct, intramolecular structural comparison of linear and bent terminal NO ligands. Clearly, the Ru—N—O angle is much smaller in the bent NO ligand, and the Ru—N distance to the linear NO ligand is significantly shorter than the Ru—N distance to the bent NO ligand. These structural features are consistent with the simple bonding descriptions shown as **6.1** and **6.4**. The use of $^{15}$N-labeled NO in preparing **6.12** allows the isotopic labeling of one of the NO ligands. Infrared spectra of this labeled complex revealed an equilibration between the bent and linear terminal NO ligands.

*Nitric sulfide*, NS, complexes are prepared indirectly from the trithiazyl trichloride cyclic trimer, [SNCl]$_3$, as shown in equation 6.13, because of the rapid polymerization of free NS.

$$Na[(\eta^5\text{-}C_5H_5)Cr(CO)_3] + \tfrac{1}{3}[SNCl]_3 \xrightarrow[-78°C]{THF} \qquad + CO + NaCl \quad (6.13)$$

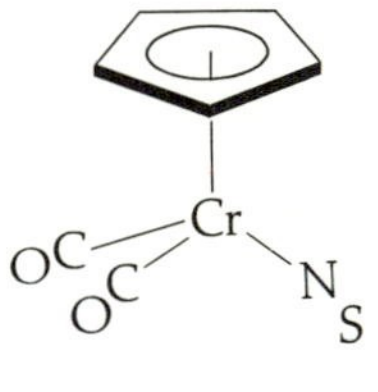

**6.13**

red-violet solid, 21%
mp 68°C

Complex **6.13** has a linear terminal *thionitrosyl ligand* with a Cr—N—S angle of 176.8(1)°. The Cr—C, Cr—N and N—S distances are 1.883(2) Å, 1.694(7) Å and 1.551(2) Å, respectively. In NS$^+$(g), the N—S distance is 1.440(5) Å, a value that may indicate some back bonding of electron density into the NS $\pi^*$ MO in **6.13**. An IR spectrum of **6.13** shows $\nu_{CO}$ bands at 2033 and 1962 cm$^{-1}$, and a $\nu_{NS}$ band at 1180 cm$^{-1}$. These $\nu_{CO}$ frequencies are slightly higher than those of the analogous nitrosyl complex. Nitric sulfide is believed to be a slightly better $\pi$-acceptor than NO because its $\pi^*$ MO is of lower energy; however, the lower electronegativity of a sulfur atom relative to an oxygen atom may make NS a better $\sigma$-donor ligand. Since thionitrosyl complexes are relatively rare, a detailed comparison of M—NO and M—NS bonding is not possible.

An example of *molecular nitrogen*, $N_2$, acting as a terminal and bridging ligand in an organometallic complex is shown in equation *6.14*.

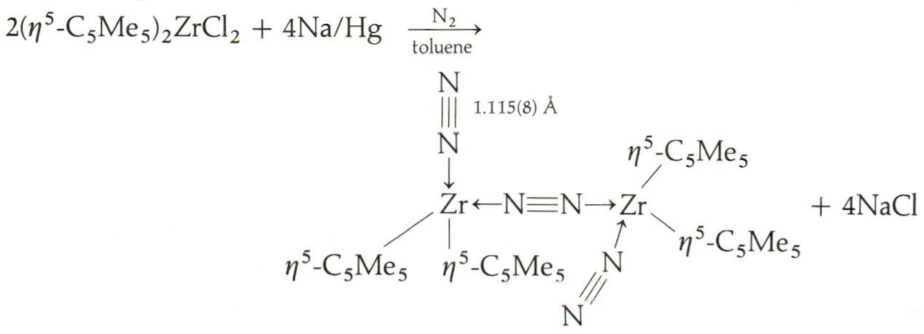

$$2(\eta^5\text{-}C_5Me_5)_2ZrCl_2 + 4Na/Hg \xrightarrow[\text{toluene}]{N_2}$$

**6.14**

$$[(\eta^5\text{-}C_5Me_5)_2Zr(N_2)]_2(\mu_2\text{-}N_2)$$
dark red solid, 30–40%

Complex **6.14** forms when the Zr(IV) reactant complex is reduced to Zr(II) under a nitrogen atmosphere. Terminal $\nu_{N_2}$ bands appear at 2041 and 2006 cm$^{-1}$, and the bridging $\nu_{N_2}$ band occurs at 1556 cm$^{-1}$. Coordination chemists have been exploring the synthesis and reaction chemistry of dinitrogen complexes intensely with the overall goal of understanding the dinitrogen fixation chemistry of natural enzymes such as the nitrogenases. The conversion of $N_2$ to $NH_3$ by these enzymes involves chemistry at metal atoms, and the development of synthetic catalysts that model this reduction is being pursued.

*In situ* generation of molybdenocene in the presence of nitriles affords $\eta^2$-*nitrile complexes* (**6.15**).

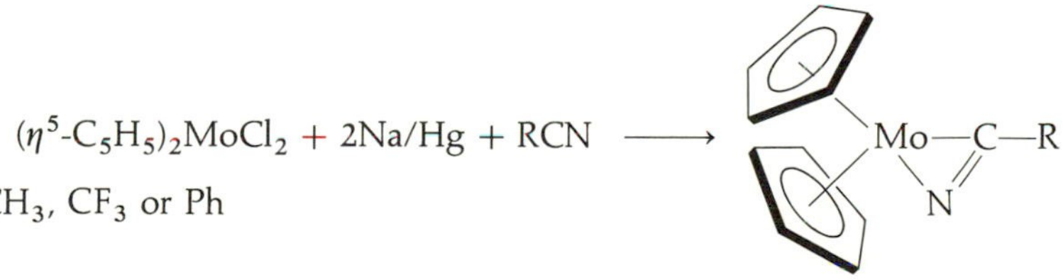

$$(\eta^5\text{-}C_5H_5)_2MoCl_2 + 2Na/Hg + RCN \longrightarrow$$

R = CH$_3$, CF$_3$ or Ph

**6.15**

red solids

The nitriles act as two-electron $\eta^2$-heteroalkyne ligands. When R is methyl, the $\nu_{CN}$ stretching frequency is 494 cm$^{-1}$ lower than that of free acetonitrile. The molybdenum atom in **6.15** obeys the EAN rule. *Dihapto*-imine and isoelectronic ketone complexes, such as **6.16** and **6.17**, respectively, are also known. In each complex, the principal coordination geometry about the metal atom is planar.

**6.16**                                    **6.17**

## Complexes of CS, CSe, CS$_2$, and Other Heterocumulenes

As with NS, complexes of *carbon monosulfide*, CS, or *carbon monoselenide*, CSe, are prepared, with one exception, via indirect methods. Carbon monosulfide is a very reactive gas, which is generated by photolysis or combustion of CS$_2$ (usually effected by using a high voltage discharge). At temperatures greater than $-160°$C, CS reacts to form an insoluble (CS)$_x$ polymer. Carbon monoselenide has not been isolated.

Over 125 complexes of CS are known, but complexes of CSe are still relatively rare. Several spectroscopic studies of M—CS complexes reveal that the $7\sigma$ MO of CS (which is equivalent to the $5\sigma$ MO in CO) has a higher energy than the $5\sigma$ MO in CO; the $\pi^*$ MO in CS, on the other hand, is slightly lower in energy than the $\pi^*$ MO in CO. Therefore, *thiocarbonyl* ligands are slightly better $\sigma$-donors and $\pi$-acceptors of electron density than CO. Metal-CX bond-disruption enthalpies, where X is S or Se, should be slightly greater than corresponding M—CO bond-disruption enthalpies. Incoming two-electron donors more readily substitute CO ligands than CS ones. Also, structural data reveal that M—CX distances are shorter than M—CO distances within the same molecule.

Thiocarbonyl complexes are prepared (1) by direct reaction between cocondensed CS(g) and metal vapor, (2) from CS$_2$ and a metal complex with cleavage of one of the C—S bonds, or (3) by reaction of a metal complex and *thiophosgene*, Cl$_2$CS, or a thiophosgene derivative such as (RO)(Cl)CS. *Selenocarbonyl* complexes are prepared from CSe$_2$ or C(S)(Se). When CS$_2$ is the source of CS, complexes of CS$_2$ may be isolable, or they may act as transient intermediates.

From these preparations, several structural types of CS coordination are observed (see **6.18−6.23** in table 6.1). Since a sulfur atom is a moderately strong Lewis base, the sulfur atom can also coordinate to metal atoms affording rather unusual bridging structures such as **6.21−6.23**.

Although there appears to be a nearly linear correlation between $\nu_{CS}$ values for terminal M—CS ligands and C—S bond order, the low frequencies of these bands (and those of the more complex structural types) occur in regions of the IR spectrum close to the stretching bands of other ligands in the molecule. This frequency proximity results in considerable mixing of the C—S (or C—Se) vibrations with other

**TABLE 6.1 Various structural types of CS coordination.**

| Structure | Designation | $\nu_{CS}(cm^{-1})$ | C—S Distance (Å) |
|---|---|---|---|
| M—CS<br><br>**6.18** | terminal | 1161–1409 | 1.51–1.54 |
| 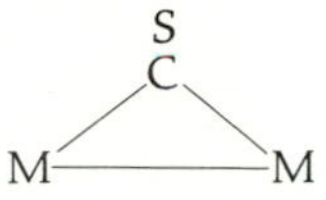 (structure **6.19**) | $\mu_2$-CS | 1106–1160 | 1.59–1.62 |
| 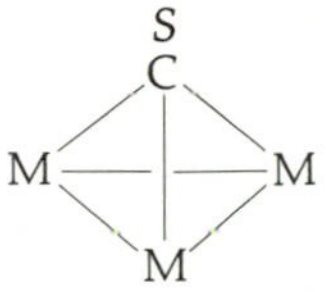 (structure **6.20**) | $\mu_3$-CS | 1040–1080 | ~1.69(5) |
| M—CS → M'<br><br>**6.21** | $\eta^2$-$(\mu_1$-C)$(\mu_1$-S) | 1048–1106 | ~1.58 (estimate) |
| 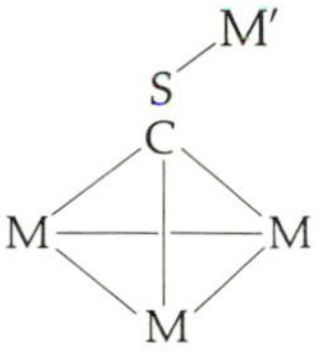 (structure **6.22**) | $\eta^2$-$(\mu_3$-C)$(\mu_1$-S) | ~960 | ~1.695(5) |
| 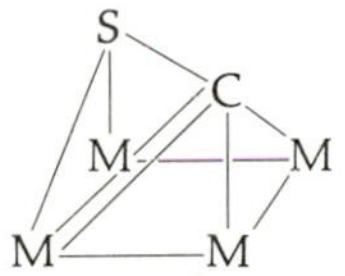 (structure **6.23**) | $\eta^2$-$(\mu_4$-C)$(\mu_2$-S) | ~921 | ~1.71(2) |

vibrations in the molecule. Such mixing removes the "purity" of the $\nu_{CS}$ bands and, thereby, complicates any correlation between $\nu_{CS}$ values and M—CS bonding interactions. Free CS and CSe have $\nu_{CX}$ values, respectively, of 1274 cm$^{-1}$ and 1036 cm$^{-1}$ (extrapolated value for CSe from its electronic spectrum). Only complexes having terminal CSe ligands are known. In these compounds, $\nu_{CSe}$ values are in the range 1063–1137 cm$^{-1}$.

Representative preparations and structures of selected thiocarbonyl complexes are shown in equations 6.16–6.18. Only Ni(CS)$_4$ has been prepared from the cocondensation of CS gas and metal vapor at 10°K. In equation 6.16, a CS$_2$ molecule is incorporated into the cluster products as a CS ligand and a $\mu_3$-*sulfido* ligand. Complex **6.24** contains a $\mu_3$-CS ligand of type **6.20**; complex **6.25** has a CS ligand of type **6.22**. In this structure, a CS ligand is a four-electron donor.

$(\eta^5\text{-}C_5H_5)Co(\eta^2\text{-}CS_2)(L) + 2(\eta^5\text{-}C_5H_5)Co(L) \xrightarrow[\substack{\Delta \\ 5\ hrs}]{C_6H_6}$

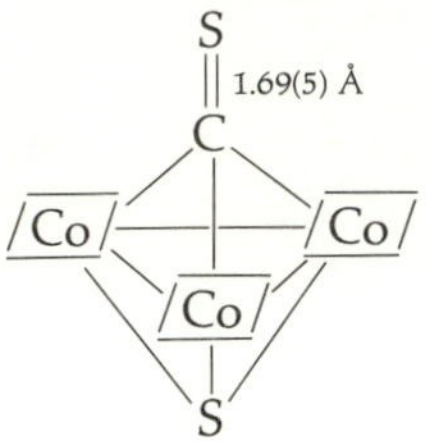

L = PMe$_3$

$\underline{/Co/} = (\eta^5\text{-}C_5H_5)Co$

**6.24**

$(\eta^5\text{-}C_5H_5)_3Co_3(\mu_3\text{-CS})(\mu_3\text{-S})$
black solid, 80%
dec. 289°C

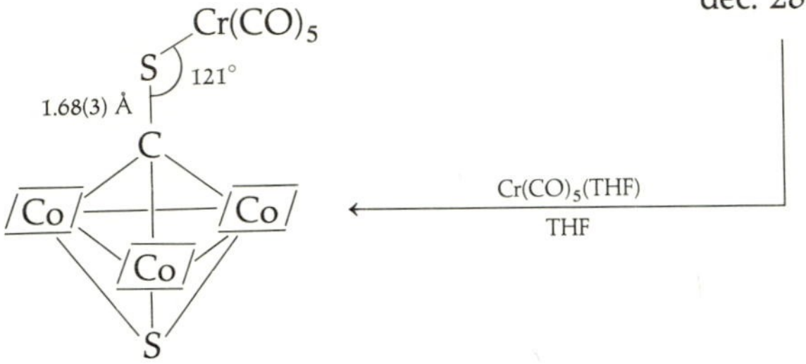

$$\xleftarrow[\text{THF}]{\text{Cr(CO)}_5\text{(THF)}} \qquad\qquad (6.16)$$

**6.25**

black solid, 35%

In equation *6.17*, CS$_2$ reacts with the Rh(I) reactant to afford an intermediate $\eta^2$-CS$_2$ complex by dissociation of a Ph$_3$P ligand. Dissolved Ph$_3$P extracts a sulfur atom from the CS$_2$ ligand affording the thiocarbonyl compound **6.26** and Ph$_3$P=S.

$$RhL_3X + CS_2 \xrightarrow[\text{PPh}_3]{\text{MeOH}} \textit{trans}\text{-Rh(CS)L}_2X + Ph_3P{=}S \qquad (6.17)$$

L = PPh$_3$; X = Cl or Br                **6.26**

orange solids, >98%
mp 250°C (X = Cl)

Equation *6.18* shows the loss of a sulfur atom from CS$_2$ by protonation of a ferra-dithioester ligand in **6.27** followed by loss of methyl mercaptan. Complex **6.27** is formed by nucleophilic attack of $[(\eta^5\text{-}C_5H_5)Fe(CO)_2]^-$ on CS$_2$ followed by subsequent alkylation of the ferra-dithiocarboxylate anion with MeI. The cationic thiocarbonyl complex (**6.28**) can be reduced to a neutral dimeric complex (**6.29**), which has two symmetrically bridging $\mu_2$-CS ligands.

Many CS$_2$ *complexes* are now known, and the formation of CS$_2$ complexes as intermediates in the syntheses of thiocarbonyl complexes was recognized many years

$$Na[(\eta^5\text{-}C_5H_5)Fe(CO)_2] + CS_2 \xrightarrow[\text{(2) MeI}]{\text{(1) THF}} (\eta^5\text{-}C_5H_5)Fe(CO)_2C(S)SMe + NaI$$

**6.27**

brown solid, 49–62%
mp 72°C

$NH_4PF_6 \Big| C_6H_6/HCl$

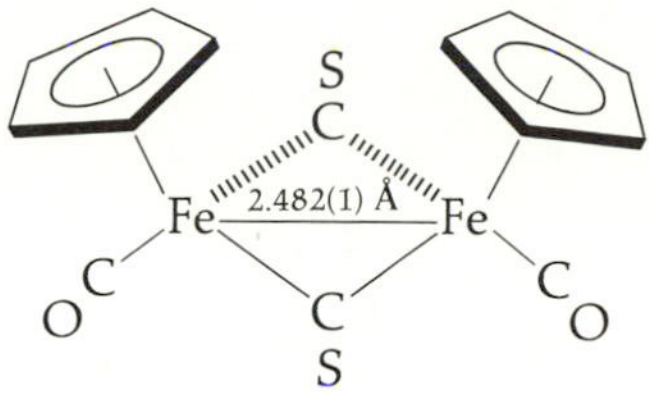

**6.29**

*cis*-$[(\eta^5\text{-}C_5H_5)Fe(CO)(\mu_2\text{-}CS)]_2$
black solid, 40%

$$\xleftarrow[\text{THF}]{\text{NaH}} [(\eta^5\text{-}C_5H_5)Fe(CO)_2(CS)]^+PF_6^- + MeSH \quad (6.18)$$

**6.28**

yellow solid, 48–63%
dec. 190°C

ago. Although some $\eta^1$-$CS_2$ complexes, $M \leftarrow S{=}C{=}S$, have been prepared, these complexes are not well characterized. The most common type of $CS_2$ coordination is as an $\eta^2$-$CS_2$ ligand, **6.30**. These complexes are prepared generally by ligand substitution reactions. Two C—S stretching vibrations are observed in the ranges of 632–653 cm$^{-1}$, $\nu_{C-S}$, and 955–1235 cm$^{-1}$, $\nu_{C=S}$. Uncomplexed $CS_2$ is linear with a C—S distance of 1.54 Å and can be regarded as a *heterocumulene*, $X{=}C{=}Y$, where X and Y are sulfur. Upon complexation, $\eta^2$-$CS_2$ ligands of type **6.30** act as two-electron donors and adopt a bent structure analogous to the coordination geometry of complexed allenes.

Because of the strong Lewis basicity of sulfur, several different types of bridging $CS_2$ geometries are known (**6.31–6.35**). Many of these structural types are quite rare, however.

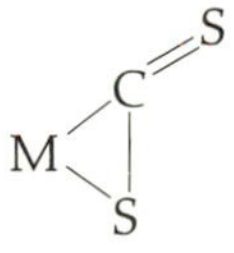

**6.30**

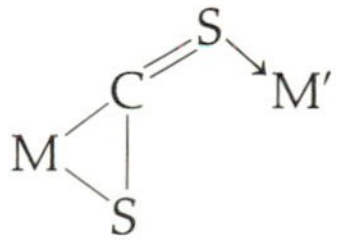

**6.31**

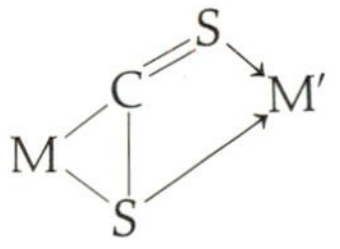

**6.32**

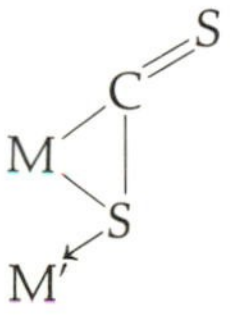

**6.33**

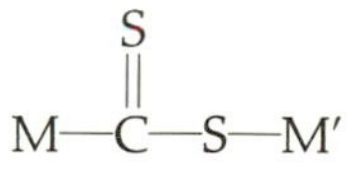

**6.34**

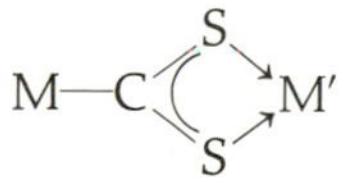

**6.35**

Selected examples of syntheses and structures of $CS_2$ complexes are shown in equations *6.19* and *6.20*. In both reactions, the products are formed via ligand substitution.

$$(\eta^2\text{-PhC}_2\text{H})\text{PtL}_2 + \text{CS}_2 \longrightarrow \quad \text{6.36} \quad + \text{PhC} \equiv \text{CH} \qquad (6.19)$$

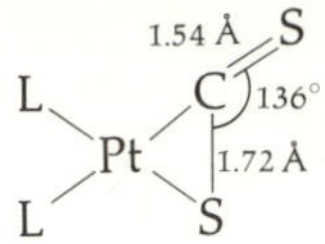

L = PPh₃

L = PPh$_3$

$$(\eta^2\text{-CS}_2)\text{Fe(CO)}_2\text{L}_2 \xrightarrow[\text{THF}]{(\eta^5\text{-C}_5\text{H}_5)\text{Mn(CO)}_2(\text{THF})}$$

L = PMe₂Ph

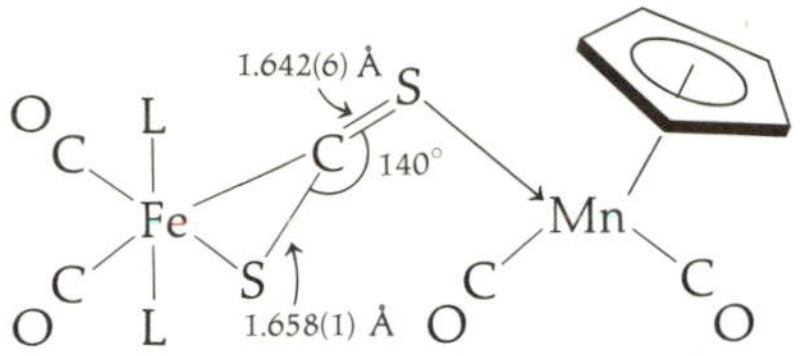

**6.37**

red solid, 56%
mp 148°C

$(6.20)$

Other heterocumulenes form complexes also, and most of these are structurally similar to those of CS$_2$. Two carbon dioxide complexes are shown as **6.38** and **6.39**. These are prepared via ligand substitution reactions using CO$_2$ (eqs. *6.21* and *6.22*). Complex **6.38** contains an $\eta^2$-CO$_2$ ligand and is structurally similar to the $\eta^2$-CS$_2$ complex **6.36**. The C—O distance in free CO$_2$ is 1.16 Å. Complex **6.39** contains a C$_2$O$_4$ ligand presumably formed by the coupling of two CO$_2$ molecules.

$$\text{L}_3\text{Ni} + \text{CO}_2 \xrightarrow[-\text{L}]{\text{toluene, 25°C}} \quad \text{6.38}$$

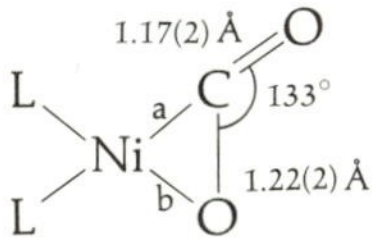

$$a = 1.84(2) \text{ Å}$$
$$b = 1.99(2) \text{ Å}$$

$(6.21)$

L = P(C₆H₁₁)₃

red-orange solid,
dec. 83°C

$$(\eta^2\text{-cis-cyclooctene})\text{IrL}_3\text{Cl} \xrightarrow[\text{CO}_2]{\text{C}_6\text{H}_6} \quad \text{6.39} \quad + \text{ olefin} \qquad (6.22)$$

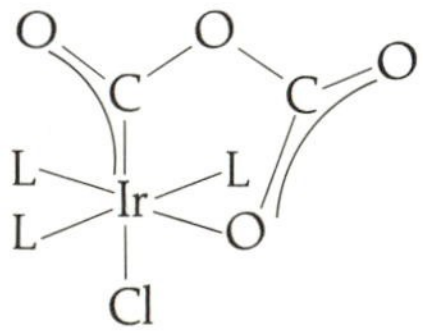

L = PMe₃

white solid
dec. 150°C

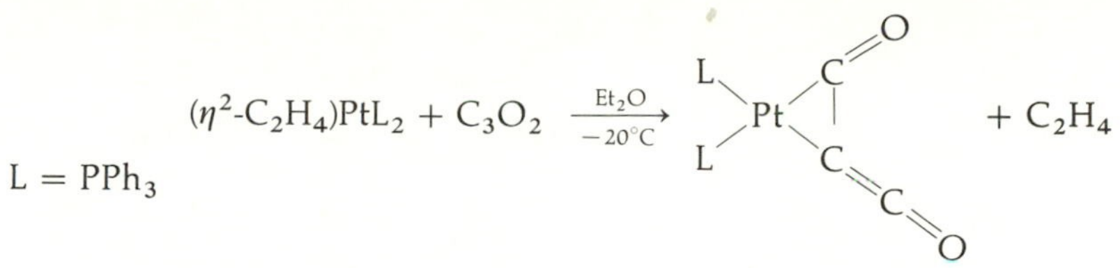
(η2-C2H4)PtL2 + C3O2
Et2O
−20°C
L
Pt
L
C
O
C
C
O
+ C2H4
L = PPh3

(6.23)

**6.40**

white solid, 37%
dec. 128°C

Equation *6.23* shows the preparation and proposed structure of a *carbon suboxide,* $C_3O_2$, complex (**6.40**). If this structure is correct, then the complex contains an $\eta^2$-diheteropentatetraene ligand.

Vanadocene also reacts with *diphenylketene* and *N-p-tolyl carbodiimide* to give representative $\eta^2$-complexes of these heterocumulene ligands (eq. *6.24*). The bent structure of the $\eta^2$-heterocumulene ligands in **6.41** and **6.42** is similar to that observed for $\eta^2$-allene ligands (see table 5.5). Since vanadocene is only a 15-electron complex, it is highly reactive toward unsaturated molecules.

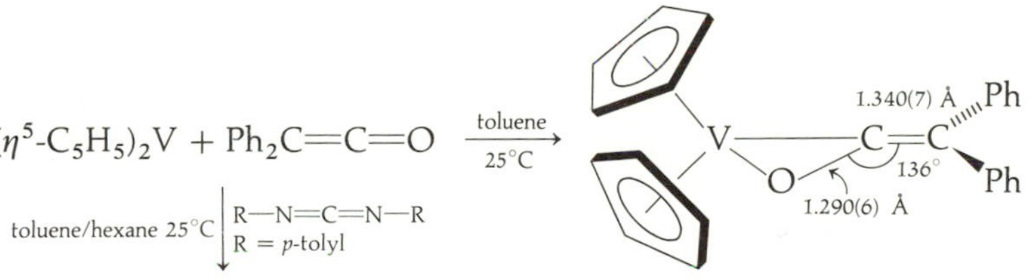
(η5-C5H5)2V + Ph2C=C=O
toluene
25°C
V
O
C=C
Ph
Ph
1.340(7) Å
136°
1.290(6) Å
toluene/hexane 25°C
R—N=C=N—R
R = p-tolyl

**6.41**

green solid, 64%

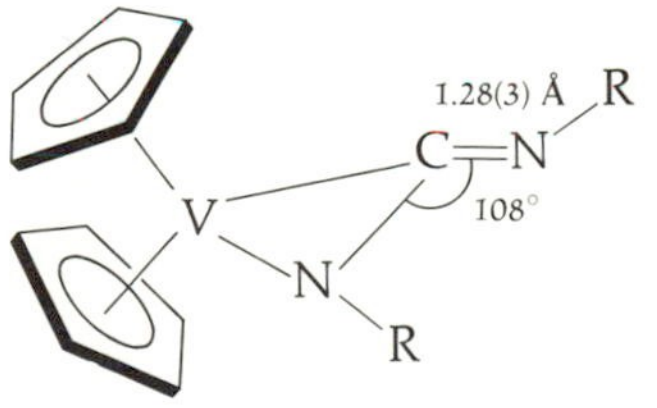
V
N
R
C=N
R
1.28(3) Å
108°

(6.24)

**6.42**

dark red solid, 65%

## Complexes of $S_2$, SO, $SO_2$, $SR_2$ and Sulfonium Ions

Organometallic complexes that have sulfur-coordinated ligands are thermally stable because of the moderately strong Lewis basicity of S donor atoms. Back bonding into empty valence *3d* orbitals on the S atom may also strengthen M—S bonds.

Early work with $Fe(CO)_5$ in basic aqueous media revealed that iron carbonyl complexes containing $S_2$ or S atoms as ligands could be prepared when inorganic sulfur reagents are present (eq. *6.25*).

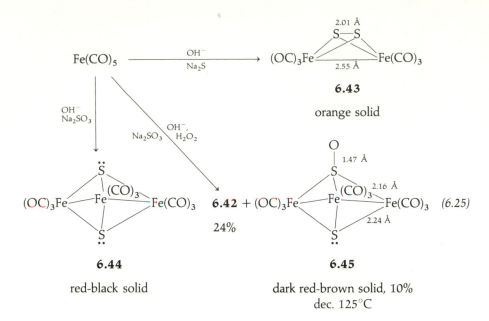

**6.43**

orange solid

**6.42** + (OC)₃Fe... (6.25)

24%

**6.44**

red-black solid

**6.45**

dark red-brown solid, 10%
dec. 125°C

With Na₂S, a dinuclear complex having an S₂ ligand (**6.43**) forms. The S—S vector is nearly perpendicular to the Fe—Fe vector. When Na₂SO₃ is present, a trinuclear cluster complex, S₂Fe₃(CO)₉, (**6.44**) is obtained. This complex has two $\mu_3$-*sulfido* ligands and an *open*, triangular Fe₃ core. If H₂O₂ is present in this latter reaction solution, then oxidation of one of the $\mu_3$-S ligands to a $\mu_3$-SO ligand occurs, affording complex **6.45**.

The selenium and tellurium analogues to **6.44** are prepared similarly. In these complexes, the $\mu_3$-chalcogen atom acts as a four-electron donor as shown in **6.46**, and each Fe atom obeys the EAN rule.

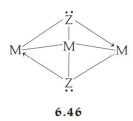

**6.46**

Complex **6.45** is derived from **6.44** by attaching a neutral oxygen atom to the lone pair of electrons on one of the $\mu_3$-S atoms. In such complexes, a $\mu_3$-SO ligand is also a four-electron donor.

*Sulfur dioxide* complexes are usually formed by ligand substitution in liquid SO₂ or in solutions containing excess dissolved SO₂. The two principal structural types of SO₂ coordination are shown in compounds **6.47** and **6.48**.

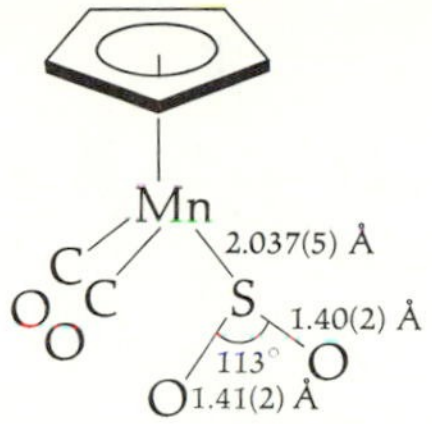

**6.47**

$(\eta^5\text{-}C_5H_5)Mn(CO)_2(\eta^1\text{-}SO_2)$

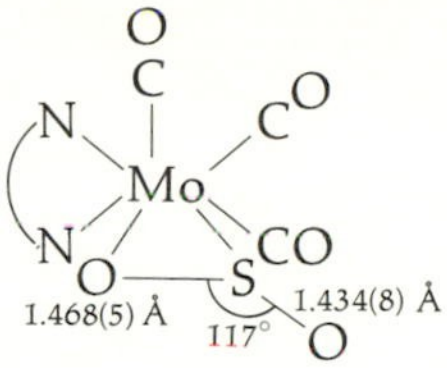

**6.48**

$Mo(CO)_3(o\text{-phen})(\eta^2\text{-}SO_2)$

Complex **6.47** has an $\eta^1\text{-}SO_2$ ligand, and in such complexes the four $MSO_2$ atoms are nearly coplanar. This geometry favors a strong $O_2S \rightarrow M$ $\sigma$-donation of electron density. Complex **6.48** contains an $\eta^2\text{-}SO_2$ ligand. This coordination mode favors strong $M \rightarrow SO_2$ $d\pi\text{-}p\pi$ back bonding because of good orbital overlap, and it is usually observed when M is relatively electron rich.

*Dialkylsulfide* complexes are prepared by thermal substitution of weakly coordinated ligands, such as THF, by $R_2\ddot{S}:$. When acting as a two-electron donor, the sulfur atom in such complexes has a lone pair of electrons in its valence shell. This lone pair can be alkylated using oxonium salts (eq. *6.26*).

$(\eta^5\text{-}C_5H_4Me)Mn(CO)_2(SMe_2) + [Et_3O]PF_6 \xrightarrow{CH_2Cl_2}$

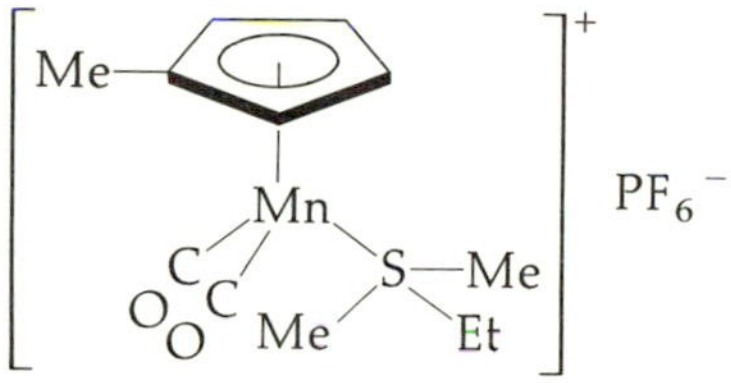

$PF_6^-$    (*6.26*)

**6.49**

yellow solid, 59%
mp 111°C

The complex formed contains a *sulfonium*, $SR_3^+$, ligand, in which the S atom remains a two-electron donor. In complex **6.49** the S atom has a nearly tetrahedral coordination geometry.

## COMPLEXES CONTAINING MONATOMIC LIGANDS

Monatomic ligands participate in the metal core unit of cluster complexes. These complexes can be classified into two structural types. In the first type, the main-group atom Z occupies a vertex position in a skeletal framework composed of metal atoms and these main-group atoms (**6.50**). In the second, the main-group atom is completely encapsulated within a metal core cluster of atoms (**6.51**).

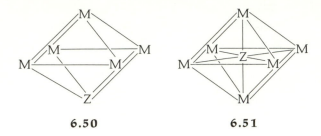

**6.50**                        **6.51**

Structures like **6.50** have "exposed" or peripheral attachment of the main-group atom to a face of a metal core polyhedron; structures like **6.51** have main-group atoms situated near the center of an $M_n$ core. These complexes are usually prepared from neutral or anionic metal carbonyl clusters.

Examples of type **6.50** clusters are shown in **6.52–6.54**, where Z is C, N and P.

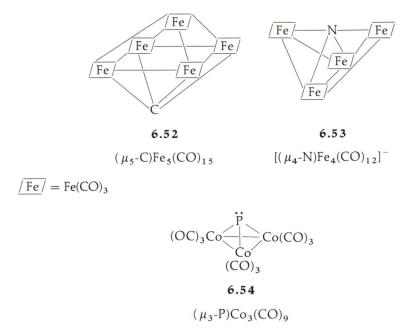

**6.52**                                **6.53**

$(\mu_5\text{-}C)Fe_5(CO)_{15}$              $[(\mu_4\text{-}N)Fe_4(CO)_{12}]^-$

$\boxed{Fe} = Fe(CO)_3$

**6.54**

$(\mu_3\text{-}P)Co_3(CO)_9$

Complexes **6.24**, **6.25**, **6.44**, and **6.45** have $\mu_3$-sulfido ligands. The *carbido* complex (**6.52**) was isolated originally in very low yield from a thermal reaction between $Fe_3(CO)_{12}$ and alkynes. The carbide-carbon atom is obtained from the reduction of CO. The *nitrido* complex (**6.53**) is obtained by reacting $Fe_3(CO)_{12}$ with $[Fe(CO)_3NO]^-$. Reduction of NO generates the nitride N atom. Complex **6.54** forms when $Co_2(CO)_8$ is treated with $PI_3$. The *phosphido* ligand acts as a one-electron donor to each Co atom. Metal-hydride complexes are also known in which a $\mu_3$-hydrogen atom is located above a triangular face of metal atoms.

Complexes of type **6.51** have complex structures because of the high nuclearity of the cluster molecules. As examples, $Ru_6(CO)_{17}C$ and $[Co_6(CO)_{15}N]^-$ have structure **6.51** where the unique carbon and nitrogen atoms are bonded equally to six metal atoms. The ruthenium complex is obtained from the thermal decomposition of $Ru_3(CO)_{12}$ or from reacting $Ru_3(CO)_{12}$ with $C_2H_4$ at 30 atm and 150°C. This complex was the first encapsulated cluster to be characterized. One CO ligand is doubly bridging and the remaining 16 CO ligands are terminal. The cobalt complex has six terminal CO and nine $\mu_2$-CO ligands.

The complexes $[Rh_9(CO)_{21}P]^{2-}$ and $[Rh_{17}(CO)_{32}(S)_2]^{3-}$ are isolated from high temperature and pressure reactions between $Rh(CO)_2(acac)$ and a mixture of CO and $H_2$ gases. Sources of the encapsulated S and P atoms are $H_2S$ or $SO_2$ and $PPh_3$, respectively. Each S atom and the P atom are bonded to *nine* Rh atoms. The complexes $[Rh_9(CO)_{21}As]^{2-}$ and $[Rh_{12}(CO)_{27}Sb]^{3-}$ are formed from similar reactions and have been shown to contain encapsulated As and Sb atoms.

Four complexes are known in which an encapsulated H atom is bonded to six metal atoms. Two examples are $[HCo_6(CO)_{15}]^-$ and $[HRu_6(CO)_{18}]^-$.

A unique carbide complex, $Rh_{12}(CO)_{25}C_2$ has an *encapsulated* $C_2$ molecule. One carbon atom is bonded to eight rhodium atoms, and the other C is bonded to six. The C—C distance of 1.48 Å is slightly shorter than a $C(sp^3)$-$C(sp^3)$ single bond.

## STUDY QUESTIONS

1. Given the procedure for preparing complex **6.11**, predict similar reactions that may occur with Vaska's complex.
2. Rationalize the presence of a $\mu_2$-CO ligand in the CS complex shown below.

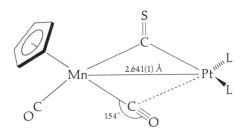

3. Explain why $SO_2$ can act both as a Lewis acid and as a Lewis base.
4. By using Wade's rules and a modification of equation 2.24, determine how many valence electrons of the carbide carbon atoms in $Fe_5(CO)_{15}C$ (**6.52**) and $Ru_6(CO)_{17}C$ must be involved in M—C bonding in order for these complexes to conform to Wade's rules.
5. The photolysis of $Co(CO)_3(NO)$, which is an 18-electron complex having a linear terminal NO ligand, gives an excited electronic state having a bent terminal NO ligand. Explain the observation by considering changes in the M—NO bonding.
6. For the two complexes shown below, predict the value of the M—M bond orders **(a)** for M = Fe and **(b)** for M = Co if both metals obey the EAN rule.

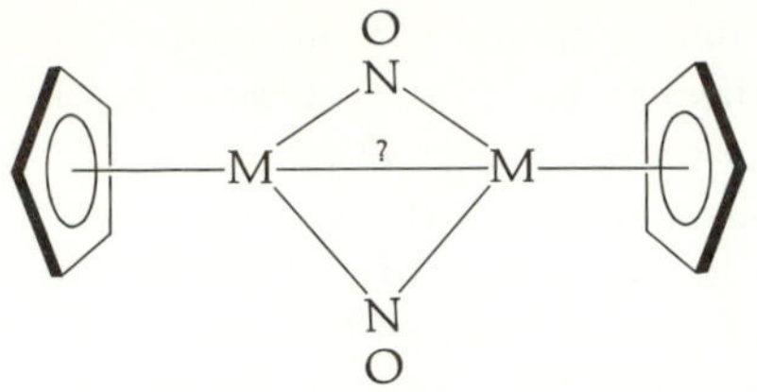

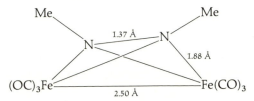

$$[(\eta^5\text{-}C_5H_5)M(\mu_2\text{-}NO)]_2$$

**7.** Azomethane, $\text{Me\ddot{N}}=\text{\ddot{N}Me}$, reacts with $Fe_2(CO)_9$ to give the complex shown below. Describe the electron-donor count of the bridging nitrogen atoms within the bridging azomethane ligand.

**8.** After reaction of $W_2(CO)_{10}^{2-}$ with $Cl_2C{=}S$, a mixture of $W(CO)_6$ and $W(CO)_5(CS)$ is isolated. Separation of the CS complex by fractional crystallization is inefficient. It was found that at 35°C the complex $W(CO)_5(CS)$ reacts with $I^-$ salts to give *trans*-$W(CO)_4(CS)I^-$, whereas $W(CO)_6$ did not undergo reaction under these conditions. The anionic complex could then easily be separated from the neutral $W(CO)_6$. What property of the neutral CS complex apparently enables such facile CO substitution, and how could the anionic CS complex be converted into $W(CO)_5(CS)$ after separation?

**9.** The product of the reaction shown below is an example of an "A-frame" molecular structure. A metal-metal bond is present in the reactant but absent in the product. Give a formal description of this reaction.

$$Pt_2Cl_2(Ph_2PCH_2PPh_2)_2 + CS_2(xs) \xrightarrow[\Delta]{CH_2Cl_2}$$

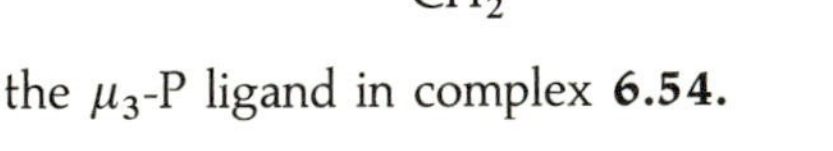

**10.** Predict possible reaction chemistry for the $\mu_3$-P ligand in complex **6.54**.

## SUGGESTED READING

### Complexes Containing NO Ligands

Bottomley, F. *Acc. Chem. Res.* **11**, 158 (1978).
Caulton, K. G. *Coord. Chem. Rev.* **14**, 317 (1975).
Connelly, N. G. *Inorg. Chim. Acta Rev.* **6**, 47 (1972).

Eisenberg, R., and Meyer, C. D. *Acc. Chem. Res.* **8**, 26 (1975).

Enemark, J. H., and Feltham, R. D. *Coord. Chem. Rev.* **13**, 339 (1974).

Frenz, B. A., and Ibers, J. A. *MTP Int. Rev. Sci. Phys. Chem. One.* **11**, 33 (1972).

Griffith, W. P. *Adv. Organometal. Chem,* **7**, 211 (1968).

Johnson, B. F. G., and McCleverty, J. A. *Progr. Inorg. Chem.* **7**, 277 (1966).

Lewis, J. *Science Progress* (London). **47**, 506 (1959).

## Complexes Containing CS, CSe, CS$_2$, and CO$_2$ Ligands

Butler, I. S. *Acc. Chem. Res.* **10**, 359 (1977).

Ito, T., and Yamanoto, A. *J. Soc. Org. Synth. Chem. Tokyo.* **34**, 308 (1976).

Vol'pin, M. E., and Kolomnikov, I. S. *Organometallic Reactions*, Vol. 5. E. I. Becker and M. Tsutsui, eds., p. 313. New York: Wiley, 1975.

Yaneff, P. V. *Coord. Chem. Rev.* **23**, 183 (1977).

## Transition Metal Cluster Complexes

Chini, P. *J. Organometal. Chem.* **200**, 37 (1980).

Chini, P., Longoni, G., and Albano, V. G. *Adv. Organometal. Chem.* **14**, 285 (1976).

Kaesz, H. D. *J. Organometal. Chem.* **200**, 145 (1980).

Muetterties, E. L. *J. Organometal. Chem.* **200**, 177 (1980).

Muetterties, E. L., Rhodin, T. N., Band, E., Bunker, C. F., and Pretzer, W. R. *Chem. Rev.* **79**, 91 (1979).

J. *Organometal. Chem.* **213.** A special issue dedicated to the late Professor Paolo Chini.

# Dynamic Intramolecular Rearrangements of Organometallic Complexes

## CHAPTER 7

Many molecules undergo fast, reversible *intramolecular* rearrangements that can be detected experimentally. Molecules that exhibit these intramolecular motions are said to be *stereochemically nonrigid*. Typical examples of time-dependent intramolecular processes include rotation about chemical bonds or permutation of like atoms among equivalent or nearly equivalent bonding sites, such as a scrambling of CO ligands over all of the metal atoms within a cluster complex. The observation of intramolecular motions, the elucidation of the mechanistic pathways by which these motions occur, and the determination of the height of the energy barriers associated with these rearrangements provide fundamental information regarding the various types of intramolecular processes and the relative energies of closely related molecular structures. These rearrangements are the simplest reactions undergone by organometallic complexes.

Most intramolecular rearrangements are detected by proton or $^{13}C$—NMR. Complexes are usually studied in the solution phase although some rearrangements, such as the rotation of the $\eta^5$-$C_5H_5$ ligands in ferrocene, are also observed in the solid state.

Intramolecular rearrangements occur either between ground-state molecular structures of identical free energies, i.e., a *degenerate rearrangement*, or between ground-state molecular structures of different free energies, a *nondegenerate rearrangement*. An example of a degenerate rearrangement is rotation about the C—N bond in an N,N-dimethyl amide (eq. 7.1).

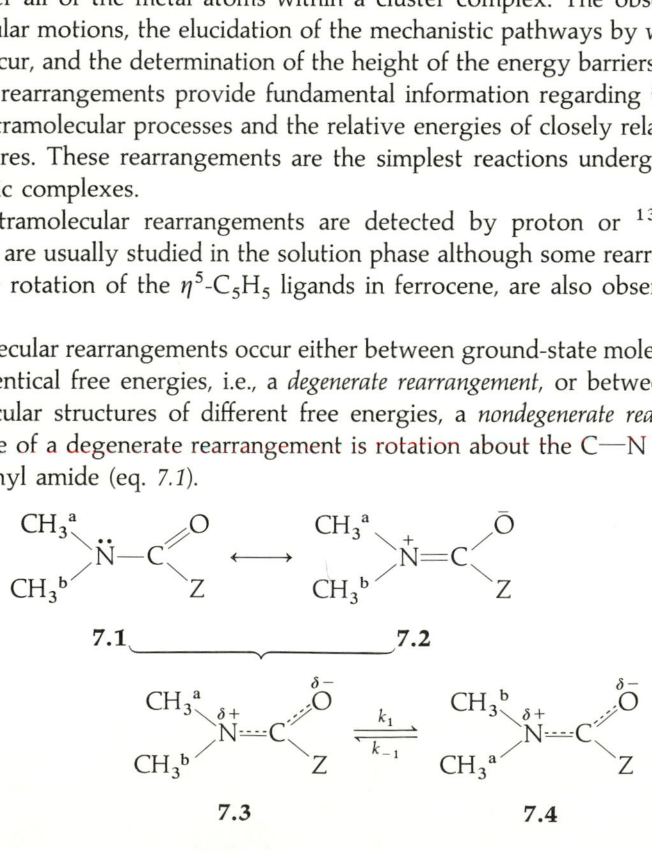

$$\text{7.1} \qquad \text{7.2}$$

$$(7.1)$$

$$\text{7.3} \qquad \qquad \text{7.4}$$

The ground-state structure of such an amide is shown as **7.3**, where Z is a general substituent. The C—N bond order is about 1.70 because of contributions to the molecular electronic structure by resonance forms **7.1** and **7.2**. The *instantaneous* molecular structure of an amide, as determined by x-ray crystallography, would reveal considerable C—N multiple bonding, essentially $sp^2$ hybridization for the N and carbonyl-carbon atoms, and the location of one methyl group *syn* to the oxygen atom and the other *syn* to the substituent, Z. A proton NMR spectrum of **7.3** in solution would be consistent with this solid-state structure. The two methyl groups would appear as singlets having different chemical shifts.

As shown in equation *7.1*, thermal activation can cause rotation about the C—N multiple bond of **7.3** to generate structure **7.4**. Structures **7.3** and **7.4** are equivalent, but *not* identical since they differ in the arbitrary labeling of the two methyl groups. Notice that the two methyl groups have exchanged positions during this intramolecular rearrangement. (To illustrate this principle, suppose that one of the $CH_3$ groups of **7.3** is changed to $CD_3$; then rotation about the C—N bond clearly indicates how **7.3** and **7.4** differ. However, in this isotopically substituted molecule, the isomers **7.3** and **7.4** do *not* have exactly the same free energy, and therefore, they are *not* truly equivalent isomers.)

An energy profile of equation *7.1* is shown as **7.6**. Since **7.3** and **7.4** have identical energy, the reaction is a *degenerate rearrangement*.

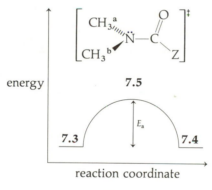

**7.6**

Conversion of **7.3** into **7.4**, and the reverse conversion, occur through a *transition state species*, **7.5**, in which there has been a 90° rotation about the amide C—N bond relative to the ground-state structure. The magnitude of the activation energy, $E_a$, in going from **7.3** to **7.5** depends on the strength of the C—N $\pi$-bond and other minor energy terms. The rate of exchange, $k$, between **7.3** and **7.4** at a given temperature depends on the height of this energy barrier according to the Arrhenius equation:

$$k = Ae^{-E_a/RT}$$

Interestingly, the appearance of the proton NMR spectrum can be influenced by exchange processes. This effect is shown in figure 7.1 for a hypothetical molecule such as **7.3**, where the difference in chemical shift of $CH_3{}^a$ and $CH_3{}^b$ at this field

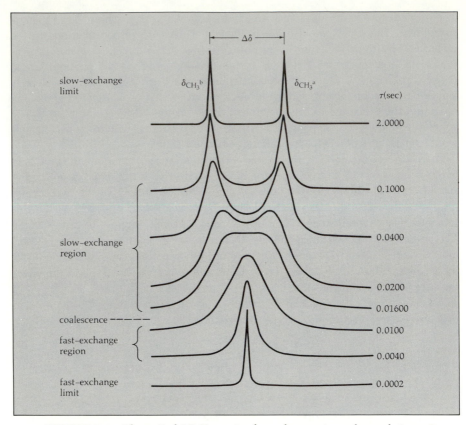

**FIGURE 7.1** Theoretical NMR spectra for a degenerate exchange between two sites as a function of the average lifetime, $\tau$, at each site. *Adapted from H. Gunther, NMR Spectroscopy—An Introduction (New York: John Wiley & Sons, Inc., 1980), p. 239.*

strength, $\Delta\delta$, is 30 Hz and where the "natural line width" at half-height of each methyl singlet is 1 Hz. The parameter $\tau$ represents the *average lifetime that a methyl group spends in each site*, a or b, as shown in equation 7.1. In this example, $\tau$ is just the inverse of the first-order rate constants, as shown in equation 7.2 for the exchange process.

$$\tau = 1/k_1 = 1/k_{-1} \tag{7.2}$$

The top spectrum of figure 7.1 shows the anticipated spectrum of **7.3** when no exchange or *very slow* exchange of the methyl groups is occurring. The lifetime of a methyl group in site a or b is 2 sec or greater. As rotation about the C—N bond increases in rate, the lifetime of a methyl group in site a or b decreases. Within a lifetime range of ~1.0 sec to ~0.0002 sec, several spectral changes are observed. Initially, the two methyl singlets begin to broaden; this is called the *slow-exchange region*. These broad resonances then coalesce into one broad resonance; this point is called *coalescence*. As the rate of exchange increases, this broad peak begins to sharpen and finally becomes a sharp singlet of "natural line width"; this is the *fast-exchange region*. Since sites a and b are equally populated in this example, the fast-exchange singlet appears at the average chemical shift of $CH_3^a$ and $CH_3^b$. Throughout the exchange process, the *total* integrated intensity of the methyl resonance remains constant.

Two questions are raised by the spectral data shown in figure 7.1:

1. Why is the appearance of the NMR spectrum dependent on the rate of chemical exchange between the two different methyl environments?

2. How can the rate of chemical exchange be varied experimentally to permit the detection of such intramolecular exchange processes?

The answer to the first question is based on the uncertainty principle as shown in equation 7.3 or the equivalent relationship shown in equation 7.4.

$$\Delta E \, \Delta t \approx \hbar \tag{7.3}$$

$$\Delta\delta(\text{in Hz}) \; \tau(\text{in sec}) \approx 1/\pi \tag{7.4}$$

For degenerate exchange processes, equation 7.4 defines a lower limit for $\tau$ where two absorptions separated in energy by $\Delta\delta$(Hz) can be observed as distinct peaks in the spectrum. In the NMR spectra of figure 7.1, $\Delta\delta$ is 30 Hz. Therefore coalescence should occur at about $(1/30\pi)$ sec, or 0.0106 sec, according to equation 7.4. This value is quite close to the lifetime that produces observed coalescence in figure 7.1.

As a consequence of the uncertainty principle, there is a lifetime associated with the absorption process for each experimental technique, such as NMR or IR. If a chemical exchange process occurs very fast relative to the lifetime required by a spectroscopic method, then chemically distinct absorptions are seen as a single absorption at an average energy. This effect is shown in figure 7.1 where rapid exchange of $CH_3^a$ and $CH_3^b$ in **7.3** affords a single proton NMR resonance at an average chemical shift. An analogous effect is observed in photography where movement of an object at a rate comparable to or faster than the shutter speed produces a blurred image.

Since observation of spectral changes like those shown in figure 7.1 indicate that a dynamic process is occurring, how does one record such a set of spectra in answer to the second question? For NMR, the answer to this question depends on the chemical shift difference, $\Delta\delta$, between exchanging sites, the magnitude of $E_a$ for the exchange process, and a little bit of luck! If a molecule exhibits an NMR spectrum at room temperature that is consistent with the static or instantaneous ground-state structure, then the sample should be heated in the NMR probe to promote the chemical exchange. Heating the sample increases the rate of the exchange process, and resonance averaging among exchanging sites may be observed. Conversely, if a fast-exchange limiting spectrum is observed at normal probe temperature, then cooling the sample solution may reduce the exchange rate to afford a spectrum consistent with the ground-state structure of the molecule. For organic amides, proton NMR samples usually need to be heated before rotation about the amide C—N bond becomes rapid relative to the proton NMR time scale.

Luck becomes important when the temperature of a sample is varied. Heating a sample may cause thermal decomposition and cooling it may initiate crystallization. Both effects can terminate the variable temperature experiment, because the amount of desired sample compound present *in the solution* is reduced. Also, if $E_a$ is quite small, then a slow-exchange limiting spectrum may not be evident even at temperatures low enough to increase the viscosity of the solvent. This effect, too, impairs good spectral resolution and instrumental sensitivity.

*Intra*molecular exchange processes are distinguished from *inter*molecular exchange processes by recording sets of variable temperature spectra at different molar concentrations of the sample complex. Intramolecular reactions show first-order kinetics; their rates are *not* dependent on reactant concentrations. Intermolecular reactions, on the other hand, have higher-order rate laws that are dependent on reactant concentrations. Sometimes uncomplexed ligand is added to a sample to measure any effect on the observed rate of exchange in a process involving that ligand.

Once a set of variable temperature spectra like that in figure 7.1 is recorded, the rate of exchange at each temperature can be calculated accurately by using a rather complicated equation or approximately by using a set of approximate relationships such as equations 7.5–7.7. These equations are applicable to a degenerate rearrangement between two sites as shown in equation 7.1.

Slow-exchange region (for each resonance):

$$1/\tau = k_1 = \pi \Delta_e \tag{7.5}$$

At the coalescence temperature, $T_c$:

$$k_{coal.} = 2.22 \, \Delta\delta \tag{7.6}$$

Fast-exchange region (for the average resonance):

$$1/\tau = k_1 = \frac{\pi(\Delta\delta)^2}{2\Delta_e} \tag{7.7}$$

In these equations $\Delta_e$ is the *observed* line width at half-height for an NMR resonance at a given temperature minus the natural line width of that resonance as recorded in the slow- or fast-exchange limiting spectra. Therefore, $\Delta_e$ is a measure of the amount of line broadening (in hertz) of a resonance caused by an exchange process.

Once the exchange rates are known at several temperatures, the value of $E_a$ can be calculated from an Arrhenius plot, or $\Delta G^\ddagger$, the free energy of activation, can be calculated from the Eyring equation. Equation 7.8 is derived from the Eyring equation.

$$\log(k_1/T) = 10.32 - (\Delta H^\ddagger/4.57T) + (\Delta S^\ddagger/4.57) \tag{7.8}$$

A plot of $\log(k_1/T)$ versus $1/T$ gives $\Delta H^\ddagger$ and $\Delta S^\ddagger$. If only the coalescence temperature, $T_c$, is known, then equation 7.9 permits calculation of $\Delta G^\ddagger$ for an exchange process *at this temperature*.

$$\Delta G^\ddagger = 0.00457T_c(9.97 + \log T_c/\Delta\delta) \text{ kcal/mole} \tag{7.9}$$

An example of a *nondegenerate intramolecular rearrangement* is shown in equation 7.10.

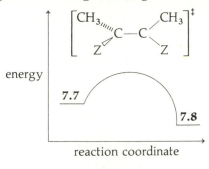

$$(7.10)$$

**7.7**                                    **7.8**

Interconversion of **7.7** and **7.8** represents a geometrical isomerization about the C—C double bond. Since the two isomers have different free energies, the energy profile for this nondegenerate rearrangement might be as shown in **7.9**.

**7.9**

If the interconversion shown in equation 7.10 occurred at a rate rapid enough to produce a set of variable temperature NMR spectra like those of figure 7.1, then the values of the rate of exchange, $E_a$ or $\Delta H^{\ddagger}$ and $\Delta S^{\ddagger}$ could be determined by a procedure similar to that described above for a degenerate exchange between two sites. However, since **7.7** and **7.8** have different free energies, the relative amounts of each isomer are not equal and vary with temperature. To account for this change in equilibrium constant with temperature, the above mathematical equations need to be modified slightly. The most obvious experimental effect arising from nondegenerate exchange processes is that the chemical shift of the exchange-averaged NMR resonance at the fast-exchange limit occurs at a weighted-average value, closer in chemical shift to the resonance of the more abundant isomer at that temperature.

## DYNAMIC INTRAMOLECULAR REARRANGEMENT
## OF $[(\eta^5\text{-}C_5H_5)Mo(CO)_3]_2$

Figure 7.2 shows several proton NMR spectra of $[(\eta^5\text{-}C_5H_5)Mo(CO)_3]_2$ in acetone solution at various temperatures. Notice that the spectrum at 30°C shows only a broad resonance for the two cyclopentadienyl ligands. This effect is *not* the result of paramagnetic impurities; rather, this temperature is close to the coalescence temperature for a dynamic intramolecular rearrangement. Heating the sample solution to 62°C produces a sharp singlet defining the fast-exchange limit. Cooling the sample to −15°C generates a spectrum showing two sharp singlets that define the slow-exchange limit. Since these two singlets have unequal intensities (which also vary with temperature and solvent polarity), a nondegenerate exchange process must be occurring.

These spectra are interpreted as reflecting an interconversion of **7.10** and **7.11** by rotation about the Mo—Mo single bond.

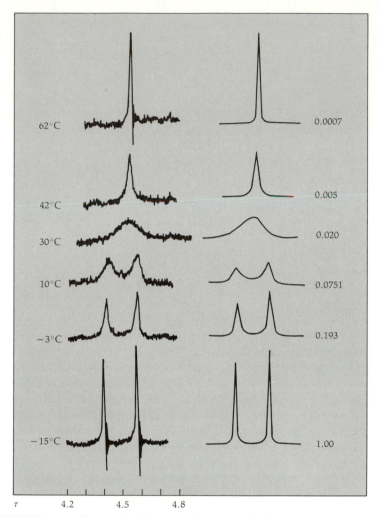

**FIGURE 7.2** Observed and computed proton NMR spectra (100 MHz) of $[(\eta^5\text{-}C_5H_5)Mo(CO)_3]_2$ in acetone solution at various temperatures. *Adapted from R. D. Adams and F. A. Cotton, Inorg. Chim. Acta,* **7,** *153 (1973).*

$$\text{7.10} \quad\quad\quad\quad\quad\quad \rightleftharpoons \quad\quad\quad\quad\quad\quad \text{7.11} \tag{7.11}$$

**7.10**                                        **7.11**

*gauche* isomer                              *anti* isomer

The *anti* isomer is the most stable rotamer in the solid state (see **4.17**) and in non-polar solvents. The $\eta^5$-$C_5H_5$ ligands within *each* rotamer are equivalent by symmetry and appear in the proton NMR spectrum as a singlet. However, the chemical shifts of these singlets are different because **7.10** and **7.11** are different compounds having $\eta^5$-$C_5H_5$ ligands in different chemical environments. Changing the solvent polarity determines that the resonance at highest field can be assigned to the *anti* isomer. As solvent polarity is decreased, the isomer having the lower dipole moment, **7.11** in this case, increases in relative abundance because of more favorable solvent stabilization. Conversely, an increase in solvent polarity increases the relative abundance of the isomer having a higher dipole moment.

As the sample temperature is increased from $-15°C$, the rate of rotation about the Mo—Mo single bond increases and causes rapid exchange between *gauche*- and *anti*-$C_5H_5$ environments relative to the $\Delta\delta$ between these resonances. Resonance broadening, coalescence, and the appearance of a single resonance at the fast-exchange limit are observed. The Arrhenius parameters for this process are $E_a = 15.3 \pm 1.0$ kcal/mole and $\log A = 13.0 \pm 1$.

## DYNAMIC INTRAMOLECULAR REARRANGEMENT OF $(\eta^5$-$C_5H_5)(\eta^1$-$C_5H_5)Fe(CO)_2$

Figure 7.3 shows proton NMR spectra of $(\eta^5$-$C_5H_5)(\eta^1$-$C_5H_5)Fe(CO)_2$ (**7.12**) in $CS_2$ solution at various temperatures.

**7.12**

The spectrum at $-80°C$ shows complex AA′ and BB′ multiplets at about 6.3 $\delta$ and 6.0 $\delta$, respectively, and a singlet at about 3.5 $\delta$. The relative intensities of these resonances are 2:2:1, respectively, and a singlet at 4.4 $\delta$ has a relative intensity of 5. This spectrum is consistent with structure **7.12** as determined by x-ray diffraction. The singlet at 4.4 $\delta$ results from the $\eta^5$-$C_5H_5$ ligand, and this resonance remains unchanged as the temperature increases. The remaining resonances are attributable to the $\eta^1$-$C_5H_5$ ligand. The complex multiplet at lowest field can be assigned to the A protons, the multiplet near 6.0 $\delta$ to the B protons, and the resonance at 3.5 $\delta$ of unit intensity is due to the X proton.

As the sample is warmed, the $\eta^1$-$C_5H_5$ resonances broaden, coalesce, and then appear as a slightly broad singlet near 5.7 $\delta$ at $+30°C$. Clearly, some dynamic process is permuting the A, B and X proton environments. The experimental spectra

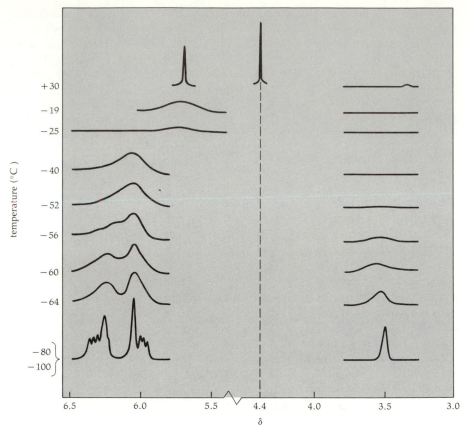

**FIGURE 7.3**    Proton NMR spectra (60 MHz) of $(\eta^5\text{-}C_5H_5)(\eta^1\text{-}C_5H_5)Fe(CO)_2$ in $CS_2$ solution at various temperatures. *Adapted from Bennett et al., J. Am. Chem. Soc., **88**, 4371 (1966).*

also reveal that the A protons broaden at a rate nearly *two times as fast* as the B protons. The Arrhenius parameters for this process are $E_a = 10.7 \pm 0.5$ kcal/mole and $\log A = 12.6 \pm 0.5$, where $A$ is the preexponential coefficient of the Arrhenius equation.

The relatively low value of $E_a$ eliminates a dissociative mechanism for the dynamic process because cleavage of the Fe—$(\eta^1\text{-}C_5H_5)$ bond would require considerably more energy than 11 kcal/mole. A mechanism involving a momentary conversion of the $\eta^1\text{-}C_5H_5$ ligand to an $\eta^5\text{-}C_5H_5$ ligand with rapid subsequent formation of an $\eta^1\text{-}C_5H_5$ ligand (in which the Fe atom is bonded to a different carbon atom of the ring) would afford a *symmetrical* broadening of the A- and B-type proton resonances. Therefore, this mechanism is not consistent with the observed asymmetrical collapse of the A and B multiplets.

The only reasonable mechanisms remaining are shown in equation 7.12. In going from a reference ground-state structure, **7.13**, to **7.14**, the Fe atom executes a *1,2-shift* relative to the $\eta^1\text{-}C_5H_5$ ligand; in going from **7.13** to **7.15**, the Fe atom executes a *1,3-shift*. All three structures are degenerate and equivalent to the solid-state structure **7.12**. Rapid rearrangement by either mechanism would permute the different hydrogen environments and lead to the observed spectral averaging of these resonances to a single resonance. However, each mechanism results in a different type of *asymmetrical collapse* of the resonances of the A- and B-type of protons.

**7.14**                **7.13**                **7.15**

$$
\begin{matrix} C1 \\ C2 \\ C3 \\ C4 \\ C5 \end{matrix}
\begin{bmatrix} A \\ X \\ A \\ B \\ B \end{bmatrix}
\xleftarrow{\text{1,2-shift}}
\begin{matrix} C1 \\ C2 \\ C3 \\ C4 \\ C5 \end{matrix}
\begin{bmatrix} X \\ A \\ B \\ B \\ A \end{bmatrix}
\xrightarrow{\text{1,3-shift}}
\begin{matrix} C1 \\ C2 \\ C3 \\ C4 \\ C5 \end{matrix}
\begin{bmatrix} B \\ A \\ X \\ A \\ B \end{bmatrix}
\qquad (7.12)
$$

Inspection of the column matrix associated with each structure reveals how these two permutation mechanisms differ. In going from **7.13** to **7.14**, *both* A proton environments permute to non-A environments, while *only one* of the B-type proton environments permutes to a non-B environment. Therefore, in a 1,2-shift the A-type protons would be permuting with non-A environments twice as fast as the B-type protons. Conversely, in going from **7.13** to **7.15**, the B protons are being permuted twice as fast as the A protons during a 1,3-shift. Only the 1,2-shift mechanism is consistent with the asymmetrical collapse observed experimentally. Molecules that exhibit *degenerate, intramolecular stereochemical nonrigidity* are frequently referred to as *fluxional molecules*. Complex **7.12** is one such complex.

## DYNAMIC INTRAMOLECULAR REARRANGEMENT OF $Rh_4(CO)_{12}$

Figure 7.4 shows $^{13}$C-NMR spectra of $Rh_4(CO)_{12}$ (**7.16**) in $CD_2Cl_2$ and $CDCl_3$ solutions at various temperatures. The spectrum recorded at $-65°C$ is consistent with the solid-state molecular structure shown in table 2.4. This spectrum represents the slow-exchange limit, and peak assignments are shown in **7.16**. Note that $^{13}$C and $^{103}$Rh nuclei have nuclear spins of $\frac{1}{2}$ just as the proton has. The three equivalent $\mu_2$-CO ligands, A, couple to *two* rhodium nuclei affording a triplet at about 229 $\delta$. The three terminal CO ligands, B, on the apical Rh atom appear as a doublet at about 183 $\delta$, and the two sets of terminal CO ligands on the basal rhodium atoms, C and D, appear as doublets near 182 and 176 $\delta$. The relative intensities of the multiplets are roughly 3:3:3:3 as predicted from structure **7.16**.

As the sample solution is warmed, the $^{13}$C-NMR resonances broaden at about the same rate with coalescence of the peaks at $-5 \pm 5°C$. The fast-exchange limit is reached at approximately $63°C$. This spectrum shows a single quintet resonance

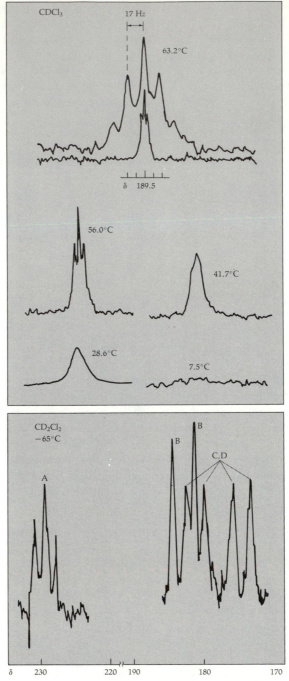

**FIGURE 7.4** $^{13}C$—NMR spectra of $Rh_4(CO)_{12}$ in $CD_2Cl_2$ and $CDCl_3$ solutions at various temperatures. Chemical shifts are in $\delta$ relative to TMS. The spectrum at $+63.2°C$ is enlarged to show the quintet structure more clearly. *Adapted from F. A. Cotton, et al., J. Am. Chem. Soc., **94**, 6191 (1972), and L. M. Jackman and F. A. Cotton, eds., Dynamic Nuclear Magnetic Resonance Spectroscopy (New York: Academic Press, 1975), p. 520.*

at near $189\,\delta$ with a $^{13}$C-Rh coupling constant of 17.1 Hz. This chemical shift is near the average chemical shift of resonances A through D, and the $^{13}$C-Rh coupling constant of about 17 Hz is the value expected if each CO ligand were permuted equally over all four rhodium nuclei.

The permutation mechanism is shown in equation 7.13. Each ground-state molecule converts to the *all-terminal isomer*, which has a slightly higher energy, and then collapses back to a structure equivalent, but *not* identical, to **7.16** in which the CO ligands have been permuted. Rapid permutation would exchange each $^{13}$C environment at the same rate. Each CO ligand would spend a certain amount of time as a terminal or $\mu_2$-CO ligand and would be coordinated for some time period to each rhodium atom. The observation of $^{13}$C-Rh coupling in the fast-exchange limiting spectrum confirms an intramolecular exchange in which dissociation of Rh—CO bonds has not occurred. Such bond dissociation would abolish $^{13}$C-Rh coupling.

Other examples of dynamic intramolecular rearrangements are provided as study questions.

## STUDY QUESTIONS

1. An organometallic complex of the type $(\eta^5\text{-}C_5H_5)M(CO)L_n$ exists in solution as two interconverting isomers. In the proton NMR spectrum, the two $C_5H_5$ singlets are separated by 25 Hz; in the $^{13}$C-NMR spectrum, the two $C_5H_5$ $^{13}$C-singlets are separated by 60 Hz; and in the IR spectrum, the two $\nu_{CO}$ bands are separated by 15 cm$^{-1}$, which is equivalent to $4.5 \times 10^{11}$ Hz.
   (a) Use equations 7.2 and 7.4 to calculate the slowest rate of interconversion necessary to cause spectral-peak averaging for *each* of the three spectroscopic methods.
   (b) How would the temperature ranges used in proton and $^{13}$C-NMR studies of this dynamic process differ? (Refer to eq. 7.6.)
   (c) What experimental effect would result from undertaking this study on a PMR spectrometer operating at three times the magnetic field strength of the one assumed in the original statement of the problem?
   (d) Why does x-ray crystallography show the "instantaneous" ground-state structure even when dynamic processes may be occurring in the solid state?

2. The complex $(\eta^5\text{-}C_5H_5)(\eta^1\text{-}C_5H_5)Fe(CO)(PF_2NMe_2)$ exhibits a "ring-whizzing" fluxionality like complex **7.12**. However, the $^{19}F$-NMR spectra reveal *nonequivalent* F atoms even at the fast-exchange limit. What does this observation indicate about retention or inversion of configuration at the Fe atom during this fluxional process?

3. Proton NMR spectra of $(\eta^1\text{-}C_5H_5)_2(\eta^5\text{-}C_5H_5)_2Ti$ in toluene solution from $-27°C$ to $62°C$ are shown in figure 7.5. Analyze the spectrum at $-27°C$, and interpret the spectral changes observed at higher temperatures in terms of a dynamic intramolecular rearrangement.

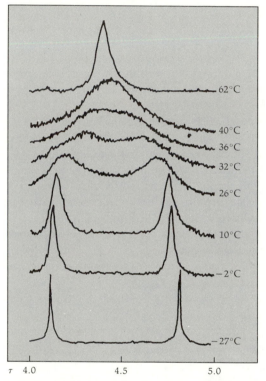

**FIGURE 7.5** Proton NMR spectra for question 3. *Adapted from L. M. Jackman and F. A. Cotton, eds.,* Dynamic Nuclear Magnetic Resonance Spectroscopy, *(New York: Academic Press, 1975), p. 420.*

4. Proton NMR spectra of $(\eta^2\text{-tetramethylallene})Fe(CO)_4$ (eq. 5.71) in $CS_2$ solution at $-60°C$ and $30°C$ are shown in figure 7.6. Rationalize the spectral changes observed in terms of a dynamic intramolecular process. For this process $E_a$ is $9.0 \pm 2.0$ kcal/mole.

5. Proton NMR spectra at 100 MHz of $(\eta^3\text{-2-methylallyl})$ $RhL_2Cl_2$, where L is $Ph_3As$, are shown in figure 7.7 for several temperatures. The resonance at about 2 ppm is due to the 2-methyl group. Resonances near 4 ppm are due to the *syn* and *anti* protons on the outer carbon atoms of the $\eta^3$-allylic ligand. Rationalize these spectral changes in terms of a dynamic intramolecular process.

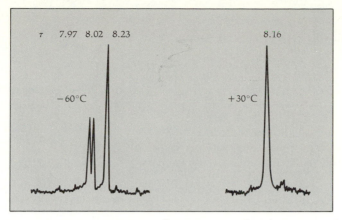

**FIGURE 7.6** Proton NMR spectra for question 4. *Adapted from R. Ben-Soshan and R. Pettit,* J. Am. Chem. Soc., **89,** *2231 (1967).*

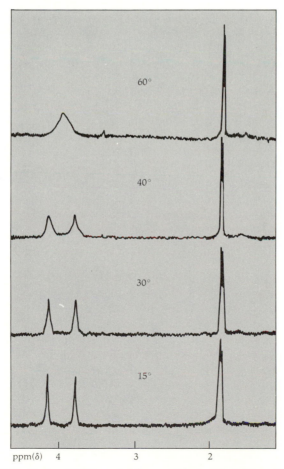

**FIGURE 7.7** Proton NMR spectra for question 5. *Adapted from K. Vrieze and H. C. Volger,* J. Organometal. Chem., **9,** *537 (1967).*

## SUGGESTED READING

Cotton, F. A. *Acc. Chem. Res.* **1**, 257 (1968).

Cotton, F. A. *J. Organometal. Chem.* **100**, 29 (1975).

Gunther, H. *NMR Spectroscopy—An Introduction*, Chapter 8. New York: John Wiley, 1980.

Jackman, L. M., and Cotton, F. A. eds., *Dynamic Nuclear Magnetic Resonance Spectroscopy.*
    New York: Academic Press, 1975.

Muetterties, E. L. *Acc. Chem. Res.* **3**, 266 (1970).

Shapley, J. R., and Osborn, J. A. *Acc. Chem. Res.* **6**, 305 (1973).

Vrieze, K., and Van Leeuwen, P. W. N. M. *Progr. Inorg. Chem.* **14**, 1 (1971).

# Complexes Containing Hydride, Alkyl, Acyl, or Related Ligands

## CHAPTER 8

Chemical reduction of metal carbonyls frequently affords anionic complexes that are relatively strong Brønsted bases or good nucleophiles. Protonation of these anionic complexes is a common preparative route to *metal-hydride complexes*, and alkylation or acylation of these anions gives *metal-alkyl* or *metal-acyl* complexes. Chemical reduction of coordinated molecules (for example, by external nucleophilic attack) also occurs, but these reactions are discussed in chap. 11.

## ANIONIC METAL CARBONYL COMPLEXES

Like carbanions, anionic organometallic complexes exhibit extensive chemical reactivity. A common method of preparing anionic complexes is by reducing neutral complexes with electropositive metals. Similar reactions with boron hydride reagents or hydroxide ion also yield anionic complexes.

Most neutral dimeric complexes, such as $[(\eta^5\text{-}C_5H_5)M(CO)_3]_2$ (where M is Cr, Mo or W), $[(\eta^5\text{-}C_5H_5)M(CO)_2]_2$ (where M is Fe, Ru or Os), $[(\eta^5\text{-}C_5H_5)Ni(CO)]_2$, $M_2(CO)_{10}$ (where M is Mn, Tc or Re) or $Co_2(CO)_8$, undergo *reductive-cleavage* of the M—M single bond to give the corresponding monoanionic monomers when they are stirred over sodium or potassium amalgam (eq. *4.46*). A frequent, but usually minor, complication with such alkali metal/amalgam reductions is formation of mercury organometallic compounds:

$$Mn_2(CO)_{10} \xrightarrow[\text{THF}]{\text{Na/Hg}} Na[Mn(CO)_5] + Hg[Mn(CO)_5]_2 \qquad (8.1)$$
$$\sim 89\% \qquad \quad \textbf{8.1}, 11\%$$

Complex **8.1** contains two covalent Hg—Mn bonds and is representative of *heteronuclear complexes*, those that contain metal-metal bonds between at least two different metals. Use of lithium or potassium trialkylborohydrides as reducing agents eliminates this difficulty as shown in equation *8.2*.

$$\left.\begin{array}{l} [(\eta^5\text{-}C_5H_5)Mo(CO)_3]_2 \\ [(\eta^5\text{-}C_5H_5)Fe(CO)_2]_2 \\ Mn_2(CO)_{10} \\ Co_2(CO)_8 \end{array}\right\} \xrightarrow[\text{THF or HMPA}]{M[BR_3H]} \left\{\begin{array}{l} 2[(\eta^5\text{-}C_5H_5)Mo(CO)_3]^- \\ 2[(\eta^5\text{-}C_5H_5)Fe(CO)_2]^- \\ 2Mn(CO)_5^- \\ 2Co(CO)_4^- \end{array}\right. \qquad (8.2)$$

nearly quantitative yields

$M = Li^+$ or $K^+$; $R = Et$ or $sec$-$C_4H_9$;
$HMPA = $ hexamethylphosphoramide, $(Me_2N)_3PO$

Metal-halogen bonds can be cleaved reductively (eq. 8.3). These reductions are usually less facile than reduction of metal-metal single bonds. The reduction appears to proceed in higher yield when one of the reduction products is a good halide acceptor, such as Zn(II) ion.

$$2(\eta^3\text{-}C_3H_5)Fe(CO)_3Br \xrightarrow[\text{Et}_2O,\ 2hrs]{Zn^0\ dust} 2[(\eta^3\text{-}C_3H_5)Fe(CO)_3]^- + Zn^{2+} + ZnBr_2 \quad (8.3)$$

Reduction of metal carbonyl or substituted metal carbonyl complexes, particularly mononuclear complexes, affords anionic complexes in which electrons provided by the reducing agent have displaced one or more ligands. Interesting features of these preparations include:

**1.** Displacement of one CO ligand by two electrons forming dianionic complexes (eq. 8.4 and 8.5).

$$(\eta^5\text{-}C_5H_5)V(CO)_4 \xrightarrow[\text{THF}]{2Na} Na_2[(\eta^5\text{-}C_5H_5)V(CO)_3] + CO \qquad (8.4)$$
$$\textbf{8.2}$$

$$M(CO)_6 + 2Na \xrightarrow[\text{liq. NH}_3]{} Na_2[M(CO)_5] + CO \qquad (8.5)$$
$$\textbf{8.3}$$

$M = $ Cr, Mo or W

**2.** Use of electron transfer agents such as sodium napthalenide in equation 8.6 and the benzophenone ketyl radical anion in equations 8.7 and 8.8.

$$M(CO)_5(NMe_3) + 2NaNp \xrightarrow[-78°C]{THF} Na_2[M(CO)_5] + NMe_3 + 2Np \quad (8.6)$$

$M = $ Mo or W

$$Fe(CO)_5 + 2Na[Ph_2CO] \xrightarrow{THF} Na_2[Fe(CO)_4] + CO + 2Ph_2CO \quad (8.7)$$
$$\textbf{8.4},\ 95\%$$

$$4Ru_3(CO)_{12} + 6K[Ph_2CO] \xrightarrow{THF} 3K_2[Ru_4(CO)_{13}] + 9CO + 6Ph_2CO$$
$$\textbf{8.5},\ 80\%$$

$$\xrightarrow{THF,\ 50°C,\ 12K[Ph_2CO]} 3K_4[Ru_4(CO)_{12}] + 12CO + 12Ph_2CO \quad (8.8)$$
$$\textbf{8.6},\ 85\%$$

**3.** Use of strong $\sigma$-donor amine ligands, such as NMe$_3$ in equation 8.6 and tetramethylethylenediamine (TMEDA) in equation 8.9, as ligands that readily dissociate from the metal atom when the negative charge on the metal increases.

$$\text{(TMEDA)M(CO)}_4 \xrightarrow[\text{liq. NH}_3]{\text{Na}^0} \text{Na}_4[\text{M(CO)}_4] + \text{TMEDA} \qquad (8.9)$$
$$\textbf{8.7, } 95\%$$

M = Cr, Mo or W

**4.** Application of photolytic activation to facilitate loss of CO (eq. *8.10*).

$$2\text{M(CO)}_6 + 2\text{Na/Hg} \xrightarrow[\text{(2) [Et}_4\text{N]Br}]{\text{(1) THF, } h\nu,\ 25°\text{C}} [\text{Et}_4\text{N]}_2[\text{M}_2\text{(CO)}_{10}] + 2\text{CO} \qquad (8.10)$$
$$\textbf{8.8, } 35{-}50\%$$

M = Cr, Mo or W

**5.** Generation of metal carbonyl anions by using base, as shown in equation *8.11*.

$$2\text{Fe(CO)}_5 \xrightarrow[\text{NH}_2\text{OH} \cdot \text{HCl}]{\text{NH}_3 \cdot \text{H}_2\text{O}} [\text{Fe(NH}_3)_6][\text{Fe}_2\text{(CO)}_8] + 2\text{CO} \qquad (8.11)$$
$$\textbf{8.9}$$

**6.** Formation of larger clusters during a reduction process (eqs. *8.8* and *8.10–8.12*).

$$4\text{Ni(CO)}_4 + 2\text{Na/Hg} \xrightarrow{\text{THF}} \text{Na}_2[\text{Ni}_4\text{(CO)}_9] + 7\text{CO} \qquad (8.12)$$
$$\textbf{8.10}$$

Such *cluster expansion* by treating neutral metal carbonyl complexes with anionic metal carbonyl complexes is a common method of preparing clusters of higher nuclearity. Mixed-metal clusters can be formed by this method, also.

**7.** Preparation of highly reduced anionic complexes, such as $\text{V(CO)}_5{}^{3-}$, $\text{Mn(CO)}_4{}^{3-}$, $\text{Co(CO)}_3{}^{3-}$, or $\text{Cr(CO)}_4{}^{4-}$, as shown in equations *8.9* and *8.13–8.15*.

$$[\text{Na(diglyme)}_2][\text{V(CO)}_6] \xrightarrow[\text{liq. NH}_3]{\text{M}^0} \text{M}_3[\text{V(CO)}_5] + \text{CO} \qquad (8.13)$$
$$\textbf{8.11, } 80{-}94\%$$

$\text{M}^0$ = K, Rb or Cs

$$\text{Na[M(CO)}_5] \xrightarrow[\text{HMPA}]{\text{Na}^0} \text{Na}_3[\text{M(CO)}_4] + \text{CO} \qquad (8.14)$$
$$\textbf{8.12}$$

M = Mn or Re

$$\text{Na[M(CO)}_4] \xrightarrow[\text{liq. NH}_3]{\text{Na}^0} \text{Na}_3[\text{M(CO)}_3] + \text{CO} \qquad (8.15)$$
$$\textbf{8.13, } 20{-}90\%$$

M = Co, Rh or Ir

Historically, the use of base to generate anionic metal carbonyl complexes was a common synthetic method. This procedure is still very useful although it is used more frequently to prepare metal carbonyl hydride complexes. Anion formation presumably occurs as shown in equation *8.16*.

$$(\text{OC})_n\text{M}{=}\text{C}{=}\text{O} + \text{OH}^- \longrightarrow (\text{OC})_n\overset{-}{\text{M}}-\overset{\overset{\displaystyle O}{\|}}{\text{C}}-\text{OH} \xrightarrow{-\text{CO}_2} (\text{OC})_n\overset{-}{\text{M}}-\text{H} \qquad (8.16)$$

$$\text{OH}^- \Big\downarrow -\text{H}_2\text{O} \qquad\qquad \text{OH}^- \Big\downarrow -\text{H}_2\text{O}$$

$$(\text{OC})_n\overset{-}{\text{M}}-\overset{\overset{\displaystyle O}{\|}}{\text{C}}-\text{O}^- \xrightarrow{-\text{CO}_2} \text{M(CO)}_n{}^{2-}$$

Nucleophilic attack at a CO ligand by a hydroxide ion generates a metallacarboxylic acid. Subsequent loss of a proton and $CO_2$ affords the metal carbonyl dianion. When this dianion is a strong base, it can acquire a proton from the protic solvent to give a metal-hydride complex.

For example, the dianion $Fe_3(CO)_{11}{}^{2-}$ can be prepared either from $Fe_3(CO)_{12}$ in $2\ M$ KOH/MeOH solution or by deprotonation of $HFe_3(CO)_{11}{}^-$ by a $1.2\ N$ KOH solution. The structure of this dianion (**8.14**) reveals a symmetrical $\mu_2$-CO ligand and an asymmetrical $\mu_3$-CO ligand.

$$\bar{a} = 2.593(2)\ \text{Å}$$
$$b = 2.603(3)\ \text{Å}$$
$$\bar{c} = 2.21(2)\ \text{Å}$$

**8.14**

$$[Fe_3(CO)_9(\mu_2\text{-CO})(\mu_3\text{-CO})]^{2-}$$

In contrast to the structures of most compounds containing representative elements, a lone electron pair in the valence shell of a transition metal has little stereochemical significance and therefore does not occupy a coordination site. Thus, anions of the type $M(CO)_4{}^{n-}$ or $M(CO)_5{}^{n-}$ have the expected tetrahedral or trigonal bipyramidal geometry, respectively. The dinuclear anions **8.8**, where M is Cr or Mo, have structures like that of $Mn_2(CO)_{10}$, where the Cr—Cr and Mo—Mo distances are 2.97(1) Å and 3.123(7) Å, respectively. In **8.9**, the $Fe_2(CO)_8{}^{2-}$ ion has a structure similar to that of the all-terminal isomer of $Co_2(CO)_8$ with an Fe—Fe distance of 2.75 Å. Notice that the mono- and dinuclear anions shown in equations 8.4–8.15 obey the EAN rule.

All anionic metal carbonyl complexes have $\nu_{CO}$ absorptions at lower frequency than those of their neutral precursors. Very highly reduced complexes, such as $Co(CO)_3{}^{3-}$ or $Cr(CO)_4{}^{4-}$, have $\nu_{CO}$ bands at 1744 and 1600 $cm^{-1}$ or 1657 and 1462 $cm^{-1}$, respectively. Terminal $\nu_{CO}$ bands at such low frequencies are consistent with high negative charge densities at the metal atoms.

## ORGANOMETALLIC COMPLEXES CONTAINING HYDRIDE LIGANDS

Many organometallic anions are moderately strong Lewis bases. Deliberate, or solvent-induced, protonation of these anions affords a facile method of preparing *metal hydride* complexes, complexes that contain an M—H bond (eqs. 8.17–8.23). Protic solvents or stronger acids are used as proton sources.

$$Na_2[(\eta^5\text{-}C_5H_5)V(CO)_3] \xrightarrow[\text{(2) H}_2\text{O}]{\text{(1) [PPN]Cl}} PPN[(\eta^5\text{-}C_5H_5)V(CO)_3H] \qquad (8.17)$$

$$\mathbf{8.2} \qquad\qquad\qquad\qquad\qquad \mathbf{8.15}$$

$$\text{yellow solid}$$

$$PPN = (Ph_3P)_2N^+$$

$$\text{Na}_2[\text{Cr(CO)}_5] \xrightarrow[\text{(2) [Et}_4\text{N]Br}]{\text{(1) MeOH}} [\text{Et}_4\text{N}][\text{HCr(CO)}_5] \qquad (8.18)$$
$$\textbf{8.16}$$

$$\text{Na}[(\eta^5\text{-C}_5\text{H}_5)\text{M(CO)}_3] + \text{HOAc} \xrightarrow{\text{THF}} (\eta^5\text{-C}_5\text{H}_5)\text{M(CO)}_3\text{H} + \text{NaOAc} \quad (8.19)$$
$$\textbf{8.17, 90\%}$$
$$\text{M = Mo or W}$$

$$\text{Na}[\text{Mn(CO)}_5] \xrightarrow{\text{H}_3\text{PO}_4} \text{HMn(CO)}_5 + \text{Na}[\text{H}_2\text{PO}_4] \qquad (8.20)$$
$$\textbf{8.18, 83\%}$$

$$\text{Na}_2[\text{Fe(CO)}_4] \xrightarrow{\text{H}_2\text{O}} \text{Na}[\text{HFe(CO)}_4] \xrightarrow{\text{H}^+} \text{H}_2\text{Fe(CO)}_4 \qquad (8.21)$$
$$\textbf{8.19} \qquad\qquad \textbf{8.20}$$

$$\text{Na}_2[\text{M(CO)}_4] \xrightarrow[\text{[PPN]Cl}]{\text{MeOH}} \text{PPN}[\text{HM(CO)}_4] \qquad (8.22)$$
$$\textbf{8.21, 60\%}$$
$$\text{M = Ru or Os}$$

$$\text{Na}[\text{Co(CO)}_4] \xrightarrow[\text{THF}]{\text{H}_2\text{SO}_4} \text{HCo(CO)}_4 + \text{Na}[\text{HSO}_4] \qquad (8.23)$$
$$\textbf{8.22}$$

Hydroxide or borohydride reactions with metal carbonyls afford anionic complexes including metal hydrides, as shown in equations 8.24–8.29. Protonation of these anionic intermediates is a common route to neutral metal carbonyl hydride complexes.

$$\text{Cr(CO)}_6 + 3\text{KOH} \longrightarrow \text{K}[\text{HCr(CO)}_5] + \text{K}_2\text{CO}_3 + \text{H}_2\text{O} \qquad (8.24)$$
$$\textbf{8.23}$$

$$\text{M(CO)}_6 + \text{Na}[\text{BH}_4] \xrightarrow{\text{THF}} \text{Na}[\text{M}_2(\text{CO})_{10}(\mu_2\text{-H})] \qquad (8.25)$$
$$\textbf{8.24}$$
$$\text{M = Cr, Mo or W}$$

$$\text{Mn}_2(\text{CO})_{10} \xrightarrow[\text{(2) H}_3\text{PO}_4]{\text{(1) KOH}} \text{H}_3\text{Mn}_3(\text{CO})_{12} \qquad (8.26)$$
$$\textbf{8.25, 70\%}$$

$$\text{Re}_2(\text{CO})_{10} \xrightarrow[\text{(2) H}_3\text{PO}_4, \, \text{C}_6\text{H}_{12}]{\text{(1) NaBH}_4, \, \text{THF}} \text{H}_3\text{Re}_3(\text{CO})_{12} \qquad (8.27)$$
$$\textbf{8.26, 50\%}$$

$$\text{Fe(CO)}_5 + 3\text{NaOH} \xrightarrow{\text{H}_2\text{O}} \text{Na}[\text{HFe(CO)}_4] + \text{Na}_2\text{CO}_3 + \text{H}_2\text{O} \quad (8.28)$$
$$\textbf{8.19}$$

$$\text{Ru}_3(\text{CO})_{12} + \text{NaBH}_4 \xrightarrow{\text{THF}} \text{Na}[\text{Ru}_3(\text{CO})_{11}\text{H}] \qquad (8.29)$$

Metal carbonyl hydrides can be used to form other carbonyl hydride complexes (eqs. 8.30 and 8.31).

$$PPN[(\eta^5\text{-}C_5H_5)V(CO)_3H] \xrightarrow{Fe(CO)_5} (\eta^5\text{-}C_5H_5)V(CO)_4 + PPN[HFe(CO)_4]$$

**8.15**                                 47%

$$\xrightarrow{Cr(CO)_6} (\eta^5\text{-}C_5H_5)V(CO)_4 + PPN[HCr(CO)_5] \quad (8.30)$$
$$+ PPN[Cr_2(CO)_{10}(\mu_2\text{-}H)]$$

$$(\eta^5\text{-}C_5H_5)Nb(CO)_4 + (\eta^5\text{-}C_5H_5)_2NbH_3 \xrightarrow[-H_2]{THF,\ h\nu} (\eta^5\text{-}C_5H_5)_3Nb_2(CO)_4(\mu_2\text{-}H) \quad (8.31)$$

**8.27**

The reactions in equation *8.30* involve CO—H exchange; equation *8.31* involves photolytic elimination of $H_2$ and a subsequent redistribution of CO ligands between the two niobium atoms.

Metal-hydride complexes can be prepared by several other routes, including oxidative-addition, reductive-elimination, and reductive-addition reactions. Specific examples of these reactions are presented in chapters 9 and 10.

Structurally, hydride ligands occupy a distinct coordination site on a metal atom. For example $HMn(CO)_5$ and $HCr(CO)_5^-$ have quasi-octahedral coordination geometries about the metal atoms. The dinuclear hydride complexes **8.24** contain M—H—M bonds that are not accompanied by direct M—M bonds or other $\mu_2$-bridging ligands, as shown in **8.28**.

**8.28**

$$M_2(CO)_{10}(\mu_2\text{-}H)$$

M = Cr, Mo or W

Considerable structural variation is observed depending on the cation present in the crystalline lattice. When M is molybdenum and the cation is $Et_4N^+$, the Mo—H—Mo bond is linear, $\theta = 180°$, and the carbonyl ligands on the Mo atoms are eclipsed. However, if the cation is $PPN^+$, then $\theta$ is only 136(3)°, and the two sets of CO ligands are staggered relative to one another. Clearly, the value of $\theta$ depends strongly on differences in interionic crystal packing forces. The M—H—M bond in **8.28** is considered to be a *three-center two-electron bond* arising from the formal protonation of the metal-metal bond of the corresponding $M_2(CO)_{10}^{2-}$ dianions (**8.8**). Considerable structural flexibility in this type of bond is likely.

Complex **8.27** has a "bent" $\mu_2$-Nb—H—Nb bond, as shown in **8.29**. The Nb $\cdots$ Nb distance is 3.738(3) Å. During the reaction, one CO ligand has transferred to the $(\eta^5\text{-}C_5H_5)_2Nb$ fragment from $(\eta^5\text{-}C_5H_5)Nb(CO)_4$.

Other types of bridging hydride structures are shown in **8.30** and **8.31**. Frequently, $\mu_2$-hydride ligands bridge M—M bonds, as in **8.30**.

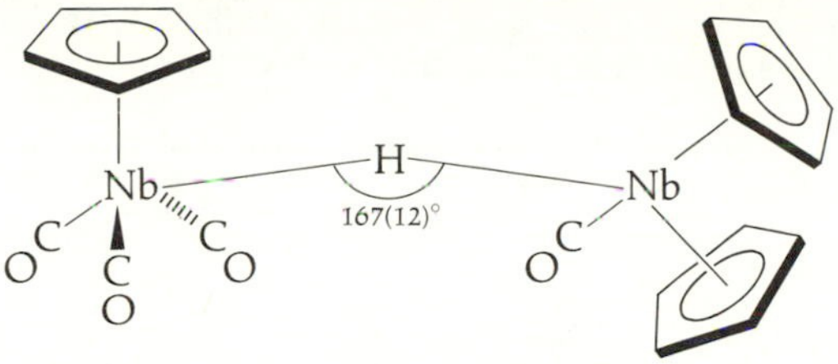

**8.29**

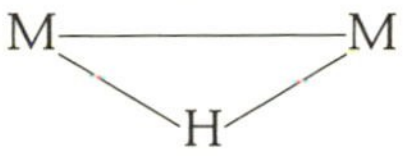

| **8.30** | **8.31** |
| --- | --- |
| $\mu_2$-hydride ligand | $\mu_3$-hydride ligand |

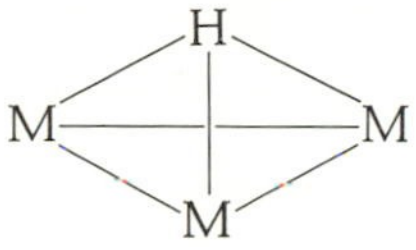

The complex, $H_3Mn_3(CO)_{12}$ (**8.25**) has three such $\mu_2$-hydride ligands, one $\mu_2$-H ligand bridging each edge of a triangular $Mn_3$ core. Each manganese atom has four terminal CO ligands, as well. The Mn—Mn distance of 3.111(2) Å in $H_3Mn_3(CO)_{12}$ is considerably longer than the Mn—Mn distance of 2.923(2) Å in $Mn_2(CO)_{10}$. Hydride ligands can also bridge three metal atoms, as shown in **8.31**. The complex $H_4Re_4(CO)_{12}$ has a tetrahedral $Re_4$ core with each rhenium atom having three terminal CO ligands. Each of the four trigonal faces of the $Re_4$ tetrahedron is capped by a $\mu_3$-H ligand.

Measurements of M—H bond energies in organometallic complexes give values in the range of 53−87 kcal/mole with an average of 68 kcal/mole. The relative ease of heterolytic cleavage of M—H bonds is reflected in the $pK_a$ values of metal-hydride complexes. The hydrides $HV(CO)_6$ and $HCo(CO)_4$ are strong acids, $pK_a \sim$ 0; $H_2Fe(CO)_4$ has only one relatively acidic proton, $pK_a^1 = 6.8$ and $pK_a^2 = 15$; $HMn(CO)_5$ is moderately acidic, $pK_a = 7.1$; and $HRe(CO)_5$, $H_2Os(CO)_4$, $pK_a =$ 12.8, and $H_2Ru_4(CO)_{13}$, $pK_a = 14.7$, are very weak acids. Clearly, the term "metal hydride" is a misnomer for a complex like $HCo(CO)_4$ where the hydrogen atom is very acidic, and thus not very hydridic. However, this term is useful in denoting the presence of a metal-hydrogen bond within a complex.

## COMPLEXES CONTAINING ALKYL, ACYL, OR RELATED LIGANDS

Convenient availability of metal carbonyl anions resulted historically in the rapid preparation of many new classes of organometallic complexes. These anions, like carbanions, act as nucleophiles and undergo a wide variety of nucleophilic displacement reactions. Alkylation affords *metal-alkyl complexes*, $L_nM$—R, and acylation gives *metal-acyl complexes*, $L_nM$—C(O)R. Although several metal-alkyl complexes were known long before the exploitation of carbonyl anion chemistry, these anions provided the first general method of forming metal-carbon $\sigma$ bonds in complexes

**TABLE 8.1** Rate constants and relative nucleophilicities of selected metal carbonyl anions in reaction with $CH_3I$ at 25°C in THF solution.

| Metal carbonyl anion | $k(M^{-1}sec^{-1})^a$ | Relative nucleophilicity[b,c] |
|---|---|---|
| $(\eta^5\text{-}C_5H_5)Fe(CO)_2^-$ | $2.8 \times 10^6$ | 64,100,000 |
| $(\eta^5\text{-}C_5H_5)Ru(CO)_2^-$ | $3 \times 10^5$ | 6,860,000 |
| $(\eta^5\text{-}C_5H_5)Ni(CO)^-$ | $2.2 \times 10^5$ | 5,030,000 |
| $Re(CO)_5^-$ | $1 \times 10^3$ | 22,900 |
| $Mn(CO)_5^-$ | 7.4 | 169 |
| $(\eta^5\text{-}C_5H_5)W(CO)_3^-$ | 2.4 | 55 |
| $(\eta^5\text{-}C_5H_5)Mo(CO)_3^-$ | 1.5 | 34 |
| $(\eta^5\text{-}C_5H_5)Cr(CO)_3^-$ | $7.5 \times 10^{-2}$ | 2 |
| $Co(CO)_4^-$ | $4.37 \times 10^{-2}$ | 1 |

[a]Data from R. G. Pearson and P. E. Figdore, *J. Am. Chem. Soc.*, **102**, 1541 (1980).
[b]As a reference to more normal organic nucleophiles, $PhS^-$ reacts with $CH_3I$ in THF at 25°C with a rate constant of $1 \times 10^5 \ M^{-1} \ sec^{-1}$ and a relative nucleophilicity of 2,290,000.
[c]Presumably these anions attack $CH_3I$ by an $S_N2$ mechanism, although a competitive free-radical mechanism might be operative also.

that were very stable to heat or to air oxidation. These alkyl complexes were stable enough to permit spectroscopic and structural examination and to be practical reagents in subsequent reaction chemistry.

Selected relative nucleophilicities for metal carbonyl anions are shown in table 8.1. Notice that the nucleophilicities listed extend over a range of nearly $10^8$. These data are useful in deciding which anion should be used to displace a particular leaving group—especially when a particular metal anion did not react properly. Since nucleophilicity is a kinetic parameter, the observed trend is difficult to rationalize. It is generally assumed that, within a series of similar metal carbonyl anions, nucleophilicity increases as the size of the metal increases and as the number of strongly $\pi$-acidic ligands on M decreases. Notice that $(\eta^5\text{-}C_5H_5)Ni(CO)^-$ is slightly more nucleophilic than is $PhS^-$.

Preparations of metal-alkyl and acyl complexes by nucleophilic addition or displacement reactions are shown in equations *8.32–8.35*.

$$Na[(\eta^5\text{-}C_5H_5)M(CO)_3] + CH_3I \xrightarrow{THF}$$

M = Mo or W

$$-NaCl \quad | \quad CH_3C(O)Cl$$

**8.32**

(8.32)

**8.33**

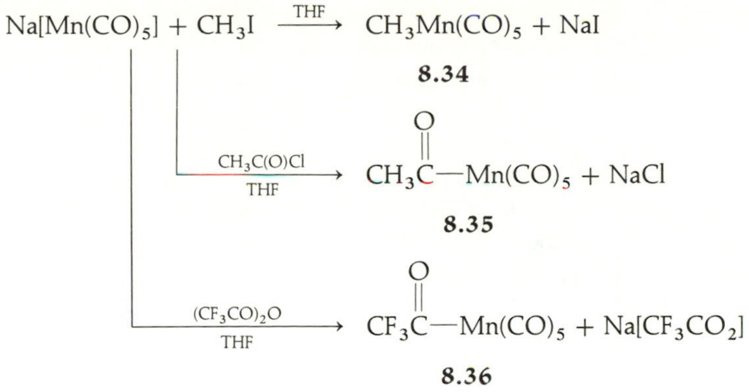

$$Na[Mn(CO)_5] + CH_3I \xrightarrow{\text{THF}} CH_3Mn(CO)_5 + NaI$$

**8.34**

$$\xrightarrow[\text{THF}]{CH_3C(O)Cl} \quad CH_3\overset{\displaystyle O}{\overset{\|}{C}}\!-Mn(CO)_5 + NaCl$$

**8.35**

$$(8.33)$$

$$\xrightarrow[\text{THF}]{(CF_3CO)_2O} \quad CF_3\overset{\displaystyle O}{\overset{\|}{C}}\!-Mn(CO)_5 + Na[CF_3CO_2]$$

**8.36**

$$Na_2[Fe(CO)_4] + R\!-\!X \xrightarrow{\text{THF}} Na^+ \left[ OC\!-\!\underset{\underset{O}{\overset{|}{C}}}{\overset{R}{\overset{|}{Fe}}}\!\!\!\!\!\overset{\displaystyle C^{O}}{\underset{\displaystyle C_O}{}} \right]^- + NaX$$

R = alkyl;
X = halide or OTs

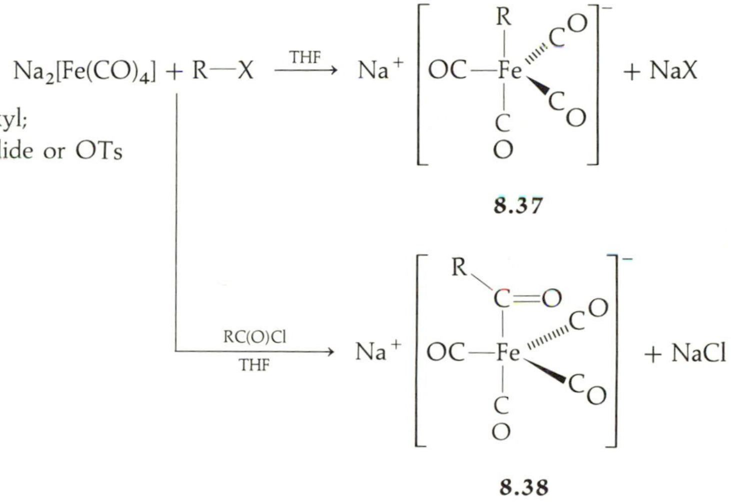

**8.37**

$$\xrightarrow[\text{THF}]{RC(O)Cl} \quad Na^+ \left[ OC\!-\!\underset{\underset{O}{\overset{|}{C}}}{\overset{\overset{R}{\diagdown}\,_{C=O}}{Fe}}\!\!\!\!\!\overset{\displaystyle C^{O}}{\underset{\displaystyle C_O}{}} \right]^- + NaCl \quad (8.34)$$

**8.38**

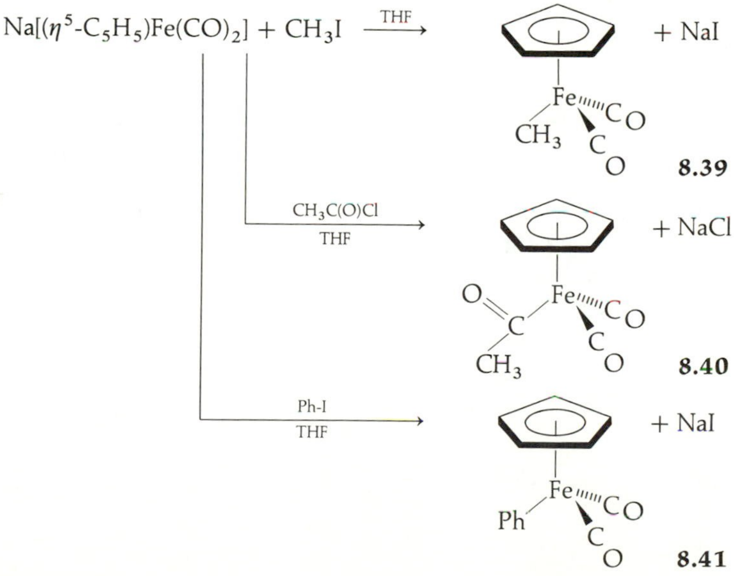

$$Na[(\eta^5\text{-}C_5H_5)Fe(CO)_2] + CH_3I \xrightarrow{\text{THF}}$$

+ NaI

**8.39**

$$\xrightarrow[\text{THF}]{CH_3C(O)Cl}$$

+ NaCl

**8.40**

$$\xrightarrow[\text{THF}]{\text{Ph-I}}$$

+ NaI    (8.35)

**8.41**

A wide range of organic alkylating or acylating reagents can be used in these re-actions. Usually, alkyl iodides and acyl chlorides are preferred reagents, but acid anhydrides (eq. 8.33) or organic tosylates (eq. 8.34) can also be used. The $(\eta^5\text{-}C_5H_5)$-$Fe(CO)_2^-$ anion is a sufficiently strong nucleophile that it reacts with iodobenzene as shown in equation 8.35. Stronger alkylating agents, such as the Meerwein oxo-nium salts, $[R_3O]BF_4$, "magic methyl," $FSO_3CH_3$, or dimethylsulfate also result in M—C bond formation provided that oxidation of the metal atom does not occur. These reagents are used more frequently to alkylate metals in neutral complexes or other basic atoms within an organometallic complex. Note that anionic alkyl and acyl complexes such as **8.37** and **8.38** can be prepared; they are isovalent to the corresponding anionic metal hydride compounds.

As shown in equations 8.32–8.35, alkyl or aryl ligands occupy normal coordina-tion sites around a metal atom. As mentioned in chap. 2, $L_nM$—$C(sp^3)$ distances provide a direct measure of the single-bond covalent radius of an $L_nM$ moiety because the M—$C(sp^3)$ bond lacks any $\pi$ character, and the single-bond covalent radius of a $C(sp^3)$ atom is known to be 0.77 Å.

Perfluoroalkyl complexes, $L_nMR_F$, are more stable thermally than corresponding alkyl complexes. The M—C distances in perfluoroalkyl complexes also are shorter than M—C distances in similar alkyl complexes:

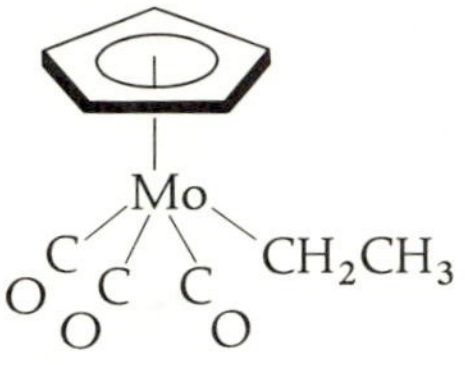

**8.42**                                  **8.43**

Mo—$CH_2$ = 2.40(3) Å          Mo—$CF_2$ = 2.28(2) Å

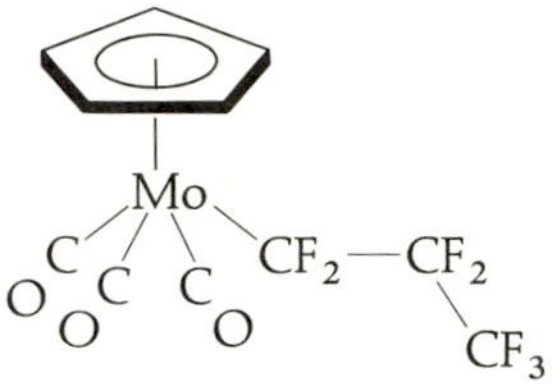

Several explanations for this increased stability of an M—$R_F$ bond have been offered. Perhaps the simplest is to assume that the highly electronegative fluorine atoms on the donor carbon atom distort the $sp^3$ hybrid orbitals of the carbon atom. This distortion would increase the $p$ character of the carbon hybrid orbitals involved in C—F bonding, thereby increasing the s character of the carbon hybrid orbital that forms the M—C bond. An M—$CF_2R_F$ bond should be shorter and stronger than an analogous M—$CH_2R$ bond since the M—$CF_2R_F$ bond has a greater amount of s-character contribution from the donor carbon atom. Thermochemical data indicate that some M—$CF_3$ bond energies are about 14 kcal/mole larger than the corresponding M—$CH_3$ bond energies.

Four complications that might be encountered when preparing metal-alkyl com-plexes are shown in equations 8.36–8.39, in which $ML_n$ represents an organo-metallic moiety and $L$ may be CO or other ligands.

$$Na[ML_n] + Me_3CCl \xrightarrow{\text{E}_2\text{elimination}} HML_n + Me_2C{=}CH_2 + NaCl \quad (8.36)$$

$$R_2CH{-}CH_2{-}ML_n \xrightarrow[\text{elimination}]{\beta\text{-hydrogen}} HML_n + R_2C{=}CH_2 \quad (8.37)$$

$$R\!-\!ML_n \xrightarrow[\text{L or CO}]{\text{"CO-insertion"}} R\!-\!\overset{\overset{\displaystyle O}{\|}}{C}\!-\!ML_n \tag{8.38}$$

$$R\!-\!ML_n \xrightarrow{\text{R isomerization}} R'\!-\!ML_n \tag{8.39}$$

When an organometallic anion is a strong base, reaction with a tertiary alkyl halide may occur by an $E_2$ elimination affording an olefin and a metal-hydride complex (eq. 8.36). In fact, reaction of $Na[(\eta^5\text{-}C_5H_5)Fe(CO)_2]$ with $Me_3CCl$ is one of the preferred methods of preparing $(\eta^5\text{-}C_5H_5)Fe(CO)_2H$. The basic dianion $[Fe(CO)_4]^{2-}$ also exhibits $E_2$ chemistry.

Metal-alkyls are frequently unstable thermally because of an intramolecular elimination reaction that also gives metal-hydride complexes. Equation 8.37 shows a β-*hydrogen elimination*. A detailed discussion of this and other elimination reactions is presented in the next chapter.

Insertion reactions may occur under the same conditions used for preparing an alkyl complex. A *"CO-insertion"* is shown in equation 8.38. Equations 8.40 and 8.41 show two specific examples of "CO-insertion."

$$Na[RFe(CO)_4] \xrightarrow{CO} Na[R\overset{\overset{\displaystyle O}{\|}}{C}\!-\!Fe(CO)_4] \tag{8.40}$$

$$RCo(CO)_4 \xrightarrow{CO} R\overset{\overset{\displaystyle O}{\|}}{C}\!-\!Co(CO)_4 \tag{8.41}$$

Alkyl cobalt tetracarbonyl complexes undergo very facile conversion to acyl complexes in the presence of CO. Insertion reactions are also discussed in the following chapter.

*Decarbonylation reactions*, the reverse of CO-insertion, which convert acyl ligands into alkyl ligands, usually occur at elevated temperatures. However, spontaneous decarbonylation may occur also, as shown in equation 8.42. The driving force of this reaction is the high stability of the $Co\!-\!R_F$ bond.

$$Na[Co(CO)_4] + C_3F_7C(O)Cl \longrightarrow C_3F_7Co(CO)_4 + CO + NaCl \tag{8.42}$$

Decarbonylation reactions provide a convenient route to alkyl complexes for anionic metal complexes which are such poor nucleophiles that they can react only with acid halides and not alkyl halides.

Alkyl ligand isomerization reactions are known (eq. 8.39). Thermal isomerization of secondary alkyl ligands to primary alkyl ligands and of primary to secondary alkyl ligands are shown in equations 8.43 and 8.44, respectively.

$$\tag{8.43}$$

$L = PPh_3$

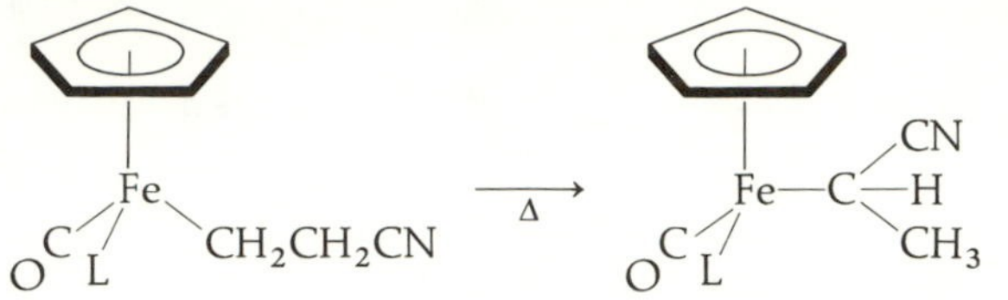

$$\text{(8.44)}$$

$L = PPh_3$

Presumably, the isomerization shown in equation 8.43 is driven by a reduction in intramolecular steric interactions; the isomerization shown in equation 8.44 is driven by an increase in the Fe—C bond energy when the very electronegative cyano group is bonded to the carbon donor atom. Both steric and electronic effects appear to be important in these isomerizations.

Because of the great interest in metal-alkyl complexes, several other methods of preparing these complexes have been developed. These synthetic pathways, with selected examples, include the following:

**1.** Reaction of a carbanion with a metal-halide.

$$Et_4N[W(CO)_5Br] + RLi \xrightarrow{\text{THF}} Et_4N[W(CO)_5R] + LiBr \qquad \text{(8.45)}$$

$R = Me$ or $Ph$

$$(\eta^5\text{-}C_5H_5)_2ZrCl_2 + 2C_6F_5Li \xrightarrow{\text{Et}_2\text{O}} (\eta^5\text{-}C_5H_5)_2Zr(C_6F_5)_2 + 2LiCl \quad \text{(8.46)}$$

**2.** Addition of an M—H or M—C bond across an olefin.

$$HCo(CO)_4 + C_2F_4 \longrightarrow HCF_2CF_2Co(CO)_4 \qquad \text{(8.47)}$$

$$CH_3Re(CO)_5 + C_2F_4 \longrightarrow CH_3CF_2CF_2Re(CO)_5 \qquad \text{(8.48)}$$

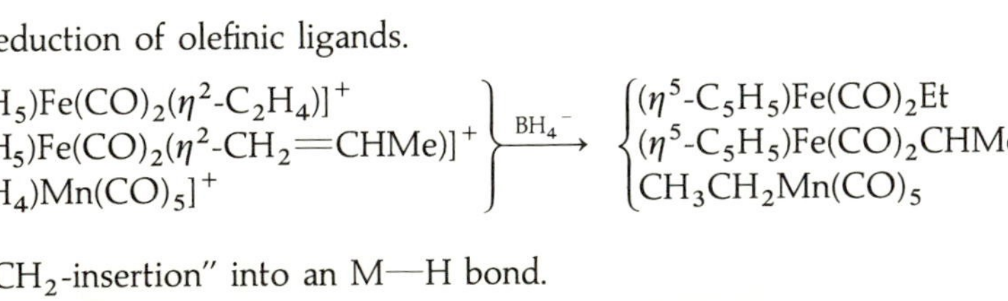

$$\text{(8.49)}$$

**3.** Reduction of olefinic ligands.

$$\left.\begin{array}{l}[(\eta^5\text{-}C_5H_5)Fe(CO)_2(\eta^2\text{-}C_2H_4)]^+ \\ [(\eta^5\text{-}C_5H_5)Fe(CO)_2(\eta^2\text{-}CH_2{=}CHMe)]^+ \\ [(\eta^2\text{-}C_2H_4)Mn(CO)_5]^+\end{array}\right\} \xrightarrow{BH_4^-} \left\{\begin{array}{l}(\eta^5\text{-}C_5H_5)Fe(CO)_2Et \\ (\eta^5\text{-}C_5H_5)Fe(CO)_2CHMe_2 \;\;\text{(8.50)} \\ CH_3CH_2Mn(CO)_5\end{array}\right.$$

**4.** "CH$_2$-insertion" into an M—H bond.

$$\left.\begin{array}{l}(\eta^5\text{-}C_5H_5)Fe(CO)_2H \\ (\eta^5\text{-}C_5H_5)Mo(CO)_3H \\ HMn(CO)_5\end{array}\right\} \xrightarrow{CH_2N_2} \left\{\begin{array}{l}(\eta^5\text{-}C_5H_5)Fe(CO)_2CH_3 \\ (\eta^5\text{-}C_5H_5)Mo(CO)_3CH_3 \quad \text{(8.51)} \\ CH_3Mn(CO)_5\end{array}\right.$$

**5.** Oxidative-addition reactions.

$$(\eta^5\text{-}C_5H_5)Co(CO)_2 + CF_3I \xrightarrow{C_6H_6} (\eta^5\text{-}C_5H_5)Co(CO)(CF_3)I + CO \quad (8.52)$$

$$(\eta^2\text{-}C_2H_4)PtL_2 + CH_2I_2 \xrightarrow[25°C]{C_6H_6}$$

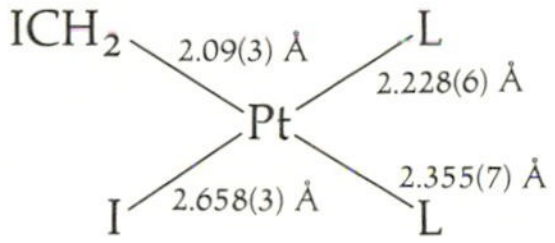

$$+ C_2H_4 \quad (8.53)$$

L = PPh₃

$$(\eta^2\text{-}C_2H_4)_2Pt_2Cl_4 +$$

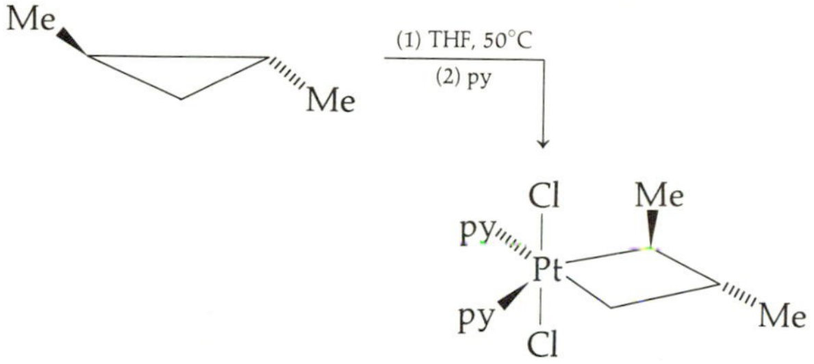

$$(8.54)$$

Although alkyl complexes are frequently prepared by using carbanionic reagents or by oxidative-addition, the other, less routine methods become more useful when common routes fail to yield a particular complex of interest. Addition across C—C double or triple bonds is also an important step in many catalytic processes. Metal vapor cocondensation with an alkyl or aryl halide (eq. 8.55), is another type of oxidative-addition reaction that has been studied recently.

$$nM^0(g) + nR\text{—}X \longrightarrow (R\text{—}M\text{—}X)_n \quad (8.55)$$

An acyl ligand may also coordinate to metals as a *dihapto* ligand, as shown in **8.44–8.46**.

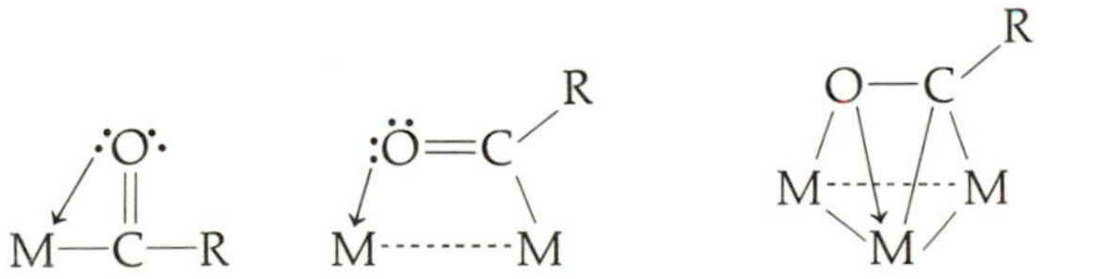

| **8.44** | **8.45** | **8.46** |
|---|---|---|
| $\eta^2$-acyl | $\mu_2$, $\eta^2$-acyl | $\mu_2(O)$, $\mu_2(C)$, $\eta^2$-acyl |

Coordination of the oxygen atom occurs when the metal atom would be electron deficient with only a *monohapto*-acyl ligand. Structural type **8.44** is found in mononuclear complexes where the metal atom has strong *oxophilicity*. These metals commonly include those of the titanium and vanadium groups as well as lanthanide and actinide metals. For transition metal complexes, such as **8.47**, the M—C distance is usually shorter than the M—O distance, a difference consistent with the adduct structure shown in **8.44**. However, the apparently stronger oxophilicity of thorium, as shown in **8.48**, affords a Th—O distance *shorter than* the Th—C distance. Presumably, the Th—O interaction is greater than the Th—C interaction; therefore an oxycarbenoid, or C-adduct, structure has been proposed to rationalize the bonding within this complex (**8.49**).

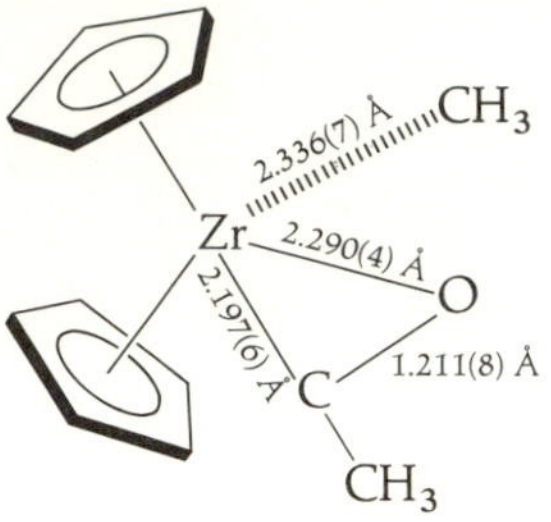

|                   |                   |                   |
|:-----------------:|:-----------------:|:-----------------:|
| **8.47**          | **8.48**          | **8.49**          |

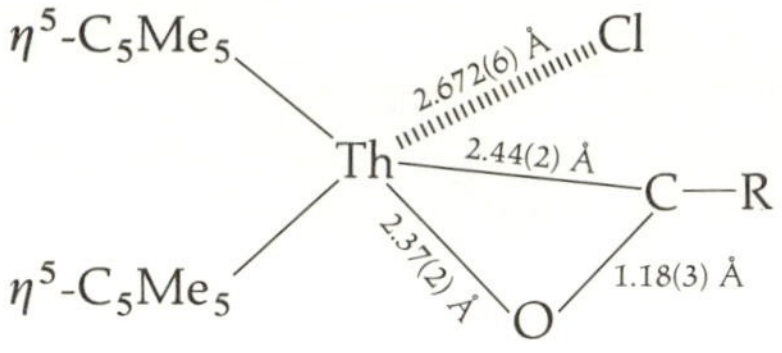

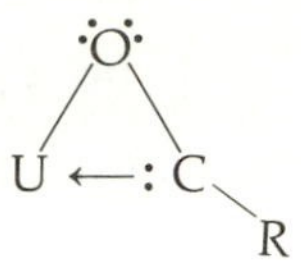

$$R = CH_2CMe_3$$

Structures **8.50** and **8.51** contain $\mu_2$, $\eta^2$-acyl ligands like those of **8.45** and **8.46**, respectively.

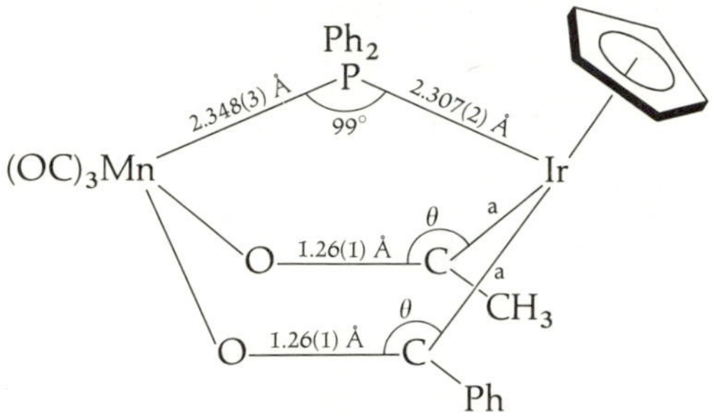

|                                                      |                                                      |
|:----------------------------------------------------:|:----------------------------------------------------:|
| **8.50**                                             | **8.51**                                             |
| $\bar{a} = 1.995(9)$ Å, $\bar{\theta} = 125(1)°$     | $\bar{a} = 1.98(1)$ Å, $\theta = 114(1)°$            |
|                                                      | $[Fe_3(CO)_9(CH_3CO)]^-$                             |

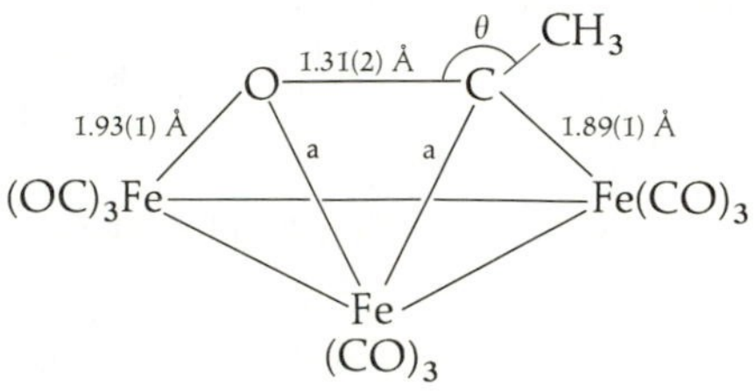

In both complexes, the acyl C—O bond retains partial double bond character, although coordination to three metal atoms, as in **8.51**, does reduce the C—O bond order considerably. The EAN rule is obeyed for complexes **8.50** and **8.51** if the acyl oxygen atom acts as a two- and three-electron donor, respectively.

A special "acyl ligand" that has generated much current interest is the *formyl ligand*, M—C(O)H. Formyl ligands have been proposed as intermediates in several catalytic processes, and they also represent the first reduction product in the reductive conversion of CO to $CH_3OH$ or related saturated hydrocarbons. Formyl complexes are usually prepared by one of three methods:

**1.** Formylation of a metal carbonyl anion (eq. *8.56*).

$$Na_2[Fe(CO)_4] + HC(O)OC(O)CH_3 \longrightarrow Na[H-\overset{\overset{\displaystyle O}{\displaystyle \|}}{C}Fe(CO)_4] + NaOAc \quad (8.56)$$

**2.** Hydride addition to a CO ligand (eqs. *8.57* and *8.58*).

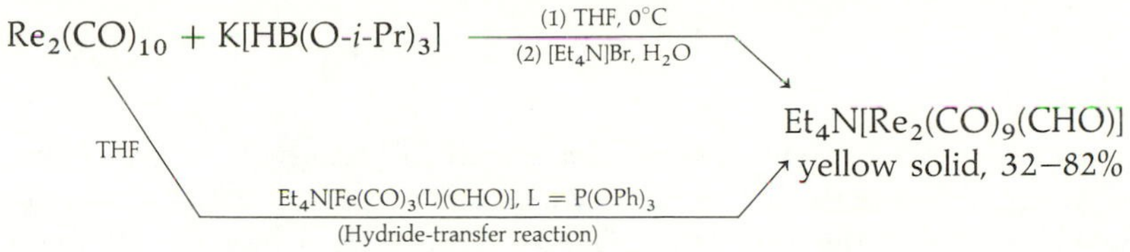

$$Re_2(CO)_{10} + K[HB(O\text{-}i\text{-}Pr)_3] \xrightarrow[\text{(2) }[Et_4N]Br, H_2O]{\text{(1) THF, 0°C}}$$

$$Et_4N[Re_2(CO)_9(CHO)] \quad (8.57)$$
yellow solid, 32–82%

THF

$$Et_4N[Fe(CO)_3(L)(CHO)], L = P(OPh)_3$$
(Hydride-transfer reaction)

$$[(\eta^5\text{-}C_5H_5)Re(CO)(L)(NO)]BF_4 + Li[HBEt_3] \xrightarrow[22°C]{THF}$$

$$L = PPh_3$$

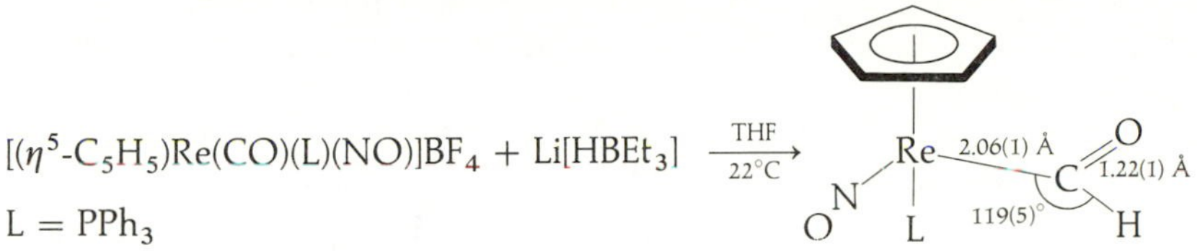

$$(8.58)$$

yellow solid, 59%

**3.** Oxidative-addition of formaldehyde (eqs. *8.59* and *8.60*).

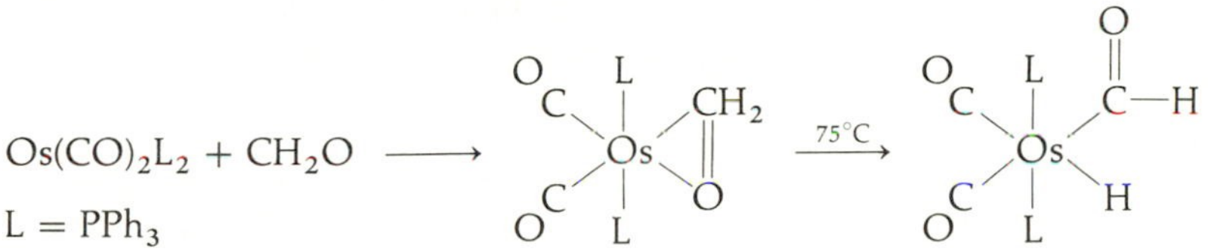

$$Os(CO)_2L_2 + CH_2O \longrightarrow$$

$$L = PPh_3$$

$\eta^2$-formaldehyde
complex

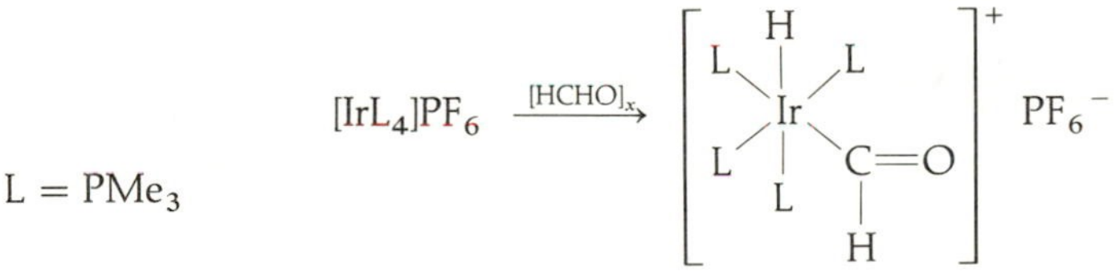

$$\xrightarrow{75°C} \qquad (8.59)$$

$$[IrL_4]PF_6 \xrightarrow{[HCHO]_x}$$

$$L = PMe_3$$

$$PF_6^- \qquad (8.60)$$

Formyl-ligand formation by insertion of CO into an M—H bond has been observed in the two reactions of equations *8.61* and *8.62*.

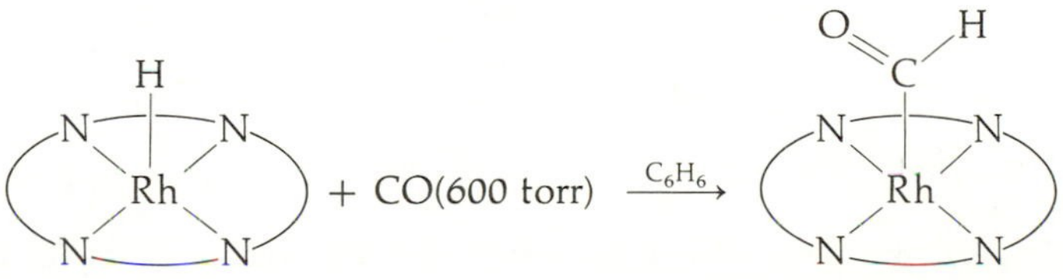

$$+ CO(600 \text{ torr}) \xrightarrow{C_6H_6}$$

$$(8.61)$$

Rh(OEP)(H)                    Rh(OEP)(CHO)

OEP = octaethylporphyrin

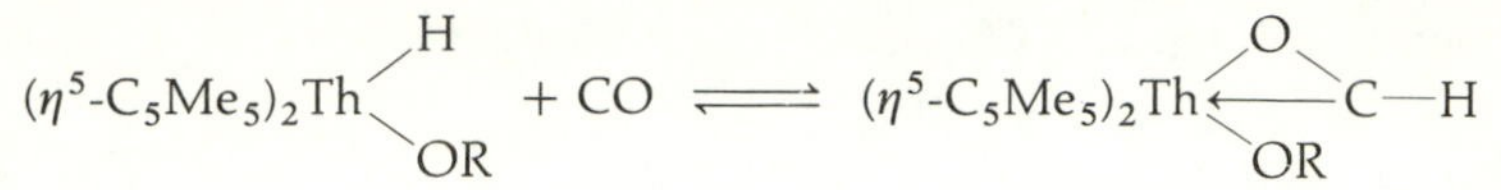

$$(\eta^5\text{-}C_5Me_5)_2Th\begin{array}{c} H \\ \diagup \\ \diagdown \\ OR \end{array} + CO \;\rightleftharpoons\; (\eta^5\text{-}C_5Me_5)_2Th\begin{array}{c} O \\ \diagdown \\ \longleftarrow C-H \\ \diagup \\ OR \end{array} \quad (8.62)$$

R = alkyl

The rhodium porphyrin hydride complex reacts with CO to give a normal *mono-hapto*-formyl ligand. However, the thorium hydride complex reacts with CO to give an $\eta^2$-formyl ligand, like the analogous acyl ligand, **8.47**. A strong Th—O bond apparently stabilizes the formyl complex. The equilibrium shown in equation 8.62 also provides a unique example of the decarbonylation of a formyl ligand to give a metal-hydride complex.

Reaction of the hydride complex $[(\eta^5\text{-}C_5Me_4Et)TaCl_2(\mu_2\text{-}H)]_2$ with CO gives complex **8.52**, which contains a $\mu_2$, $\eta^2$-formyl ligand. The C—O distance of 1.50(2) Å indicates very little C—O multiple bonding, as expected for coordination to two electron-deficient metal atoms. This formyl complex is structurally similar to acyl complexes **8.46** and **8.47**.

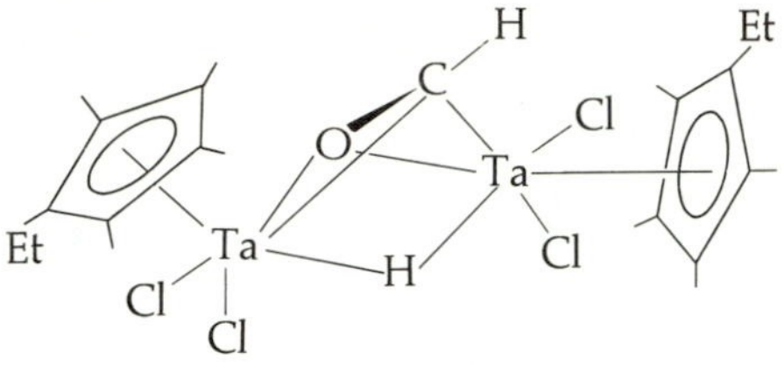

**8.52**

$(\eta^5\text{-}C_5Me_4Et)_2Ta_2Cl_4(\mu_2\text{-}H)(\mu_2\text{-}CHO)$

## COMPLEXES CONTAINING OTHER TYPES OF M—C σ BONDING

Four types of $\sigma$-bonding between the metal and the $sp^3$ hybrid orbitals of carbon are shown as **8.53–8.56**.

| M—CH$_2$R | M- - -M | M- - -M | (with M, M, M) |
|:---:|:---:|:---:|:---:|
| **8.53** | **8.54** | **8.55** | **8.56** |
| $\eta^1$-alkyl | $\mu_2$-alkyl | $\mu_2$-alkylidene | $\mu_3$-alkylidyne |

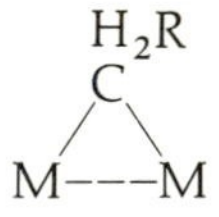

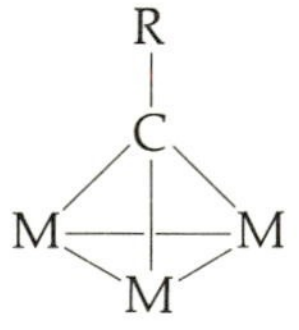

Ordinary M-alkyl coordination is represented as **8.53**. Several examples of doubly bridging alkyl ligands (**8.54**) are now known. In these compounds, a $\mu_2$-alkyl ligand contributes one electron to the *three-center two-electron* M—C—M bond, as in Al$_2$Me$_6$. Selected examples of this type of structure are shown as **8.57** and **8.58**. A formal M—M bond does not always accompany a $\mu_2$-alkyl ligand.

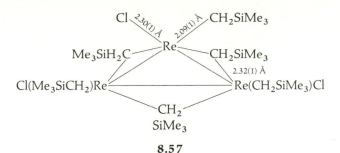

**8.57**

Re$_3$Cl$_3$($\mu_2$-CH$_2$SiMe$_3$)$_3$(CH$_2$SiMe$_3$)$_3$

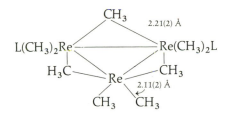

L = PEt$_2$Ph

**8.58**

Re$_3$($\mu_2$-CH$_3$)$_3$(CH$_3$)$_6$L$_2$

Doubling bridging alkylidene ligands (**8.55**) are possibly important intermediates in metal promoted CH$_2$ transfer reactions, in the reduction of CO to alkanes, and in olefin metathesis. Representative preparations and structures are shown in equations *8.63–8.66.*

Na[($\eta^5$-C$_5$H$_5$)$_2$Co$_2$(CO)$_2$]$^-$ $\xrightarrow[\text{THF}]{\text{CH}_2\text{I}_2}$

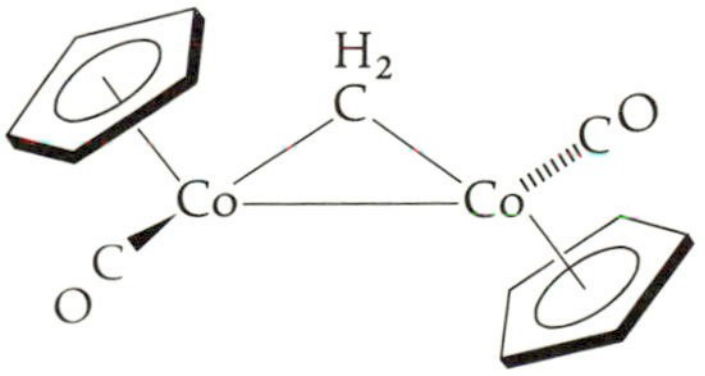

[structure diagram] $\quad + [(\eta^5$-C$_5$H$_5$)Co($\mu_2$-CO)]$_2 + 2$NaI    (8.63)

**8.59**

dark red solid, 48%

Na$_2$[Fe$_2$(CO)$_8$] + CH$_2$I$_2$ $\xrightarrow[\text{0°C}]{\text{acetone}}$ (OC)$_3$Fe[structure, 2.507(1) Å]Fe(CO)$_3$ + 2NaI    (8.64)

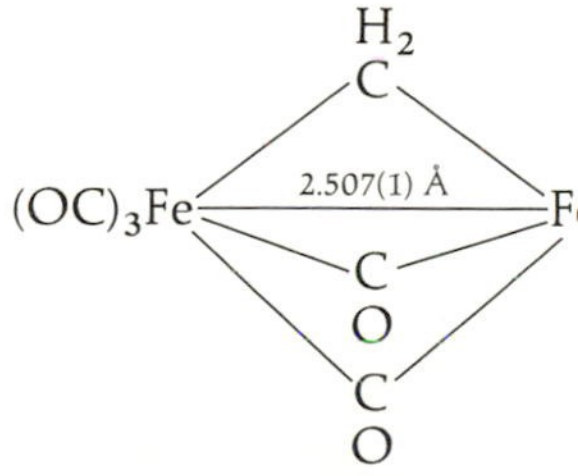

**8.60**

yellow solid, 60%

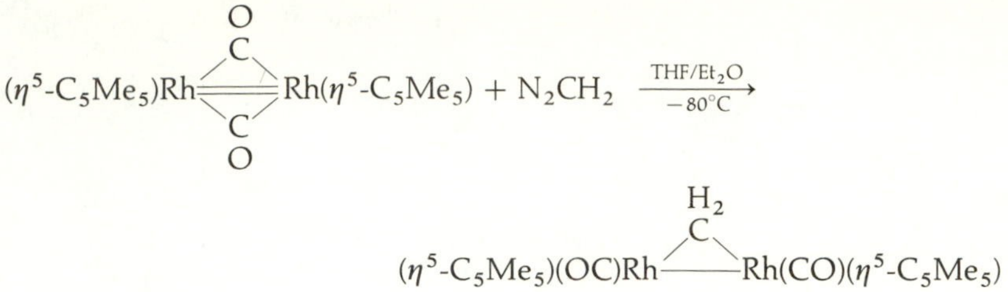

$$(\eta^5\text{-}C_5Me_5)Rh\overset{\overset{\displaystyle O}{\underset{\displaystyle C}{\|}}}{\underset{\overset{\displaystyle C}{\underset{\displaystyle O}{\|}}}{}}Rh(\eta^5\text{-}C_5Me_5) + N_2CH_2 \xrightarrow[-80°C]{THF/Et_2O}$$

$$(\eta^5\text{-}C_5Me_5)(OC)Rh\overset{\overset{\displaystyle H_2}{\displaystyle C}}{\underline{\qquad}}Rh(CO)(\eta^5\text{-}C_5Me_5) \quad (8.65)$$

**8.61**

red solid, 94%

$$[Ru_3(\mu_3\text{-}O)(OAc)_6 \cdot 3H_2O](OAc) + Me_2Mg + L(xs) \xrightarrow[-CH_4]{THF}$$

L = Me₃P

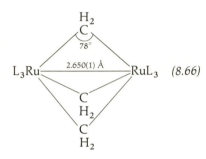

$$(8.66)$$

**8.62**

Doubly bridging alkylidene ligands formally donate one electron to each metal atom, resulting in formal M—C(sp³) single bonds.

Triply bridging alkylidyne ligands (**8.56**) are well known, and the chemistry of the alkylidynetricobalt nonacarbonyl complexes (**8.63**) has been studied extensively. Preparations of several such trinuclear cluster complexes are shown in equations *8.67–8.69*.

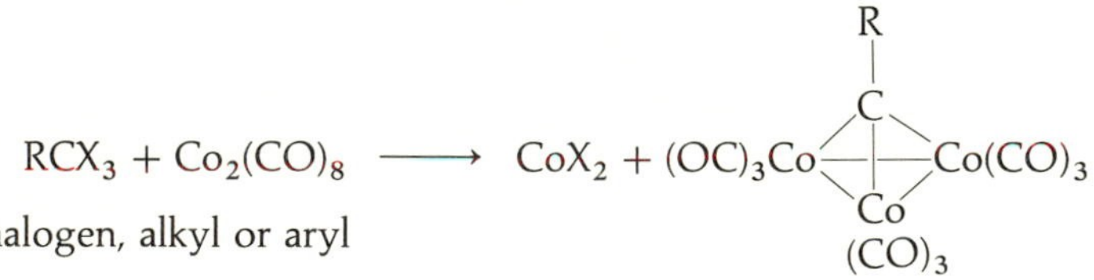

$$RCX_3 + Co_2(CO)_8 \longrightarrow CoX_2 + (OC)_3Co\overset{\overset{\displaystyle R}{\underset{\displaystyle C}{|}}}{\underset{\overset{\displaystyle Co}{(CO)_3}}{}}Co(CO)_3 \quad (8.67)$$

R = H, halogen, alkyl or aryl

**8.63**

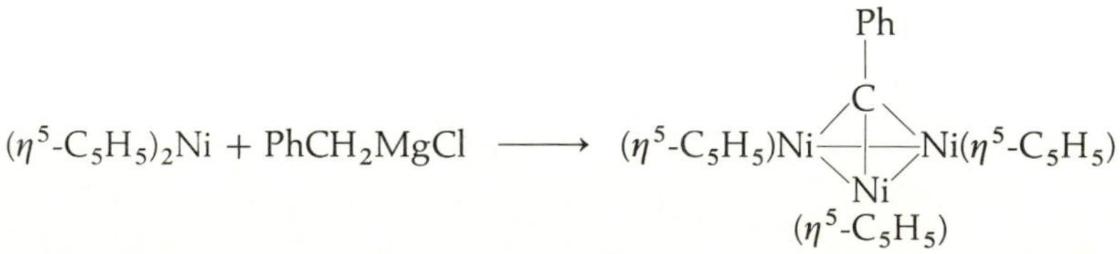

$$(\eta^5\text{-}C_5H_5)_2Ni + PhCH_2MgCl \longrightarrow (\eta^5\text{-}C_5H_5)Ni\overset{\overset{\displaystyle Ph}{\underset{\displaystyle C}{|}}}{\underset{\overset{\displaystyle Ni}{(\eta^5\text{-}C_5H_5)}}{}}Ni(\eta^5\text{-}C_5H_5) \quad (8.68)$$

**8.64**

$$H_4Ru_4(CO)_{12} + CH_2{=}CH_2 \xrightarrow[70°C]{C_6H_{12},\ 1\ atm} (OC)_3Ru\ \underset{\underset{(CO)_3}{\overset{|}{Ru}}}{\overset{\overset{Me}{\overset{|}{C}}}{\diagdown}}Ru(CO)_3 \qquad (8.69)$$

**8.65**

In each case the $\mu_3$-alkylidyne ligand acts formally as a one-electron donor to each metal atom. There is some evidence that the hybridization at the methine carbon atom is closer to $sp$ than to $sp^3$. Recently, a $\mu_2$-alkylidene ligand was converted into a $\mu_3$-alkylidyne ligand upon protonation with trifluoroacetic acid (TFA) (eq. 8.70). Although the initial site of attack by the proton is not known, a methyl complex intermediate is presumed. Complex **8.66** has two asymmetrical $\mu_2$-CO ligands.

$$2(\eta^5\text{-}C_5H_5)_2Rh_2(CO)_2(\mu_2\text{-}CH_2) \xrightarrow[(2)\ KPF_6,\ MeOH]{(1)\ TFA,\ CH_2Cl_2}$$

$$\left[ (\eta^5\text{-}C_5H_5)Rh \underset{\underset{(\eta^5\text{-}C_5H_5)}{\overset{|}{\underset{O}{\overset{b}{C}}}}}{\overset{\overset{H}{\overset{|}{C}}}{\diagdown}} Rh(\eta^5\text{-}C_5H_5) \right] PF_6 + (\eta^5\text{-}C_5H_5)Rh(CO)_2 + CH_4 \quad (8.70)$$

**8.66**

brown solid, 61%

$$\overline{Rh\text{—}Rh} = 2.685(4)\ \text{Å}$$
$$\bar{a} = 1.96\ \text{Å}$$
$$b = 1.918(5)\ \text{Å}$$
$$c = 2.267(5)\ \text{Å}$$

Several classes of organometallic complexes contain M—C bonds that have not only a strong $\sigma$-bond component but some degree of M—C $\pi$ bonding as well. Ligands of this type are shown in table 8.2. In complexes **8.67–8.78**, the carbon donor atoms have formal $sp^2$ or $sp$ hybridization.

Metal-carbon bonding in structures **8.67–8.74** can be represented formally as a pure M—C $\sigma$ bond, although some amount of M—C $\pi$ bonding may occur to enable electronic delocalization between M and the unsaturated ligand. For example, M—C bonding to terminal acyl ligands (**8.71**) is usually described by two resonance forms:

$$\underset{\textbf{8.79}}{M\text{—}\overset{\overset{O}{\parallel}}{C}\text{—}R} \quad\longleftrightarrow\quad \underset{\textbf{8.80}}{\overset{+}{M}{=}\overset{\overset{O^-}{|}}{C}\text{—}R}$$

In neutral acyl complexes, acyl C—O stretching frequencies are usually in the range 1590–1650 cm$^{-1}$, considerably lower than carbonyl stretching frequencies

**TABLE 8.2  Ligands in which the M—C bond contains a strong σ component and a possible π component.**

| Structure | Name | Structure | Name |
|---|---|---|---|
| 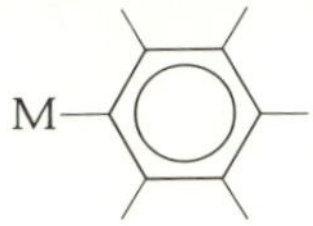 M—⟨benzene ring⟩ **8.67** | aryl |  ⟨benzene ring bridging⟩ M  M **8.68** | $\mu_2$-aryl |
| 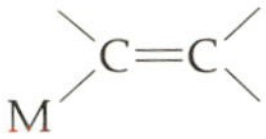 M\C=C/ **8.69** | $\eta^1$-alkenyl | 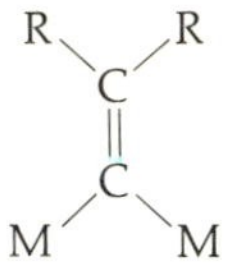 R\ /R  C‖C  M  M **8.70** | $\mu_2$-vinylidene |
| 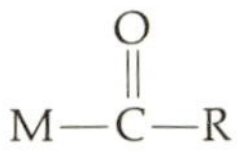 O‖ M—C—R **8.71** | acyl | M—C≡CR **8.72** | $\eta^1$-acetylide |
| 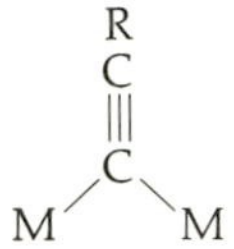 R  C‖C  M  M **8.73** | $\mu_2$-acetylide | 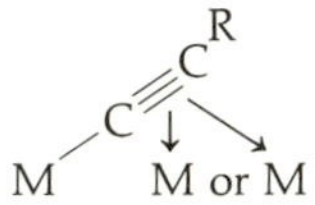 R—C≡C  M  M or M **8.74** | $\mu_n, \eta^2$-acetylide |
| M=CR$_2$ **8.75** | alkylidene (carbene) | M=C=CR$_2$ **8.76** | vinylidene |
| M≡CR **8.77** | alkylidyne (carbyne) | 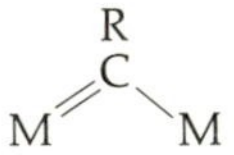 R—C  M  M **8.78** | $\mu_2$-alkylidyne |

in ketones or esters. Also, M—C distances to acyl ligands are usually 0.04–0.15 Å shorter than expected for M—C($sp^2$) single-bond distances.

Within a similar series of complexes of the type $(\eta^5\text{-}C_5H_5)Fe(CO)_2(R)$, the observed Fe—C(R) distances *as a function of R* are: Fe—C($sp^3$) *alkyl*, 2.098(2) Å; Fe—C($sp^2$) *vinylic*, 1.987(5) Å; Fe—C($sp^2$) *acyl*, 1.960(3) Å; and Fe—C($sp$) $\eta^1$-*acetylide*, 1.920(6) Å. Based on differences in single bond covalent radii, Fe—C($sp^x$) *single-bond* lengths are expected to be: Fe—C($sp^3$), 2.10 Å; Fe—C($sp^2$), 2.07 Å; Fe—C($sp$), 2.03 Å. Since the observed Fe—C lengths to $sp^2$- and $sp^3$-hybridized carbon atoms are significantly less than these estimated Fe—C($sp^x$) single-bond distances, the Fe—C bonds in these complexes presumably have some double-bond character. A very rough estimate for a formal Fe—C($sp^2$) *double-bond* length in this type of complex is 1.78 Å.

Metal-carbon bonding in structures **8.75–8.78** must involve formal M—C multiple bonding to satisfy the covalency of the carbon atom. In general, M—C distances to these ligands are considerably shorter than those expected for M—C single bonds when the R substituents are alkyl groups. The alkylidene and alkylidyne complexes are particularly important classes of compounds. The synthesis, structure, and reaction chemistry of many of the types of compounds shown in table 8.2 appear throughout the remaining chapters of this text, particularly in chapters 11 through 14.

## STUDY QUESTIONS

1. Coordination of Lewis acids to carbonyl ligands has been observed for several anionic metal carbonyls. When this coordination occurs at the carbonyl oxygen atom, as in M—CO → M′, the carbonyl group is referred to as an *isocarbonyl ligand*. Predict changes in the $\nu_{CO}$ stretching frequencies for the isocarbonyl ligand and for other terminal carbonyl ligands upon formation of such a complex.
2. Complexes **8.28**, **8.57** and **8.58** have $\mu_2$-ligands that are involved in three-center two-electron bonds. Describe this type of bond by using overlapping atomic orbitals. Why is considerable structural flexibility in **8.28** "reasonable"?
3. By using Lewis structures or Wade's rules, show how the complexes $Mn_3(CO)_{12}$ $(\mu_2\text{-H})_3$ (**8.50**) and $Re_4(CO)_{12}(\mu_3\text{-H})_4$ (**8.51**) obey the EAN rule.
4. Rationalize the observation that complex **8.66** contains two asymmetrical $\mu_2$-CO ligands.
5. By consulting the literature, locate $^1$H—NMR chemical shift data for hydride, alkyl, acyl, and formyl ligands. Compare these chemical shifts with those of analogous organic compounds.
6. Draw reasonable mechanisms showing appropriate shifts of electron pairs for the reactions shown in equations *8.16*, *8.33*, *8.56*, and *8.58*.

## SUGGESTED READING

### Anionic Organometallic Complexes

Collman, J. P. *Acc. Chem. Res.* **8**, 342 (1975).
Ellis, J. E. *J. Organometal. Chem.* **86**, 1 (1975).
Jonas, K. *Adv. Organometal. Chem.* **19**, 97 (1981).
King, R. B. *Acc. Chem. Res.* **3**, 417 (1970).
King, R. B. *Adv. Organometal. Chem.* **2**, 157 (1964).
King, R. B. *J. Organometal. Chem.* **100**, 111 (1975).

### Metal-Hydride Complexes

Chatt, J. *Adv. Organometal. Chem.* **12**, 1 (1974).
Ginsberg, A. P. *Trans. Metal Chem.* **1**, 111 (1965).
Humphries, A. P., and Kaesz, H. D. *Progr. Inorg. Chem.* **25**, 145 (1979).
Kaesz, H. D. *J. Organometal. Chem.* **200**, 145 (1980).

**Metal-Alkyl and Related Complexes**

Bergman, R. G. *Acc. Chem. Res.* **13,** 113 (1980).

Churchill, M. R. *Perspectives Struct. Chem.* **3,** 91 (1970).

Churchill, M. R., and Mason, R. *Adv. Organometal. Chem.* **5,** 93 (1967).

Davidson, P. J., Lappert, M. F., and Pearce, R. *Chem. Rev.* **76,** 219 (1976).

Nesmeyanov, A. N. *Adv. Organometal. Chem.* **10,** 1 (1972).

Parshall, G. W., and Mrowca, J. J. *Adv. Organometal. Chem.* **7,** 157 (1968).

Puddephatt, R. J. *Coord. Chem. Rev.* **33,** 149 (1980).

Schrock, R. R., and Parshall, G. W. *Chem. Rev.* **76,** 243 (1976).

Seyferth, D. *Adv. Organometal. Chem.* **14,** 97 (1976).

Treichel, P. M., and Stone, F. G. A. *Adv. Organometal. Chem.* **1,** 143 (1964).

Wilkinson, G. *Science,* **185,** 109 (1974).

# Insertion, Elimination, and Abstraction Reactions

## CHAPTER 9

Two types of addition reactions that are commonly observed in organometallic chemistry are shown in equations 9.1 and 9.2.

$$L_nMR + XY \longrightarrow L_nM\overset{\overset{\displaystyle Y}{|}}{-}X-R \quad (or \quad L_nM-X-Y-R) \qquad (9.1)$$

$$L_nMR + XY \longrightarrow L_nM\overset{\overset{\displaystyle X}{|}}{\underset{\underset{\displaystyle Y}{|}}{-}}R \quad (or \quad L_nM\overset{\overset{\displaystyle X}{|}}{\underset{\underset{\displaystyle R}{|}}{-}}Y) \qquad (9.2)$$

In the first, a substrate molecule, XY, has topologically inserted into the M—R bond, although the precise mechanism of this *insertion reaction* is not specified. *Insertion reactions are usually characterized by addition of a metal-ligand bond to a substrate molecule such that the coordination number and formal oxidation state of M does not change.* The reverse of equation 9.1 is termed an *elimination* or *extrusion* reaction. The molecule that is inserted, XY, must be unsaturated because an X—Y bond is maintained in the product. The "coordination number" of X and/or Y increases upon insertion.

In equation 9.2, the substrate molecule, XY, has added to M leading to two new M—X and M—Y bonds *without* inserting into a preexisting metal-ligand bond. Both the coordination number and the formal oxidation state of M increase in this addition reaction. This type of addition is, therefore, referred to as an *oxidative-addition reaction*. The presence or absence of an X—Y bond in the addition product depends on the degree of saturation within the free XY molecule. The reverse of equation 9.2 is termed a *reductive-elimination reaction*. Addition reactions that involve a change in the formal oxidation state of M are discussed in chap. 10.

Another type of elimination reaction that is now being observed more frequently is the *abstraction reaction:*

$$
\begin{array}{c}
\text{R} \\
| \\
\text{L}_n\text{M}
\end{array}
\cdots
\begin{array}{c}
\text{H} \quad \text{R}' \\
\diagdown \diagup
\end{array}
\quad \xrightarrow{\ \Delta\ } \quad
\text{L}_n\text{M}\!\!\bigcirc^{\text{R}'} \ + \ \text{R}\!-\!\text{H} \tag{9.3}
$$

Mechanistically, loss of RH can occur either by a concerted process or by a stepwise process probably involving a combination of oxidative-addition and reductive-elimination reactions. However, since the formal oxidation state and coordination number (or the "valency") of M does not change during abstraction of RH, this reaction type is discussed along with elimination reactions. Depending on the number of atoms between M and H in the reactant complex, these reactions are classified as $\alpha$-, $\beta$-, $\gamma$-, $\delta$-, or 1,1-, 1,2-, 1,3-, 1,4-abstractions.

Insertion reactions are fundamentally important in organometallic chemistry. They permit the direct incorporation of substrate molecules into an organometallic complex. The newly formed metal-ligand bond is a chemically reactive functional group that can be used to insert other substrate molecules or to effect metal-ligand cleavage affording substituted derivatives of the original substrate molecules. Such chemical transformations are crucial steps in the application of transition metal organometallic chemistry to organic synthesis (see chapters 12–14).

In this chapter we shall discuss insertion reactions, elimination reactions, and selected abstraction reactions. The practical importance of these reactions will become evident in Part Two of this text.

## INSERTION REACTIONS

Most commonly, insertion reactions involve addition of an M—R bond to a substrate molecule with complete scission of the original M—R bond. The majority of these insertions are either **1,1-additions** (eq. 9.4), or **1,2-additions** (eq. 9.5).

$$
\text{L}_n\text{MR} + \text{XY} \xrightarrow[\text{1,1-insertion}]{\text{1,1-addition or}}
\begin{array}{c}
\text{Y} \\
| \\
\text{L}_n\text{M}\!-\!\text{X}\!-\!\text{R}
\end{array} \tag{9.4}
$$

$$
\text{L}_n\text{MR} + \text{XY} \xrightarrow[\text{1,2-insertion}]{\text{1,2-addition or}} \text{L}_n\text{M}\!-\!\text{X}\!-\!\text{Y}\!-\!\text{R} \tag{9.5}
$$

Other types of insertions, such as 1,4-addition, are also known.

Table 9.1 gives a list of selected known insertion reactions of the types shown in equations 9.4 and 9.5 and one example of a 1,4-insertion. Specific examples of these insertions are shown in equations 9.6–9.25, as noted in the table.

In equations 9.6–9.9, molecules insert into metal-metal single bonds to give dinuclear products that lack direct metal-metal bonding.

**TABLE 9.1** Selected insertion reactions of organometallic molecules.

| Bond type | Inserting molecule | Insertion type | Product | Equation |
|---|---|---|---|---|
| M—M | $SnCl_2$ | 1,1 | Cl<br>\|<br>M—Sn—M<br>\|<br>Cl | 9.6 |
| | CZ (Z = O or NR) | 1,1 | 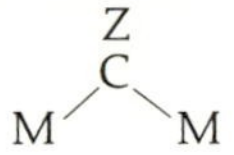 Z<br>‖<br>C<br>M⟋ ⟍M | 9.7 |
| | $SO_2$ | 1,1 | 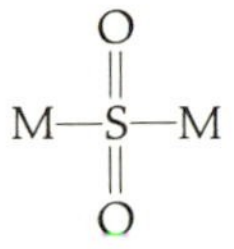 O<br>‖<br>M—S—M<br>‖<br>O | 9.8 |
| | $CF_3C{\equiv}CCF_3$ | 1,2 | 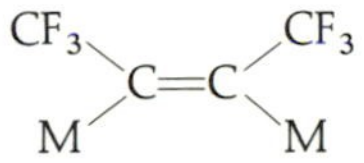 $CF_3$⟍ ⟋$CF_3$<br>C=C<br>M⟋ ⟍M | 9.9 |
|  M—C⟨ | CO | 1,1 | 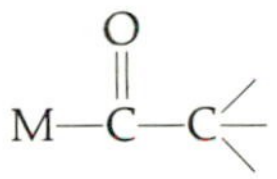 O<br>‖<br>M—C—C⟨ | 8.48 |
| | CZ (Z = O or NR) | 1,2 | 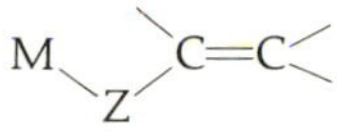 M⟍ C=C⟨<br>  Z⟋ | 9.10 |
| | RNC | 1,1 | 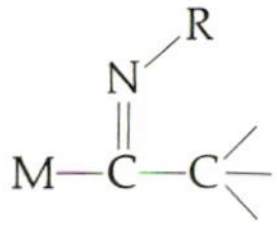 R<br>N⟋<br>‖<br>M—C—C⟨ | 9.11 |
| | RNC | 1,2 | 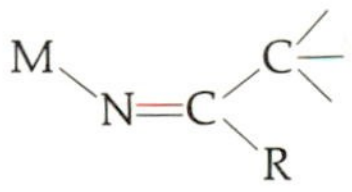 M⟍ C⟨<br>  N=C<br>    ⟍R | 9.10 |
| | $SO_2$ | 1,1 | 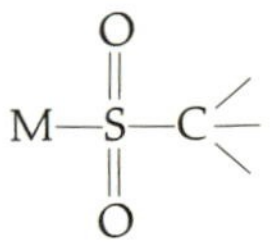 O<br>‖<br>M—S—C⟨<br>‖<br>O | 9.12 |
| | | 1,2 | 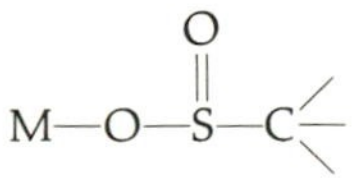 O<br>‖<br>M—O—S—C⟨ | 9.12 |
| | $EO_2$ (E = S, Se or Te) | 1,1 | 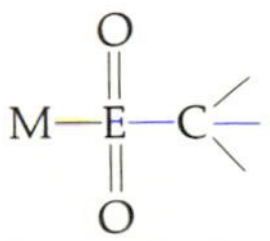 O<br>‖<br>M—E—C⟨<br>‖<br>O | 9.13 |

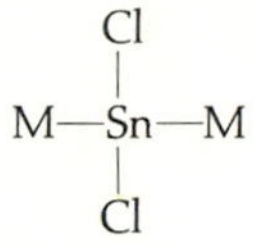

*(continued)*

**TABLE 9.1**  (continued)

| Bond type | Inserting molecule | Insertion type | Product | Equation |
|-----------|-------------------|----------------|---------|----------|
| $M-C\!\!\stackrel{/}{\diagdown}$ | $SO_3$ | 1,2 | $M-S(O)_2O-C\!\!\stackrel{/}{\diagdown}$ | 9.14 |
| | $CO_2$ | 1,2 | $M-O-\overset{\overset{O}{\|\|}}{C}-C\!\!\stackrel{/}{\diagdown}$ | 9.15 |
| | $CS_2$ | 1,2 | 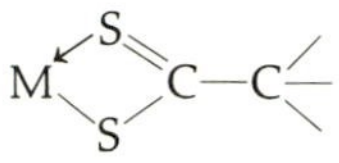 $M-S-\overset{\overset{S}{\|\|}}{C}-C\!\!\stackrel{/}{\diagdown}$ | 9.16 |
| | | | $M\!\!\stackrel{\diagup S}{\diagdown S}\!\!C-C\!\!\stackrel{/}{\diagdown}$ | |
| | NO | 1,1 | $M-N-\overset{\overset{O}{\|\|}}{\phantom{N}}C\!\!\stackrel{/}{\diagdown}$ | 9.17 |
| | $M'Cl_2$ (M' = Ge or Sn) | 1,1 | $M-\overset{\overset{Cl}{\|}}{\underset{\underset{Cl}{\|}}{M'}}-C\!\!\stackrel{/}{\diagdown}$ | 9.18 |
| | $CF_2{=}CF_2$ | 1,2 | $M-CF_2CF_2C\!\!\stackrel{/}{\diagdown}$ | 9.19 |
| | $RC{\equiv}CR$ | 1,2 | $M-C(R){=}C(R)C\!\!\stackrel{/}{\diagdown}$ | 9.20 |
| M—H | CO | 1,1 | $M-\overset{\overset{O}{\|\|}}{C}-H$ | 8.68 |
| | $CF_2{=}CF_2$ | 1,2 | $M-CF_2CF_2H$ | 9.21 |
| | $CF_3C{\equiv}CCF_3$ | 1,2 | $M-C(CF_3){=}C(CF_3)H$ | 9.22 |
| | 1,3-diene | 1,4 | $M-CR_2C(R){=}C(R)CR_2H$ | 9.23 |
| $M-N\!\!\stackrel{/}{\diagdown}$ | CO | 1,1 | $M-\overset{\overset{O}{\|\|}}{C}-N\!\!\stackrel{/}{\diagdown}$ | 9.24 |
| M—X | RNC | 1,1 | $M-\overset{\overset{N^{\diagup R}}{\|\|}}{C}-X$ | 9.25 |

$$[(\eta^5\text{-}C_5H_5)Fe(CO)_2]_2 + SnCl_2 \rightarrow (\eta^5\text{-}C_5H_5)(OC)_2Fe\!-\!\underset{\underset{\displaystyle Cl}{|}}{\overset{\overset{\displaystyle Cl}{|}}{Sn}}\!-\!Fe(CO)_2(\eta^5\text{-}C_5H_5) \quad (9.6)$$

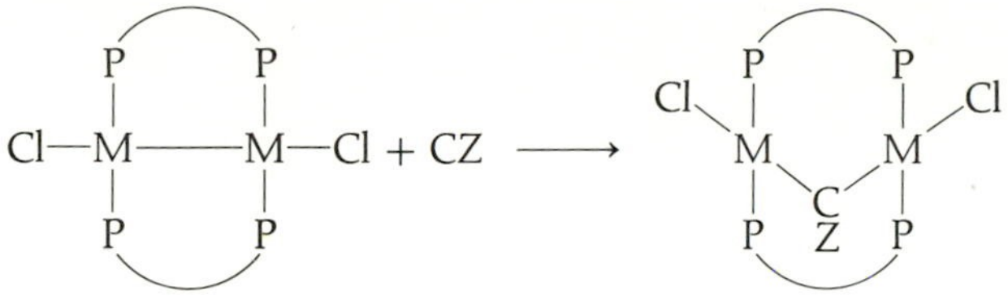

$$ \text{Cl}\!-\!\text{M}\!-\!\!-\!\text{M}\!-\!\text{Cl} + \text{CZ} \longrightarrow \qquad\qquad (9.7)$$

M = Pt(I) or Pd(I)     P⌒P = H₂C(PPh₂)₂ = DPM
CZ = CO or CNR

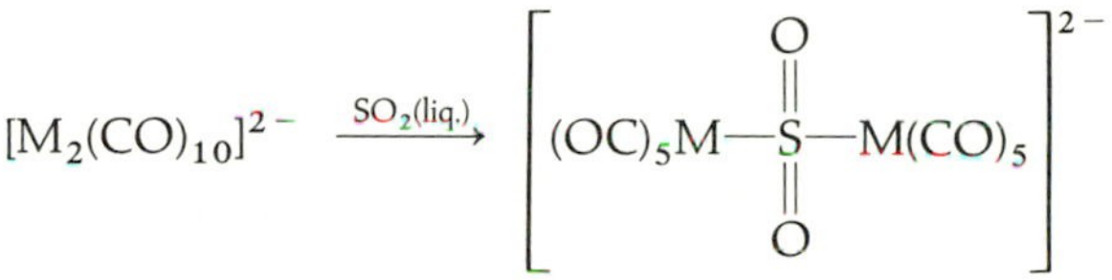

$$[M_2(CO)_{10}]^{2-} \xrightarrow{\ SO_2(\text{liq.})\ } \left[ (OC)_5M\!-\!\overset{\overset{\displaystyle O}{\|}}{\underset{\underset{\displaystyle O}{\|}}{S}}\!-\!M(CO)_5 \right]^{2-} \quad (9.8)$$

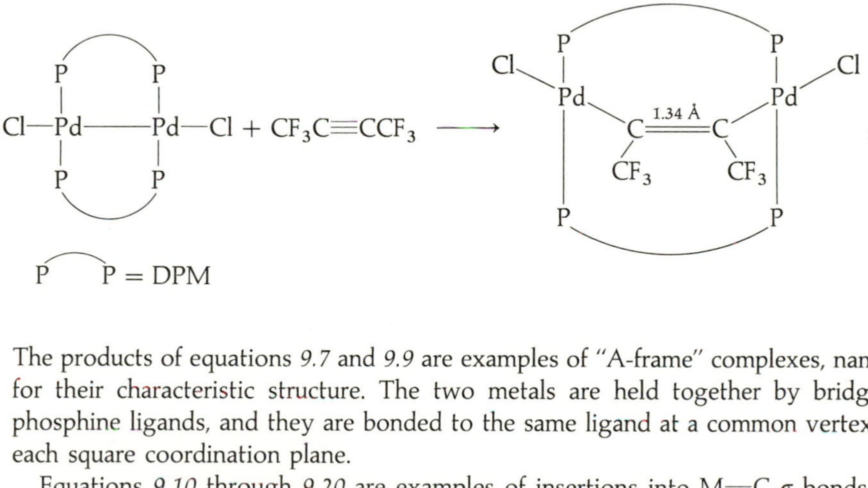

$$\text{Cl}\!-\!\text{Pd}\!-\!\!-\!\text{Pd}\!-\!\text{Cl} + \text{CF}_3\text{C}\!\equiv\!\text{CCF}_3 \longrightarrow \qquad\qquad (9.9)$$

1.34 Å

P⌒P = DPM

The products of equations 9.7 and 9.9 are examples of "A-frame" complexes, named for their characteristic structure. The two metals are held together by bridging phosphine ligands, and they are bonded to the same ligand at a common vertex of each square coordination plane.

Equations 9.10 through 9.20 are examples of insertions into M—C σ bonds. In equations 9.10 and 9.11, the tendency of titanium and actinide metals to coordinate to highly electronegative elements, such as oxygen and nitrogen, is revealed in the unusual structures of the products.

Sulfur dioxide presumably inserts into M—C σ bonds to give an O-sulfinato 1,2-addition product. This complex rearranges to the more stable S-sulfinato 1,1-addition product when M is a transition metal. Examples of this and related reactions are shown in equations 9.12–9.14.

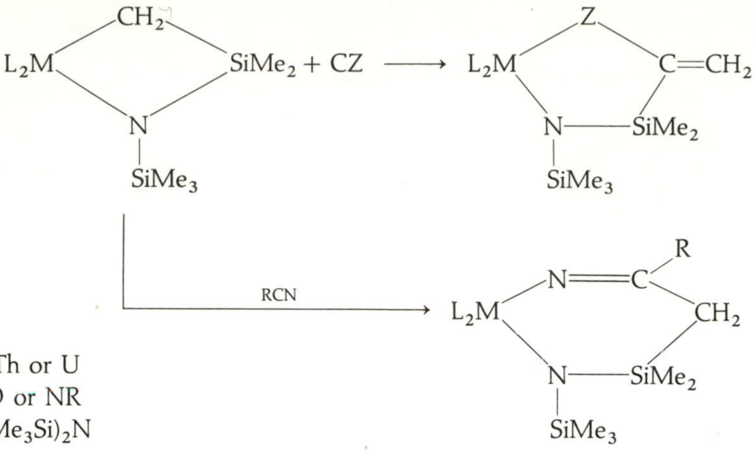

$$L_2M \underset{\underset{SiMe_3}{|}}{\overset{\overset{CH_2}{\diagup}}{\diagdown}} SiMe_2 + CZ \longrightarrow L_2M \begin{array}{c} Z \\ \diagdown \\ N-SiMe_2 \\ | \\ SiMe_3 \end{array} C=CH_2$$

$$\xrightarrow{RCN} L_2M \begin{array}{c} -N=C \diagdown \overset{R}{} \\ \diagdown CH_2 \\ N-SiMe_2 \\ | \\ SiMe_3 \end{array} \quad (9.10)$$

M = Th or U
Z = O or NR
L = $(Me_3Si)_2N$

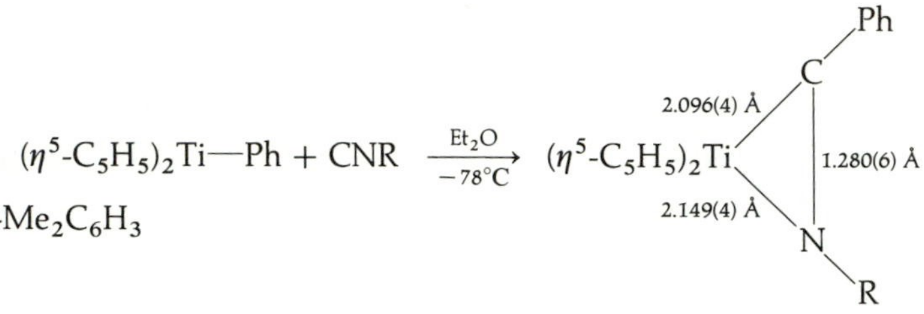

$$(\eta^5\text{-}C_5H_5)_2Ti\text{—}Ph + CNR \xrightarrow[-78°C]{Et_2O} (\eta^5\text{-}C_5H_5)_2Ti \begin{array}{c} \overset{Ph}{\diagup} \\ C \\ 2.096(4) \text{ Å} \quad | \quad 1.280(6) \text{ Å} \\ 2.149(4) \text{ Å} \\ N \\ \diagdown R \end{array} \quad (9.11)$$

R = $2,6\text{-}Me_2C_6H_3$

$\eta^2$-acylimidoyl complex

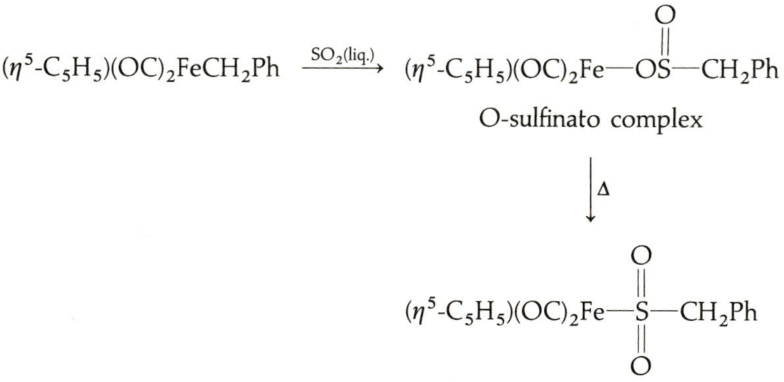

$$(\eta^5\text{-}C_5H_5)(OC)_2FeCH_2Ph \xrightarrow{SO_2(liq.)} (\eta^5\text{-}C_5H_5)(OC)_2Fe\text{—}O\overset{\overset{O}{\parallel}}{S}\text{—}CH_2Ph$$

O-sulfinato complex

$$\downarrow \Delta$$

$$(\eta^5\text{-}C_5H_5)(OC)_2Fe\text{—}\overset{\overset{O}{\parallel}}{\underset{\underset{O}{\parallel}}{S}}\text{—}CH_2Ph \quad (9.12)$$

S-sulfinato complex

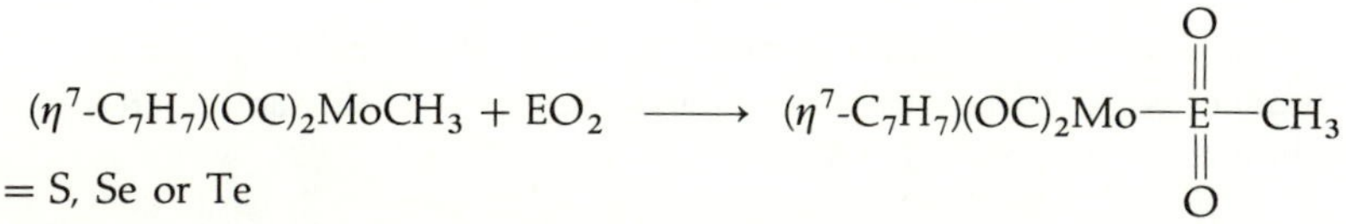

$$(\eta^7\text{-}C_7H_7)(OC)_2MoCH_3 + EO_2 \longrightarrow (\eta^7\text{-}C_7H_7)(OC)_2Mo\text{—}\overset{\overset{O}{\parallel}}{\underset{\underset{O}{\parallel}}{E}}\text{—}CH_3 \quad (9.13)$$

E = S, Se or Te

$$(OC)_5ReCH_3 + SO_3 \xrightarrow[0^\circ C]{CCl_4} (OC)_5Re{-}O(O)_2SCH_3 \qquad (9.14)$$

Insertions of $CO_2$ and $CS_2$ into M—C bonds are shown in equations *9.15* and *9.16.* A benzyne intermediate has been proposed in the titanium reaction. The tetracarbonyl manganese product of equation *9.16* is formed when the thiocarbonyl sulfur atom displaces a *cis* carbonyl ligand.

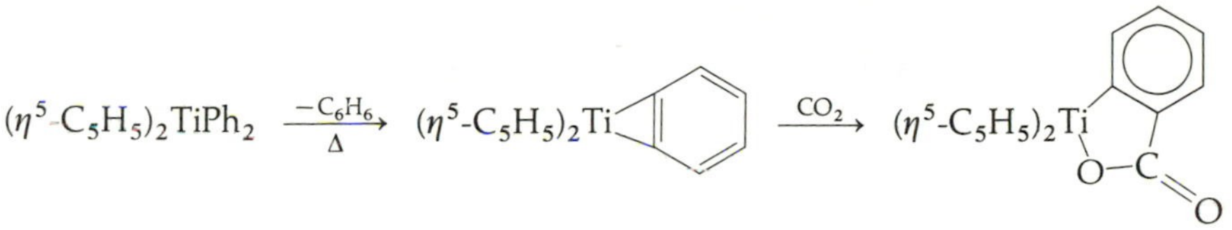

(η⁵-C₅H₅)₂TiPh₂
−C₆H₆
Δ
(η⁵-C₅H₅)₂Ti
CO₂
(η⁵-C₅H₅)₂Ti
O
C
O

(9.15)

(OC)₅MnCH₃ + CS₂
~100°C
pressure
(OC)₄Mn
S
S
C—CH₃ + CO

(9.16)

Equation *9.17* is an "NO-insertion" into a M-alkyl bond. This reaction is analogous to CO-insertion. Both types of reactions are discussed later in this chapter.

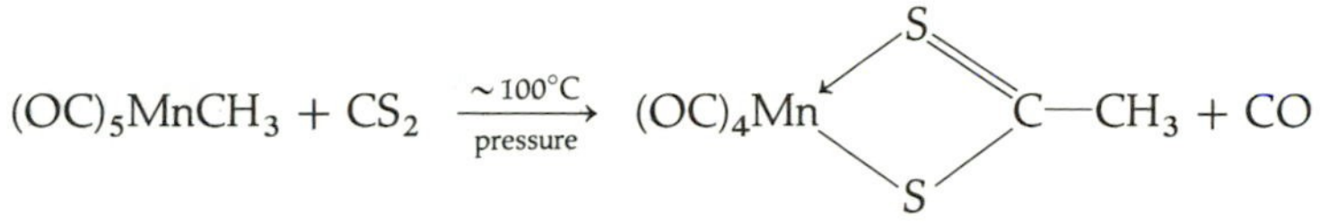

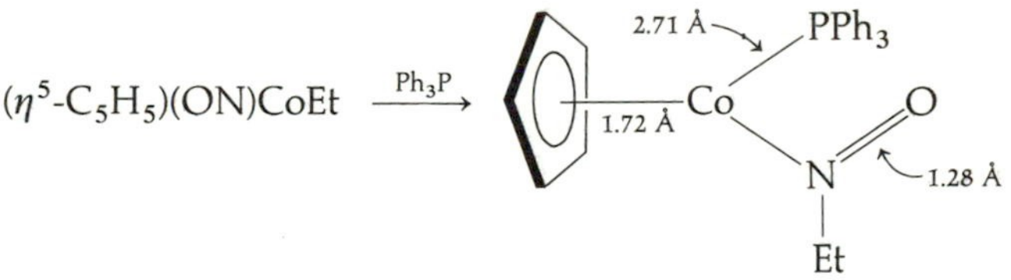

(η⁵-C₅H₅)(ON)CoEt
Ph₃P
2.71 Å
PPh₃
Co
1.72 Å
O
N
1.28 Å
Et

(9.17)

Insertion of $GeCl_2$ or $SnCl_2$ into an Fe—$CH_3$ bond is shown in equation *9.18.* This reaction is related to equation *9.6* in that Group IV metal dihalide reactants act as carbenoid species and insert into the relatively reactive M—M or M—C bonds.

$$(\eta^5\text{-}C_5H_5)(OC)_2FeCH_3 + M'Cl_2 \longrightarrow (\eta^5\text{-}C_5H_5)(OC)_2Fe{-}M'Cl_2{-}CH_3 \qquad (9.18)$$

M′ = Ge or Sn

Equations *9.19* through *9.22* show 1,2-additions of M-alkyl or M—H bonds to unsaturated organic molecules. In these reactions and in the similar 1,4-addition of equation *9.23,* the bond order of the unsaturated C—C bond is reduced formally by one unit in each addition.

$$(OC)_5ReCH_3 + CF_2{=}CF_2 \longrightarrow (CO)_5ReCF_2CF_2CH_3 \qquad (9.19)$$

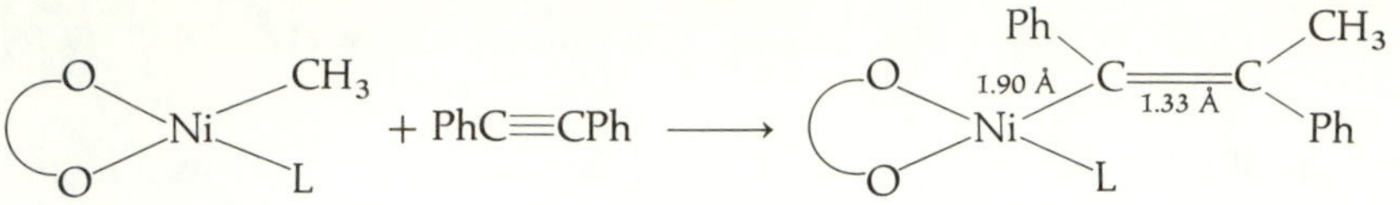

$$
\text{(acac)Ni(CH}_3\text{)(L)} + \text{PhC}\equiv\text{CPh} \longrightarrow \text{(acac)Ni(L)}\left[\text{C(Ph)}=\text{C(CH}_3\text{)(Ph)}\right] \qquad (9.20)
$$

L = Ph$_3$P, O    O = acac$^-$

$$
(\text{OC})_4\text{CoH} + \text{CF}_2\text{=CF}_2 \xrightarrow{\text{pentane}} (\text{OC})_4\text{CoCF}_2\text{CF}_2\text{H} \qquad (9.21)
$$

$$
(\text{OC})_5\text{MnH} + \text{CF}_3\text{C}\equiv\text{CCF}_3 \longrightarrow (\text{OC})_5\text{MnC(CF}_3\text{)}=\text{C(CF}_3\text{)(H)} \qquad (9.22)
$$

$$
\xrightarrow{\text{1,3-butadiene}} (\text{OC})_5\text{MnCH}_2\text{CH}=\text{C(H)(CH}_3\text{)} \qquad (9.23)
$$

*cis* and *trans* isomers

Insertions into metal-heteroatom bonds are shown in equations *9.24* and *9.25*. As also observed in equations *9.10* and *9.11*, the titanium and uranium atoms of these products coordinate to oxygen and nitrogen atoms within the ligand.

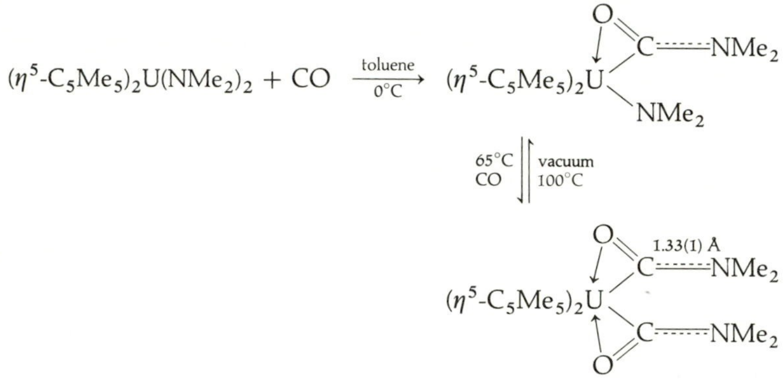

$$
(\eta^5\text{-C}_5\text{Me}_5)_2\text{U(NMe}_2)_2 + \text{CO} \xrightarrow[0°C]{\text{toluene}} (\eta^5\text{-C}_5\text{Me}_5)_2\text{U}\left[\text{OC}\text{-----NMe}_2\right]\text{NMe}_2
$$

$$
\Big\Updownarrow \begin{array}{l}65°C\\ \text{CO}\end{array} \quad \begin{array}{l}\text{vacuum}\\ 100°C\end{array} \qquad (9.24)
$$

$$
(\eta^5\text{-C}_5\text{Me}_5)_2\text{U}\left[\text{(OC-----NMe}_2\text{)}_2\right] \quad 1.33(1)\ \text{Å}
$$

$$
\text{TiCl}_4 + 4\text{RNC} \xrightarrow[20°C]{\text{CH}_2\text{Cl}_2} \text{Cl}_3(\text{RNC})\text{Ti}\left[\text{(ClC=NR)}_2\right]\text{Ti(CNR)Cl}_3 \qquad (9.25)
$$

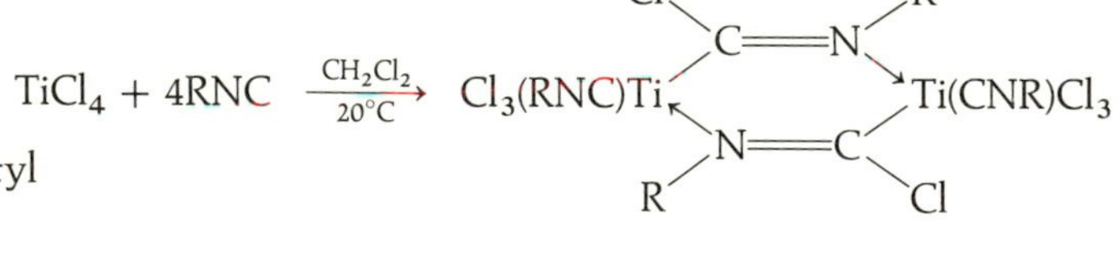

R = *t*-butyl

Some types of insertion reactions are better documented than others. Insertion of unsaturated organic molecules into M—C and M—H bonds (eqs. *9.19–9.23*) has received considerable attention because these reactions are fundamentally important steps in stoichiometric or catalytic applications of organometallic chemistry to organic synthesis. Insertion into M—M bonds is being studied more intensely at present.

Acylimidoyl ligands (eq. *9.11*) coordinate to metals as *mono-* or *dihapto* ligands in analogy to acyl ligands. Examples of $\eta^1$- and $\eta^2$-acylimidoyl, or iminoacyl, ligands in similar complexes are shown below as **9.1** and **9.2**, respectively.

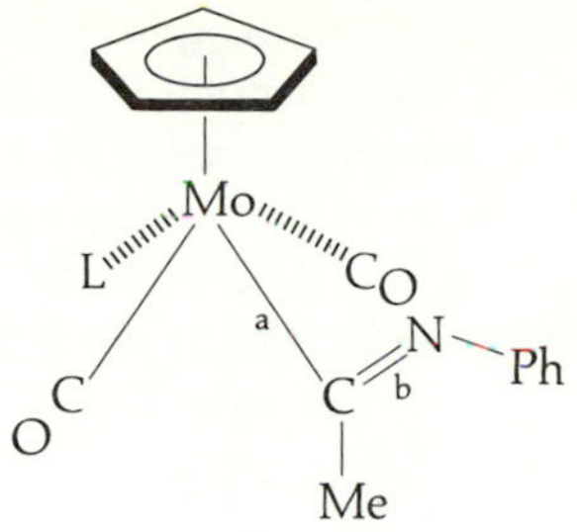

a = 2.27(1) Å
b = 1.27(2) Å

**9.1**

$(\eta^5\text{-}C_5H_5)Mo(CO)_2[P(OMe)_3](\eta^1\text{-}MeCNPh)$

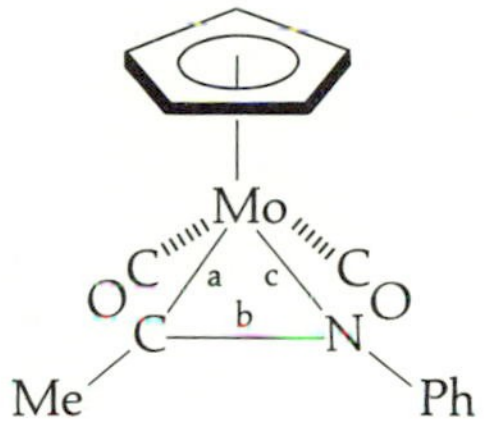

a = 2.143(4) Å
b = 1.233(6) Å
c = 2.106(5) Å

**9.2**

$(\eta^5\text{-}C_5H_5)Mo(CO)_2(\eta^2\text{-}MeCNPh)$

The molybdenum atom obeys the EAN rule in both complexes. In complex **9.1**, the Mo $\cdots$ N distance of 3.062(9) Å represents a nonbonding interaction. The corresponding Mo—N bonding distance in **9.2** is 2.106(5) Å.

The $\eta^1$-alkylnitroso complex formed in equation 9.17 appears to result from a *formal* insertion of NO into a Co—Et bond. Reaction of paramagnetic alkyl complexes with NO usually affords $\eta^2$-alkylnitroso complexes, as shown in **9.3** and **9.4**.

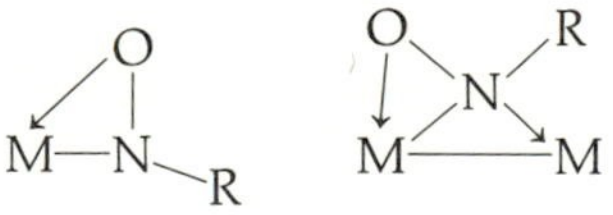

**9.3**  **9.4**

However, the reactant of the reaction of equation 9.17 is a diamagnetic alkyl complex. Such compounds usually react with NO to give an NO coupled N-alkyl-N-nitrosohydroxylaminato ligand as shown in equation 9.26.

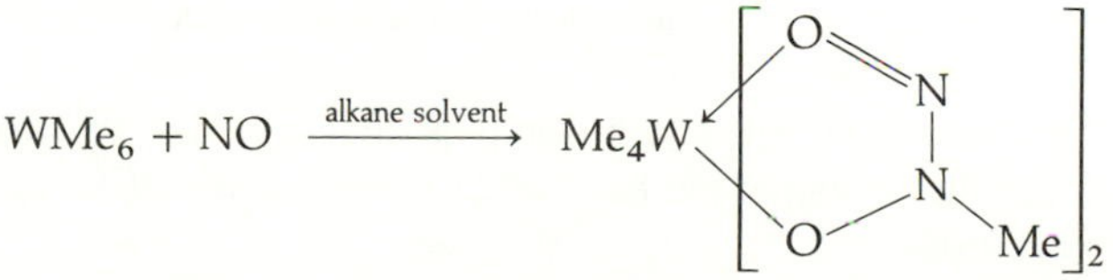

$$WMe_6 + NO \xrightarrow{\text{alkane solvent}} Me_4W \qquad\qquad (9.26)$$

Of course, the changes in topology observed in going from reactants to products may not define the actual mechanism of the reaction. The question of reaction mechanism for several of these insertion reactions is discussed below.

A related type of addition reaction is observed when the initial M—R bond is *multiple* (eqs. 9.27 and 9.28). A 1,1-addition of XY to an M=Z bond or two 1,1-additions to an M≡Z triple bond afford metallacyclic products that retain a formal M—Z single bond.

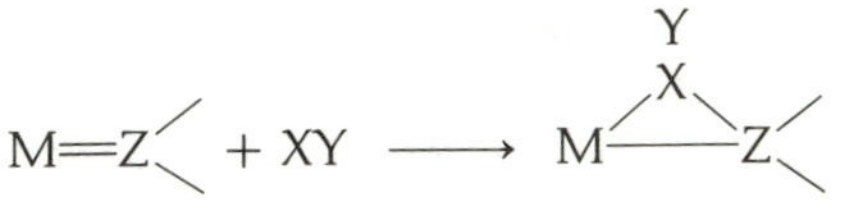

$$M{=}Z{\big\langle} + XY \longrightarrow M{\underset{X}{\overset{Y}{\diagup}}}Z{\big\langle} \tag{9.27}$$

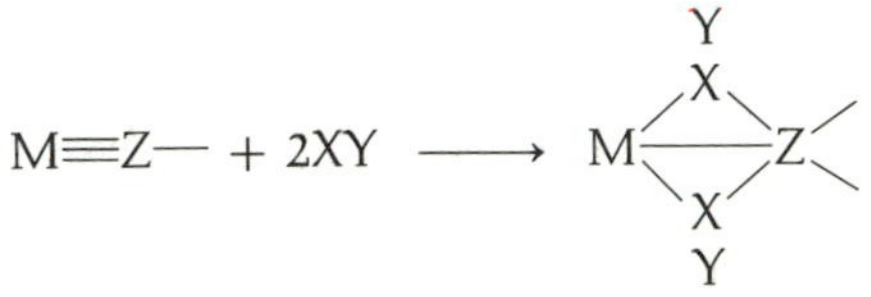

$$M{\equiv}Z{-} + 2XY \longrightarrow M{\Big\langle}{\overset{\overset{Y}{|}}{\underset{\underset{Y}{|}}{X}}}{\Big\rangle}Z{\big\langle} \tag{9.28}$$

Table 9.2 shows several such additions when Z is either a metal or a carbon atom. Specific examples cited in Table 9.2 are shown in equations 9.29–9.33.

Compound **9.5** is the first intermediate in diazoalkane-metal reactions (eq. 9.29) to be structurally characterized. The $\mu_2$-diazodiphenylmethane ligand acts as an unsymmetrically bridging four-electron donor. A semi-bridging $\mu_2$-CO ligand helps equalize the unequal charge density on the two Mo atoms.

$$[(\eta^5\text{-}C_5H_5)Mo(CO)_2]_2 + Ph_2CN_2 \longrightarrow$$

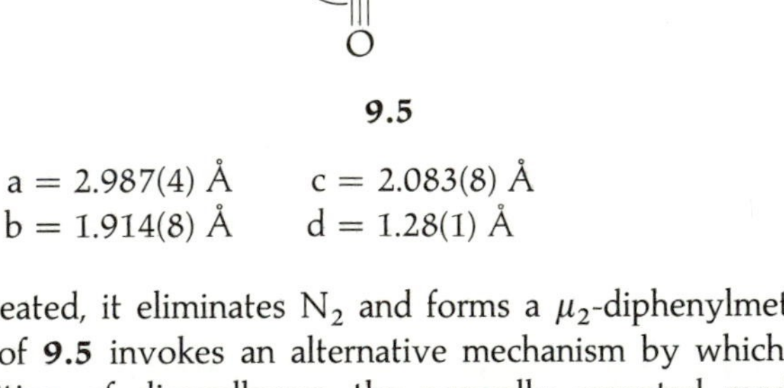

**9.5**

$$a = 2.987(4)\ \text{Å} \qquad c = 2.083(8)\ \text{Å}$$
$$b = 1.914(8)\ \text{Å} \qquad d = 1.28(1)\ \text{Å}$$

When complex **9.5** is heated, it eliminates $N_2$ and forms a $\mu_2$-diphenylmethylene complex. The isolation of **9.5** invokes an alternative mechanism by which atoms catalyze the decomposition of diazoalkanes; the normally accepted mechanism for this reaction is attack by the methylene carbon at a metal atom, thereby displacing $N_2$.

Complex **9.6** forms from a reactant complex that contains an Mo—Mo triple bond, and it is the first example we have encountered of a $\mu_2$-CO ligand bridging an M=M double bond.

**TABLE 9.2  Selected insertions across M—Z multiple bonds.**

| M—Z bond | Inserting molecule | Product | Equation |
|---|---|---|---|
| M=M | $N_2CH_2$ | 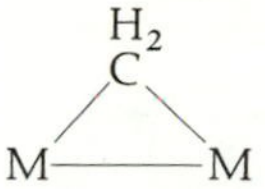 $\begin{array}{c} H_2 \\ C \\ \diagdown \\ M————M \end{array}$ | 8.65 |
| M≡M | $N_2CR_2$ | 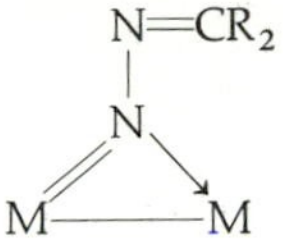 $\begin{array}{c} N{=}CR_2 \\ \mid \\ N \\ \diagdown \\ M————M \end{array}$ | 9.29 |
| | CO | 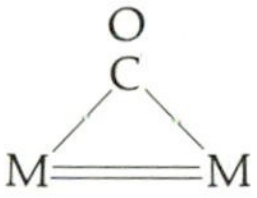 $\begin{array}{c} O \\ C \\ \diagdown \\ M{=}{=}M \end{array}$ | 9.30 |
| M=C< | $PtL_2$ | 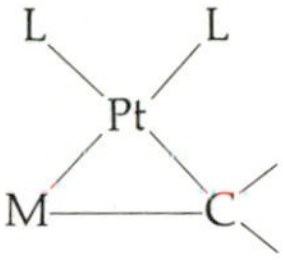 $\begin{array}{c} L \quad L \\ \diagdown \quad / \\ Pt \\ \diagdown \\ M————C< \end{array}$ | 9.31 |
| M≡C— | AgCl | 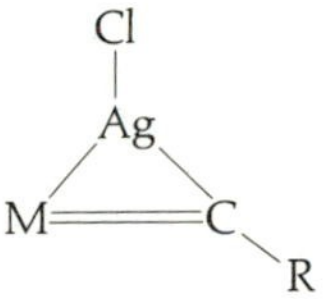 $\begin{array}{c} Cl \\ \mid \\ Ag \\ \diagdown \\ M{=}{=}C{\diagdown}R \end{array}$ | 9.32 |
| | RNC | 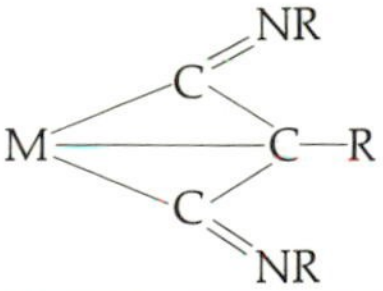 $\begin{array}{c} NR \\ C{<} \\ M————C{-}R \\ C{<} \\ NR \end{array}$ | 9.33 |

$$2Mo_2(OR)_6 \xrightarrow[25°C,\ 1\ atm]{6CO} $$

2.02(1) Å   2.498(1) Å   2.02(1) Å

**9.6**

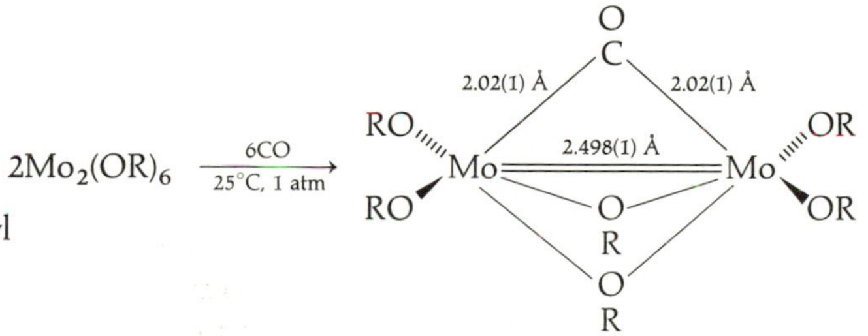

R = *t*-butyl

(9.30)

Complexes **9.7** and **9.8** represent addition of a metal fragment across an M—Z multiple bond to give dimetalla-cyclopropane and cyclopropene rings, respectively, as shown in equations *9.31* and *9.32*. Formal insertion of two RNC molecules into an M≡CR bond affords only an M—CR single bond in the product (eq. 9.33). Notice that the product complex **9.9** as written, is the "all-sigma" representation of an $\eta^3$-allyl coordination of the newly formed ligand.

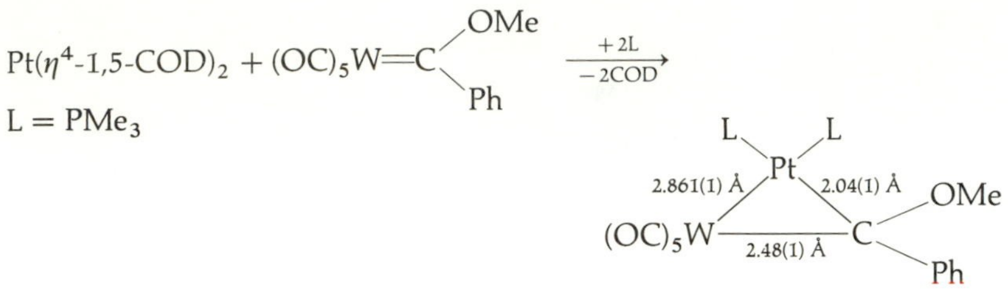

$$Pt(\eta^4\text{-}1,5\text{-COD})_2 + (OC)_5W{=}C\overset{OMe}{\underset{Ph}{\diagdown}} \xrightarrow[-2COD]{+2L}$$

L = PMe₃ ... *L = PMe$_3$*

(9.31)

**9.7**

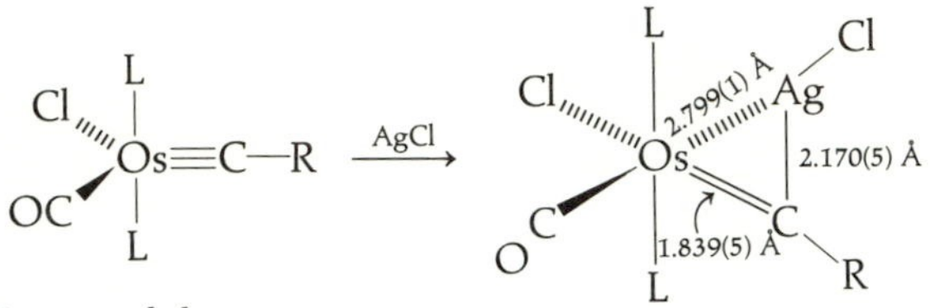

(9.32)

L = Ph₃P and R = p-tolyl ... *L = Ph$_3$P and R = p-tolyl*

**9.8**

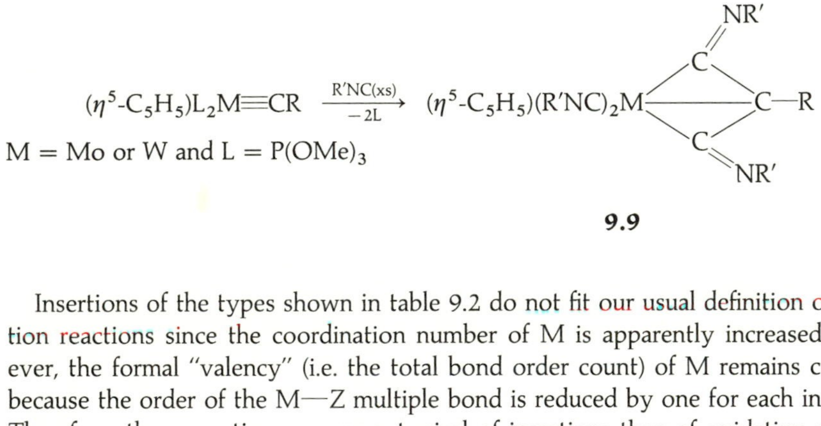

$$(\eta^5\text{-C}_5\text{H}_5)\text{L}_2\text{M}{\equiv}\text{CR} \xrightarrow[-2L]{R'NC(xs)} (\eta^5\text{-C}_5\text{H}_5)(R'NC)_2\text{M}$$

(9.33)

M = Mo or W and L = P(OMe)₃ ... *M = Mo or W and L = P(OMe)$_3$*

**9.9**

Insertions of the types shown in table 9.2 do not fit our usual definition of insertion reactions since the coordination number of M is apparently increased. However, the formal "valency" (i.e. the total bond order count) of M remains constant because the order of the M—Z multiple bond is reduced by one for each insertion. Therefore, these reactions are more typical of insertions than of oxidative-addition reactions.

A 1,2-insertion into an M=Z double bond is shown in equation 9.34. An $\eta^2$-alkene adds across the M—C $\pi$ bond of an alkylidene ligand forming a metallacyclobutane intermediate (**9.10**). Subsequent decomposition of **9.10** affords a new alkene and a different alkylidene complex. The overall reaction has effected *olefin metathesis* in that the original olefin, CB$_2$=CD$_2$, and alkylidene complex, M=CA$_2$, have interchanged methylene fragments.

$$\text{M=C} \underset{\text{A}}{\overset{\text{A}}{<}} + \text{CB}_2\text{=CD}_2 \longrightarrow \overset{\text{CB}_2\text{=CD}_2}{\underset{\downarrow}{\text{M=CA}_2}}$$

$$\text{M=C} \underset{\text{B}}{\overset{\text{B}}{<}} + \text{CA}_2\text{=CD}_2 \longleftarrow \quad \text{M} \underset{\underset{\text{A}_2}{\text{C}}}{\overset{\overset{\text{B}_2}{\text{C}}}{<}} \text{CD}_2 \qquad (9.34)$$

9.10

A specific example of such a conversion is shown in equation 9.35. The mechanism of olefin metathesis is discussed in more detail in chap. 14.

$$(\text{OC})_5\text{W=CPh}_2 + \text{CH}_2\text{=C} \underset{\text{OPh}}{\overset{\text{Ph}}{<}} \longrightarrow (\text{OC})_5\text{W=C(Ph)(OPh)} + \text{CH}_2\text{=CPh}_2 \quad (9.35)$$

## Mechanism of CO-Insertion Into M—R Bonds

Most stoichiometric or catalytic applications of organometallic chemistry to organic synthesis involve formation of new C—C bonds. Building more complex organic molecules from readily available simple feedstock chemicals is of particular interest. A fundamental reaction of this type is CO-insertion into an M-alkyl bond to form an acyl complex. This reaction increases the carbon chain of the alkyl ligand by one carbon atom and introduces a carbonyl function group (eq. 8.38).

Mechanistic understanding of CO-insertion comes from extensive study of CO-insertion into $CH_3Mn(CO)_5$, as shown in equation 9.36. This reaction is the best-studied of all insertion reactions.

$$\text{CH}_3\text{Mn(CO)}_5 + \text{CO} \longrightarrow \overset{\overset{\text{O}}{\|}}{\text{CH}_3\text{C}}\text{—Mn(CO)}_5 \qquad (9.36)$$

The three mechanisms shown in figure 9.1 are considered possible. Mechanism A requires that the *incoming CO molecule insert directly into the original Mn—CH$_3$ bond.* In mechanism B, a CO ligand *cis* to the methyl ligand *inserts directly* into the *original* Mn—CH$_3$ bond while the incoming CO occupies the vacated coordination site. Mechanism C entails *migration of the CH$_3$ ligand* from its original site *onto an adjacent CO ligand* with the incoming CO ligand occupying the coordination site where the methyl ligand was coordinated initially.

The mechanism is elucidated by conducting three experiments. When reaction 9.36 is carried out using isotopically labeled CO gas, the *only* reaction product is This result rules out mechanism A and establishes that thermal scrambling or permutation of CO ligands does not occur in this complex under these reaction conditions

Mechanism A: Intermolecular direct CO-insertion

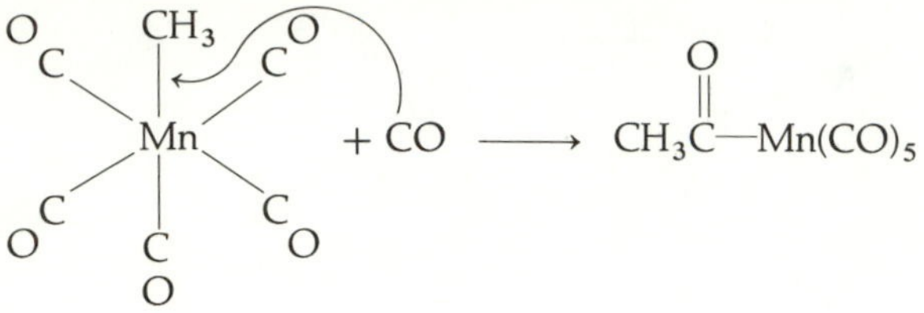

Mechanism B: Intramolecular direct CO-insertion

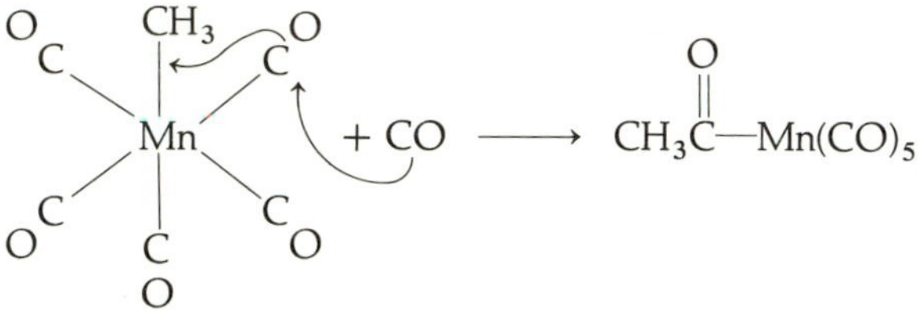

Mechanism C: Methyl group migration to an adjacent CO ligand

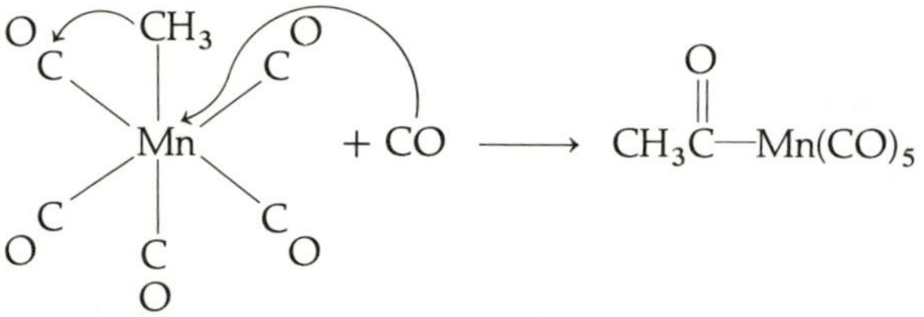

**FIGURE 9.1**  Postulated mechanisms for CO-insertion of $CH_3Mn(CO)_5$.

(otherwise about 20% *trans*-labeled complex would be observed). Therefore, **9.11** is the kinetically controlled product. Mechanisms B and C *cannot* be distinguished solely on the basis of equation *9.37*. Both mechanisms would afford **9.11** as product, as shown in equation *9.38*.

A second reaction verifies microscopic reversibility in a *decarbonylation* reaction (eq. *9.39*). Complex **9.12** is formed as the sole product from thermal decarbonyla-

$cis\text{-}CH_3\overset{\text{O}}{\overset{\|}{C}}\text{—}Mn(*CO)(CO)_4$:

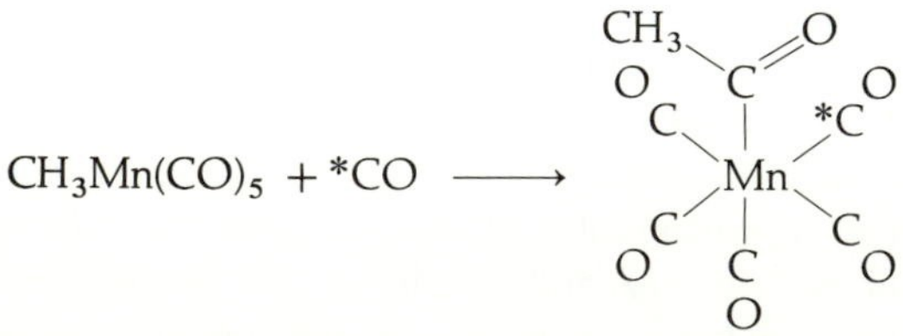

$$CH_3Mn(CO)_5 + {}^*CO \longrightarrow$$

(9.37)

**9.11**

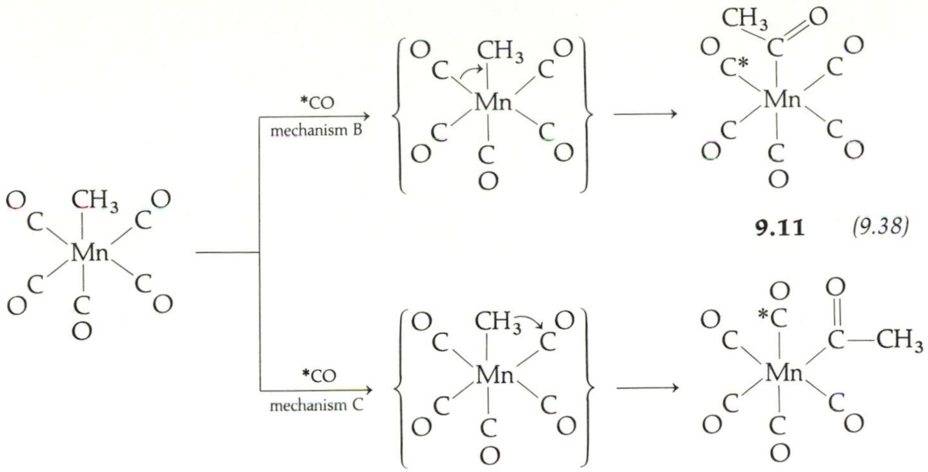

$\xrightarrow[\text{mechanism B}]{\text{*CO}}$

**9.11**          (9.38)

**9.11**

tion of acetyl-labeled acetylpentacarbonylmanganese. This result indicates loss of a *cis*-CO ligand during the decarbonylation reaction, i.e., the reverse of equation 9.36.

$$CH_3-{}^*\overset{\displaystyle O}{\overset{\|}{C}}-Cl + Na[Mn(CO)_5] \longrightarrow CH_3{}^*\overset{\displaystyle O}{\overset{\|}{C}}-Mn(CO)_5 \xrightarrow{\Delta}$$

$$cis\text{-}CH_3Mn({}^*CO)(CO)_4 + CO \quad (9.39)$$

**9.12**

A third reaction makes use of this decarbonylation to differentiate between mechanisms B and C. Decarbonylation of the labeled complex **9.13** affords a mixture of *trans*- and *cis*-$CH_3Mn(^*CO)(CO)_4$ in a 1:2 *ratio* (eq. 9.40). Figure 9.2 reveals that this ratio could be obtained *only from mechanism C*. In fact, mechanism B would not afford any *trans* isomer. Therefore, the term "CO-insertion" in this case is a misnomer for equation 9.36. This reaction is rather an *alkyl-group migration*.

$$cis\text{-}CH_3\overset{\displaystyle O}{\overset{\|}{C}}-Mn(^*CO)(CO)_4 \xrightarrow{\Delta} trans\text{- and } cis\text{-}CH_3Mn(^*CO)(CO)_4 \quad (9.40)$$

**9.13** (same as **9.11**)                    1    :    2

(*trans*- and *cis*- **9.12**)

Product analysis of isomeric mixtures like that required in equation 9.40 can be obtained from solution IR spectra of the $\nu_{CO}$ region. Isotopically labeled complexes of different geometries have different symmetries; therefore the number and relative intensities of the $\nu_{CO}$ bands are different for each complex. With $^{13}CO$, isotopic shifts of specific $\nu_{CO}$ bands are also observed (eq. 2.23).

Mechanisms B (reverse)

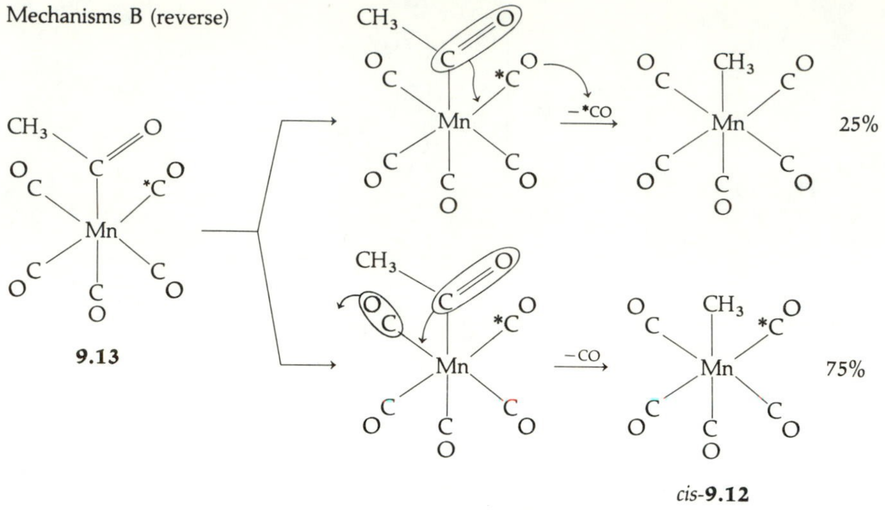


Mechanisms C (reverse)

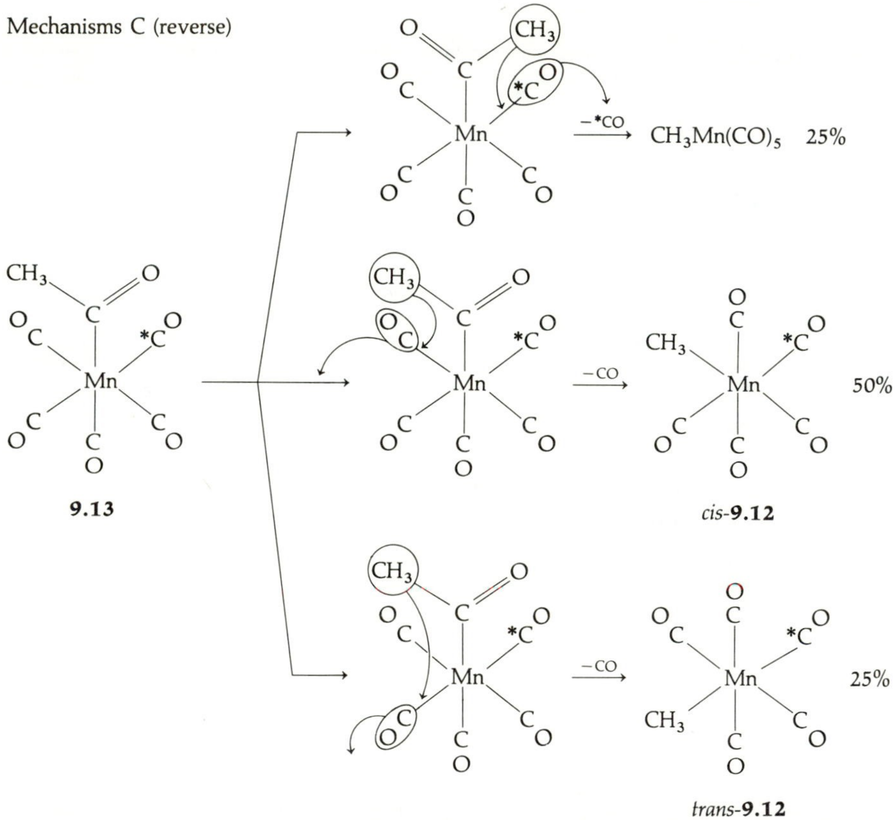


**FIGURE 9.2** Predicted product distribution of equation *9.40* according to mechanisms B and C.

Other two-electron donors, such as phosphines, isocyanides, or amines, also induce alkyl migration, as shown in equation *9.41*.

$$CH_3Mn(CO)_5 + Ph_3P \xrightarrow{\text{fast}}$$

$$cis\text{-}CH_3\overset{\overset{\textstyle O}{\|}}{C}\text{—}Mn(CO)_4(Ph_3P) \xrightleftharpoons{\text{slow}} \textit{trans} \text{ isomer} \quad (9.41)$$

The *cis* isomer is the kinetically controlled product, as expected from mechanism C, and this isomer undergoes slow isomerization to and eventual equilibration with the *trans* isomer. Frequently, the *trans* isomer is thermodynamically more stable.

Detailed kinetic studies reveal that the rate of alkyl migration is dependent on the concentration of alkyl manganese complex (a first-order dependence) but *independent* of the concentration of the incoming ligand L. Alkyl migration is highly stereospecific at manganese, giving an initial *cis* product, and probably involves retention of configuration of the migrating alkyl group. A mechanism consistent with these data is shown in equation *9.42*, and a probable *transition state species* for the migration is depicted as **9.15**. (The alkyl ligand presumably "slides over" onto an adjacent CO ligand.)

$$RMn(CO)_5 \xrightleftharpoons[k_{-1}]{k_1} [R\overset{\overset{\textstyle O}{\|}}{C}Mn(CO)_4] \xrightleftharpoons[k_{-2}(-L)]{k_2(+L)} cis\text{-}R\overset{\overset{\textstyle O}{\|}}{C}\text{—}Mn(CO)_4L \quad (9.42)$$

$$\textbf{9.14}$$

$$\textbf{9.15}$$

The first step of equation *9.42* is rate determining, affording a coordinately unsaturated intermediate (**9.14**), which then reacts rapidly with L to give the product complex. When L is a phosphine or amine, the reverse step, $k_{-2}$, can usually be neglected at ambient temperature. Values of $k_1$ when $CH_3Mn(CO)_5$ is treated with CO, $Ph_3P$, $(PhO)_3P$, or $C_6H_{11}NH_2$ in ether solvents are essentially the same, about $6 \times 10^{-4} \text{ sec}^{-1}$ at 26°C. This rate constant represents the rate of formation of the intermediate **9.14**. As **9.14** forms, the Mn—R bond is broken, and a terminal CO ligand is converted into an acyl ligand. One might expect the rate constant $k_1$ to be smaller for Mn—R bonds that have a high bond-disruption enthalpy. Experimental evidence supports this prediction.

Recently, methyl migration of $cis\text{-}CH_3Mn(CO)_4(^{13}CO)$ was examined using $^{13}$C-NMR at −115°C. Several solvents were used, and the incoming ligand L was either $^{12}CO$ or $P(OCH_2)_3CMe$. Experimental determination of the identity and relative abundances of the various product complexes verified an alkyl migration mechanism as shown in figure 9.2 (mechanism C). Furthermore, product distributions were consistent *only* with the intermediate **9.14** (R = $CH_3$) having a rigid square-pyramidal geometry in which the acetyl ligand occupies a basal site (**9.16**). The $\Delta G^*$ for structural rearrangement of **9.16** is $\geq 9$ kcal/mole, so this structure is quite stable at this low temperature.

**9.16** (one C is labeled)     **9.17** (one C is labeled)

Since **9.16** is coordinatively unsaturated, an $\eta^2$-acetyl structure (**9.17**) has been proposed for this intermediate. However, since postulating **9.17** is not necessary to explain the observed product distributions and since recent theoretical calculations do not reveal any additional stabilization upon forming an $\eta^2$-acetyl mode of coordination, intermediate **9.16** is assumed to be the form present in this system. Coordinatively unsaturated structures of the types **9.14–9.16** might be stabilized by solvent coordination. Geometrical isomerization of the kinetic *cis* product to the *trans* isomer (eq. *9.41*) could occur through five-coordinate intermediates such as **9.16** or through other geometrical isomers of **9.16.**

Similar studies of CO-insertion into M—R bonds in organometallic complexes of metals other than manganese reveal diverse stereochemistry. Some CO-insertions seem to be alkyl migrations to an adjacent CO ligand; others apparently involve direct CO-insertion. A few complexes appear to react by even more complicated mechanisms. In some cases, a polar solvent that is also a good donor, such as $CH_3CN$, can greatly accelerate the rate of CO-insertion. Clearly, more detailed mechanistic study is warranted.

CO-insertion is accelerated by the presence of Lewis acids. Molecular Lewis acids, such as $M'X_3$ compounds ($BF_3$, $AlCl_3$, or $AlBr_3$, for example) induce rapid alkyl migration in a variety of complexes even in the absence of an external ligand (eq. *9.43*). A halide atom of the $M'X_3$ reagent acts as the incoming ligand, which coordinates *cis* to the newly formed acyl ligand as in **9.18.**

$$(9.43)$$

**9.18**

Methyl migration of $CH_3Mn(CO)_5$ under CO gas (eq. *9.36*) is also accelerated by protic acids. The addition of trifluoroacetic acid increases the observed rate 7.2-fold. Rate enhancement by molecular Lewis acids and protic acids probably results from a lowering of the energy of the transition state species **9.15** by coordination to the latent acyl-oxygen atom, as shown in **9.19** and **9.20.**

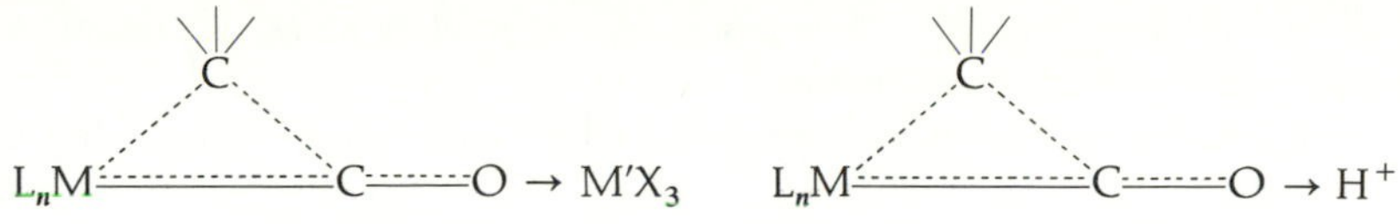

**9.19**                                   **9.20**

Lewis acid coordination to acyl-oxygen atoms is known and such observations support the supposition of intermediate adducts like **9.19** and **9.20**.

CO-insertions in $(\eta^5\text{-}C_5H_5)Fe(CO)(L)R$ complexes are particularly interesting because these complexes are common organometallic reagents. Furthermore, when L is other than CO, the resulting acyl complex is chiral at the pseudo-tetrahedral iron atom. Changes in carbon stereochemistry of the donor atom can be established, (eq. 9.44), as well as changes in stereochemistry at the iron atom.

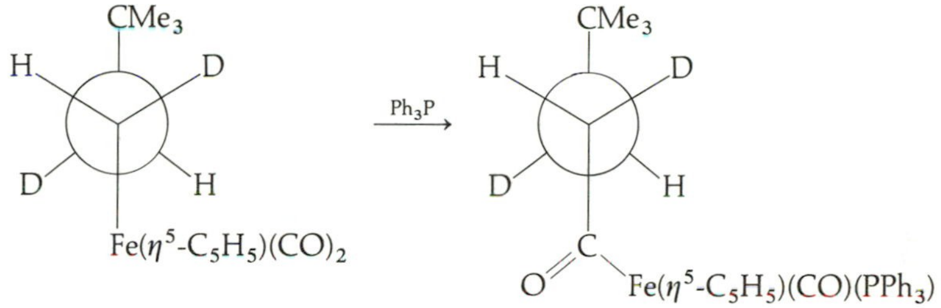

**9.21**                                   **9.22**

Triphenylphosphine-induced CO-insertion of the *erythro*-$(Me_3CCHDCHD)Fe(\eta^5\text{-}C_5H_5)(CO)_2$ alkyl complex **9.21** affords the *erythro*-acyl derivative **9.22** with $95+\%$ *retention of configuration at carbon*. The reason for using such an unusual alkyl ligand is that the configuration at the alkyl carbon atom can be determined by measuring the magnitude of $J_{H-H}$ coupling constants.

Stereochemical changes at the iron atom should provide additional information about the mechanism of CO-insertion as shown in equation *9.45*.

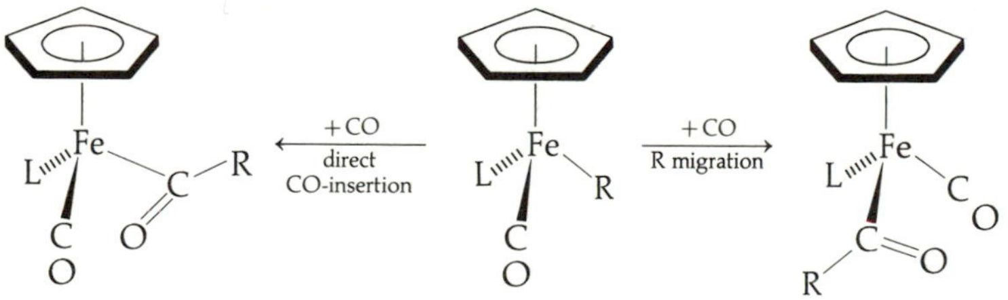

**9.24**                **9.23**                **9.25**

If an enantiomerically pure Fe-alkyl complex such as **9.23** undergoes direct CO-insertion to give **9.24**, then the stereochemistry at Fe would be retained. Alkyl migration, however, would lead to inversion of configuration at Fe in the acyl product, **9.25**.

When enantiomerically pure $(\eta^5\text{-}C_5H_5)Fe(CO)(L)Et$ complexes in which L is $Ph_3P$ or $P(OCH_2)_3CMe$ are treated with CO at 60 psi at 25°C in nitroethane solvent, acyl complexes form with 90 + % *inversion of configuration at Fe.* Therefore, alkyl migration is the presumed mechanism.

Two intermediate species have been proposed for insertion reactions of this type of iron complex:

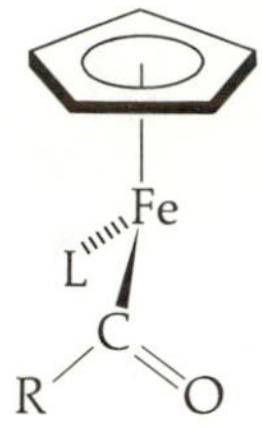

**9.26** 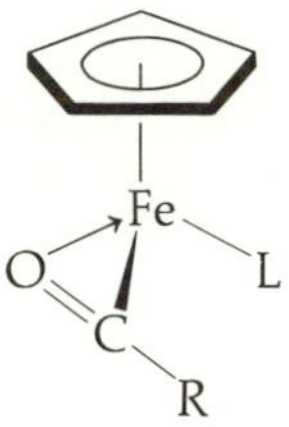          **9.27**

These structures differ only in containing $\eta^1$- or $\eta^2$-acyl ligands, respectively.

Photochemical *decarbonylation* of **9.28** in THF solution yields **9.29** with inversion of configuration at Fe. This result is also consistent with alkyl migration:

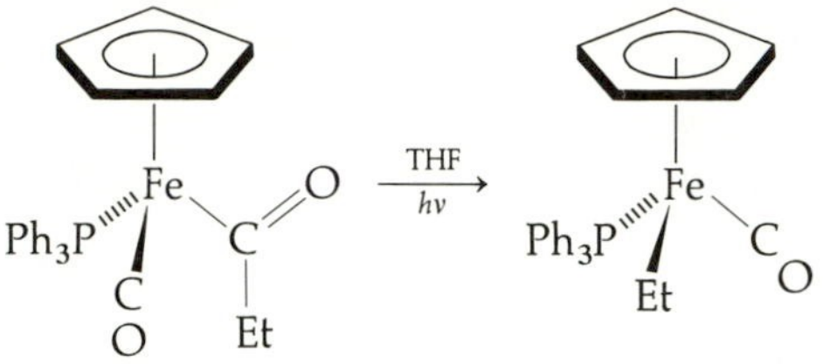

(9.46)

**9.28**                          **9.29**

Recently, thermal carbonylation of **9.30** was reported to give **9.31**, as shown in equation *9.47*, with 90 + % *retention of configuration at Fe.* This result implies a *direct CO-insertion* into the original Fe—CH$_3$ bond and thus is inconsistent with the results discussed above for the reaction of $(\eta^5$-C$_5$H$_5$)Fe(CO)(L)Et with CO in nitroethane solution. Clearly, a rationalization of these mechanistic differences is needed.

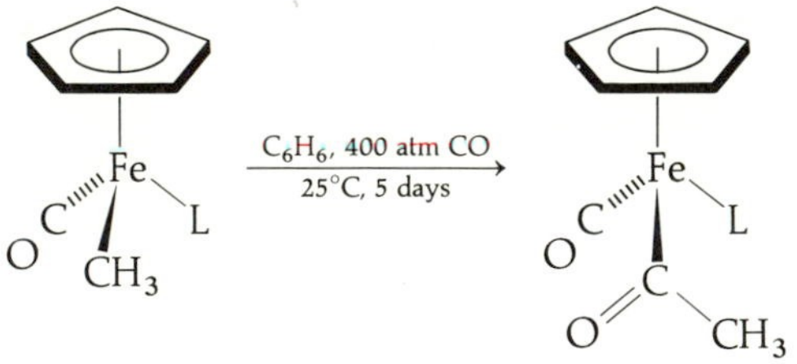

(9.47)

**9.30**                          **9.31**

L = PPh$_2$NR$_2$

Phosphine-induced CO-insertion in $(\eta^5$-C$_5$H$_5$)Mo(CO)$_3$R complexes (**9.32**) have also been studied. Kinetic data are explained best by assuming a *cis*-acyl kinetic product that isomerizes essentially completely to the *trans* isomer as shown in equation *9.48*. When X is Br, the rate of insertion is $4.3 \times 10^{-3}$ sec$^{-1}$ at 25°C, and the value of $\Delta H^{\ddagger}$ is $12 \pm 3$ kcal/mole. Again, alkyl migration appears to be the mechanism.

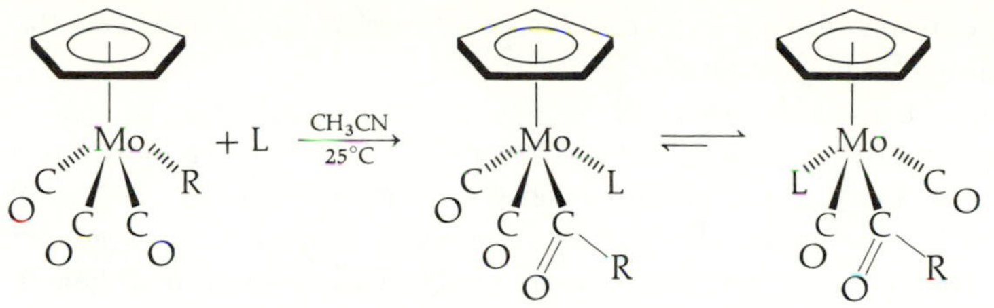

**9.32**         *cis*-acyl         *trans*-acyl      (9.48)

$R = CH_2CH_2CH_2X; L = Ph_3P$

Migratory insertion occurs when the osmium complex **9.33** is treated with iso-cyanide (eq. 9.49). In forming the propionyl complex **9.34**, an isocyanide displaces the very labile perchlorate ligand.

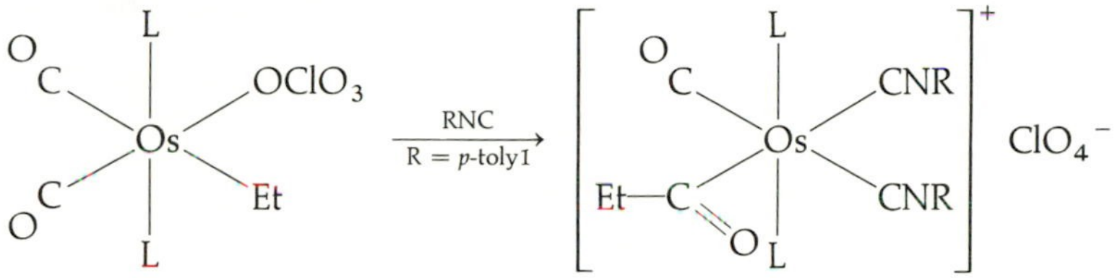

**9.33**                   **9.34**

$L = Ph_3P$

In analogy to CO-insertion, *alkyl migration to alkylidene ligands* has been observed. An example is shown in equation 9.50. Electrophilic attack on the methoxymethyl ligand in **9.35** generates the presumed methylidene complex intermediate **9.36** by abstraction of methoxide anion. Presumably methyl migration occurs spontaneously

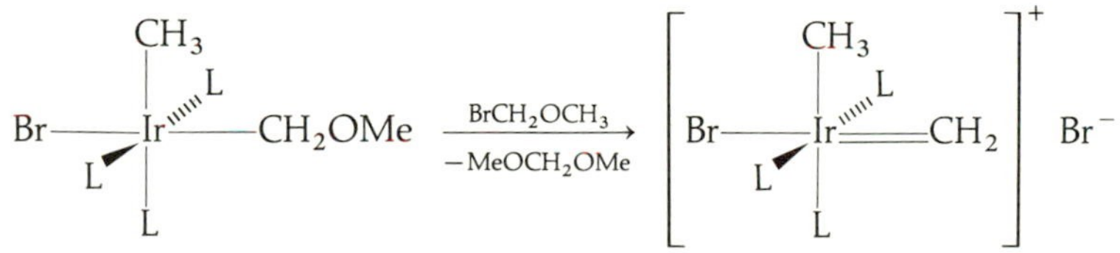

**9.35**                   **9.36**

$L = Me_3P$

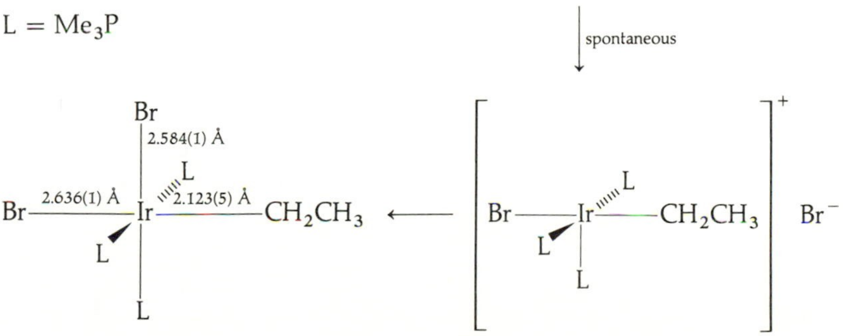

spontaneous

**9.38**                   **9.37**     (9.50)

to give a five-coordinate ethyl complex **9.37**, which then coordinates the bromide counterion affording the product **9.38**.

Like CO, isocyanides insert into M—C bonds with retention of configuration at the migrating carbon atom. These insertions have not been studied extensively.

A kinetic study of the NO-insertion shown in equation *9.17* reveals that a co-ordinatively unsaturated alkylnitroso complex is formed as an intermediate. This species, which might be solvated, reacts rapidly with the incoming ligand L. These observations are consistent with an NO-*migratory insertion* reaction.

### Mechanism of SO₂-Insertion Into M—R Bonds

Sulfur dioxide insertion into metal-alkyl or aryl bonds is now recognized to be much different mechanistically from CO-insertion. One possible complicating factor is that $SO_2$ can act as either a nucleophile or a strong electrophile. Thus, treating an (alkyl)metal-carbonyl complex such as **9.39** with $SO_2$ might induce CO-insertion:

$$RM(CO)_n + SO_2 \quad \xrightarrow{\quad\times\quad} \quad (OC)_{n-1}M\overset{\displaystyle O}{\underset{\displaystyle \underset{SO_2}{\uparrow}}{\overset{\|}{\underset{}{C}}}}R \qquad (9.51)$$

**9.39**

This type of reaction has not been observed, however. Another possible complicating feature is the variety of possible structures for SO₂-insertion products, as shown by **9.40–9.43**.

$$\underset{\textbf{9.40}}{M-\overset{\displaystyle O}{\underset{\displaystyle O}{\overset{\|}{\underset{\|}{S}}}}-R} \qquad \underset{\textbf{9.41}}{M-O-\overset{\displaystyle O}{\overset{\|}{S}}-R} \qquad \underset{\textbf{9.42}}{M-\overset{\displaystyle O}{\overset{\|}{S}}-OR} \qquad \underset{\textbf{9.43}}{M\overset{O}{\underset{O}{\diamondsuit}}S-R}$$

| **9.40** | **9.41** | **9.42** | **9.43** |
| S-sulfinate | O-sulfinate | O-alkyl-S sulfoxylate | O,O'-sulfinate |

For alkyl or aryl transition metal complexes, the most commonly observed structure for an insertion product is the *S-sulfinate.* Oxophilic metals, such as Ti and Zr, give stable *O-sulfinate* structures also. Insertion products containing *O,O'-sulfinate ligands* are observed less frequently because to form them either the coordination number of M must be increased by one relative to that of the reactant complex, **9.39**, or an ancillary ligand must be lost.

The observed preference of an S-sulfinate structure appears to be thermodynam-ically controlled. Both mechanistic and spectroscopic data support intermediate formation of O-sulfinate complexes (eq. *9.52*).

$$L_nM-R + SO_2 \quad \xrightarrow{k_1} \quad M-O-\overset{\displaystyle O}{\overset{\|}{S}}-R \quad \xrightarrow{k_2} \quad M-\overset{\displaystyle O}{\underset{\displaystyle O}{\overset{\|}{\underset{\|}{S}}}}-R \qquad (9.52)$$

When M obeys the EAN rule in the reactant complex, attack by $SO_2$ is believed to be direct, i.e., *not* involving prior coordination of $SO_2$ to M. In complexes where M is electron deficient, prior coordination of $SO_2$ is possible. Infrared spectroscopy is used to identify S-sulfinate ligands because they have two strong $v_{SO_2}$ bands in the ranges of $1250-1100$ cm$^{-1}$ and $1100-1000$ cm$^{-1}$. Other structural types have $v_{SO_2}$ bands around $1085-1050$ cm$^{-1}$ or lower.

Mechanistic information about $SO_2$-insertion reactions is scarce; however, some studies have been reported—particularly on $(\eta^5\text{-}C_5H_5)Fe(CO)_2R$, $RMn(CO)_5$, $(\eta^5\text{-}C_5H_5)Mo(CO)_3R$, or related complexes. Although the mechanism may differ for reactions conducted in neat $SO_2$ and those in solutions of $SO_2$ with organic solvents, the basic features of $SO_2$-insertion are the following:

1. A radical mechanism appears unlikely because free-radical scavengers do not affect the rate of insertion.
2. For R ligands of the type $R = CH_2X$, the rate of insertion increases as X becomes more electron releasing.
3. As R becomes bulkier, the rate of insertion decreases.
4. As ancillary ligands on M become better $\sigma$-donors, the rate of insertion increases.

These observed trends are consistent with electrophilic cleavage of the M—R bond by the incoming $SO_2$ molecule. Stereochemical results support a $S_E2$ (inversion or back-side) mechanism. As shown in equation 9.53, conversion of the alkyl complex (**9.44**) to the S-sulfinate complex (**9.45**) occurs with *inversion of configuration at the carbon donor atom.*

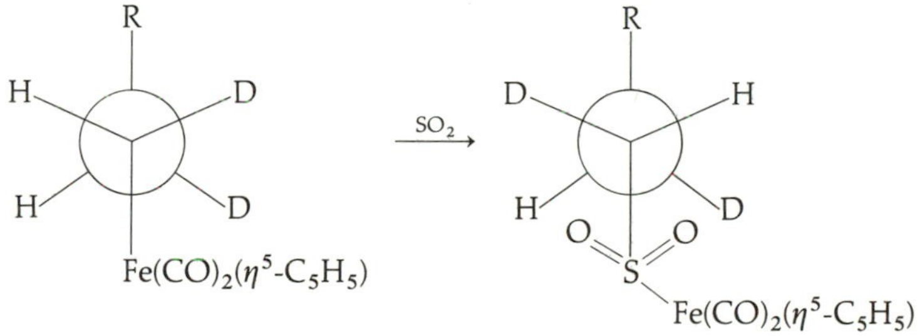

(9.53)

**9.44**                                             **9.45**

$R = CMe_3$ or Ph

X-ray diffraction data confirm that the conversion of alkyl complex **9.46** to the S-sulfinate product **9.47** *occurs with retention of configuration at Fe:*

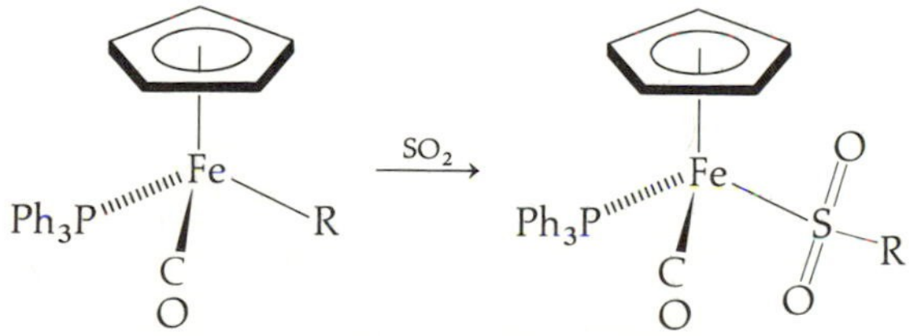

(9.54)

**9.46**                                             **9.47**

$R = $ *iso*-butyl

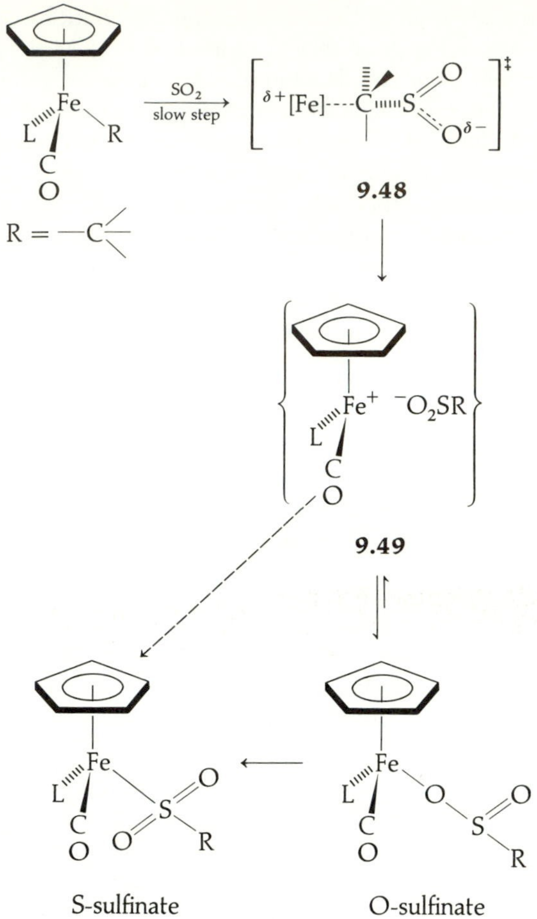

R = —C⟨

9.48

9.49

S-sulfinate                                    O-sulfinate

**FIGURE 9.3**   Accepted $S_E2$ mechanism for $SO_2$-insertion.

Since O-sulfinate complexes are observed spectroscopically as intermediates in several $SO_2$-insertion reactions, the currently accepted mechanism for $SO_2$-insertion is as shown in figure 9.3 for a pseudo-tetrahedral Fe-alkyl complex. An incoming $SO_2$ molecule attacks the Fe—C bond in an $S_E2$ mechanism, giving transition state species **9.48**. Inversion at the carbon donor atom results in a close ion-pair, **9.49**. The latter rearranges to give primarily the O-sulfinate product, which then isomerizes to the thermodynamically more stable S-sulfinate product complex. Direct conversion of **9.49** to the S-sulfinate complex might also occur. The configurational stability of the ion-pair must be quite high to explain the high degree of retention of configuration that is observed at Fe.

### Mechanism of Alkene- and Alkyne-Insertion Into M—H or M—R Bonds

Addition of M—H or M—R bonds across an olefinic bond (eq. 9.55) are fundamentally important reactions.

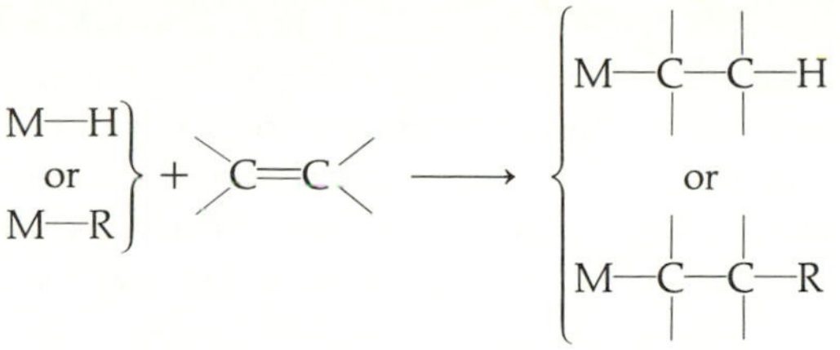

$$
\left.\begin{array}{c} M-H \\ or \\ M-R \end{array}\right\} + \begin{array}{c} \diagdown \\ \diagup \end{array} C = C \begin{array}{c} \diagup \\ \diagdown \end{array} \longrightarrow \left\{ \begin{array}{c} M-\underset{|}{\overset{|}{C}}-\underset{|}{\overset{|}{C}}-H \\ or \\ M-\underset{|}{\overset{|}{C}}-\underset{|}{\overset{|}{C}}-R \end{array}\right. \qquad (9.55)
$$

These additions provide a direct method of incorporating alkenes into organometallic complexes. Alkene feedstock chemicals can be activated toward further chemical reactivity by metal atoms. Subsequent reactions, such as CO-insertion, become facile, allowing the olefinic carbon chain to be increased. In some systems, these additions occur catalytically and can be reversible. Metal-promoted olefin hydrogenation, isomerization, or polymerization proceed through these insertion reactions. Elimination reactions, which are the reverse of these insertions, are discussed below, as well.

Kinetic and spectroscopic data indicate that M—H additions to olefinic bonds proceed through an $\eta^2$-alkene complex to afford an alkyl complex (eq. 9.56).

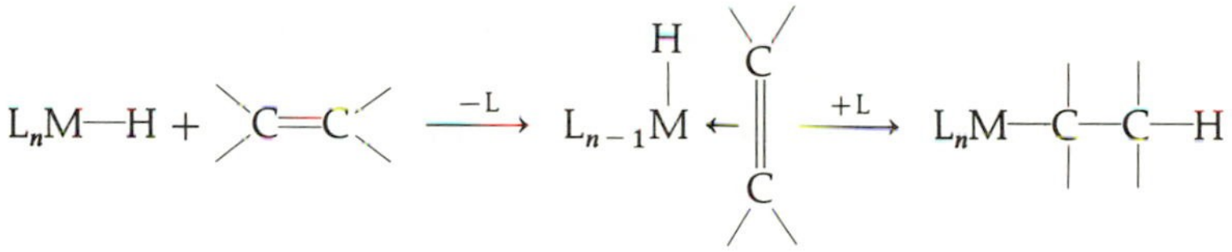

$$
L_nM-H + \begin{array}{c} \diagdown \\ \diagup \end{array} C = C \begin{array}{c} \diagup \\ \diagdown \end{array} \xrightarrow{-L} L_{n-1}M \leftarrow \begin{array}{c} H \\ | \\ \parallel \\ \end{array} \overset{\diagdown C \diagup}{\underset{\diagup C \diagdown}{\parallel}} \xrightarrow{+L} L_nM-\underset{|}{\overset{|}{C}}-\underset{|}{\overset{|}{C}}-H \quad (9.56)
$$

When L is CO, for example, an increase in CO pressure slows the rate of reaction since $\eta^2$-alkene complexation is inhibited. Loss of an ancillary ligand, L, is usually required when the initial $L_nMH$ complex obeys the EAN rule.

Labeling studies suggest a 1,2-*syn-addition* of an M—H bond to an alkene, as shown in equation 9.57 for a zirconium complex. Most analogous M—H additions are assumed to be *syn*-additions. Equation 9.57 is an example of a *hydrozirconation reaction*.

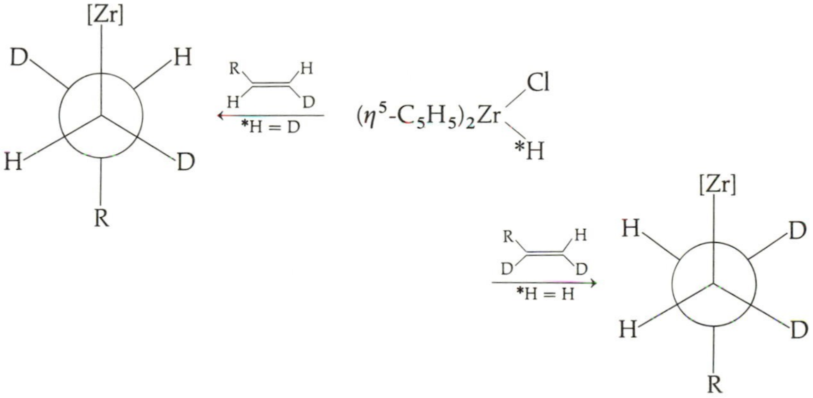

$$[Zr] = (\eta^5\text{-}C_5H_5)_2ZrCl; \; R = t\text{-butyl}$$

Similarly, many studies confirm that M—R bonds, in which R is alkyl or aryl, also undergo *syn*-addition to alkenes, and an $\eta^2$-alkene intermediate is proposed. These insertions are important to polymerization of olefins because repeated alkene insertions increase the length of the alkyl ligand. Retention of configuration at the migrating carbon center of R is observed in several reactions. This result is consistent with a 1,2-*syn*-addition and is presumed to characterize most M—R additions to alkenes.

Alkyne insertions into M—H or M—R bonds are also well known (eq. *9.58*).

$$L_nM-R + -C\equiv C- \xrightarrow{-L} L_{n-1}M \leftarrow \| \xrightarrow{+L} L_nM-C \qquad (9.58)$$

R = H, alkyl, or aryl

An $\eta^2$-alkyne intermediate undergoes M—R addition to afford an $\eta^1$-alkenyl product. The relative orientation of M and R in the $\eta^1$-alkenyl product reveals the stereochemistry of the M—R addition provided that isomerization does not occur under the reaction conditions.

Alkyne insertions into M—H bonds occur by both *syn* and *anti* mechanisms and frequently by mixed *syn*- and *anti*-mechanisms, although some complexes give high stereoselectivity. The Zr—H complex shown in equation *9.59* adds to 2-pentyne to give only the two *E*-isomers of the $\eta^1$-alkenyl product. In general, insertion of alkynes into M—H bonds appears to be less stereospecific than is insertion of alkenes.

$$(\eta^5\text{-}C_5H_5)_2Zr\overset{H}{\underset{Cl}{\diagdown}} + Me-C\equiv C-Et \longrightarrow$$

$$\underset{Me}{\overset{[Zr]}{\diagdown}}C=C\overset{H}{\underset{Et}{\diagup}} + \underset{Et}{\overset{[Zr]}{\diagdown}}C=C\overset{H}{\underset{Me}{\diagup}} \qquad (9.59)$$

$$55 \quad : \quad 45$$

Alkyne insertion into M—R bonds is presumed to follow a *syn*-addition pathway. Radical or some other more complicated reaction mechanisms result in product mixtures. As shown in equation *9.60*, a Ni—Ph bond adds to dimethylacetylene to afford a *syn*-addition product.

$$Br-\underset{\underset{L}{|}}{\overset{\overset{L}{|}}{Ni}}-Ph + Me-C\equiv C-Me \longrightarrow Br-\underset{\underset{L}{|}}{\overset{\overset{L}{|}}{Ni}}-C\overset{\overset{Ph}{\diagdown}C-Me}{\underset{Me}{\diagup}} \qquad (9.60)$$

L = Ph$_3$P

In a related reaction, a Ni-alkyl or Ni-aryl bond adds to alkynes to give *both syn-* and *anti*-addition products *as kinetically observed products* (eq. *9.61*).

$$\text{(structure)} + R^2-C\equiv C-R^3 \longrightarrow$$

$$\text{9.50} \quad + \quad \text{9.51} \quad\quad (9.61)$$

where L = $Ph_3P$; $R^1$ = $CH_3$, $CD_3$ or Ph; $R^2$ = Ph or *t*-butyl;
$R^3$ = H, $CH_3$, $CD_3$ or Ph

By using proton NMR to follow selected reactions indicated in equation *9.61*, one can measure the ratio of the initially formed products **9.50** and **9.51** and observe their slow thermal equilibration to the final product ratio.

A reaction scheme consistent with the observed results is shown in figure 9.4. An $\eta^2$-alkyne complex is formed by dissociation of L. A *syn*-1,2-addition of the Ni—$CH_3$ bond occurs to form the unsaturated *cis*-$\eta^1$-alkenyl intermediate, **9.52.** This compound can undergo $Ph_3P$-catalyzed isomerization (possibly through an ylide intermediate) to and equilibration with *trans*-$\eta^1$-alkenyl intermediate, **9.53.** Subsequent recombination of **9.52** and **9.53** with L gives the kinetically observed products **9.50** and **9.51**. These complexes undergo very slow thermal equilibration. An important result of this study is that insertion may occur by *only syn*-addition even though the kinetically observed products appear to reflect at least partial *anti*-addition.

Metal complexes frequently catalyze acetylene insertion to form oligomers, co-oligomers, or polymers. In some cases, cyclic oligomeric organic products are obtained, as shown earlier in equation 5.71. In most instances, a *syn*-addition of the M—R bond to an alkyne appears to be the preferred insertion pathway. Use of these reactions in organic synthesis is discussed in chap. 12.

## ELIMINATION REACTIONS

The reverse reaction of an insertion like that of equation 9.1 is referred to as an *elimination* or *extrusion* reaction. The most thoroughly studied elimination reactions involve the thermal decompositions of metal-alkyl or metal-aryl complexes because of the great interest in transition metal hydrogenation and polymerization reactions. A general representation of elimination reactions is shown in equations 9.62–9.64.

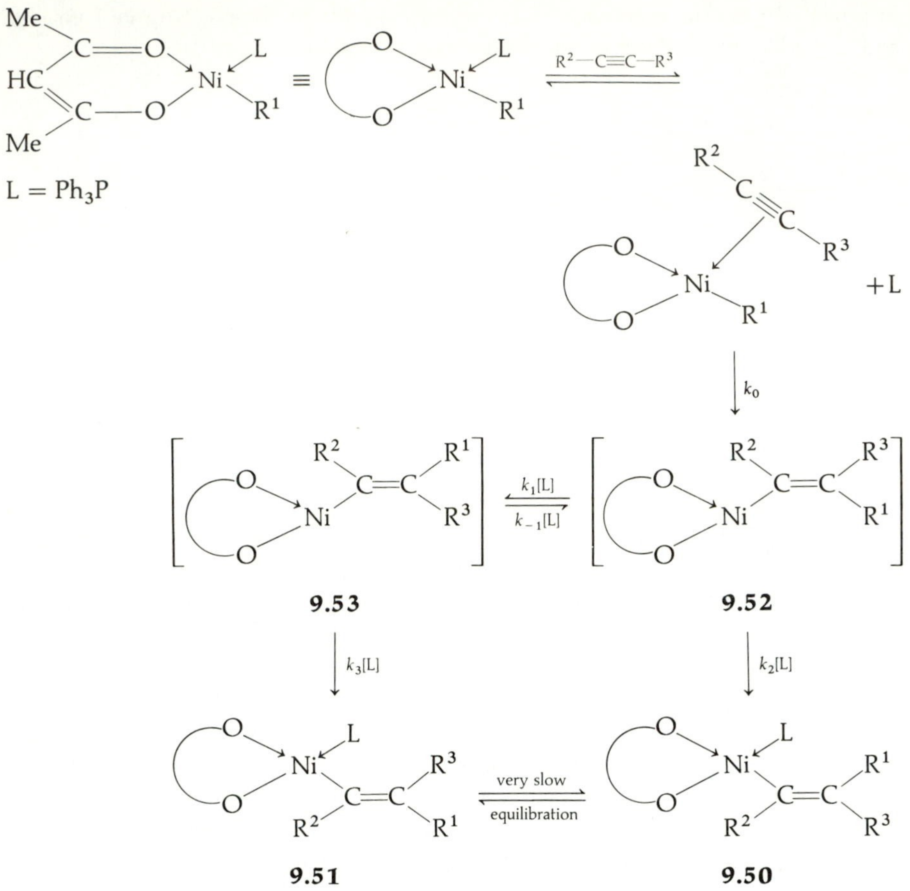

L = Ph$_3$P

**9.53**          **9.52**

**9.51**          **9.50**

**FIGURE 9.4**   Reaction scheme for Ni—R addition to an alkyne.

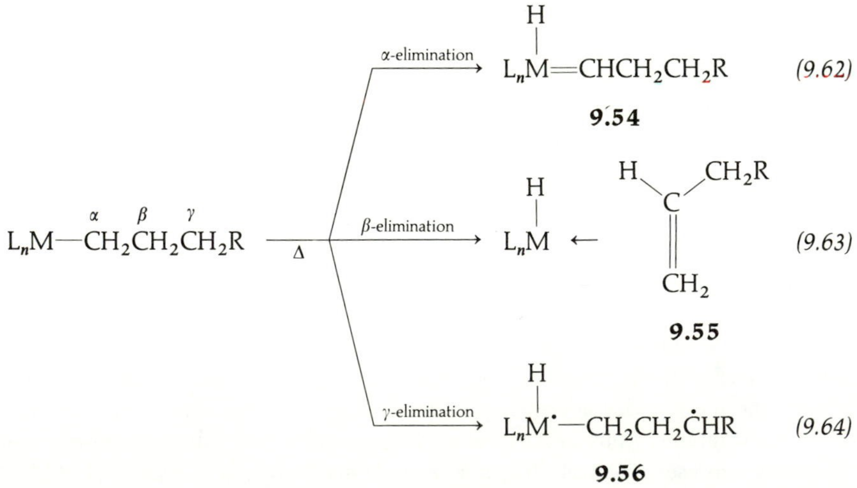

$\alpha$-elimination

$$L_nM=CHCH_2CH_2R \qquad (9.62)$$

**9.54**

$$L_nM\text{—}CH_2CH_2CH_2R$$

$\alpha \quad \beta \quad \gamma$

$\beta$-elimination

$(9.63)$

**9.55**

$\gamma$-elimination

$$L_nM^{\cdot}\text{—}CH_2CH_2\overset{\cdot}{C}HR \qquad (9.64)$$

**9.56**

Equation *9.62* is the reverse of a 1,1-addition (eq. *9.4*), in which a carbene formally inserts into an M—H bond (see eq. *8.51*). The alkylidene-hydride complex thus formed (**9.54**) may not be stable toward further decomposition although a tantalum complex of this type (**9.57**) has recently been isolated.

**9.57**                          **9.58**

$(\eta^5\text{-}C_5Me_5)Ta(CHCMe_3)(H)(PMe_3)Cl$       $(\eta^2\text{-}C_2H_4)Co(PMe_3)_3H$

Equation *9.63* is the reverse of a 1,2-addition by an M—H bond across an alkene (eq. *9.56*). *Dihapto*-alkene-hydride complexes, such as **9.55**, usually dissociate the alkene ligand to give free alkene and a metal-hydride complex. A cobalt complex of this type (**9.58**) is known, however.

More distal eliminations, such as $\gamma$-elimination (eq. *9.64*) are possible; however, radical intermediates like **9.56** should be very reactive. Cycloalkanes and metal-hydride complexes might be reasonable products. High-energy intermediates like **9.56** make such distant eliminations less likely reaction pathways.

Notice from equations *9.62–9.64* that the initial elimination products, as shown, require an increase in coordination number of the metal atom. If the original alkyl complex obeys the EAN rule, then these initial elimination products would be of exceptionally high energy. In such instances, either elimination is preceded by dissociation of an ancillary ligand L (the reverse of eq. *9.56*), or else thermal decomposition proceeds by homolytic cleavage of the M-alkyl bond, as shown in equation *9.65*.

$$L_nM{-}R \xrightarrow{\Delta} L_nM\cdot + R\cdot \longrightarrow L_nM{-}ML_n + R{-}R \qquad (9.65)$$

Relative thermal stabilities of various types of alkyl complexes provide considerable information regarding the mechanism of elimination reactions. Figure 9.5 shows the relative rates of thermal decomposition of $R_2Mn$ complexes (which are probably solvent stabilized) as a function of R. It is obvious that ethyl, *i*-propyl, and *t*-butyl complexes decompose thermally much faster than do methyl, benzyl, or neopentyl complexes.

Similar comparisons of the thermal stabilities of metal-alkyl complexes reveal the following trends:

1. Alkyl complexes in which the alkyl ligand *lacks* hydrogen substituents on the $\beta$-carbon atom are more stable thermally than those with $\beta$-hydrogens. For example,

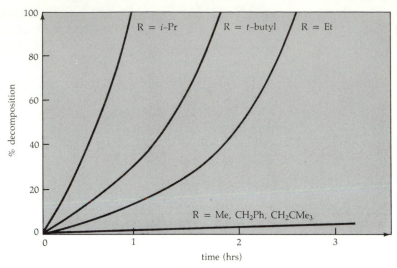

**FIGURE 9.5** Rates of thermal decomposition of selected $R_2Mn$ complexes in THF solution at 2°C. *Source: M. Tamura and J. Kochi, J. Organometal. Chem.,* **29**, *111 (1971).*

$WMe_6$ and $Cr(CH_2SiMe_3)_4$ are more stable thermally than the respective $WEt_6$ and $CrEt_4$ complexes.

2. Coordinatively saturated alkyl complexes are more stable thermally than coordinatively unsaturated complexes. A related observation is that binary alkyl complexes (referred to as *peralkyl* complexes) containing M in a high formal oxidation state are usually more stable than those having M in a low formal oxidation state. For example, thermal stabilities follow the trends ($\eta^5$-$C_5H_5)Mo(CO)_3Pr >$ $Pr_2Mo$; $EtMn(CO)_5 > Et_2Mn$; $WMe_6PR_3 > WMe_6$; $TiMe_4(dipy) > TiMe_4$; **9.59** $>$ **9.60**; and $WMe_6 \gg WMe_2$. In each pair of complexes, the more coordinatively saturated complex is the more thermally stable.

$$\left[ \underset{\underset{Me_2}{N}}{\overset{CH_2}{\bigcirc}} Cr \right]_3 \qquad (PhCH_2)_3Cr$$

**9.59**                                **9.60**

3. Complexes that are kinetically inert to ligand substitution, such as octahedral complexes of Cr(III), Co(III) and Rh(III), afford unusually stable alkyl complexes. For example, $[Rh(NH_3)_5Et]^{2+}$, vitamin $B_{12}$ coenzyme (which contains a Co(III)-alkyl bond) and **9.59** are stable complexes.

For quite some time, the observed thermal instability of transition metal-alkyl complexes compared to that of main-group metal-alkyl compounds was assumed to be the result of a very low transition metal-alkyl bond energy. However, thermo-chemical data, as shown in table 9.3, reveal that transition metal-alkyl bond disruption enthalpies are very similar to those of the main group metal-alkyls, even though the latter, are much more stable. Clearly, the relative thermal instability of transition metal-alkyls must be rationalized on a kinetic rather than a thermodynamic basis.

**TABLE 9.3   M—CH$_3$ bond-disruption enthalpies, D$_{M-Me}$, where M is a main-group or transition metal.**

| M | D$_{M-Me}$ (kcal/mole) | M | D$_{M-Me}$ (kcal/mole) |
|---|---|---|---|
| Mn | 28 | Hg | 29 |
| W | 38 | Zn | 42 |
| Re | 53 | Ge | 62 |
| Ti, Ta | 62 | Al | 66 |
| Zr | 74 | Si | 74 |
| Hf | 79 | B | 87 |

A mechanism postulated as the lowest-energy decomposition pathway of metal-alkyl complexes is the *β-hydrogen elimination* mechanism shown in equation *9.66*.

$$L_nM—\overset{\alpha}{C}H_2\overset{\beta}{C}H_2R \rightleftharpoons \left\{ \begin{matrix} H^{\cdots\cdots}CHR \\ | \quad | \\ M^{\cdots\cdots}CH_2 \end{matrix} \right\}^{\ddagger} \rightleftharpoons L_nM \leftarrow \begin{matrix} H \\ H \end{matrix}\diagdown\!\!\begin{matrix} R \\ C \end{matrix} \longrightarrow$$

transition state

intermediate

$$L_nM—H + H_2C{=}C(R)H \quad (9.66)$$

In a concerted process, a hydrogen atom substituent on the β-carbon atom of an alkyl ligand forms a four-centered transition state, which proceeds to an ($\eta^2$-alkene)metal-hydride complex. This complex frequently releases free alkene to give a metal-hydride coproduct. A key feature of this reaction is that formation of the ($\eta^2$-alkene)metal-hydride intermediate increases the coordination number of M by one. Therefore, any electronic or steric factors that hinder such an increase in the coordination number raise the energy of this intermediate (and of the transition state species as well) and so enhance the thermal stability of the initial alkyl complex.

This mechanism rationalizes the observed trends in thermal stability of alkyl complexes. First, aryl ligands and alkyl ligands that *lack* β-hydrogen atoms, such as CH$_2$SiR$_3$, CH$_2$CR$_3$, CH$_2$Ph, or CH$_3$, cannot undergo β-hydrogen elimination. These complexes decompose thermally by α-hydrogen elimination or by homolytic M—R bond rupture mechanisms, which are higher-energy pathways. Second, coordinatively saturated complexes (in which M may or may not obey the EAN rule) can undergo β-hydrogen elimination only through high-energy intermediates. For example, if the initial alkyl complex were octahedral, then the intermediate would be a higher-energy complex. Loss of an ancillary ligand would normally be required to afford facile β-hydrogen elimination. Prior dissociation of Ph₃P from complex **9.61** leads to a β-hydrogen elimination of alkene for precisely this reason (9.67).

$$(\eta^5\text{-}C_5H_5)Fe(CO)(Ph_3P)(alkyl) \xrightarrow{\Delta} (\eta^5\text{-}C_5H_5)Fe(CO)(Ph_3P)(H) + alkene \quad (9.67)$$

**9.61**

Thirdly, if ancillary ligands in coordinatively saturated complexes cannot dissociate easily, then these alkyl complexes should be quite stable thermally. This effect is observed for the types of kinetically inert octahedral complexes mentioned above. It is now possible to choose appropriate alkyl ligands to prepare relatively stable alkyl complexes of transition metals and even of lanthanide and actinide metals.

The thermolysis of *cis*-(Et₃P)₂PtEt₂ in cyclohexane solution illustrates the salient features of a β-hydrogen elimination reaction.

L = Et₃P

The original four-coordinate 16-electron alkyl complex, **9.62**, dissociates a Et₃P ligand to give the three-coordinate 14-electron T-shaped complex **9.63**. Beta-hydrogen elimination converts **9.63** to **9.64**, which is a four-coordinate 16-electron (η²-ethylene) Pt-hydride complex. The stereochemistry of this intermediate is not known. Reductive-elimination of ethane affords the two-coordinate Pt(0) complex, **9.65**, which then recombines with the dissociated ligand, L, to give the three-coordinate Pt(0) product, **9.66**.

One difficulty in ascertaining the stereochemistry of β-hydrogen elimination is that hydrogen scrambling frequently occurs at a rapid rate. For example, in equation 9.68 if β-hydrogen elimination (**9.63** to **9.64**) is rapidly reversible relative to the rate of conversion of **9.64** to **9.65**, then any deuterium labeling of the ethyl groups in **9.62** will lead to random labeling in the ethylene and ethane products. By using CH₂CD₃ ligands in **9.62**, the extent of D—H scrambling has been measured. With no added L, very little D—H scrambling occurred, an observation consistent with the slowest step being dissociation of L from **9.62**. However, with an added L concentration of 0.3 $M$ or greater, extensive D—H scrambling occurred. This result is consistent with a *reversible* β-hydrogen elimination and a *slow* subsequent reductive-elimination step.

An example of a successful stereochemical study of β-hydrogen elimination is shown in equation *9.69*.

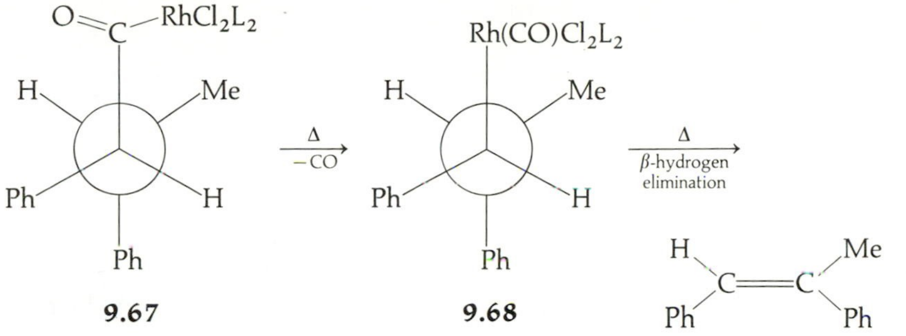

L = Ph₃P

The *threo*-Rh-acyl complex is decarbonylated thermally to the presumed *threo*-Rh-alkyl complex (**9.68**). Subsequent β-hydrogen elimination affords 90% *cis*- and only 10% *trans*-1,2-diphenylpropene. Similar thermal decomposition of the *erythro*-Rh-acyl derivative of **9.67** gives only *trans*-1,2,-diphenylpropene. These results are consistent with a *syn*-β-hydrogen elimination. Recall that 1,2-addition of an M—H bond to an alkene, the reverse of β-hydrogen elimination, proceeds by a *syn*-addition mechanism.

## ABSTRACTION REACTIONS

As depicted in equation *9.3*, abstractions are a type of elimination reaction in which the overall coordination number or valency of the metal atom remains unaltered. The term *abstraction* refers to a type of reaction rather than a precise reaction mechanism. Both inter- and intramolecular abstractions are known.

### Intermolecular Abstraction

Examples of intermolecular abstractions are shown in equations *9.70* and *9.71*.

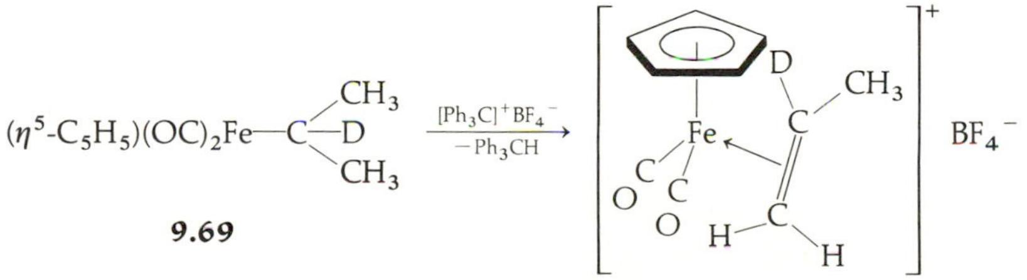

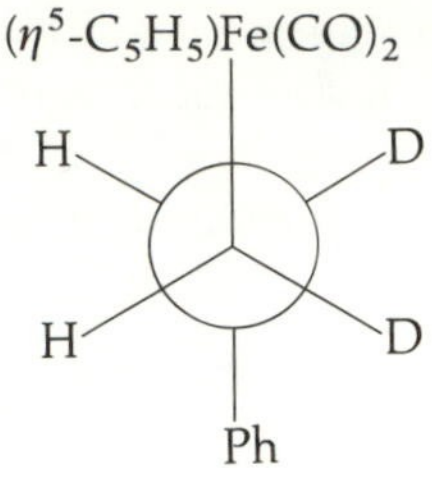

$(\eta^5\text{-}C_5H_5)Fe(CO)_2$

$$\xrightarrow[\text{(2) Ph}_3P]{\text{(1) [Ph}_3\text{C]}^+\text{BF}_4^-} \quad [(\eta^5\text{-}C_5H_5)Fe(CO)_2(Ph_3P)]^+ BF_4^-$$

**9.70**

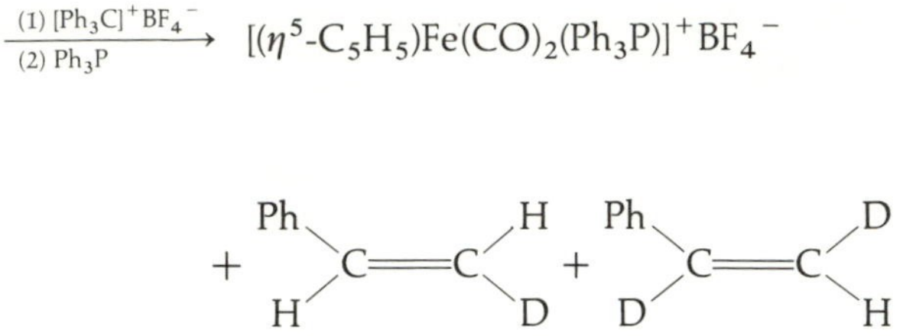

(9.71)

In each equation $Ph_3C^+$ abstracts a hydride anion from the $\beta$-carbon of an alkyl ligand affording cationic ($\eta^2$-alkene)Fe complexes. Displacement of the alkene ligand in equation 9.71 by addition of $Ph_3P$ permits the isolation of the free alkene products. The stereochemistry of the free alkenes indicates that *β-abstraction* occurred *anti* to the iron moiety in **9.70**.

Several examples of intermolecular *α-abstraction* reactions are shown in equations 9.50 and 9.72–9.76. In each case intermolecular abstraction of a substituent on the α-carbon atom of an alkyl or alkylidene ligand generates a metal-carbon bond of in-

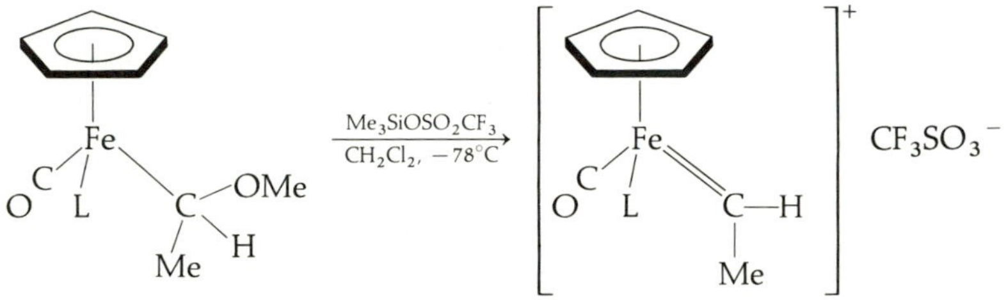

$$\xrightarrow[\text{CH}_2\text{Cl}_2,\ -78°C]{\text{Me}_3\text{SiOSO}_2\text{CF}_3}$$

L = Ph₃P or CO

$CF_3SO_3^-$

+ Me₃SiOMe   (9.72)

$$\xrightarrow[-78°C]{\text{HCl}}$$

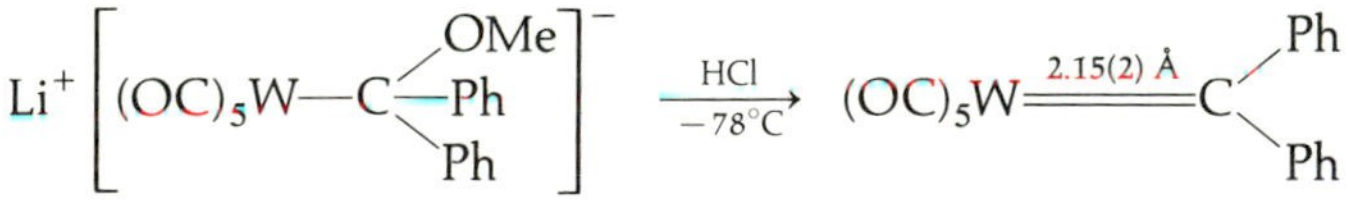

+ MeOH + LiCl   (9.73)

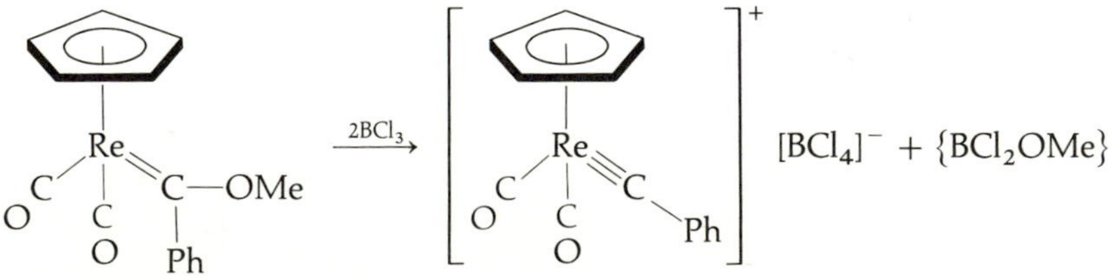

$$\xrightarrow{2\text{BCl}_3}$$

$[BCl_4]^- + \{BCl_2OMe\}$   (9.74)

$$[(\eta^5\text{-}C_5H_5)_2TaMe_2]^+BF_4^- \xrightarrow{\text{NaOMe}} \ \underset{136°}{} Ta \underset{2.03(1)\,Å}{\overset{2.25(1)\,Å}{\Longleftarrow}} \begin{matrix} CH_3 \\ H \\ C \\ \vdots \\ H \end{matrix} + MeOH + NaBF_4 \quad (9.75)$$

$$(OC)_5Cr=C\underset{CH_3}{\overset{OMe}{<}} \xrightarrow[\text{pentane}]{BI_3} \ I \overset{2.792(2)\,Å}{\underset{a}{\overbrace{\phantom{xx}}^{a}}} Cr \overset{1.69(1)\,Å}{\Longleftarrow} C \overset{1.49(2)\,Å}{\longrightarrow} CH_3 + \{BI_2OMe\} \quad (9.76)$$
$$\bar{a} = 1.95(1)\,Å$$

creased bond order. Notice that the electron count of the metal atom is the same in the reactant and product complexes. Examples of other types of apparent overall inter-molecular abstractions have been presented earlier, such as those involving $\pi$-bonded ligands (e.g., eqs. 5.17, 5.79, 5.84, and 5.85).

### Intramolecular Abstractions

Many intramolecular abstraction reactions are known, but their mechanisms are still unclear. Either a concerted mechanism (eq. 9.77) or a stepwise oxidative-addition and reductive-elimination sequence (eq. 9.78) would be a reasonable route to an overall abstraction reaction.

$$\begin{matrix} R & H \\ | & \\ L_nM-C-R' \\ | \\ R' \end{matrix} \longrightarrow \left[ \begin{matrix} R\cdots H \\ L_nM\cdots C-R' \\ R' \end{matrix} \right]^{\ddagger} \longrightarrow L_nM=C\overset{R'}{\underset{R'}{<}} + R-H \quad (9.77)$$

$$\begin{matrix} R & H \\ | & \\ L_nM-C-R' \\ | \\ R' \end{matrix} \xrightarrow[\text{addition}]{\text{oxidative-}} L_nM\overset{R\ H}{\underset{C}{\diagdown}}R' \xrightarrow[\text{elimination}]{\text{reductive-}} L_nM=C\overset{R'}{\underset{R'}{<}} + R-H \quad (9.78)$$

In the absence of more direct evidence, the presumed mechanism is postulated based on the known chemistry of the particular metal atom involved. Several types of in-tramolecular abstractions are known, depending upon the number of atoms between the metal atom and the center at which abstraction occurs. Equations 9.77 and 9.78 show intramolecular *α-hydrogen abstractions*. Most abstractions studied involve removal of hydrogen substituents from hydrocarbon ligands because of the great interest in activating ligand C—H bonds toward further reaction.

An interesting example of an intramolecular α-hydrogen abstraction is shown in equation 9.79. Complex **9.71** is an osmium cluster complex with a terminal methyl ligand. One of the C—H bonds in **9.71** is weakly bonded to one of the Os atoms as shown.

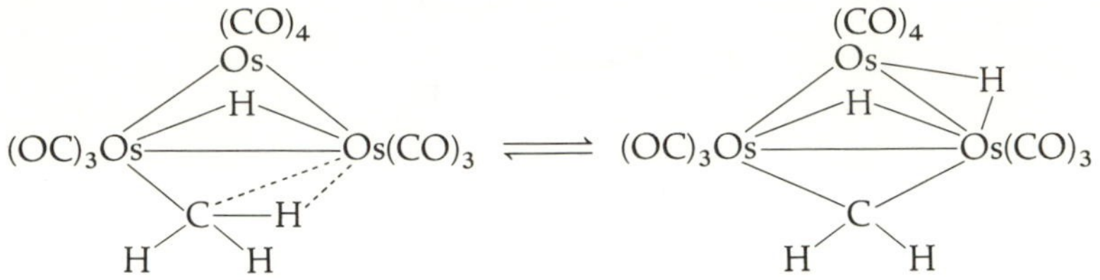

**9.71**                                                    **9.72**

$(\mu_2\text{-H})\text{Os}_3(\text{CO})_{10}(\text{CH}_3)$                    $(\mu_2\text{-H})_2\text{Os}_3(\text{CO})_{10}(\mu_2\text{-CH}_2)$

This alkyl complex is in thermal equilibrium with the ($\mu_2$-methylene)Os$_3$-hydride cluster (**9.72**), which is formed by abstraction of a methyl hydrogen atom to give an additional Os—C single bond and a second $\mu_2$-H ligand.

Tantalum alkylidene and alkylidyne complexes are also prepared by apparent intramolecular α-hydrogen abstractions (eqs. *9.80* and *9.81*).

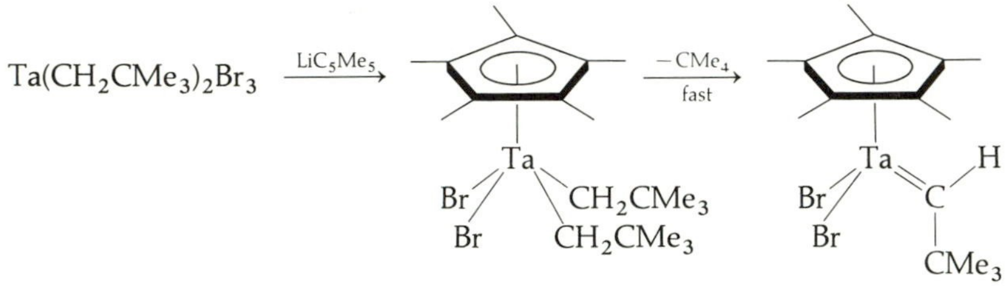

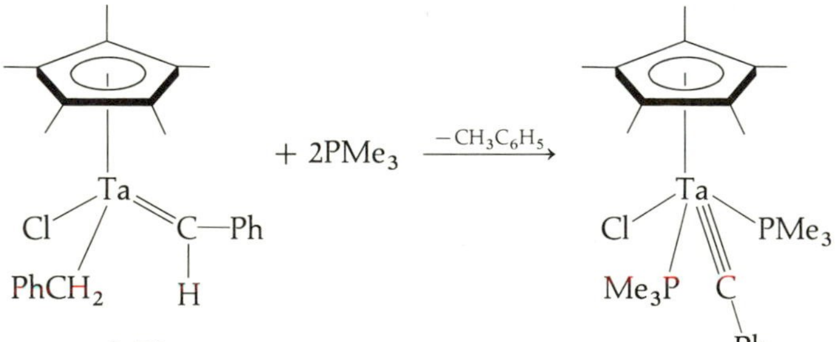

**9.73**

A concerted α-abstraction is postulated for these reactions based on the general lack of known oxidative-addition chemistry for Ta(V) or Ta(III). Considerable spectroscopic and structural data indicate that electron density in the $C_\alpha$-H bond of Ta-alkylidene complexes, such as **9.73**, interacts weakly with the metal orbitals of the highly electron-deficient (14-electron) Ta(III) atom, thus facilitating α-hydrogen abstraction. Interligand steric repulsion also enhances this reaction. Introduction of bulky phosphine ligands or $\eta^5$-C$_5$Me$_5$ ligands induces the facile loss of R—H for this reason. Furthermore, theoretical calculations show that a *complete* transfer of an α-hydrogen atom from the carbon donor atom to the metal atom (as in oxidative-addition) is difficult in five-coordinate 14-electron complexes because of an exceptionally large energy barrier.

An intramolecular $\beta$-hydrogen abstraction is shown in equation 9.82. A methyl ligand abstracts an *ortho*-hydrogen atom from the Ph ligand in **9.74** to generate the $\eta^2$-benzyne complex (**9.75**). An equivalent bonding description of **9.75** is as an *ortho*-phenylene ligand having two formal Ta—C $\sigma$ bonds as shown in **9.76**. A similar $\beta$-hydrogen abstraction is observed in equation 9.15.

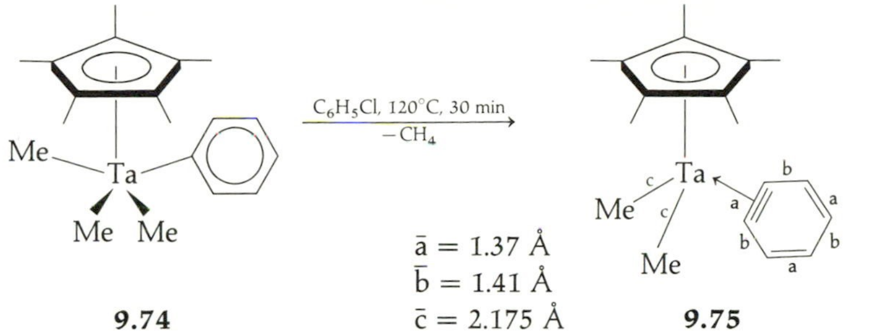

$$\xrightarrow[-CH_4]{C_6H_5Cl,\ 120°C,\ 30\ min}$$

$$\bar{a} = 1.37\ \text{Å}$$
$$\bar{b} = 1.41\ \text{Å}$$
$$\bar{c} = 2.175\ \text{Å}$$

**9.74**

$(\eta^5\text{-}C_5Me_5)TaMe_3(Ph)$

**9.75**

$(\eta^5\text{-}C_5Me_5)TaMe_2(\eta^2\text{-}C_6H_4)$

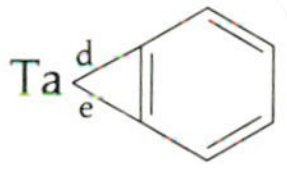

$$d = 2.059(4)\ \text{Å}$$
$$e = 2.091(4)\ \text{Å}$$

**9.76**

(9.82)

One type of intramolecular abstraction which yields cyclometallated products is a *concerted ortho-metallation reaction*. Triphenylphosphine frequently undergoes *ortho*-metallation, as shown schematically in equation 9.83.

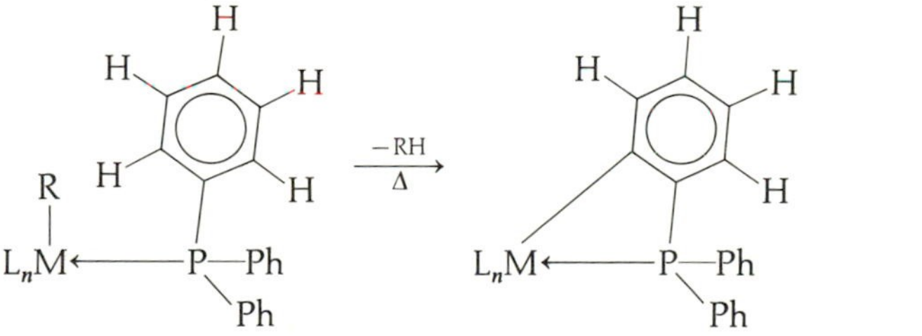

$$\xrightarrow[\Delta]{-RH}$$

(9.83)

**9.77**

The cyclometallated product, **9.77**, can be formed by a concerted abstraction or a stepwise oxidative-addition, reductive-elimination mechanism. Since most *ortho*-metallations are presumed to occur by a stepwise process, this reaction type is discussed more completely in the next chapter. However, an *ortho*-metallation that is believed to occur by a concerted abstraction process is shown in equation 9.84. The octahedral Ru(II) complex (**9.78**) would presumably not undergo initial oxidative-addition through a high-energy seven-coordinate intermediate, but predissociation of one of the ancillary ligands followed by a rapid stepwise process cannot be ruled out in the formation of **9.79**.

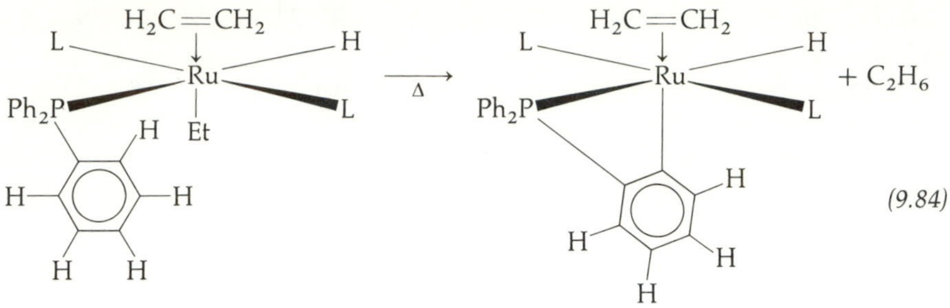

(9.84)

**9.78**

**9.79**

$L = Ph_3P$

## STUDY QUESTIONS

1. Compare the essential features of "CO-insertion" of $MeMn(CO)_5$ to $SO_2$-insertion of $(\eta^5\text{-}C_5H_5)Fe(CO)(L)Me$.

2. Consider the phosphine (L)-induced *alkyl migration* of *cis*-$MeMn(CO)_4(*CO)$. For *each* type of intermediate—a square pyramid like **9.16**; a trigonal bipyramid (axial acyl); and a trigonal bipyramid (equatorial acyl)—determine the structures of the various possible *cis*-acetyltetracarbonylphosphinemanganese products and determine the relative amounts of all possible products for each mechanism. [See T. C. Flood, J. E. Jenson, and J. A. Statler, *J. Am. Chem. Soc.* **103**, 4410 (1981).]

3. The acetyl $\nu_{CO}$ stretching frequency of $CH_3\overset{\overset{\displaystyle O}{\|}}{C}Mn(CO)_5$ is $\sim 1663$ cm$^{-1}$ in solution. When trifluoroacetic acid is added to this solution, the acetyl $\nu_{CO}$ frequency drops to $1581$ cm$^{-1}$. Explain this result.

4. Locate a literature reference pertaining to a hydrosilation reaction. How is this reaction type related to insertion reactions discussed in this chapter?

5. What feature of the mechanism shown in figure 9.4 is supported by the experimental data shown below. [See J. M. Huggins and R. G. Bergman, *J. Am. Chem. Soc.* **101**, 4410 (1979).]

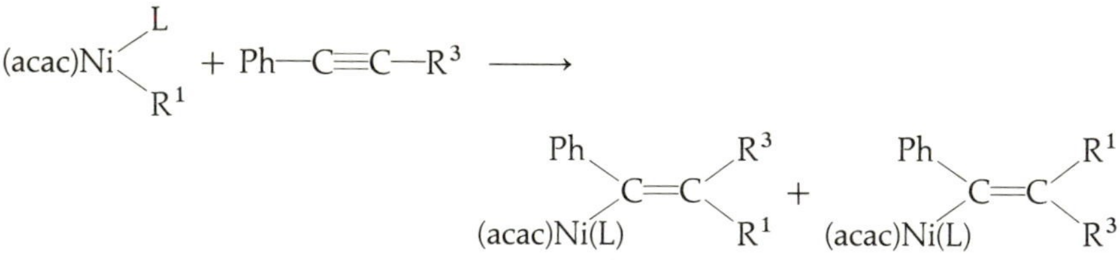

| $R^1$ | $R^3$ | Kinetic products (%) | | Thermodynamic products (%) | |
|---|---|---|---|---|---|
| | | cis | trans | cis | trans |
| $CH_3$ | $CD_3$ | 65 | 35 | 50 | 50 |
| $CD_3$ | $CH_3$ | 61 | 39 | 50 | 50 |

**6.** Characterize the apparent reaction types for the following equations:

**(a)** $(\eta^5\text{-}C_5Me_5)Ta(CHCMe_3)(H)(PMe_3)Cl \xrightarrow[\substack{60°C}]{Me_3P}$

$$H_2 + (\eta^5\text{-}C_5Me_5)Ta(CCMe_3)(PMe_3)_2Cl$$

**(b)** $MeMn(CO)_5 + Ph\text{—}C\equiv C\text{—}H \longrightarrow$

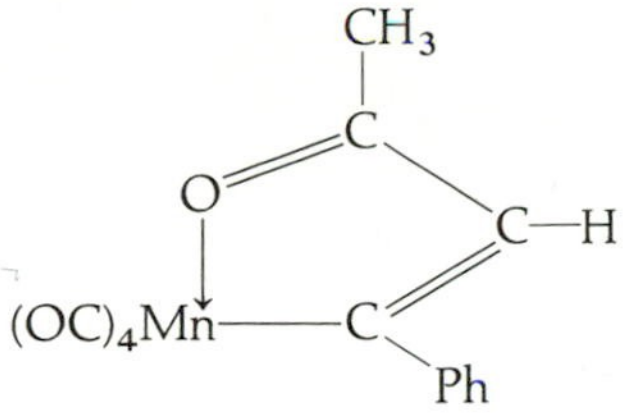

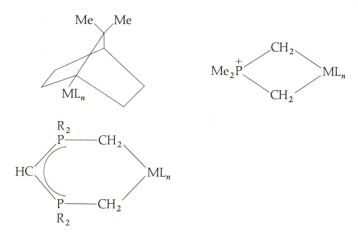

**(c)** $MeRh(PPh_3)_3 \xrightarrow{\Delta} CH_4\uparrow +$

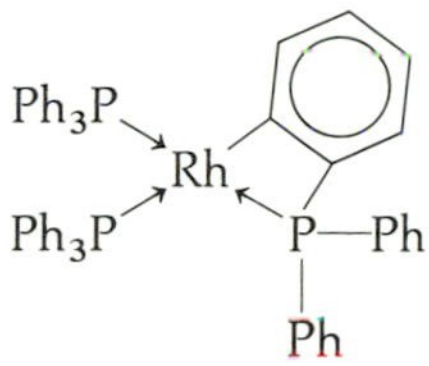

**7.** Explain why complexes containing the following alkyl ligands are quite stable thermally:

**8.** Use equation 2.23 to calculate the isotopic shift of a $\nu_{CO}$ band for an $M\text{—}^{13}CO$ ligand when assuming that the "natural" $M\text{—}^{12}CO$ ligand $\nu_{CO}$ band appears at 2000 cm$^{-1}$. Assume that the CO force constant is unaffected by the isotopic labeling.

**9.** For the reaction shown below, identify the product complex, and discuss how the differences in the $\nu_{CO}$ spectrum between reactant complex (spectrum A) and product complex (spectrum B) are consistent with your answer.

$$(\eta^5\text{-}C_5H_5)Mo(CO)_3Me \xrightarrow[C_6H_6, RT]{Ph_3P} product$$

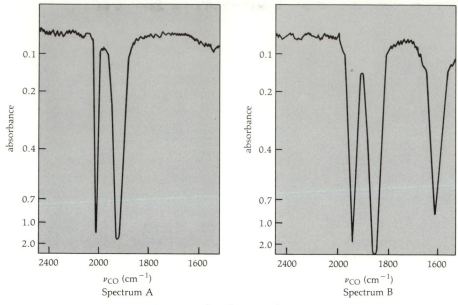

FIGURE 9.6   Infrared spectra for question 9.

10. For the reaction shown below, identify the product complex from the observed changes in the $\nu_{CO}$ region of the IR spectrum. Explain your reasoning. (*Note*: $CH_3Mn(CO)_5$ is stable in THF solution under these conditions.)

$$CH_3Mn(CO)_5(\text{spectrum A}) + LiBr \xrightarrow[\Delta]{\text{THF}} \text{product (spectrum B)}$$

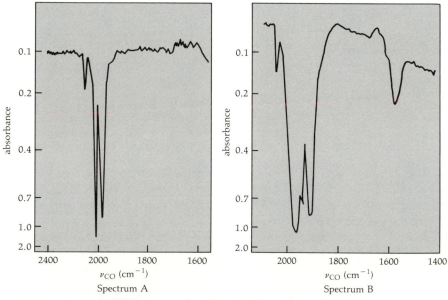

FIGURE 9.7   Infrared spectra for question 10.

## SUGGESTED READING

### Insertion Reactions

Calderazzo, F. *Angew. Chem. Internal. Ed. Engl.* **16**, 299 (1977).
Flood, T. C. *Top. Stereochem.* **12**, 37 (1981).
Heck, R. F. *Adv. Chem. Ser.* **49**, 181 (1965).
Heck, R. F. *Acc. Chem. Res.* **2**, 10 (1969).
Kuhlmann, E. J., and Alexander, J. J. *Coord. Chem. Rev.* **33**, 195 (1980).
Lappert, M. F., and Prokai, B. *Adv. Organometal. Chem.* **5**, 225 (1967).
Wojcicki, A. *Acc. Chem. Res.* **4**, 344 (1971).
Wojcicki, A. *Adv. Organometal. Chem.* **11**, 87 (1973).
Wojcicki, A. *Adv. Organometal. Chem.* **12**, 31 (1974).

### Elimination and Abstraction Reactions

Braterman, P. S., and Cross, R. J. *Chem. Soc. Rev.* **2**, 271 (1973).
Davidson, P. J., Lappert, M. F., and Pearce, R. *Chem. Rev.* **76**, 219 (1976).
Schrock, R. R., and Parshall, G. W. *Chem. Rev.* **76**, 243 (1976).
Schrock, R. R. *Acc. Chem. Res.* **12**, 98 (1979).
Wilkinson, G. *Science.* **185**, 109 (1974).

# Oxidative-Addition
# and Reductive-Elimination Reactions

## CHAPTER 10

In the preceding chapter, insertion reactions (and the reverse elimination reactions) were discussed as important routes for incorporating substrate molecules into organometallic complexes. This overall incorporation did not alter the coordination number or formal oxidation state of the metal atom. A second important method of accomplishing this incorporation is by *oxidative-addition*, as defined in equations *1.5* and *9.2*. The reverse reaction, *reductive-elimination*, removes ligand moieties from a metal atom.

The term *oxidative-addition reactions* (like insertion reactions; see chap. 9) refers to a particular type of reaction rather than to a specific reaction mechanism. For some complexes, a reaction normally classified as an oxidative-addition may in fact entail very little *actual* oxidation at the metal atom. This problem of ambiguous terminology is discussed below. However, an example of a classical oxidative-addition (O-A), reductive-elimination (R-E) reaction is shown in equation *10.1*.

$$L_nM + X-Y \underset{R-E}{\overset{O-A}{\rightleftharpoons}} L_nM \begin{smallmatrix} X \\ \diagdown \\ Y \end{smallmatrix}$$

(10.1)

In this equation, $L_nM$ represents a stable (and usually isolable) organometallic complex, and XY is a substrate molecule that adds to M with an overall complete dissociation of the X—Y bond and formation of two new bonds, M—X and M—Y. Examples of common substrate molecules, XY, are shown in table 10.1. The organometallic complex may be neutral, anionic, or cationic (and even polynuclear). The substrate molecule *usually* contains a highly polarized X—Y bond (e.g., an H—X or C—X bond) or a very reactive, low-energy bond between highly electronegative atoms (e.g., halogen-halogen or S—S bonds). However, oxidative-additions of H—H, C—H, or C—C bonds to metal atoms are known, and these reactions are fundamentally important in effecting hydrogenation or functionalization of hydrocarbons.

Notice that the coordination number of M increases upon addition of the X—Y bond—in fact, by two coordination sites in equation *10.1*. Since X and Y are usually

**TABLE 10.1   Common substrate molecules X—Y in oxidative-addition reactions.**

| Substrate molecule X—Y | Substrate molecule X—Y |
|---|---|
| $H_2$ | R—X (R = alkyl or aryl) |
| $X_2$ (X = Cl, Br, I) | |
| NC—CN | $$\begin{matrix}O\\\parallel\end{matrix}$$ |
| H—X (X = Cl, Br, I, CN, $ClO_4$) | R—C—X (R = alkyl or aryl) |
| H—OR (R = H, Me, $C_6F_5$) | R—H (R = aryl, usually) |
| H—SR (R = H, Ph) | M—X ($R_3Sn$—Cl, $R_3Si$—Cl, $R_3Ge$—Cl, |
| H—NH($C_6F_5$) | $HgCl_2$, MeHg—Cl) |
| $$\begin{matrix}O\\\parallel\end{matrix}$$ | |
| H—OCR (R = alkyl or aryl) | |
| H—$SiR_3$ | |

electronegative atoms and coordination chemists usually prefer to think of ligands as two-electron donors, the *formal* oxidation state of M is also increased upon addition of the X—Y bond. Therefore, the term *oxidative-addition* is used to describe the forward process of equation *10.1*. The reverse of equation *10.1* is consequently defined as a *reductive* (formal oxidation state of M decreases)-*elimination* (molecule X—Y is eliminated from M).

A crude, but useful, thermodynamic analysis of equation *10.1* is summarized as follows:

$$\Delta H_{rx}(\text{O-A}) \approx D_{X-Y} + E^M_{VSP} - D_{M-X} - D_{M-Y} + E_{steric} \qquad (10.2)$$

For an exothermic oxidative-addition, the bond energy gained by formation of two new bonds, $D_{M-X}$ and $D_{M-Y}$, must exceed the energy expended to break the X—Y bond, $D_{X-Y}$, to effect valence-state promotion of M, $E^M_{VSP}$, and to account for additional interligand steric repulsions in the product complex, which now contains two additional ligands.

The valence-state promotion energy reflects the hybridization and electronic changes necessary to convert the valence state of M to the correct coordination geometry for product formation. A qualitative representation of these $E^M_{VSP}$ energy contributions is exemplified in equation *10.3* for a presumed four-coordinate reactant complex (**10.1**).

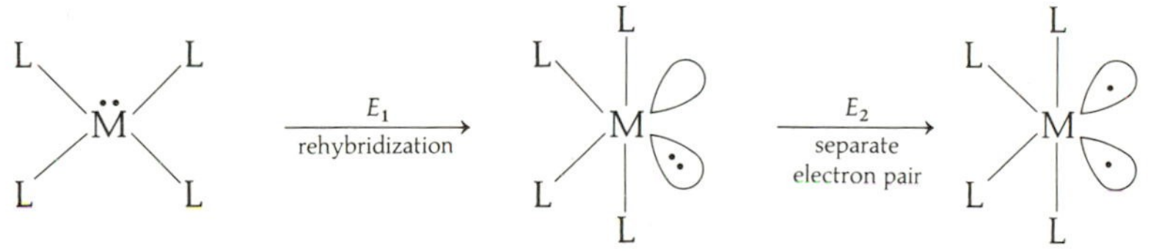

$$(10.3)$$

| **10.1** | **10.2** | **10.3** |
|---|---|---|
| CN = 4 | CN = 4 | valence state |
| square planar, $dsp^2$ | "octahedral," $d^2sp^3$ | |

Frequently, reactant complexes are low-spin and diamagnetic since all valence $d$ electrons are present as pairs. In using hybridization theory, the coordination geometry of **10.1** is defined by four $dsp^2$ hybrid orbitals. Since the final complex after adding X—Y is six-coordinate with octahedral geometry, two atomic valence orbitals on **10.1** must be used to generate octahedral hybrid orbitals on M, as shown in **10.2**. Finally, a valence electron pair on M must be separated so that each of the newly generated hybrid orbitals on M contains one electron. Complex **10.3** has the appropriate valence state to combine with the homolytic fission products of X—Y, X· and Y·, and produce the oxidative-addition product. The concept of valence-state promotion energy is encountered in organic chemistry when an $sp^2$-$\ddot{C}R_2$ singlet carbene species reacts to give an $sp^3$-$CR_4$ product.

Equations 10.2 and 10.3 suggest several factors that would facilitate oxidative-addition: (1) the presence of nonbonded $d$-electron density in the valence shell of M; (2) a low ionization potential of the valence electrons on M (minimizes the endothermic $E_1$ term); (3) a low X—Y bond energy; (4) high M—X and M—Y bond energies; and (5) the presence of ancillary ligands, L, that include strong $\sigma$-donors (to stabilize the higher oxidation state of M in the product complex) and of relatively nonbulky ligands (to minimize interligand steric repulsion in the product complex). Unfortunately, *a priori* thermochemical analysis of a particular oxidative-addition reaction is thwarted by a lack of specific bond energy and steric repulsion data. However, observed trends in a large number of oxidative-addition reactions support these general predictions.

In the remainder of this chapter we shall discuss oxidative-addition reactions and reductive-elimination reactions in more detail, including nondissociative oxidative-addition and reductive-elimination reactions, as well as *reductive-additions* and *ortho-metallation* reactions. We shall also address questions relating to nomenclature.

## OXIDATIVE-ADDITION REACTIONS

Three major classes of oxidative-addition reactions are distinguished by the type of reactant complex. These classes are (1) five-coordinate $d^7$-complexes, (2) four-coordinate $d^8$-complexes, and (3) two-coordinate $d^{10}$-complexes.

### Five-Coordinate $d^7$-Complexes

A general oxidative-addition reaction of this type is shown in equation 10.4. Two molecules of a five-coordinate $d^7$-complex react with a substrate molecule X—Y to afford two six-coordinate 18-electron complexes.

$$2ML_5 \quad \xrightarrow{+ X—Y} \quad ML_5X \quad + \quad ML_5Y \quad (10.4)$$

$$\begin{array}{ccc} d^7(17\text{ electron}) & d^6(18\text{ electron}) & d^6(18\text{ electron}) \\ CN = 5 & CN = 6 & CN = 6 \end{array}$$

Each metal has undergone a *formal* one-electron oxidation and an increase of one in coordination number. Specific examples of this reaction type are shown in equations 10.5 and 10.6.

$$2[Co^{II}(CN)_5]^{3-} + X{-}Y \longrightarrow [Co^{III}(CN)_5X]^{3-} + [Co^{III}(CN)_5Y]^{3-} \quad (10.5)$$

$X{-}Y = H_2, Br_2, H{-}OH, HO{-}OH, R{-}I, I{-}CN, H_2N{-}OH,$ and others

$$2[Co^{II}(DH)_2L] + R{-}X \longrightarrow [Co^{III}(DH)_2(L)X] + [Co^{III}(DH)_2(L)R] \quad (10.6)$$

R = alkyl; X = halide
L = amine or phosphine (R$_3$P)
DH = dimethylglyoximato anion,

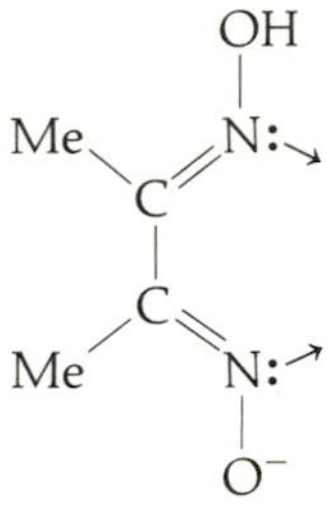

Since the reactant is an odd-electron (paramagnetic) species, it might be expected to react with X—Y via a radical mechanism. Considerable experimental work confirms a stepwise free-radical mechanism for nearly all of these oxidative-additions. In each such case, the reactant complex extracts a radical from X—Y in the rate-determining step, as shown with $H_2O_2$ and R—X substrate molecules in equations 10.7–10.9. The rate equation is overall second order, first order in [reactant complex] and in [XY].

$$[Co(CN)_5]^{3-} + HO{-}OH \xrightarrow{\text{slow}} [Co(CN)_5(OH)]^{3-} + HO\cdot$$
$$\underline{[Co(CN)_5]^{3-} + HO\cdot \longrightarrow [Co(CN)_5(OH)]^{3-}}$$
$$2[Co(CN)_5]^{3-} + HO{-}OH \longrightarrow 2[Co(CN)_5(OH)]^{3-} \quad (10.7)$$

$$[Co(CN)_5]^{3-} + Me{-}I \xrightarrow{\text{slow}} [Co(CN)_5I]^{3-} + Me\cdot$$
$$\underline{[Co(CN)_5]^{3-} + Me\cdot \longrightarrow [Co(CN)_5Me]^{3-}}$$
$$2[Co(CN)_5]^{3-} + MeI \longrightarrow [Co(CN)_5I]^{3-} + [Co(CN)_5Me]^{3-} \quad (10.8)$$

$$[Co(DH)_2L] + R{-}X \xrightarrow{\text{slow}} [Co(DH)_2(L)X] + R\cdot$$
$$\underline{[Co(DH)_2L] + R\cdot \longrightarrow [Co(DH)_2(L)R]}$$
$$2[Co(DH)_2L] + RX \longrightarrow [Co(DH)_2(L)X] + [Co(DH)_2(L)R] \quad (10.9)$$

The hydroxyl radical formed in equation 10.7 can be trapped with added iodide ion. The rate of oxidative-addition of organic halides decreases in the following order: RI > RBr > RCl and PhCH$_2$I > Me$_3$CI > Me$_2$CHI > EtI > MeI. These trends are consistent with a free-radical mechanism first because C—I bonds would be the most easily cleaved of the carbon-halogen bonds and second because the more stable carbon free radical affords faster reaction rates.

When an ethyl or higher alkyl halide is used in the reaction described by equation 10.8, a competing hydrogen abstraction reaction is observed:

$$[Co(CN)_5]^{3-} + C_2H_5I \xrightarrow{\text{slow}} [Co(CN)_5I]^{3-} + C_2H_5\cdot$$

$$[Co(CN)_5]^{3-} + C_2H_5\cdot \longrightarrow [Co(CN)_5Et]^{3-} \tag{10.10}$$

$$\longrightarrow [Co(CN)_5H]^{3-} + C_2H_4$$

As would be expected from the general predictions summarized earlier, reaction rates for equation $10.9$ are faster for more basic ancillary ligands L but are apparently slower when L is sterically bulky.

### Four-Coordinate $d^8$-Complexes

A four-coordinate 16-electron complex (which is therefore coordinatively unsaturated) reacts with a substrate molecule X—Y to afford a six-coordinate 18-electron complex:

$$ML_4 \xrightarrow[+ X—Y]{} ML_4(X)(Y) \tag{10.11}$$

$$d^8 (16 \text{ electron}) \qquad\qquad d^6 (18 \text{ electron})$$
$$CN = 4 \qquad\qquad\qquad CN = 6$$

Notice that the metal atom undergoes a *formal* two-electron oxidation and increases its coordination number by two. The product is electronically and structurally similar to those formed in equation $10.4$.

The reactivity of $d^8$-metals in oxidative-additions decreases in the following order: $Os^0 > Ru^0 > Fe^0 \gg Ir^I > Rh^I > Co^I \gg Pt^{II} > Pd^{II} \gg Ni^{II}$, $Au^{III}$. This trend parallels the expected decrease in relative ease of metal oxidation. Similarly, ancillary ligands that increase the amount of electron density at M stabilize the product complex relative to the reactant complex. This effect is evident from the value of the equilibrium constant of equation $10.12$ as X and L are varied. The value of $K$ decreases as $X = I > Br > Cl$ and $L = PMe_3 > PMe_2Ph > PMePh_2 > PPh_3$, as expected.

$$\textit{trans-}Ir^IX(CO)L_2 + RCO_2H \underset{}{\overset{K}{\rightleftharpoons}} Ir^{III}(H)(X)(O_2CR)(CO)L_2 \tag{10.12}$$

Several specific examples of this class of oxidative-addition reactions are shown in equations $10.13{-}10.17$. By far the most intensive synthetic and mechanistic study has centered on Vaska-type $\textit{trans-}Ir^IX(CO)L_2$ complexes. Nearly all possible XY substrate molecules undergo oxidative-addition with these compounds.

$$\textit{trans-}IrCl(CO)L_2 + H_2 \longrightarrow \tag{10.13}$$

$L = PMe_2Ph$

**10.4**

$$\textit{trans-}IrCl(CO)L_2 + R—I \xrightarrow{CH_2ClCH_2Cl} \tag{10.14}$$

$L = PMe_3$
$R = I, Me, Et, Pr, \textit{i}-Pr, CH_2Ph, CH_3C(O), PhC(O)$

**10.5**

$$trans\text{-}IrCl(CO)L_2 + H\text{---}I \xrightarrow{CH_2ClCH_2Cl} \begin{array}{c} H \\ L\text{---}\overset{|}{Ir}\text{---}I \\ C\overset{|}{\diagup}\overset{|}{\diagdown}L \\ O \quad Cl \end{array} \qquad (10.15)$$

L = PMe$_3$

**10.6**

$$[(\eta^2\text{-}C_2H_4)PtCl(\mu_2\text{-}Cl)]_2 + 2 \overset{Me}{\underset{Me}{\triangleleft\!\!\!\!\cdots}} \xrightarrow[\substack{(2) \text{ py} \\ (L)}]{(1) \text{ THF}}$$

$$2 \quad \begin{array}{c} Cl \quad Me \\ L\diagdown \overset{|}{\phantom{.}} \diagup \\ Pt \diagup\!\!\!\!\triangleright\!\!\!\!\cdots Me \\ L\diagup \overset{|}{\phantom{.}} \\ Cl \end{array} + 2C_2H_4 \quad (10.16)$$

**10.7**

$$FeL_5 + PhCH_2Br \xrightarrow{-L} \begin{array}{c} L \\ L\diagdown\overset{|}{\phantom{.}}\diagup CH_2Ph \\ Fe \\ L\diagup\overset{|}{\phantom{.}}\diagdown Br \\ L \end{array} \qquad (10.17)$$

L = CN(t-But)

**10.8**

Oxidative-addition of molecular hydrogen usually generates a cis-MH$_2$ product (such as **10.4**) as would be expected from a concerted reaction. Halogens, alkyl halides, and acyl halides generally add oxidatively to Ir(I) complexes to give a trans-Ir(R)(X) product as shown in **10.5**. Hydrogen halides, such as HI, oxidatively add to afford a cis-product (eq. 10.15). Equation 10.16 shows oxidative-addition of a C—C bond to a four-coordinate Pt(II) atom, forming a platinacyclobutane. Complex **10.8** represents the formal overall oxidative-addition of benzyl bromide to a four-coordinate d$^8$-Fe$^0$ complex generated by dissociation of one of the isocyanide ligands from the pentacoordinate reactant complex. Ligand dissociation from coordinatively saturated complexes is required if an oxidative-addition product is to obey the EAN rule.

Values of $\Delta H_{rx}$ for equations 10.14 and 10.15 are shown in table 10.2. The exothermicity decreases as I$_2$ > HI > RC(O)I > RI. This order follows the increasing chemical stability of the X—I bond: I$_2$ is a stronger oxidizing agent than HI; the HI bond energy is less than an average C—I bond energy; and acyl iodides are stronger electrophiles than are alkyl iodides. These data have been used to estimate Ir—R bond energies. The trend in Ir—X bond energies is H > I $\approx$ alkyl $\approx$ acetyl > PhCH$_2$.

Calorimetric study of the oxidative-addition of acetyl chloride to trans-IrCl(CO)L$_2$ complexes for a variety of phosphine ligands gives the $\Delta H_{rx}$ values shown in table 10.3. A strong dependence on the steric bulk of L is observed. As L becomes bulkier, the exothermicity of the reaction decreases significantly, presumably because of greater steric strain in the six-coordinate product complexes. It is assumed that any electronic differences between the various phosphine ligands affect the $\Delta H_{rx}$ values only slightly.

**TABLE 10.2** $\Delta H_{rx}$ for the
oxidative-addition of R—I reagents
to *trans*-IrCl(CO)(PMe$_3$)$_2$ in
1,2-dichloroethane solution.

| R—I | $-\Delta H_{rx}$ (kcal/mole) |
|---|---|
| I$_2$ | 44.2 ± 1.7 |
| HI | 38.2 ± 0.7 |
| CH$_3$C(O)I | 30.1 ± 1.0 |
| PhC(O)I | 28.9 ± 0.7 |
| CH$_3$I | 28.0 ± 1.7 |
| EtI | 26.3 ± 0.7 |
| PrI | 24.6 ± 1.0 |
| PhCH$_2$I | 22.7 ± 1.7 |
| *i*-PrI | 21.0 ± 2.4 |

Source: G. Yoneda and D. M. Blake, *Inorg. Chem.*, **20**, 67 (1981).

**TABLE 10.3** $\Delta H_{rx}$ for the oxidative-addition of
acetyl chloride to *trans*-IrCl(CO)L$_2$ in 1,2-dichloroethane
at 25°C for a variety of phosphine ligands.

| L | $-\Delta H_{rx}$ (kcal/mole) | Cone angle $\theta$ (°) |
|---|---|---|
| PMe$_3$ | 29.2 ± 0.7 | 118 |
| P(CH$_2$Ph)$_3$ | 17.9 ± 1.0 | 165 |
| PMePh$_2$ | 27.0 ± 1.2 | 136 |
| P(*t*-But)Ph$_2$ | 21.7 ± 0.5 | 157 |

Source: G. Yoneda, S.-M. Lin, L.-P. Wang, and D. M. Blake, *J. Am. Chem. Soc.*, **103**, 5768 (1981).

Mechanisms of oxidative-addition to four-coordinate $d^8$-complexes are surprisingly varied. Observed reaction mechanisms include "three-center" concerted processes, two-step S$_N$2-type pathways, and a radical chain mechanism. Unfortunately, it is usually not possible to predict reaction mechanism based upon type of complex. The reaction path depends on the specific metal ion and its ancillary ligands, the type of substrate molecule X—Y, and the experimental conditions, such as solvent polarity, temperature, and the presence of trace amounts of oxidizing impurities.

Oxidative-addition of molecular hydrogen, as shown in equation *10.13*, gives a *cis*-dihydride complex (**10.4**) as the kinetically controlled product. This product is usually also the most stable thermodynamic product. Oxidative-addition of H$_2$ occurs in solution phase or even in the solid state with surprising facility considering the relatively high H—H bond energy, 104 kcal/mole.

Kinetic data for H$_2$ oxidative-additions suggest a "three-center" concerted process with a nonpolar transition state. A small deuterium isotope effect indicates that Ir—H bond formation is more important than H—H bond cleavage. A mechanism that accounts for these observations and that generates a *cis*-dihydride product is shown in equation *10.18*. Electron density in a filled valence $d$ orbital on M flows into the empty $\sigma^*$ MO of $H_2$ (**10.9**), thus forming two M—H bonding interactions while weakening the H—H bond in the transition state. In addition, electron density in the occupied $\sigma$ MO that defines the H—H single bond flows into an empty valence orbital on the metal atom. The H—H bond is eventually cleaved, and two M—H single bonds form. Interestingly, for M = Ir, as in Vaska complexes, this reaction might be slightly exothermic because of the high Ir—H bond energies.

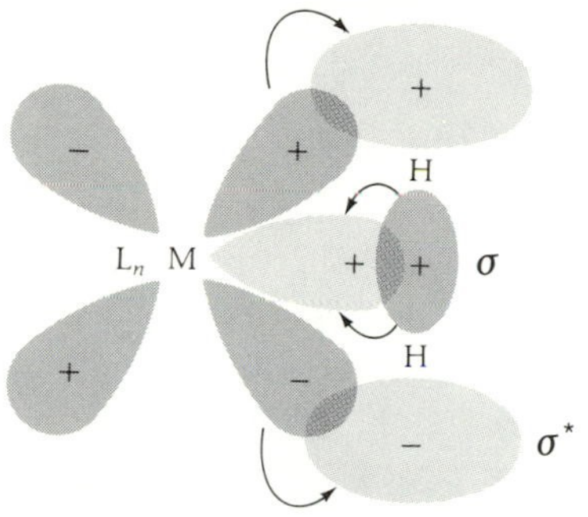

$$(10.18)$$

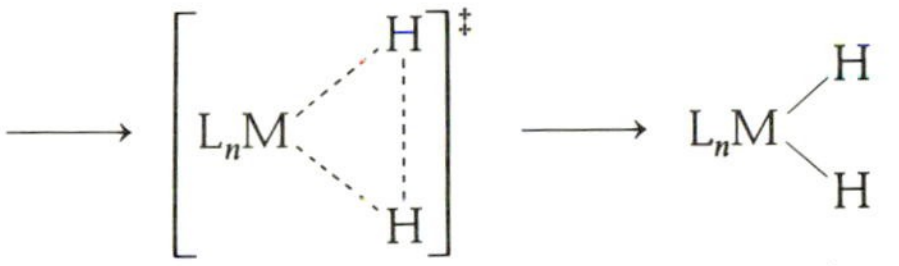

**10.9**

A classic example of an apparent overall heterolytic cleavage of an H—H bond is shown in equation *10.19*. Although the $H_2$ oxidative-addition may occur by a concerted process, the external base is eventually protonated, thus placing a formal hydride ($H^-$) ligand on the metal atom.

$$RuCl_2L_3 + H_2 \xrightarrow[C_6H_6]{Et_3N} HRuClL_3 + [Et_3NH]^+X^- \qquad (10.19)$$

$L = Ph_3P$

Oxidative-addition of alkyl halides to Vaska complexes (eq. *10.14*) is postulated to proceed by one of two mechanisms. Methyl, benzyl, or allyl halides and $\alpha$-halo ethers appear to react by $S_N2$ attack of a lone electron pair on iridium. With methyl halides, a second-order rate law is observed (first order in both [Ir complex] and [$CH_3X$]). The reaction is faster in polar solvents and shows a large negative activation entropy. Radical trapping agents do not affect the rate of reaction. The postulated two-step $S_N2$ mechanism is shown in equation *10.20*.

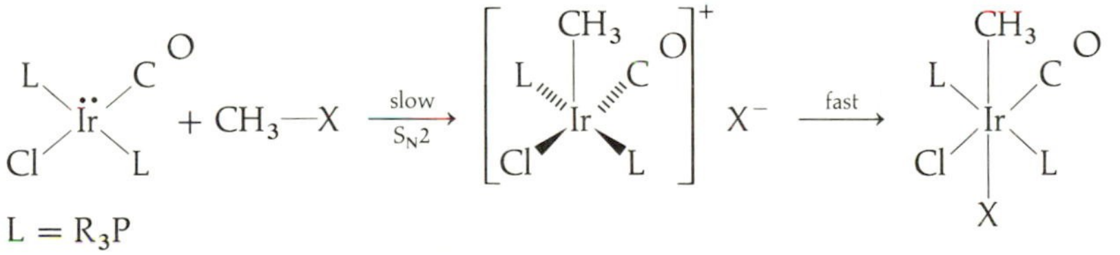

$$(10.20)$$

$L = R_3P$

The observed halide dependence of the rate of oxidative-addition and the inability to induce polymerization of acrylonitrile presumably exclude a nonchain radical mechanism. A very short-lived radical cage mechanism cannot be ruled out, however.

All other alkyl halides (e.g., ethyl halides or higher simple alkyl halides), vinyl and aryl halides, and α-halo esters apparently undergo oxidative-addition to Vaska complexes by a radical chain mechanism. Radical initiators and inhibitors greatly affect the reaction rate, and radical intermediates can be trapped by addition of acrylonitrile. Furthermore, optically-active alkyl halides, like **10.10** and **10.11**, undergo oxidative-addition with *complete loss* of stereochemistry at the reactive carbon, as shown in equation *10.21*.

**10.10**

**10.11**

*(10.21)*

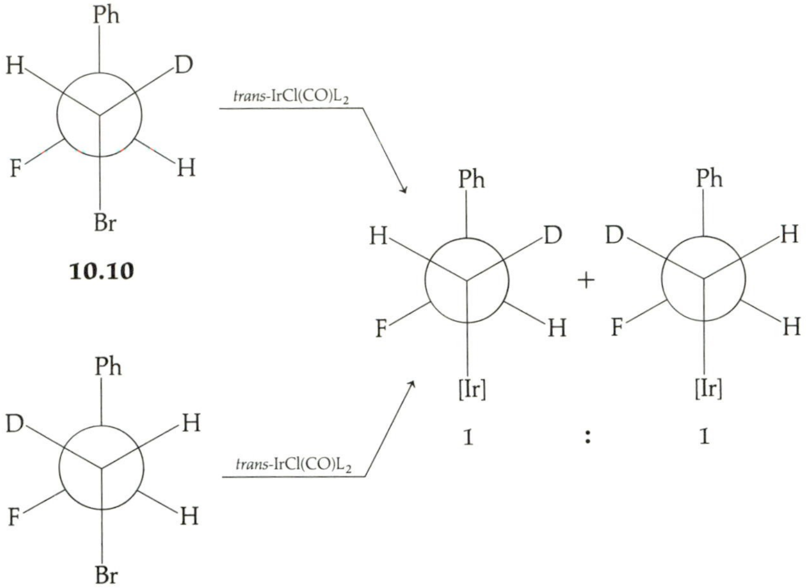

Initiation of a radical chain reaction might arise from the presence of trace quantities of Ir(II) species or of molecular oxygen. Experiments reveal that 20 mole-% of $O_2$ (relative to [Ir]) effected essentially complete reaction of *trans*-2-bromofluorocyclohexane and *trans*-$IrCl(CO)(PMe_3)_2$ in $CH_2Cl_2$ solution after two hours, whereas no significant reaction occurred within five hours at the same temperature in the absence of molecular oxygen. A proposed radical chain mechanism is shown in equation *10.22* where $[Ir] =$ *trans*-$IrX(CO)L_2$.

$$\left.\begin{aligned}&[\mathrm{Ir^{I}}] + \mathrm{Q}\cdot \text{ (trace radical)} \longrightarrow [\mathrm{Ir^{II}}] - \mathrm{Q}\cdot\\ &[\mathrm{Ir^{II}}] - \mathrm{Q}\cdot + \mathrm{R{-}X} \longrightarrow \mathrm{X{-}}[\mathrm{Ir^{III}}] - \mathrm{Q} + \mathrm{R}\cdot\end{aligned}\right\}\ \text{initiation steps}$$

$$\left.\begin{aligned}&[\mathrm{Ir^{I}}] + \mathrm{R}\cdot \longrightarrow \mathrm{R{-}}[\mathrm{Ir^{III}}]\cdot\\ &\mathrm{R{-}}[\mathrm{Ir^{II}}]\cdot + \mathrm{R{-}X} \longrightarrow \mathrm{R{-}}[\mathrm{Ir^{III}}]\mathrm{{-}X} + \mathrm{R}\cdot\end{aligned}\right\}\ \text{propagation steps}$$

$$[\mathrm{Ir^{I}}] + \mathrm{R{-}X} \longrightarrow \mathrm{R{-}}[\mathrm{Ir^{III}}]\mathrm{{-}X} \qquad \text{net reaction} \qquad (10.22)$$

Termination probably occurs by radical coupling or disproportionation.

Acyl halides undergo oxidative-addition to Vaska complexes at a much higher rate than do alkyl halides. A *trans*-addition is observed, and nucleophilic attack by the metal atom at the acyl carbon atom is postulated, as shown in equation *10.23*.

$$
\begin{array}{c}
\text{[Ir complex]} + R\!-\!\overset{\overset{\displaystyle O}{\|}}{C}\!-\!X \longrightarrow
\left[ \text{[five-coordinate intermediate]} \right] \longrightarrow
\text{[product]}
\end{array}
\qquad (10.23)
$$

Solvent effects are quite pronounced in the oxidative-additions of hydrogen halides to Vaska complexes (eq. *10.15*). Addition of HCl gas to solid *trans*-$IrCl(CO)(PPh_3)_2$ or addition of HCl or HBr to solutions of this complex in non-polar solvents such as benzene gives *cis*-HX addition products (**10.6**). A concerted oxidative-addition mechanism like that of equation *10.18* might be operative in these reactions.

Highly polar solvents, such as DMF or $C_6H_6$/MeOH mixed-solvent systems, yield a mixture of *cis*- and *trans*-HX addition products. In these solvents, ionic intermediates might be present. Initial protonation of the reactant complex would give a five-coordinate intermediate that might undergo intramolecular rearrangement thus affording both product isomers:

$$
\begin{array}{c}
\text{[Ir complex]} + H\!-\!X \longrightarrow
\left[ \text{[intermediate]} \right]^{+} \rightleftharpoons
\left[ \text{[intermediate]} \right]^{+}
\end{array}
\qquad (10.24)
$$

Intermolecular exchange of halide ligands is also more probable in highly polar media. More mechanistic study is needed, however, including determination of whether the isomeric mixture is a kinetic or thermodynamic effect (due to isomerization under the reaction conditions). The *cis*-$CH_3Br$ oxidative-addition product of *trans*-$IrCl(CO)(PPh_2Me)_2$ is known to isomerize to the *trans*-$CH_3Br$ isomer in polar solvents. Alternatively, halide ion might add first to the initial Vaska complex. This mode of addition is observed for the reaction of HCl with the Ir(I) complex shown in equation *10.25*.

$$
IrCl(COD)L_2 \xrightarrow[\text{MeOH}]{Cl^-} [IrCl_2(COD)L_2]^- \xrightarrow[\text{MeOH}]{H^+} Ir(H)(Cl)_2(COD)L_2 \qquad (10.25)
$$

$L = Ph_3P, MePh_2P, \text{ or } EtPh_2P$

Usually, halogens and HX reagents undergo oxidative-addition to Vaska complexes too rapidly for convenient kinetic study.

## Two-Coordinate $d^{10}$-Complexes

A general representation of oxidative-addition of two-coordinate $d^{10}$ complexes is shown in equation *10.26*. Here a highly coordinatively unsaturated two-coordinate 14-electron complex oxidatively adds X—Y to afford a four-coordinate, 16-electron complex.

$$ML_2 \xrightarrow[+ X—Y]{} ML_2(X)(Y) \left[\xrightarrow[+ X—Y]{} ML_2(X)_2(Y)_2\right] \quad (10.26)$$

$$d^{10}(14 \text{ electron}) \qquad d^8(16 \text{ electron}) \qquad d^6(18 \text{ electron})$$
$$CN = 2 \qquad\qquad CN = 4 \qquad\qquad CN = 6$$

Frequently, a second X—Y substrate molecule adds as shown in equation *10.11* to give a coordinatively and electronically saturated six-coordinate 18-electron complex.

Complexes of $Ni^0$, $Pd^0$ and $Pt^0$ undergo this type of oxidative-addition. Specific examples are shown below. These reactions proceed by various mechanisms, although much detailed mechanistic information is lacking. Kinetic data reveal that coordinatively saturated three- or four-coordinate complexes undergo ligand dissociation prior to oxidative-addition:

$$ML_4 \underset{+L}{\overset{\overset{K_1}{-L}}{\rightleftharpoons}} ML_3 \underset{+L}{\overset{\overset{K_2}{-L}}{\rightleftharpoons}} ML_2 \qquad (10.27)$$

A general comment about the determination of reaction mechanisms is appropriate here. The choice of a radical or nonradical mechanism depends on experimental observations, such as (1) retention or loss of stereochemistry at carbon, (2) effects attributable to radical initiators or inhibitors, (3) trapping of radical species, (4) spectroscopic evidence of radical species, and (5) a tendency for the reactant complexes to undergo one-electron redox processes.

Radical cage mechanisms differ from radical chain mechanisms (eq. *10.22*) in that the generated radicals are held within a solvent cage and they recombine *with each other* to form the products. In a radical chain process, a generated radical reacts with an external metal complex. However, as observed with oxidative-additions to Vaska compounds, generalization of reaction mechanisms for oxidative-additions to $d^{10}$-complexes is difficult at best and is probably not justified. Some specific systems are discussed below.

Nickel(0) reactions, as shown in equation *10.28*, appear to involve a radical mechanism.

Complex **10.12** reacts with benzyl halides at 25°C to give a Ni(I) complex (**10.13**). The expected Ni(II) product (**10.14**) is obtained at low temperature but at higher temperatures it is subject to disproportionation or to reduction (in the presence of added L) to **10.13**. However, **10.14** can be trapped at low temperature by "CO-insertion" to give **10.15**. Solvolysis of the acyl ligand by MeOH affords the corresponding methyl ester. Other alkyl Ni(II) complexes react similarly with CO and

$$\text{NiL}_4 + \text{PhCH}_2\text{X} \xrightarrow[25°C]{C_6H_6} \text{Ni}^I\text{L}_3\text{X} + \tfrac{1}{2}\text{PhCH}_2\text{CH}_2\text{Ph} + \text{L}$$

**10.12**  **10.13**

$\text{L} = \text{Ph}_3\text{P}; \text{X} = \text{Cl or Br}$

$$\text{PhCH}_2\text{Ni}^{II}\text{L}_2\text{X} \xrightarrow{\quad CO \quad} \text{PhCH}_2\overset{\displaystyle O}{\overset{\displaystyle \|}{\text{C}}}\text{NiL}_2\text{X}$$

(with $\xrightarrow{-20°C \; Et_2O}$ and $\xrightarrow{+L}$ arrows) (10.28)

**10.14**  **10.15**

$$\tfrac{1}{2}\text{Ni}^{II}\text{L}_2\text{X}_2 + \tfrac{1}{2}\text{PhCH}_2\text{CH}_2\text{Ph} + \tfrac{1}{2}\text{Ni}^0 + \text{L} \qquad \text{PhCH}_2\text{CO}_2\text{Me}$$

(left branch: $\Delta \mid C_6H_6$; right branch: MeOH)

undergo solvolytic cleavage to form organic esters. When optically active (benzyl-$\alpha$-$d$) halides are used, the resulting methyl $\alpha$-deuteriophenylacetate is racemic. Racemization presumably occurs during oxidative-addition.

A radical cage mechanism is postulated for oxidative-addition of aryl halides to Ni(0) complexes, as shown in equation 10.29. Electron transfer from Ni(0) to the aryl halide generates the Ni(I) radical pair, which can afford either Ni(I) or Ni(II) products.

$$\text{NiL}_3 + \text{RX} \longrightarrow \{[\text{NiL}_3{}^+\cdot][\text{RX}^-\cdot]\} \longrightarrow \text{Ni}^{II}\text{L}_2(\text{R})(\text{X}) + \text{L}$$

$\text{L} = \text{Et}_3\text{P}; \text{R} = \text{aryl}$

(10.29)

$$\text{Ni}^I\text{XL}_3 + \text{R}\cdot$$

Phosphine complexes of Pd(0) and Pt(0) react with organic halides to give *trans*-addition products, as shown in equations 10.30 and 10.31.

$$\text{PdL}_4 + \text{RX} \xrightarrow{\text{R = alkyl}} \textit{trans}\text{-PdL}_2(\text{R})(\text{X}) + 2\text{L} \qquad (10.30)$$

$\text{L} = \text{Ph}_3\text{P}$

$$\downarrow \text{RI} \mid \text{R = aryl}$$

$$\textit{trans}\text{-PdL}_2(\text{R})(\text{I}) + 2\text{L}$$

$$\text{PtL}_2{}'\text{L} + \text{R}—\text{X} \longrightarrow \textit{trans}\text{-PtL}_2(\text{R})(\text{X}) + \text{L}' \qquad (10.31)$$

$\text{L} = \text{Ph}_3\text{P}; \text{L}' = \text{Ph}_3\text{P or } \text{C}_2\text{H}_4$

Oxidative-addition of optically-active (benzyl-$\alpha$-$d$) halides to Pd(0) complexes occurs with *inversion* of configuration at the $\alpha$-carbon atom with no evidence of radical intermediates. These reactions presumably proceed by an $S_N2$-type mechanism.

Other organic halides appear to react with Pd(0) or Pt(0) complexes by radical mechanisms.

Equation *10.32* shows the oxidative-addition of allyl acetate to a two-coordinate Pd(0) complex to afford the *trihapto*-allyl Pd(II) complex. Palladium (II) $\eta^3$-allyl complexes are useful substrates in alkylations of organic molecules, as we shall discuss in chap. 12.

$$PdL_2 + allyl(OAc) \xrightarrow[25°C]{12\ hrs} (\eta^3\text{-}C_3H_5)Pd\begin{smallmatrix}L\\ \\OAc\end{smallmatrix} + L \qquad (10.32)$$

$L = (C_6H_{11})_3P$

Complex **10.16** is a dinuclear Pd(I) complex that contains a Pd—Pd single bond and three bridging dpm ligands. Oxidative-addition of $CH_2I_2$ or $CH_3I$ affords a molecular A-frame complex **10.17** or **10.19**, respectively, as shown in equation *10.33*. These reactions are formally *two-center three-fragment oxidative-additions*. For example, oxidative-addition of $CH_2I_2$ occurs on *both* palladium centers and involves cleavage of both C—I bonds to form three fragments, $CH_2$, I and I.

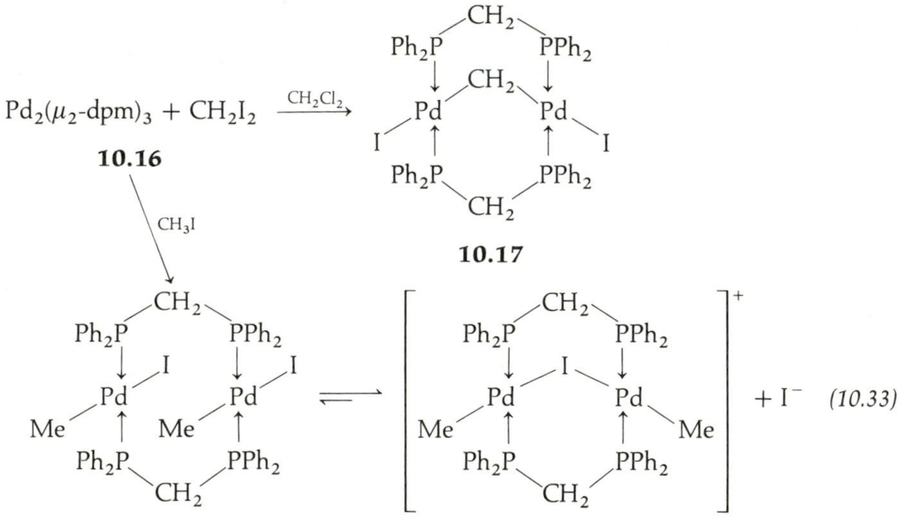

$$Pd_2(\mu_2\text{-}dpm)_3 + CH_2I_2 \xrightarrow{CH_2Cl_2}$$

**10.16**

**10.17**

$CH_3I$

**10.18**                                   **10.19**

$+ I^- \qquad (10.33)$

dpm $= CH_2(PPh_2)_2$

Oxidative-addition of hydrogen halides to Pt(0) complexes (eq. *10.34*) apparently occurs by protonation of the Pt(0) atom affording hydride complexes like **10.21** as intermediates. When $X^-$ is a good ligand, a *trans*-addition product (**10.22**) is formed by loss of L. Intermediate complexes such as **10.21** have been isolated from these reactions.

$$PtL_3 \underset{KOH}{\overset{HX}{\rightleftharpoons}} [Pt^{II}(H)L_3]^+X^- \underset{+L}{\overset{-L}{\rightleftharpoons}} trans\text{-}Pt^{II}L_2(H)(X) \qquad (10.34)$$

**10.20**                     **10.21**                        **10.22**

$L = Ph_3P$

Molecular hydrogen undergoes oxidative-addition with selected Pt(0) complexes to give *trans*-dihydride adducts (eq. *10.35*). Chelating phosphines afford *cis*-dihydride products.

$$PtL_2 + H_2 \xrightarrow[25°C]{C_6H_6} trans\text{-}PtL_2(H)_2 \qquad (10.35)$$

$$\textbf{10.23} \qquad\qquad\qquad \textbf{10.24}$$

$L = (C_6H_{11})_3P$ or $(i\text{-}Pr)_3P$

A recent theoretical calculation of the geometry of the transition state species of the reaction between $H_2$ and $Pt(PH_3)_2$ suggests the structure shown as **10.25.**

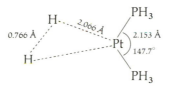

$$\textbf{10.25}$$

The two $PH_3$ ligands are displaced from the linear geometry of the reactant complex, and the H—H bond is only 4% longer than the H—H distance in free $H_2$. This transition state is reached early in the reaction, a fact consistent with the reaction's exothermicity. The calculated barrier is just 8 kcal/mole. Only a *cis*-dihydride addition product is predicted theoretically; therefore the observed predominance of *trans*-products might be the consequence of subsequent isomerization.

## Oxidative-Additions to Other $d^n$-Complexes

Many complexes that contain metals with valence electronic configurations other than $d^7$, $d^8$ or $d^{10}$ also undergo oxidative-addition reactions. These reactions usually involve either *one-* or *two-electron formal oxidations* of the metal atom and are, therefore, analogous to the oxidative-addition chemistry of the $d^7$- and $d^8$-metals, as shown in equations *10.4* and *10.11*, respectively.

Mononuclear $d^4$-Cr(II) complexes undergo one-electron oxidative-addition reactions in which alkyl halides are converted into alkanes by $[Cr(en)_2]^{2+}$:

$$R—X + [Cr^{II}(en)_2]^{2+} \longrightarrow R\cdot + [Cr^{III}(en)_2X]^{2+}$$

$$\downarrow {\scriptstyle [Cr^{II}(en)_2]^{2+}}$$

$$[RCr^{III}(en)_2]^{2+} \xrightarrow{H_2O} R—H + [Cr^{III}(en)_2(OH)]^{2+} \quad (10.36)$$

Notice that the reaction requires two equivalents of Cr(II) reactant and produces two equivalents of Cr(III) product. An alkyl radical mechanism is proposed; alkane formation occurs on hydrolysis of an alkyl-Cr(III) intermediate, as shown.

A $5f^3$-U(III) complex reacts with organo halides by a similar mechanism giving U(IV) products (eq. *10.37*). Some of the organic radical undergoes coupling, abstraction of a hydrogen atom from another species, or the loss of a hydrogen atom to give olefinic products.

$$(\eta^5\text{-}C_5Me_5)_2UCl(THF) \rightleftharpoons (\eta^5\text{-}C_5Me_5)_2UCl + THF$$

$$\downarrow R\!-\!X$$

$$R\!-\!R + R\!-\!H + alkenes \longleftarrow R\cdot + (\eta^5\text{-}C_5Me_5)_2UCl(X)$$

$$\downarrow (\eta^5\text{-}C_5Me_5)_2UCl$$

$$(\eta^5\text{-}C_5Me_5)_2UCl(R) \qquad\qquad (10.37)$$

Two dinuclear complexes that undergo oxidative-additions in which each metal is oxidized formally by one electron are shown as equations *10.38* and *10.39*.

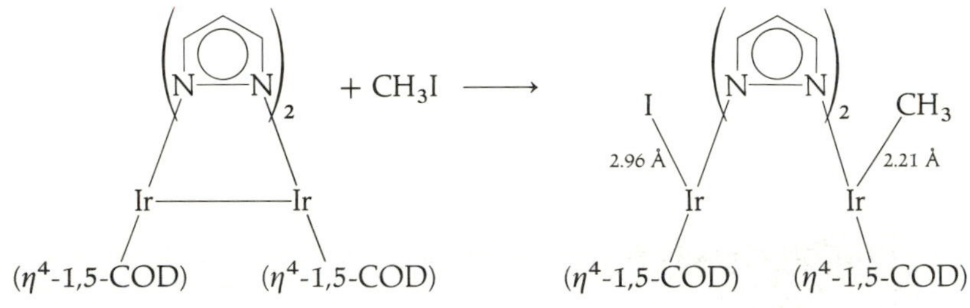

$$(10.38)$$

**10.26**                                        **10.27**

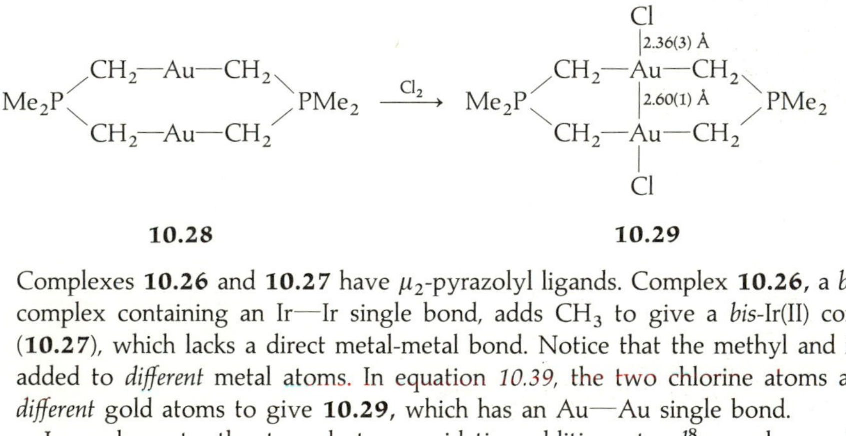

**10.28**                                        **10.29**

Complexes **10.26** and **10.27** have $\mu_2$-pyrazolyl ligands. Complex **10.26**, a *bis*-Ir(I) complex containing an Ir—Ir single bond, adds $CH_3$ to give a *bis*-Ir(II) complex (**10.27**), which lacks a direct metal-metal bond. Notice that the methyl and iodine added to *different* metal atoms. In equation *10.39*, the two chlorine atoms add to *different* gold atoms to give **10.29**, which has an Au—Au single bond.

In analogy to the two-electron oxidative-additions to $d^8$-complexes, $d^2$-$(\eta^5\text{-}C_5H_5)_2Zr^{II}(PR_3)_2$ complexes react with alkyl halides to give Zr(IV) complexes:

$$(\eta^5\text{-}C_5H_5)_2ZrL_2 + R\!-\!X \longrightarrow (\eta^5\text{-}C_5H_5)_2ZrRX + 2L \qquad (10.40)$$

L = $PMe_2Ph$ or $PPh_2Me$

An alkyl radical mechanism is supported by relative reaction rate data, racemization of the $\alpha$-carbon stereochemistry, and spectroscopic detection of effects induced by radical intermediates. Presumably, the two-electron oxidation occurs as two one-electron steps involving $(\eta^5\text{-}C_5H_5)_2Zr^{III}RL$ and $(\eta^5\text{-}C_5H_5)_2Zr^{III}XL$ intermediates. The complex $(\eta^5\text{-}C_5H_5)_2ZrX_2$ is a by-product of the reaction.

## Ortho-Metallation Reactions

As shown in equations 9.83 and 9.84, ortho-metallation is one type of cyclometallation reaction. Two ortho-metallations that occur by formal oxidative-additions are shown in equations 10.41 and 10.42. In equation 10.41, an ortho-C—H bond of a $Ph_3P$ ligand adds to the Ir(I) atom affording the cyclometallated Ir(III)-hydride complex. A similar addition occurs in the first step of equation 10.42 giving 10.30 as an observed Rh(III) intermediate. This intermediate reductively eliminates $CH_4$ to afford the Rh(I) cyclometallated product. A combination of sequential oxidative-addition and reductive-elimination steps is quite common.

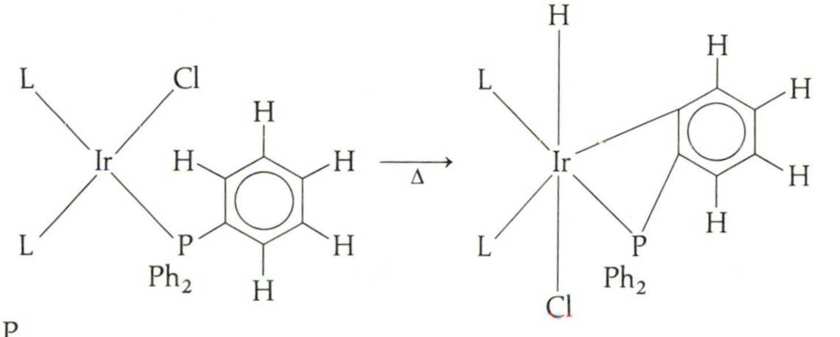

(10.41)

$L = Ph_3P$

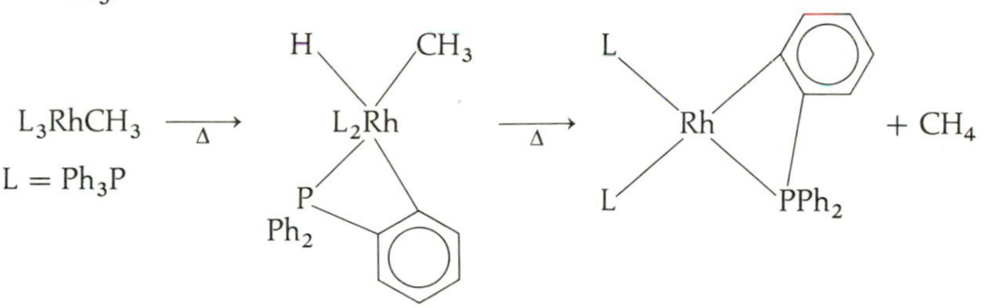

(10.42)

$L = Ph_3P$

10.30

Ortho-metallation occurs most frequently with $Ph_3P$ or $(PhO)_3P$ complexes, although azobenzene, aryl ketones, and even $\eta^5$-$C_5H_5$ ligands also undergo similar C—H oxidative-addition (see 4.3 and 4.4). This latter reaction can be prevented by using $\eta^5$-$C_5Me_5$ ligands. An example of ortho-metallation of a dialkylarylphosphine ligand is shown in equation 10.43. The x-ray structure of the product has been determined.

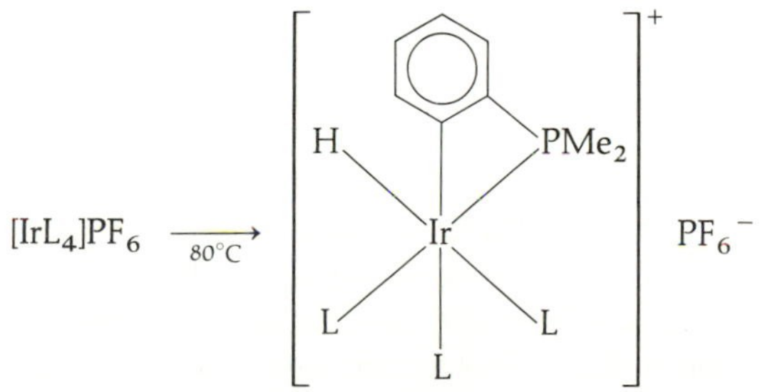

(10.43)

$L = Me_2PhP$

A particularly interesting reaction that indicates the apparent facility of *ortho*-metallation is shown in equation *10.44*. Complex **10.31** is in equilibrium with **10.32** and molecular $H_2$. Complex **10.32** is stable by itself, but it reverts to **10.31** in the presence of $H_2$.

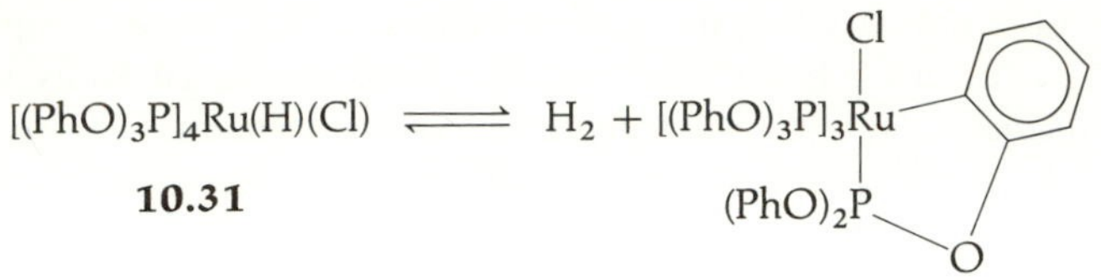

$$[(PhO)_3P]_4Ru(H)(Cl) \rightleftharpoons H_2 + [(PhO)_3P]_3Ru \qquad (10.44)$$

**10.31**

**10.32**

When $D_2$ is used, all 24 *ortho*-C—H bonds of **10.32** are deuterated at room temperature. A sequence of oxidative-addition and reductive-elimination reactions is proposed as the mechanism for this D—H exchange.

## REDUCTIVE-ELIMINATION REACTIONS

Just as oxidative-addition is an important method for attaching a substrate molecule to a metal atom, its reverse, *reductive-elimination,* is equally important for removing organic moieties from a complex. Frequently, ligands undergo one or more chemical reactions while attached to a metal atom, so that the eliminated molecule is usually different from the initial oxidative-addition product.

Four common reductive-elimination reactions are shown in equations *10.45–10.48*, in which R is an alkyl or aryl group.

$$\underset{\underset{L_nM—H}{|}}{R} \longrightarrow L_nM + R—H \qquad (10.45)$$

$$\underset{\underset{L_nM—R'}{|}}{R} \longrightarrow L_nM + R—R' \qquad (10.46)$$

$$\underset{\underset{L_nM—X}{|}}{R} \longrightarrow L_nM + R—X \qquad (10.47)$$

$$\underset{\underset{L_nM—H}{|}}{H} \longrightarrow L_nM + H—H \qquad (10.48)$$

Equation *10.45* represents C—H bond formation, an important termination step for many catalytic cycles. This type of elimination is usually very facile and often occurs rapidly and under mild conditions. An example of R—H elimination that follows $\beta$-hydrogen elimination is shown in equation *9.68*. A *cis* orientation of the R and H ligands is presumably required, as shown for the Pt(II) complex in equation *10.49*. *Trans* isomers convert to *cis* isomers before R—H elimination occurs.

$$\begin{array}{c} L \diagdown \qquad H \\ \phantom{xx} Pt \phantom{xx} \\ L \diagup \qquad CH_3 \end{array} \xrightarrow[\text{slow}]{-25°C} PtL_2 + CH_4 \qquad (10.49)$$

$$L \downarrow \text{fast}$$

$$PtL_3$$

$$L = Ph_3P$$

Equation *10.46* represents C—C bond formation, which is of fundamental importance to organic synthesis. A well-studied example of this type of reductive-elimination is shown in the following equation:

$$\begin{array}{c} L \diagdown \qquad CH_3 \\ \phantom{xx} Pd \phantom{xx} \\ L \diagup \qquad CH_3 \end{array} \underset{\text{fast}}{\rightleftarrows} \begin{array}{c} L \diagdown \qquad CH_3 \\ \phantom{xx} Pd \phantom{xx} \\ \phantom{xx} CH_3 \end{array} + L$$

$$L \downarrow \text{slow}$$

$$PdL_2 + C_2H_6 \qquad (10.50)$$

The *cis*-dialkyl Pd(II) complex eliminates ethane to give a formal Pd(0) complex. Because kinetic studies reveal that the reaction is retarded by added phosphine, predissociation of a phosphine ligand is postulated. In addition, only *cis*-$(CH_3)_2PdL_2$ complexes undergo elimination of ethane. The analogous *trans* isomers must isomerize to the *cis* form before this reaction can occur.

Equation *10.47* illustrates the reductive-elimination of organo halides. Eliminations of this kind are not particularly well studied and frequently occur as the last step of a sequence of reactions. An example of this elimination is shown in equation *10.51*. The organo halide product has an important functional group, the C—X bond, which can be used for subsequent chemistry.

$$\begin{array}{c} PhCH_2 \diagdown \quad \overset{L}{\underset{|}{}} \quad \diagup Cl \\ \phantom{xxx} Rh \\ O \diagdown \overset{C}{\phantom{x}} \diagup \overset{|}{\underset{L}{}} \quad \diagdown Cl \end{array} \longrightarrow PhCH_2Cl + \textit{trans}\text{-}Rh(CO)L_2Cl \qquad (10.51)$$

$$L = Ph_3P$$

Reductive-elimination of molecular hydrogen (eq. *10.48*) is observed for mono-, di-, and polynuclear complexes. Either thermal or photochemical activation can initiate loss of $H_2$. The principle of microscopic reversibility dictates that complexes which add $H_2$ in a concerted fashion will also eliminate $H_2$ by a concerted process. For other compounds, more complex mechanisms may be operative.

Reductive-elimination reactions are as mechanistically diverse as oxidative-addition reactions. Although the mechanisms of several reductive-elimination reactions have been studied in detail, it is not yet possible to generalize these results. Concerted, stepwise, and radical mechanisms have been postulated or observed. The

most common types of reductive-elimination reactions include both intramolecu-
lar, mononuclear, or *1,1-reductive-eliminations*, and intermolecular, dinuclear, or *1,2-reductive-eliminations*. Intra- and intermolecular processes are distinguished by using
isotopically labeled ligands and by observing the absence or presence of cross-over
products.

It is possible to generalize about some of the factors important in effecting intramolecular, mononuclear reductive-eliminations. Among them are (1) a *cis* relative orientation between the two ligands being eliminated, (2) the presence of bulky ancillary
ligands, (3) the possible formation of an electronically stable product complex, and (4)
a high formal charge on the metal atom. The first condition places the eliminating
ligands at positions of greatest proximity, while the second contributes a thermodynamic driving force to the elimination process: The coordination number of the
product complex decreases by two units upon elimination, affording a product with
considerably less interligand steric repulsion. The third criterion appears to be very
important, although a comprehensive discussion of this aspect is beyond the scope
of this text and our present understanding, as well. For general reductive-elimination
reactions, as shown in equations *10.45–10.48*, the electronic stability of the product
complex, $L_nM$, is critical. If this complex is quite stable, then reductive-elimination
is favored. The kinetically-controlled product is of particular interest. This complex
might undergo subsequent reaction chemistry.

The electronic stability of a complex $L_nM$ usually depends on the electronic properties of the ligands L, the coordination number, $n$, the geometry of the complex,
and the $d^m$ electronic configuration of M. For a *specific* complex, the dependence of
its electronic stability on these factors can be estimated by using crystal or ligand
field theory, the angular overlap model, or more sophisticated theoretical calculations. In some instances, the spin-multiplicity of the metal atom (i.e., the number of
unpaired *d* electrons) in the reactant and product complexes is important. Reductive-
elimination is usually favored when the number of unpaired electrons on M does
not change during the reaction. In other instances, the most stable product complex
has a different number of unpaired electrons on M than the reactant complex has.
Reductive-eliminations under this circumstance may be very slow (or unallowed)
thermally but more rapid with photochemical activation that excites M in the reactant complex to a spin state similar to that in the preferred product complex.
A brief examination of selected reductive-elimination reactions of four-coordinate
$d^8$-Pd(II) complexes illustrates the importance of these features. Organopalladium
complexes undergo 1,1-reductive-elimination with retention of configuration at carbon bound to palladium; thus, free-radical mechanisms are not operative in these
reactions.

A concerted, thermal 1,1-reductive-elimination from a *cis* four-coordinate square-
planar $d^8$-complex, can be shown to be allowed according to orbital symmetry:

$$L_2Pd(R)(R) \xrightarrow{\Delta} \{L_2Pd(R)(R)\}^\ddagger \longrightarrow L_2Pt + R{-}R \qquad (10.52)$$

**10.33**

If complex **10.33** were tetrahedral or even a *cis*-$L_3PdR_2$ trigonal-bipyramidal complex, a 1,1-reductive-elimination would be allowed also. However, this type of elimination is not allowed according to orbital symmetry for a *trigonal planar* $LPdR_2$ complex.

Equation *10.53* shows a 1,1-reductive-elimination of methyl and styryl ligands that proceeds predominantly by the concerted mechanism shown in equation *10.52*. Both *E*- and *Z*-geometrical isomers are observed for the $\eta^2$-olefin ligand. A possible driving force for this reaction is the formation of the relatively stable $d^{10}$-Pd(0) olefin complex (**10.34**). The olefin can act as a $\pi$-acidic ligand, thereby stabilizing the electron-rich Pd(0) atom.

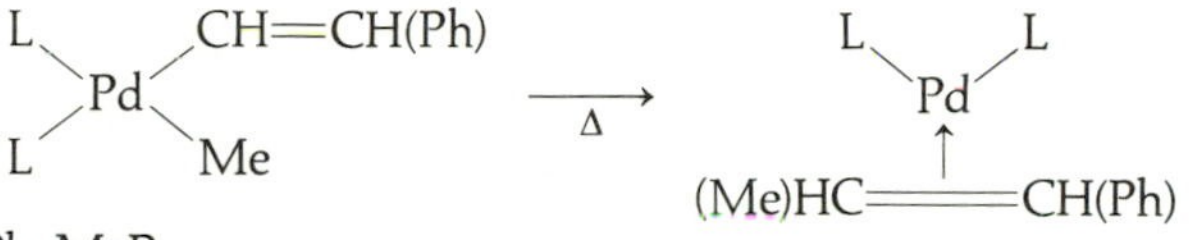

$$\underset{L}{\overset{L}{\diagdown}}Pd\underset{Me}{\overset{CH=CH(Ph)}{\diagup}} \quad \xrightarrow{\Delta} \quad \underset{(Me)HC=\!\!=CH(Ph)}{\overset{L\diagdown\,\diagup L}{Pd}} \qquad (10.53)$$

L = $Ph_3P$ or $Ph_2MeP$

**10.34**

Because of a preferred *trans* geometry, complex **10.35** does not undergo 1,1-reductive-elimination of ethane even at 100°C. However, the addition of $CD_3I$ causes immediate 1,1-reductive-elimination of $CH_3CD_3$ at 25°C:

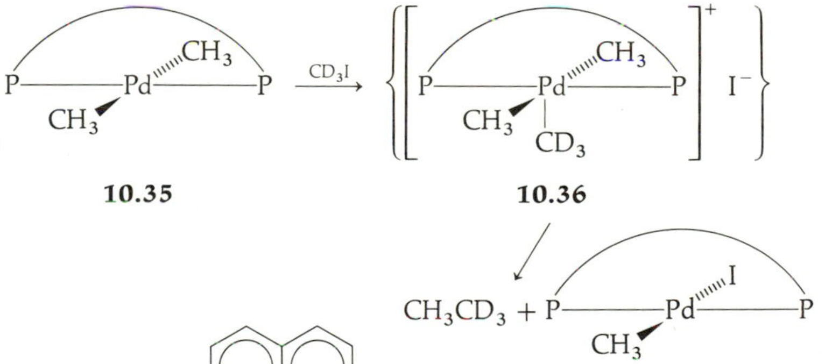

$$P\!-\!\!-Pd\overset{CH_3}{\underset{CH_3}{\cdots}}\!\!-P \quad \xrightarrow{CD_3I} \quad \left\{\left[P\!-\!\!-Pd\overset{CH_3}{\underset{\overset{|}{CD_3}}{\cdots}}\!\!-P\right]^+ I^-\right\}$$

**10.35**                             **10.36**

$$CH_3CD_3 + P\!-\!\!-Pd\overset{I}{\underset{CH_3}{\cdots}}\!\!-P \qquad (10.54)$$

where P⁀P =

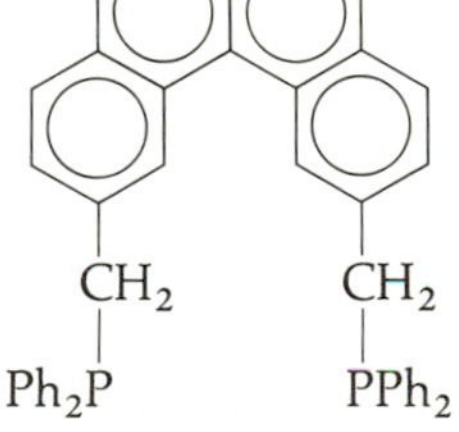

TRANSPHOS

These results indicate that a five-coordinate Pd(IV) complex, like **10.36**, is an intermediate in this reaction. Such a species contains two alkyl ligands in adjacent coordination sites. Equation *10.54* is also an example of an *oxidatively-induced* reductive-elimination reaction. Oxidation of a metal atom to a higher formal oxidation state frequently facilitates reductive-elimination. One reason for this effect is that changing the oxidation state of M often causes a structural change that places the ligands to be eliminated in adjacent coordination sites (**10.36**). Consistent with this view, rates of reductive-elimination from Pd(IV) and Pd(II) complexes containing *cis*-alkyl ligands are nearly equal in nonpolar solvents. However, there is one known system in which *cis*-dialkyl Pt(II) and Pt(IV) complexes undergo reductive-elimination at very different rates. Even though the Pt(IV) complex undergoes more rapid reductive-elimination, as expected, it is evident that the *cis*-dialkyl geometry is not the only critical factor affecting the rate of reductive-elimination.

When intramolecular 1,1-reductive-eliminations are *not* favored, dinuclear reductive-eliminations can occur. Two conditions appear to be required for these dinuclear eliminations: One metal center must contain a hydride ligand, and the other metal center must possess both a hydride, alkyl, or acyl ligand (to eliminate with the hydride ligand on the adjacent metal atom) and a vacant adjacent coordination site. This site can be present in the ground-state complex or generated by ligand dissociation or by "CO-insertion." It is still uncertain if the dinuclear eliminations occur from both metals or if migration of the groups being eliminated to one metal atom must occur prior to the elimination.

An example of a dinuclear reductive-elimination is shown in equation *10.55*.

$$cis\text{-}Os(CO)_4H_2 \xrightarrow[-CO]{slow} \{Os(CO)_3H_2\}$$

$$\textbf{10.37} \qquad\qquad\qquad\qquad \textbf{10.38}$$

fast | *cis*-Os(CO)$_4$H$_2$

$$\{Os_2(CO)_7H_2\} + H_2$$

$$\textbf{10.39}$$

fast | CO

$$\textbf{10.40}$$

(10.55)

Complex **10.37** is very stable toward *intramolecular* reductive-elimination of H$_2$, presumably because the Os(CO)$_4$ product would have two unpaired electrons on the osmium atom while complex **10.37** has all electron spins paired. However, at 125°C, complex **10.37** decomposes by *intermolecular* dinuclear reductive-elimination of H$_2$ as shown. The rate-determining step is CO dissociation from **10.37** to give **10.38**, which is coordinately unsaturated. This complex reacts with a second molecule of **10.37** to give a dinuclear intermediate that eliminates H$_2$ and forms **10.39**. Complex **10.37** recombines with CO to give the observed product **10.40**. The rate of H$_2$ elimination is decreased under external CO pressure. By using isotopically labeled complexes the dinuclear cross-coupling mechanism has been confirmed.

The apparent requirement that at least one of the eliminating ligands be a hydride suggests that a $\mu$-hydride ligand might be present in the intermediate species that undergoes dinuclear reductive-elimination. An example of a $\mu_2$-hydrido complex that apparently exhibits this behavior is shown in equation *10.56*.

$$\begin{bmatrix} & CH_2 & \\ Ph_2P & & PPh_2 \\ & H & \\ Pt & & Pt \\ H & & H \\ Ph_2P & & PPh_2 \\ & CH_2 & \end{bmatrix}^+ \xrightarrow{\;L\;} \begin{bmatrix} & CH_2 & \\ Ph_2P & & PPh_2 \\ & & \\ H-Pt & \text{———} & Pt-L \\ & & \\ Ph_2P & & PPh_2 \\ & CH_2 & \end{bmatrix}^+ + H_2 \quad (10.56)$$

**10.41**                                      **10.42**

$L = Me_2PhP$, $MePh_2P$ or $Ph_3P$

The Pt(II)-$\mu_2$-hydrido compound (**10.41**) is an A-frame complex that lacks a direct Pt—Pt bond. When **10.41** is treated with a phosphine ligand, $H_2$ is eliminated affording the dinuclear Pt(I) species (**10.42**). Since each platinum is formally reduced by one electron, a Pt—Pt single bond forms in the product complex.

## NONDISSOCIATIVE OXIDATIVE-ADDITION AND REDUCTIVE-ELIMINATION REACTIONS

When a substrate molecule XY contains a multiple bond and undergoes oxidative-addition to a metal center, the product complex retains some direct bonding between X and Y. As shown in equation *10.57*, the XY bond order is formally reduced by one upon oxidative-addition.

$$L_nM + X{=}Y \;\underset{}{\overset{K}{\rightleftharpoons}}\; L_nM\begin{array}{c} X \\ | \\ Y \end{array} \qquad (10.57)$$

**10.43**

Frequently, the oxidative-addition product formed has thermodynamic stability similar enough to that of the reactants that an equilibrium is established. Large values of the equilibrium constant $K$ indicate a more stable oxidative-addition product.

In previous chapters, selected *dihapto* complexes of olefins (**5.43**), alkynes (**5.51**), and ketones or ketone derivatives (**6.16** and **6.17**) have been described as possible oxidative-addition products rather than as $\pi$-complexes. This distinction is usually based on whether the molecular structure is described best by a metalla-cyclopropane or metalla-cyclopropene model or by the Dewar–Chatt–Duncanson model of metal-ligand bonding. Even diatomic molecules such as $O_2$ form stable adducts such as **10.43**.

A classic example of nondissociative oxidative-addition is illustrated by olefin adducts of Vaska's complex, *trans*-Ir(CO)(PPh$_3$)$_2$Cl, as shown in equation *10.58* with equilibrium and $\nu_{CO}$ data.

$$trans\text{-}Ir(CO)(PPh_3)_2Cl + \text{olefin} \overset{K}{\rightleftharpoons} (\eta^2\text{-olefin})Ir(CO)(PPh_3)_2Cl \quad (10.58)$$

**10.44**

| Olefin | $K$ $(M^{-1})$ | $\nu_{CO}^{product} - \nu_{CO}^{reactant}$ $(cm^{-1})$ |
|---|---|---|
| H$_2$C=CH$_2$ | <1 | — |
| H$_2$C=C(H)(CN) | 1.2 | 57 |
| (NC)(H)C=C(H)(CN) | 1,500 | 70 |
| (NC)$_2$C=C(CN)$_2$ | 140,000 | 96 |

Data source: L. Vaska, *Acc. Chem. Res.*, **1**, 335 (1968).

As the olefin becomes more electrophilic, complex **10.44** becomes more stable relative to the reactants. For the very electrophilic olefin, C$_2$(CN)$_4$, the $\nu_{CO}$ frequency shifts to much higher energy—an observation consistent with oxidative-addition to Vaska's four-coordinate Ir(I) complex affording a formally six-coordinate Ir(III) product.

Usually, basic ancillary ligands tend to stabilize adducts like **10.44** when the substrate molecule is strongly electrophilic. According to the Dewar–Chatt–Duncanson model of ($\eta^2$-olefin)M bonding (figure 5.2) and structural data, coordination of an olefin to a metal weakens the olefinic bond and might activate the olefin toward subsequent chemical reactivity. Investigation of nondissociative oxidative-additions of olefins or other substrate molecules (eq. *10.58*) is based on the assumption that complexes like **10.44** might either accelerate certain known reactions of the substrate molecule or confer chemical reactivity unknown for the free reactant species.

## REDUCTIVE-ADDITION REACTIONS AND NOMENCLATURE

The term *oxidative-addition* is a very useful formalism for classifying reactions of a general type. Although the addition of a molecule to a complex can easily be verified, a change in the actual charge density at the metal atom is difficult to determine. For example, oxidative-addition of H$_2$ to the Ir(I) atom in Vaska's complex represents a *formal* oxidation to Ir(III) in the product:

$$Ir(CO)(PPh_3)_2Cl + H_2 \longrightarrow Cl(Ph_3P)_2(OC)Ir\overset{\delta+}{\underset{}{\diagdown}}\begin{smallmatrix} H^{\delta-} \\ \\ H^{\delta-} \end{smallmatrix} \quad (10.59)$$

Although spectroscopic data can sometimes confirm an actual increase in the positive charge on a metal atom after such a reaction, it is important to note that the oxidative character of these reactions depends critically on the relative electronegativities of both the metal atom *in the reactant complex* and the donor atoms of the adduct molecule.

Addition of $H_2$ to the *cationic* Ir(I) complex (**10.45**) also gives a *cis*-dihydrido complex (**10.46**):

$$[(\eta^4\text{-1,5-COD})Ir(PPh_3)_2]PF_6 + H_2 \xrightarrow{-80°C} \left[ (\eta^4\text{-1,5-COD})(Ph_3P)_2Ir \begin{array}{c} {}^{\delta\delta+} H \\ {}^{\delta+} \\ H \end{array} \right]^+ PF_6^- \quad (10.60)$$

**10.45**                                             **10.46**

However, in contrast to Vaska's complex, in this case hydrogen addition is *retarded* by stronger electron donating ligands. This complex also reacts only very slowly with $CH_3I$ to give nonoxidative-addition products. This addition of $H_2$ to a *cationic* Ir(I) complex containing ancillary ligands of moderately strong $\pi$-acidity suggests that a *reductive-addition* might be occurring.

Presumably, the nearly equal Pauling electronegativities of iridium, $\chi = 2.2$, and hydrogen, $\chi = 2.1$, mean that the electronic nature of the ancillary ligands and the overall charge of a complex greatly affect the *actual* electronegativity of an Ir(I) ion *in a specific complex*. In Vaska's complex, which is neutral, the Ir atom is apparently less electronegative than a hydrogen atom; therefore, an actual oxidative-addition occurs. In **10.45**, which is cationic, the Ir atom is apparently more electronegative than a hydrogen atom, so an actual *reductive-addition* takes place.

Reductive-addition is formally similar to the hydrogenation of alkenes (eq. *10.61*), which is normally regarded as a reduction process.

$$R_2C{=}CR_2 + H_2 \longrightarrow \begin{array}{c} H\ \ H \\ |\ \ \ | \\ R_2C{-}CR_2 \end{array} \quad (10.61)$$

This type of reaction might be observed for other cationic complexes of the heavy metals.

Obviously, the observation of "oxidative-addition" reactions that appear to involve a *reduction* of the metal atom confuses the oxidative-addition nomenclature. The use of formal oxidation states is always a poor approximation to actual atomic charges. Notice, however, that the oxidative-additions described in this chapter could be classified as either one- or two-electron additions at the metal atom. For example, in equations *10.59* and *10.60*, the iridium atom gains two electrons in electron count in going from reactant to product regardless of the *real* change in atomic charge at Ir. One suggestion for a new nomenclature system regards these reactions as {3,2} additions to signify that three centers (Ir + 2H atoms) are involved and that the metal atom gains two electrons in its valence shell. A more realistic nomenclature might be adopted eventually.

## STUDY QUESTIONS

1. Describe the reactions shown below as formal oxidative-addition and/or reductive-elimination reactions:

   (a) $Ni(CO)_4$ + allyl chloride $\longrightarrow$ $(\eta^1\text{-}C_3H_5)Ni(CO)_3Cl + CO$

(b) $Fe(CO)_5 + I_2 \longrightarrow cis\text{-}Fe(CO)_4I_2 + CO$

(c) $Os(CO)_5 + H_2 \longrightarrow cis\text{-}H_2Os(CO)_4 + CO$

(d) $Os(CO)_3L_2$ (L = Ph$_3$P and in *axial* sites) + Br$_2$ $\longrightarrow$ isolated ionic
intermediate (octahedral) $\xrightarrow{\Delta}$ $cis,cis,trans\text{-}Os(CO)_2(Br)_2L_2 + CO$

(e) $trans\text{-}Pt(PEt_3)_2(GePh_3)_2 + H_2 \longrightarrow trans\text{-}Pt(PEt_3)_2(GePh_3)(H) +$
$HGePh_3$

2. A mixture of $\overset{\displaystyle O}{\overset{\displaystyle \|}{R}C}Rh(PPh_3)_2(Cl)_2$ and $Rh(R)(CO)(PPh_3)_2(Cl)_2$, where R is $p$-

$MeC_6H_4$, in $o$-xylene at $144°C$ catalytically converts $p\text{-}MeC_6H_4\overset{\displaystyle O}{\overset{\displaystyle \|}{C}}Cl$ to $p$-chlorotoluene. Propose a reaction cycle for this process. [J. A. Kampmeier, R. M. Rodehorst, and J. B. Philip, Jr., *J. Am. Chem. Soc.*, **103**, 1847 (1981).]

3. Acyl-oxygen bonds of aryl carboxylates can be cleaved by Ni$^0$ as shown:

$Ni(1,5\text{-}COD)_2 + 3Ph_3P + EtCO_2Ph \longrightarrow$
$Ni(CO)(PPh_3)_3 + C_2H_4 + PhOH + 2\ 1,5\text{-}COD$

Propose a sequence of reactions to rationalize the formation of the observed products. [See T. Yamamoto, J. Ishizu, T. Kohara, S. Komiya, and Y. Yamamoto, *J. Am. Chem. Soc.*, **102**, 3758 (1980).]

4. Propose a single intermediate for the following reaction, and determine changes in the formal oxidation states of the metals in going from reactants to products (L = Me$_2$PhP).

$Rh(CO)L_2(Cl)_3 + trans\text{-}Ir(CO)L_2Cl \longrightarrow$
intermediate $\longrightarrow$ $trans\text{-}Rh(CO)L_2Cl + Ir(CO)L_2(Cl)_3$

5. The dinuclear complex $[(\eta^5\text{-}C_5H_5)Co(CH_3)(\mu_2\text{-}CO)]_2$ decomposes thermally to produce an 85% yield of acetone and cobalt cluster complexes. Acetone is formed from a mononuclear intermediate, $(\eta^5\text{-}C_5H_5)Co(CH_3)_2(CO)$. Thermal decomposition of a 1:1 mixture of the above dinuclear complex and the analogous CD$_3$ complex affords acetone, $d^3$-acetone and $d^6$-acetone in relative yields of 26%, 47%, and 27%, respectively. Write a mechanism that explains these observations

6. Explain why a metal-metal bond is *lost* in going from **10.26** to **10.27** and *gained* in going from **10.28** to **10.29** and from **10.41** to **10.42**.

7. For the reaction shown below, addition of phosphine increases the rate of elimination of methylcyclohexane 1,000-fold. Addition of THF or Et$_3$N does not increase the rate of reaction. Propose a mechanism that explains these results. [See K. I. Gell and J. Schwartz, *J. Am. Chem. Soc.*, **103**, 2687 (1981).]

$(\eta^5\text{-}C_5H_5)_2Zr(H)(R) + 2L \xrightarrow[-30°C]{toluene} (\eta^5\text{-}C_5H_5)_2ZrL_2 + R\text{—}H$
$80\text{—}90\%$

R = cyclohexylmethyl; L = Ph$_2$MeP or PhMe$_2$P

8. Suggest an isotopic labeling experiment to confirm the dinuclear cross-coupling mechanism shown in equation *10.55*.

**9.** Preliminary structural data for the complex $W(CO)_3(PR_3)_2(\eta^2\text{-}H_2)$, in which R is *i*-Pr, indicates the geometry shown below. Relate this observation to the discussion in this chapter of the mechanism of oxidative-addition of $H_2$. The H—H distance in free $H_2$ is 0.740 Å.

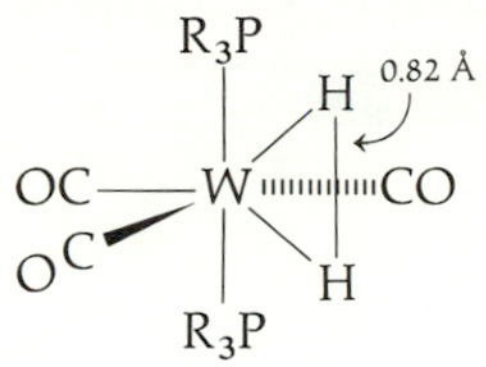

$$
\begin{array}{c}
R_3P \\
\mid \quad\quad H \quad 0.82\,\text{Å} \\
OC \text{——} W \text{'''''''''} CO \\
O^C \quad \mid \quad\quad H \\
R_3P
\end{array}
$$

**10.** Give three examples of substrate molecules (other than those listed in table 10.1) that should undergo oxidative-additions. Try to avoid trivial derivatives of those molecules listed in the table.

## SUGGESTED READING

Bergman, R. G. *Acc. Chem. Res.* **13**, 113 (1980).

Collman, J. P., and Roper, W. R. *Adv. Organometal. Chem.* **7**, 53 (1968).

Crabtree, R. *Acc. Chem. Res.* **12**, 331 (1979).

Davidson, J. L. *Inorg. React. Mech.* **5**, 398 (1977).

Davidson, J. L. *Inorg. React. Mech.* **6**, 414 (1979).

Deeming, A. J. *MTP Internat. Review of Science, Series I.* **9**, 117 (1972).

Halpern, J. *Acc. Chem. Res.* **3**, 386 (1970).

Kemmitt, R. D. W., and Burgess, J. *Inorg. React. Mech.* **2**, 350 (1972).

Kemmitt, R. D. W., and Smith, M. A. R. *Inorg. React. Mech.* **3**, 451 (1974).

Kemmitt, R. D. W., and Smith, M. A. R. *Inorg. React. Mech.* **4**, 349 (1976).

Norton, J. *Acc. Chem. Res.* **12**, 139 (1979).

Osborn, J. A. *Organotransition-Metal Chemistry.* Y. Ishü and M. Tsutsui, eds. p. 65. New York: Plenum Press, 1975.

Parshall, G. W. *Acc. Chem. Res.* **3**, 139 (1970).

Stille, J. K., and Lau, K. S. Y. *Acc. Chem. Res.* **10**, 434 (1977).

Ugo, R. *Coord. Chem. Rev.* **3**, 319 (1968).

Vaska, L. *Acc. Chem. Res.* **1**, 335 (1968).

# Electrophilic
# and Nucleophilic Attack
# on Organometallic Complexes

## CHAPTER 11

Most reactions of organometallic complexes can be classified as either electrophilic or nucleophilic attack by a reagent molecule. In this chapter we introduce several examples of each reaction type in preparation for Part II where these reactions are used to form specific organic molecules of interest. Electrophilic and nucleophilic attack on a complex occurs at the metal atom, at a ligand donor atom, or at a ligand site farther from the metal. Sometimes the initial site of attack is different from the final location of the electrophilic or nucleophilic reagent in the product. There is great interest in determining the sites of kinetic attack and the thermodynamically controlled products of these reactions for applications to organic synthesis and for better understanding these reaction mechanisms.

## ELECTROPHILIC ATTACK ON ORGANOMETALLIC COMPLEXES

Common electrophilic reagents include protic acids, alkylating and acylating reagents, and strong Lewis acids such as coordinatively unsaturated boron and aluminum compounds or other metal ions. Several examples of protonation, alkylation, and acylation reactions have been discussed earlier as preparative methods for forming hydride, alkyl and acyl complexes of metals.

### Electrophilic Attack at the Metal Atom

Only three examples of this reaction type are mentioned here because protonation, alkylation, and acylation of metals have been discussed in chap. 8. Equation *11.1* shows $CH_3I$ alkylation of a cobalt atom in a *neutral* complex (**11.1**) to afford the cationic methyl complex **11.2.**

The $d^8$-Co(I) atom in **11.1** should possess a chemically accessible lone pair of electrons in its valence shell, much like the $d^8$-Ir(I) atom in Vaska's complex. Both **11.1** and **11.2** obey the EAN rule, although complex **11.1** attains the preferred coordination number of four upon alkylation.

$$(\eta^5\text{-}C_5H_5)Co(PEt_3)_2 + CH_3I \longrightarrow \left[ \begin{array}{c} \text{Co} \\ Et_3P\overset{\displaystyle |}{\underset{\displaystyle PEt_3}{\diagdown}}CH_3 \end{array} \right]^+ I^- \qquad (11.1)$$

**11.1**

**11.2**

Neutral diphenylcarbene complexes can be prepared from metal carbonyl dianions by using two electrophilic reagents:

$$[M(CO)_5]^{2-} + \underset{\underset{Ph}{\diagup}\overset{\displaystyle C}{\underset{\displaystyle \|}{\phantom{.}}}\overset{\displaystyle +N}{\overset{Me\diagdown \diagup Me}{\phantom{.}}}Ph \longrightarrow \left[ (OC)_5\overset{-}{M}-\underset{Ph\ \ Ph}{\overset{\displaystyle NMe_2}{\underset{\displaystyle |}{C}}} \right] \xrightarrow[-Me_2NH_2^+]{2H^+}$$

$$M = Cr, Mo \text{ or } W$$

**11.3**

$$(OC)_5M{=}C\overset{\diagup Ph}{\underset{\diagdown Ph}{\phantom{.}}} \qquad (11.2)$$

Iminium ion alkylation of a Group VI $[M(CO)_5]^{2-}$ complex gives an alkyl metallate intermediate (**11.3**). Protonation of the nitrogen atom of **11.3** eliminates dimethylamine and forms the carbene complexes.

Lewis acid-base complexes are also known, as shown in equation *11.3*. The $[Re(CO)_5]^-$ anion coordinates to the strong Lewis acid, $BH_3$.

$$Re(CO)_5^- + BH_3 \longrightarrow [(OC)_5Re \rightarrow BH_3]^- \qquad (11.3)$$

### Electrophilic Attack at an M—C σ Bond

Electrophilic cleavage of M—C σ bonds is often used to remove a ligand from a metal atom. Two types of mechanisms appear to be important in these reactions: $S_E2$ bimolecular attack directly on an M—C bond and some type of electron transfer, metal oxidation, or oxidative-addition reaction at the metal atom prior to M—C bond cleavage. An $S_E2$ attack can occur from the back side to give inversion at the carbon donor atom (eq. *11.4*) or from the front side to give retention at C (eq. *11.5*).

$$L_nM{-}C{\leqslant} + E^+ \longrightarrow \left\{ L_nM^{\delta+}\text{----}\overset{\displaystyle |}{C}\text{---}E \right\}^{\ddagger} \longrightarrow L_nM^+ + {\geqslant}C{-}E \quad (11.4)$$

$$L_nM{-}C{\leqslant} + E^+ \longrightarrow \left\{ \underset{L_nM^{\delta+}}{\overset{\displaystyle C}{\underset{\diagdown}{\diagup}\!\!\!\diagdown E}} \right\}^{\ddagger} \longrightarrow L_nM^+ + E{-}C{\leqslant} \quad (11.5)$$

Unfortunately, electrophilic M—C bond cleavage may also occur by other pathways, and these reactions are frequently not stereochemically clean. The coordination number of M, the steric bulk of the ligands, solvent polarity, the hardness or softness of the electrophile, and the locations of maximum electron density in the HOMO of the complex all influence the course of these reactions.

Commonly used electrophiles include $H^+$, halogens, and metal ions such as Hg(II). An example of electrophilic cleavage by protonolysis is shown in equation 11.6. When the benzyl-Fe complex **11.4** is treated with trifluoroacetic acid, the Fe—$CH_2$Ph bond is cleaved to give toluene and a trifluoroacetate complex:

$$(\eta^5\text{-}C_5H_5)Fe(CO)_2CH_2Ph + CF_3CO_2H \longrightarrow$$

**11.4**
$$(\eta^5\text{-}C_5H_5)Fe(CO)_2(OCOCF_3) + CH_3\text{—}Ph \quad (11.6)$$

Kinetic studies of this reaction including the measurement of $H^+/D^+$ isotope effects are interpreted as indicating initial proton attack on the Fe atom followed by reductive-elimination of toluene. The coordinatively unsaturated iron product, $[(\eta^5\text{-}C_5H_5)Fe(CO)_2]^+$, picks up a $CF_3CO_2^-$ ion to give the neutral product.

Electrophilic cleavage of M—C bonds by halogens occurs via diverse pathways giving retention, inversion, or racemization of carbon stereochemistry. Radical mechanisms are frequently competing processes. An example of clean inversion of carbon stereochemistry in a halogen-induced cleavage reaction is shown in equation 11.7. Oxidative-addition of $X_2$ to the iron atom of **11.5** generates the ion pair **11.6**. Displacement of the iron complex by back-side attack of the halide counterion affords inversion of carbon stereochemistry. Likewise, the *erythro* isomer of complex **11.5** gives the *threo*-alkyl halide product.

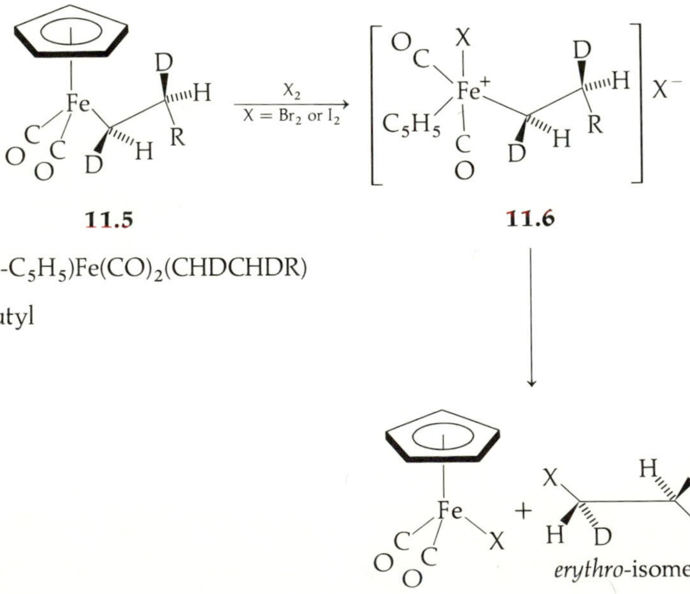

**11.5**                                                      **11.6**

*threo*-$(\eta^5\text{-}C_5H_5)Fe(CO)_2$(CHDCHDR)

R = *t*-butyl

(11.7)

*erythro*-isomer

In complex **11.5** (when R = *phenyl*), electrophilic attack by halogen proceeds with *retention* of carbon stereochemistry. Isotopic labeling studies indicate that a phenonium salt (**11.7**) is the intermediate formed by elimination of the Fe-halide complex. Coupling of the phenonium ion and the halide counterion gives *threo*-XCHDCHDPh.

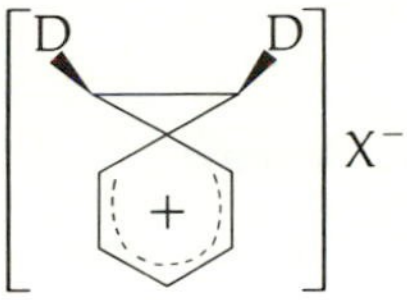

**11.7**

Electrophilic cleavage of an M—C bond by Hg(II) halides occurs with both retention and inversion of configuration at the carbon atom depending on the organometallic complex. The type of products formed sometimes depends on the nature of the alkyl ligand, as shown in equation *11.8*.

$$(\eta^5\text{-}C_5H_5)Fe(CO)_2R + HgX_2 \Bigg\langle \begin{array}{l} \xrightarrow[\text{alkyl}]{\text{primary}} (\eta^5\text{-}C_5H_5)Fe(CO)_2X + RHgX \\ \\ \xrightarrow[\text{tertiary alkyl}]{\text{benzylic or}} (\eta^5\text{-}C_5H_5)Fe(CO)_2HgX + RX \end{array}$$

$$(11.8)$$

Primary alkyls react with $HgX_2$ to give the corresponding iron-halide complex and the organomercurial, RHgX. This reaction is a formal transmetallation reaction. Benzylic or tertiary alkyl complexes give heteronuclear bimetallic complexes and alkyl halide.

For primary iron-alkyl complexes, the reaction rate with $HgX_2$ is third order, rate = $k[Fe\text{—}R][HgX_2]^2$, and the configuration at carbon is retained. The proposed mechanism for this reaction is as follows:

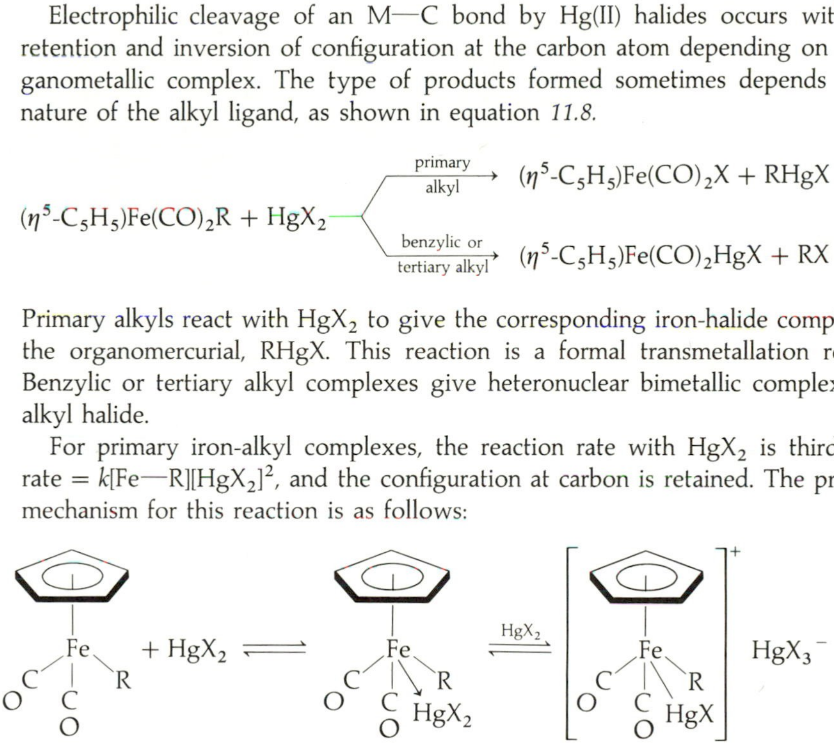

$$(11.9)$$

**11.8**          **11.9**

$$(\eta^5\text{-}C_5H_5)Fe(CO)_2X + HgX_2 \longleftarrow [(\eta^5\text{-}C_5H_5)Fe(CO)_2]^+ + HgX_3^- + RHgX$$

**11.10**

Electrophilic attack by $HgX_2$ on the $d^8$-Fe(0) atom gives the Lewis acid-base adduct **11.8**. Extraction of a halide ligand by a second molecule of $HgX_2$ affords a cationic heteronuclear complex **11.9**. Reductive-elimination from **11.9** gives RHgX with retention of configuration at the carbon atom and the cationic coordinatively unsaturated iron complex **11.10.** Because this compound is a stronger Lewis acid than $HgX_2$, it extracts a halide ion from the $HgX_3^-$ counterion to form the neutral product. Facile reductive-elimination from **11.9** seems reasonable since **11.9** is an oxidized and sterically more crowded derivative of the starting material (as discussed in chap. 10).

## Electrophilic Attack at a Donor Atom

Electrophilic attack at a donor atom occurs only when the donor is relatively electron-rich because, by definition, donor atoms are already supplying electron density to a metal atom. Electrophilic attack occurs at monatomic ligands that possess at least one lone electron pair or at unsaturated carbon donor atoms that are involved in some type of $d\pi$-$p\pi$ or $p\pi$-$p\pi$ multiple bonding to the metal atom.

Equations *11.10* and *11.11* are examples of alkylation of $\mu_2$-chalcogen atoms by $CH_3I$.

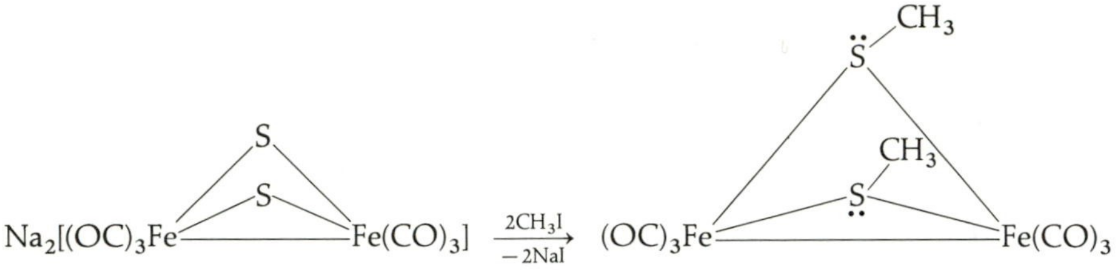

Na2[(OC)3Fe
Fe(CO)3]
S
S
2CH3I
−2NaI
(OC)3Fe
Fe(CO)3
CH3
CH3


**11.11** **11.12** *(11.10)*

(η5-C5H5)Co
Co(η5-C5H5)
Te
P
Me2
P
Me2
CH3I
CH3
I−


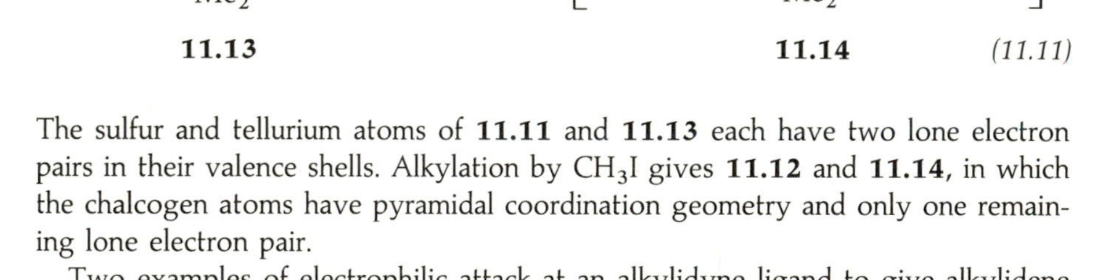

**11.13** **11.14** *(11.11)*

The sulfur and tellurium atoms of **11.11** and **11.13** each have two lone electron pairs in their valence shells. Alkylation by $CH_3I$ gives **11.12** and **11.14,** in which the chalcogen atoms have pyramidal coordination geometry and only one remaining lone electron pair.

Two examples of electrophilic attack at an alkylidyne ligand to give alkylidene complexes are shown in equations *11.12* and *11.13.*

$$HC{\equiv}WL_4Cl + CF_3SO_3H \longrightarrow [H_2C{=}WL_4Cl]^+CF_3SO_3^- \qquad (11.12)$$

**11.15** **11.16**

$L = Me_3P$

$$\underset{\textbf{11.17}}{\overset{\displaystyle O}{\underset{\displaystyle Cl}{\overset{\displaystyle \underset{L}{C}}{\overset{\displaystyle L}{Os\!\equiv\!C\!-\!R}}}}} \quad \xrightarrow{\text{HCl}} \quad \underset{\textbf{11.18}}{\overset{\displaystyle O}{\underset{\displaystyle Cl}{\overset{\displaystyle \underset{L}{C}}{\overset{\displaystyle L}{Os}}}}}\overset{R}{\underset{}{C\!-\!H}} \qquad (11.13)$$

L = Ph$_3$P; R = p-tolyl

Triflic acid protonates the electrophilic methylidyne ligand of **11.15** to afford the cationic methylidene complex **11.16**. The five-coordinate alkylidyne complex **11.17** incorporates one molecule of HCl to give the six-coordinate alkylidene complex **11.18**. These reactions represent electrophilic attack on a metal-carbon triple bond. The alkylidyne carbon atoms in these complexes are nucleophilic.

Alkylidene carbon atoms can also be nucleophilic, as shown in equations *11.14* and *11.15*. Protonation or silylation of the alkylidene α-carbon atom affords the corresponding alkyl complexes. The alkylidene carbon atoms in these *electron-deficient* complexes are nucleophilic because of particularly effective M → C $d\pi$-$p\pi$ back bonding.

$$(Me_3CCH_2)_3M\!=\!C\overset{H}{\underset{CMe_3}{}} \quad + \text{ HCl} \quad \longrightarrow \quad M(CH_2CMe_3)_4Cl \qquad (11.14)$$

M = Nb or Ta

$$(\eta^5\text{-}C_5H_5)_2Ta\overset{CH_3}{\underset{CH_2}{}} \quad + Me_3SiBr \quad \longrightarrow \quad [(\eta^5\text{-}C_5H_5)_2Ta(CH_2SiMe_3)(CH_3)]^+Br^- \qquad (11.15)$$

The most common type of electrophilic attack is protonation or acylation of the unsaturated carbon atoms of polyene ligands. Several examples of such protonation reactions are discussed in previous chapters (a well-known example being the protonation of an $\eta^4$-1,3-diene ligand to give an $\eta^3$-allylic ligand as shown in eq. *5.82*). In most of these, initial attack is at the metal atom followed by proton transfer to the polyene ligand.

Friedel–Crafts alkylation and acylation of $\eta^n$-carbocyclic ligands is also a well-known reaction. Acylation of $\eta^4$-cyclobutadiene ligands is quite facile (eq. *11.16*), presumably because the $\eta^4$-C$_4$H$_4$ ligand is moderately electron-rich. Formylation

$$\underset{\underset{(CO)_3}{Fe}}{\square} \quad \xrightarrow[\text{AlCl}_3]{\overset{O}{\overset{\|}{CH_3CCl}}} \quad \underset{\underset{(CO)_3}{Fe}}{\square}\overset{O}{\overset{\|}{-C\!-\!CH_3}} \qquad (11.16)$$

and even chloromethylation of cyclobutadiene ligands are possible using similar conditions.

Examples of acylation and sulfonylation of an $\eta^5$-$C_5H_5$ ligand in ferrocene are shown in equation 11.17. Electrophilic substitution of ferrocene can give acylation at both cyclopentadienyl ligands. Neutral metallocenes are susceptible to electrophilic attack, but cationic metallocenes are more susceptible to nucleophilic attack.

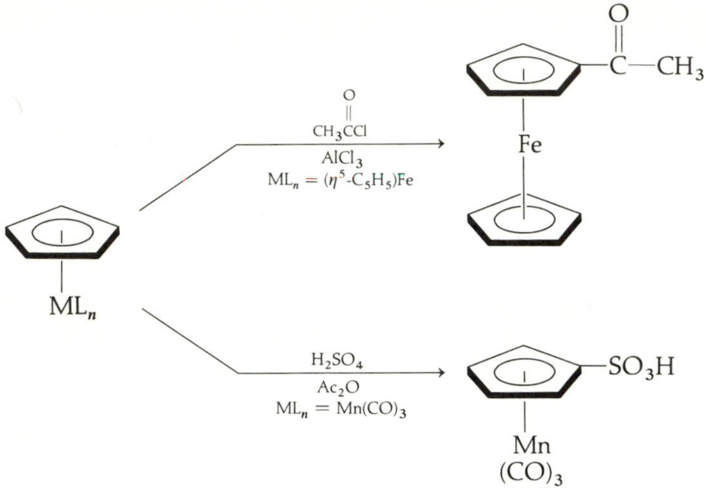

$$(11.17)$$

*Hexahapto*-arene ligands can be acylated similarly:

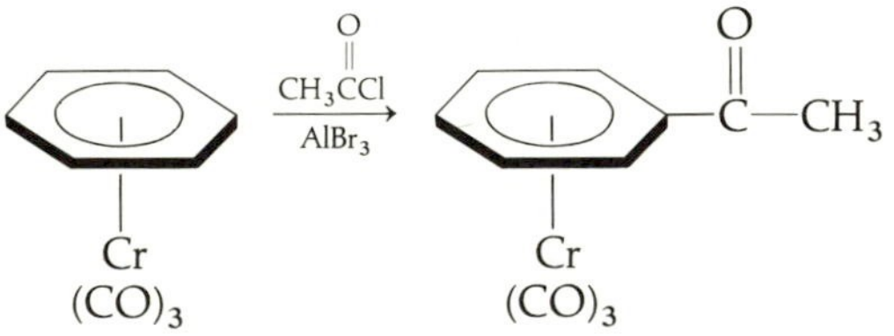

$$(11.18)$$

The $(\eta^6$-arene)Cr(CO)$_3$ complexes usually require very acidic Friedel–Crafts conditions because the metal moiety withdraws considerable $\pi$-electron density from the arene molecule. For this reason, $(\eta^6$-arene)Cr(CO)$_3$ complexes undergo facile nucleophilic attack.

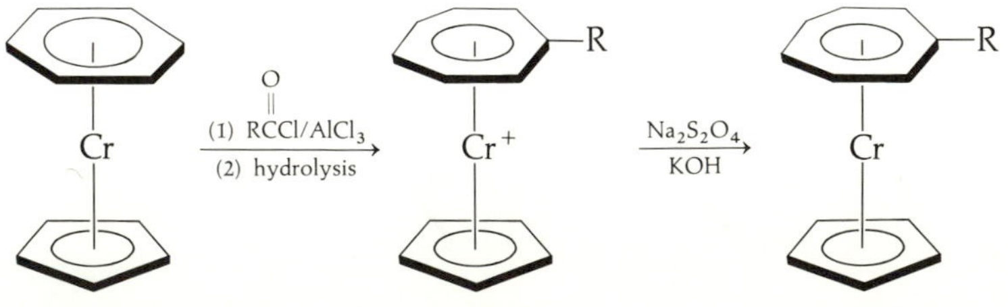

$$(11.19)$$

11.19                                            11.20                                            11.21

An interesting ring-expansion occurs upon Friedel–Crafts acylation of the arene ligand in the mixed-sandwich complex **11.19**. Acetyl or benzoyl chloride gives ring-expanded products (**11.20**) in approximately 10% yield. These complexes are probably formed from metal-stabilized carbonium ion intermediates. Reduction with dithionite ion gives the neutral analogues (**11.21**).

Direct alkylation of $\eta^3$-allylic ligands affords new alkene derivatives, as shown in equation *11.20* for a nickel complex. The allyl ligand can be regarded as an anionic four-electron donor. In this formalism, the completely filled HOMO is $2\pi$ (see figure 5.5).

$$CH_3-\!\!\diagdown\!\!\diagup\!\!\left(-Ni\diagup\!\!\diagdown\substack{Br\\Br}\right)_2 + 2RX \xrightarrow{\text{DMF}} 2 \;\; \substack{R\\ \diagup\!\!=\!\!\diagdown \\ CH_3} + \{Ni(II)salts\} \quad (11.20)$$

RX = MeI, MeBr, $C_6H_{11}I$, $Me_3CI$, PhI, vinyl Br, or $PhCH_2Br$

Electrophilic alkylation takes place only at the terminal carbon atoms because these are the only ones contributing to the $2\pi$ orbital in the uncomplexed allyl anion. Although experimental evidence indicates that this alkylation probably proceeds by a radical process, the unpaired electron of an allyl radical also occupies the $2\pi$ orbital, and therefore a similar analysis would follow.

## Electrophilic Attack at Nondonor Atoms of Ligands

Electrophilic attack at nondonor atoms of ligand molecules usually occurs at electron-rich sites, such as lone pairs or regions of $\pi$-electron density. Frequently, attack at basic sites on atoms adjacent to a donor atom significantly alters the electronic nature of the metal-ligand bond; for example, it can increase the hapticity of the ligand. Hydride abstraction by $Ph_3C^+$ from an $\eta^6$-cycloheptatriene ligand affords an $\eta^7$-$C_7H_7$ complex, as shown in equation *5.17*. Protonation of an $\eta^6$-cyclooctatetraene ligand gives an $\eta^7$-cyclooctatrienyl ligand (eq. *11.21*).

$$\underset{\substack{\text{Mo}\\(CO)_3}}{\diagup} \xrightarrow{H_2SO_4} \left[ \underset{\substack{\text{Mo}\\(CO)_3}}{\overset{H\quad H}{\diagup}} \right] HSO_4^- \qquad (11.21)$$

Electrophilic attack at carbon, oxygen, nitrogen, and sulfur atoms adjacent to donor atoms is very common. Protonation or alkylation of the propynyl ligand of complex **11.22** affords the vinylidene complexes **11.23** and **11.24**. The proton and methyl electrophiles attack the $\beta$-carbon atom of the propynyl ligand.

The oxygen atom of a CO ligand is susceptible to electrophilic attack because M—CO $d\pi$-$p\pi$ back bonding tends to increase its negative charge density. Lewis acid adducts of terminal CO, $\mu_2$-CO and $\mu_3$-CO ligands have been prepared according to equations *11.23–11.25*.

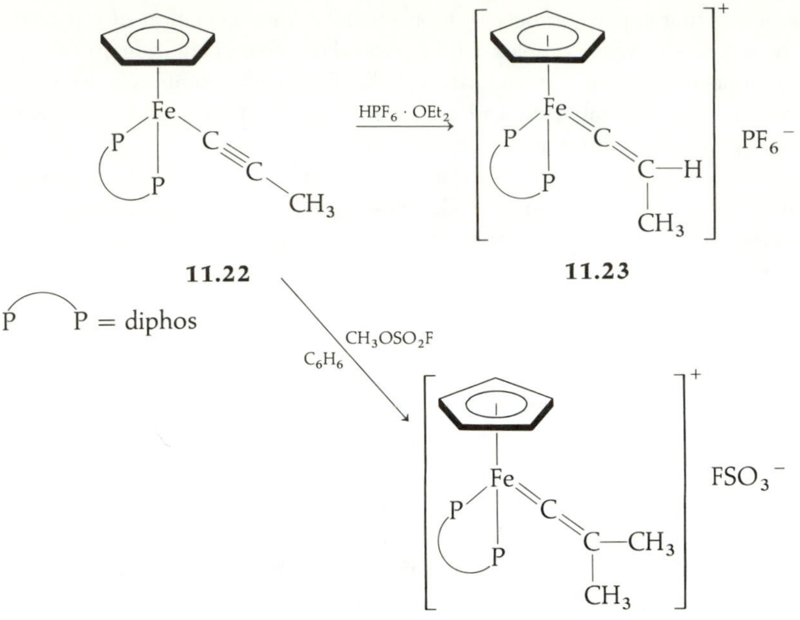

**11.22**                                                    **11.23**

P⌒P = diphos

$(\eta^5\text{-}C_5H_5)Mo(CO)_3H + (\eta^5\text{-}C_5Me_5)(\eta^5\text{-}C_5Me_4{=}CH_2)Ti(CH_3)\longrightarrow$

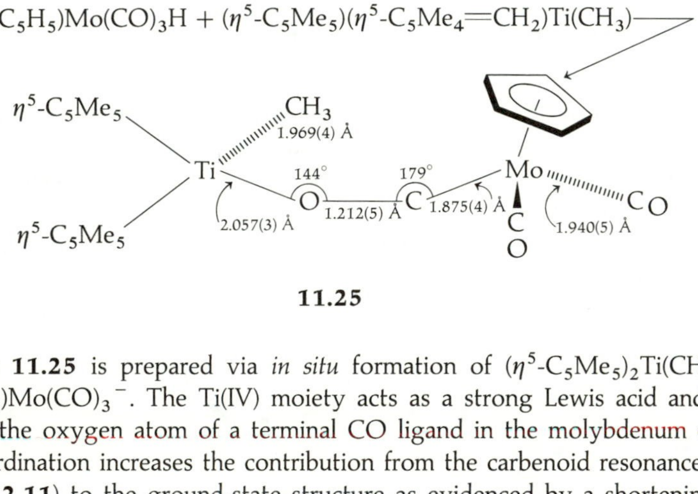

**11.24**

FSO$_3^-$    (11.22)

(11.23)

**11.25**

Complex **11.25** is prepared via *in situ* formation of $(\eta^5\text{-}C_5Me_5)_2Ti(CH_3)^+$ and $(\eta^5\text{-}C_5H_5)Mo(CO)_3^-$. The Ti(IV) moiety acts as a strong Lewis acid and coordinates to the oxygen atom of a terminal CO ligand in the molybdenum complex. This coordination increases the contribution from the carbenoid resonance form of CO (see **2.11**) to the ground-state structure as evidenced by a shortening of the Mo—C distance to the isocarbonyl CO ligand relative to the Mo—C distance to the other terminal CO groups. The C—O distance of the unique CO is about 0.085 Å longer than that of a normal terminal CO ligand.

Complex **11.26** reveals coordination of Et$_3$Al to a $\mu_2$-CO ligand. Triethylaluminum, a strong Lewis acid, coordinates even more strongly to $\mu_3$-CO ligands, as in complex **11.27**.

Molecular weight and IR data for solutions of **11.27** indicate very little dissociation of Et$_3$Al. In complexes **11.26** and **11.27**, the $\nu_{CO}$ stretching frequencies of the $\mu_2$- and $\mu_3$-CO ligands decrease by 112 cm$^{-1}$ and 125 cm$^{-1}$, respectively, upon adduct formation.

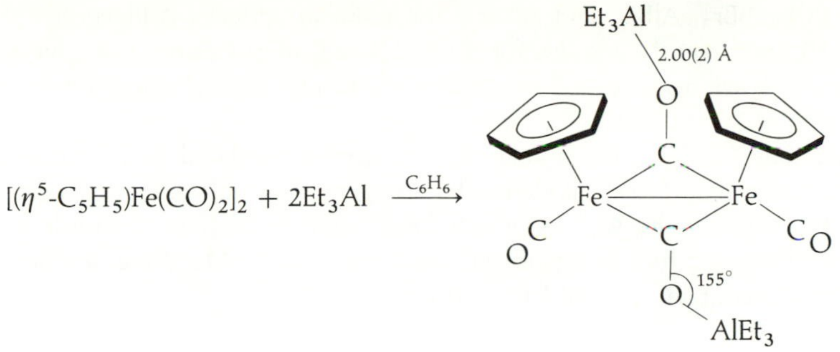

$$[(\eta^5\text{-}C_5H_5)Fe(CO)_2]_2 + 2Et_3Al \xrightarrow{C_6H_6}$$ (11.24)

**11.26**

$$[(\eta^5\text{-}C_5H_5)Fe(\mu_3\text{-}CO)]_4 + 4Et_3Al \longrightarrow [(\eta^5\text{-}C_5H_5)Fe(\mu_3\text{-}CO \cdot AlEt_3)]_4 \quad (11.25)$$

**11.27**

Protonation of the tetrairon butterfly cluster (**11.28**) affords complex **11.29**, which contains an O-protonated CO ligand:

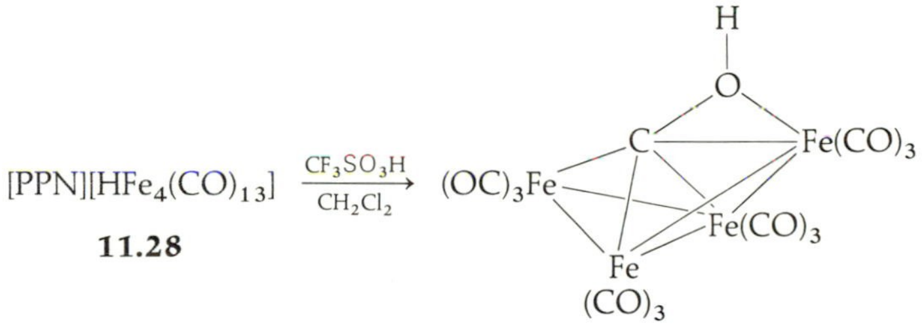

$$[PPN][HFe_4(CO)_{13}] \xrightarrow[CH_2Cl_2]{CF_3SO_3H}$$ (11.26)

**11.28**

**11.29**

Further protonation and a concomitant two-electron reduction (probably involving another iron complex) cleaves the C—O bond to give water and a C—H ligand. This reaction illustrates the activation of CO toward C—O bond scission via $\eta^2$-CO coordination to metal atoms.

The oxygen atoms of acyl ligands have moderately basic lone electron pairs. Alkylation or adduct formation at these sites is usually possible, as shown in equations *11.27–11.30*. Complex **11.30** is a Lewis acid adduct of an acetyl ligand.

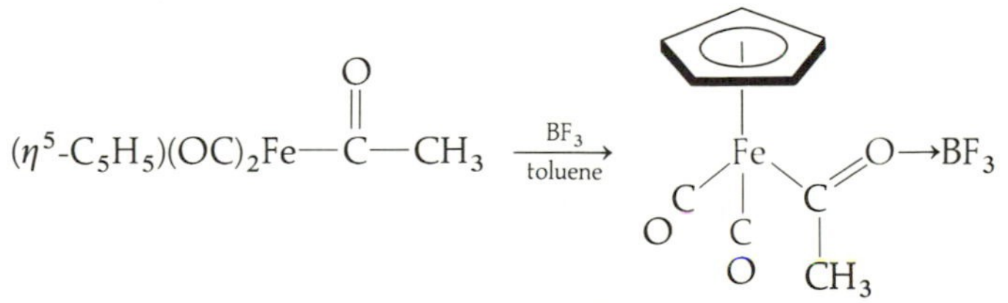

$$(\eta^5\text{-}C_5H_5)(OC)_2Fe-\overset{\overset{\displaystyle O}{\|}}{C}-CH_3 \xrightarrow[\text{toluene}]{BF_3}$$ (11.27)

**11.30**

The acetyl carbonyl stretching frequency is shifted 161 cm$^{-1}$ to lower energy relative to the acetyl stretching frequency of the starting material. Other Lewis acids such as $BCl_3$, $BBr_3$, $AlBr_3$, and $AlMe_3$, form similar adducts. Adducts like **11.25** and **11.30** are molecular models for the proposed intermediates that give rise to the observed Lewis acid-induced acceleration of the rate of alkyl migration, as shown in structures **9.19** and **9.20**.

In equations *11.28–11.30*, an acyl oxygen atom is alkylated. Complex **11.31** is formed by an *intermolecular* alkylation by using methyl triflate. Complexes **11.32** and **11.33** are formed by $Ag^+$ ion-assisted *intramolecular* alkylations affording cyclic products. The newly formed ligands in complexes **11.31–11.33** are *carbene ligands* (and therefore contain a formal M—C double bond).

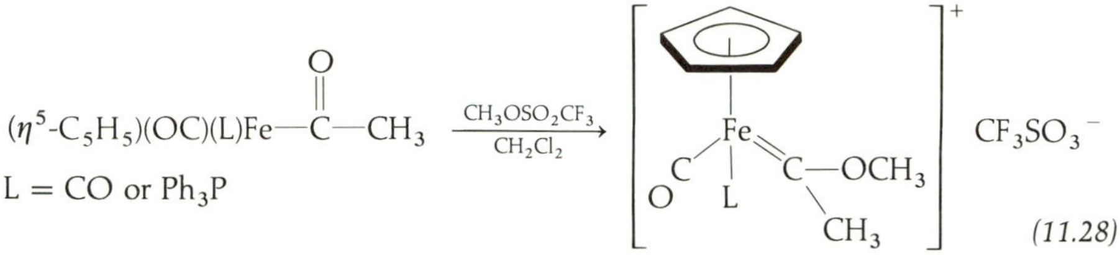

$$(\eta^5\text{-}C_5H_5)(OC)(L)Fe\overset{\overset{\displaystyle O}{\|}}{-C}-CH_3 \xrightarrow[CH_2Cl_2]{CH_3OSO_2CF_3} \qquad \mathbf{11.31} \qquad CF_3SO_3^{-}$$

L = CO or $Ph_3P$

(11.28)

$$\underset{\substack{L \\ | \\ C \\ \| \\ O}}{Mo}\overset{\overset{\displaystyle O}{\|}}{-C}-CH_2CH_2CH_2CH_2Br \xrightarrow[-AgBr]{AgBF_4} \qquad \mathbf{11.32} \qquad BF_4^{-}$$

L = $Ph_3P$

(11.29)

$$(OC)_5Mn\overset{\overset{\displaystyle O}{\|}}{-C}-OCH_2CH_2Cl \xrightarrow[-AgCl]{AgBF_4} [(OC)_5Mn{=}C] \qquad BF_4^{-}$$

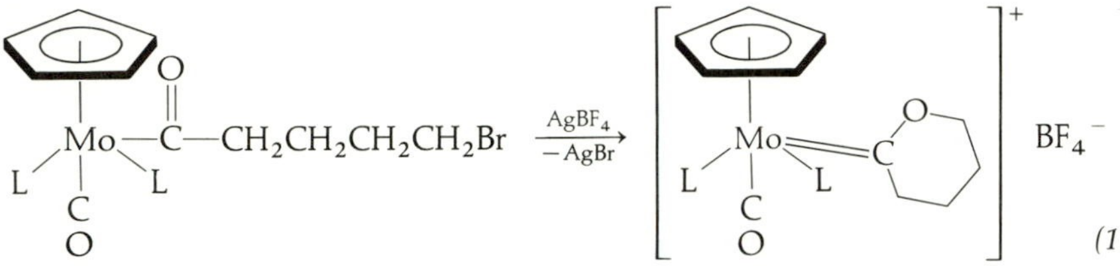

**11.33**

(11.30)

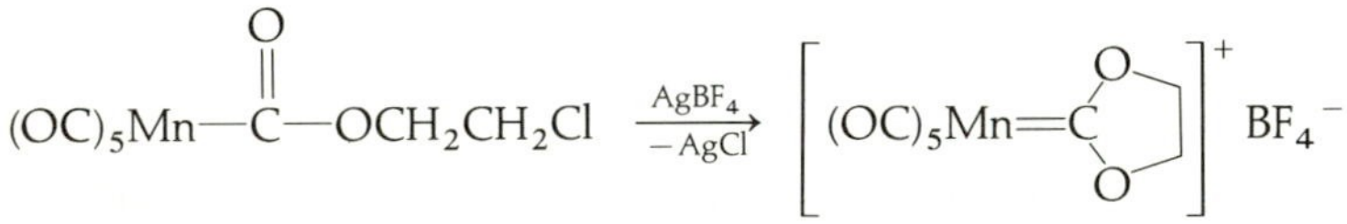

Protonation of the oxygen atom of the ethoxy methyl ligand of **11.34** gives a cationic methylidene complex (**11.35**) upon elimination of ethanol:

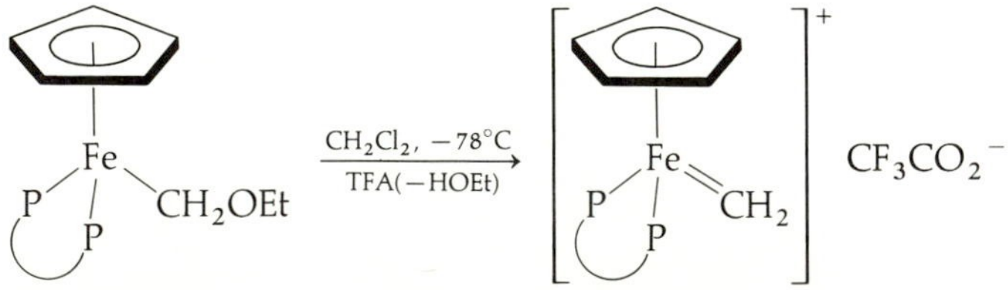

$$\underset{\substack{P \\ \diagdown}}{\overset{Fe}{\diagup}}\underset{CH_2OEt}{} \xrightarrow[TFA(-HOEt)]{CH_2Cl_2, \; -78°C} \underset{\substack{P \\ \diagdown P}}{\overset{Fe}{\diagup}}\underset{CH_2}{} \qquad CF_3CO_2^{-} \qquad (11.31)$$

**11.34**                                            **11.35**

P          P = diphos

Cationic complexes like **11.35** are stabilized best when they contain $\sigma$-donor ligands such as phosphines. Equation *11.31* represents an intermolecular abstraction, as discussed in chap. 9 (see eq. *9.72*).

Alkylation of an oxygen atom of the $CO_2$ ligand in $Ir(dmpe)_2Cl \cdot CO_2$ generates a cationic methoxycarbonyl complex (**11.36**):

$$Ir(dmpe)_2Cl \cdot CO_2 + CH_3OSO_2F \xrightarrow{\text{toluene}} \qquad (11.32)$$

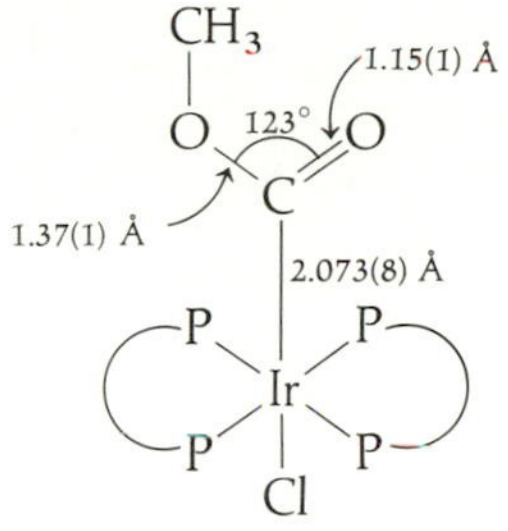

$$dmpe = Me_2PCH_2CH_2PMe_2$$

**11.36**

A similar alkylation of the oxygen atom of an S-sulfinate ligand gives a cationic complex containing an ethyl methanesulfinato ligand as shown in equation *11.33*. In equation *11.34*, an alkyl sulfito ligand is alkylated to give a dialkylsulfito ligand.

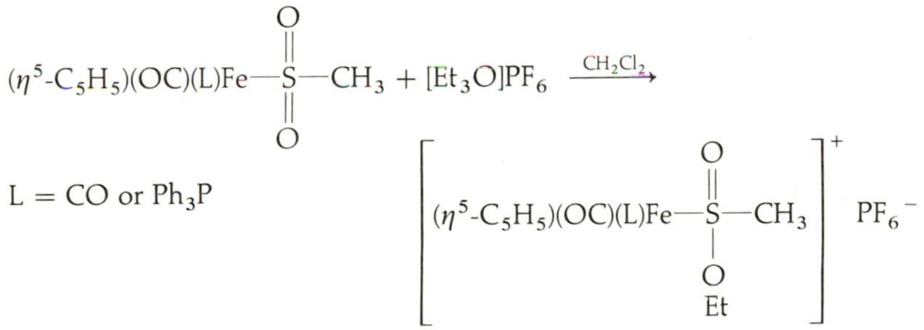

$$(\eta^5\text{-}C_5H_5)(OC)(L)Fe\text{—}\!\!\overset{\overset{\displaystyle O}{\|}}{\underset{\underset{\displaystyle O}{\|}}{S}}\!\!\text{—}CH_3 + [Et_3O]PF_6 \xrightarrow{CH_2Cl_2}$$

$$L = CO \text{ or } Ph_3P$$

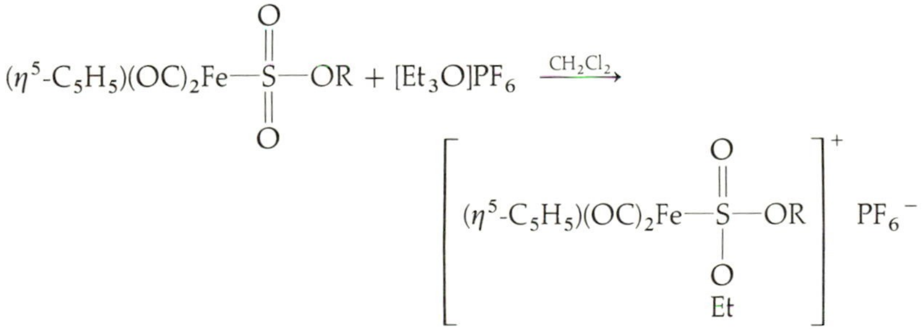

$$\left[ (\eta^5\text{-}C_5H_5)(OC)(L)Fe\text{—}\!\!\overset{\overset{\displaystyle O}{\|}}{\underset{\underset{\displaystyle O}{|}}{S}}\!\!\text{—}CH_3 \atop Et \right]^+ PF_6^- \qquad (11.33)$$

$$(\eta^5\text{-}C_5H_5)(OC)_2Fe\text{—}\!\!\overset{\overset{\displaystyle O}{\|}}{\underset{\underset{\displaystyle O}{\|}}{S}}\!\!\text{—}OR + [Et_3O]PF_6 \xrightarrow{CH_2Cl_2}$$

$$\left[ (\eta^5\text{-}C_5H_5)(OC)_2Fe\text{—}\!\!\overset{\overset{\displaystyle O}{\|}}{\underset{\underset{\displaystyle O}{|}}{S}}\!\!\text{—}OR \atop Et \right]^+ PF_6^- \qquad (11.34)$$

Four examples of electrophilic attack at nitrogen atoms bonded to donor atoms are shown in equations *11.35–11.38*. Alkylation of the cyano ligand in **11.37** gives the cationic ethyl isonitrile complex **11.38**.

$$(\eta^6\text{-}C_6H_6)Mn(CO)_2CN + [Et_3O]BF_4 \longrightarrow$$

**11.37**

$$[(\eta^6\text{-}C_6H_6)Mn(CO)_2(CNEt)]^+ BF_4^- \qquad (11.35)$$

**11.38**

$$(\eta^5\text{-}C_5H_5)_2Fe_2(CO)_3(CNEt) + Me\overset{\overset{\displaystyle O}{\|}}{C}Cl \xrightarrow{\text{THF}}$$

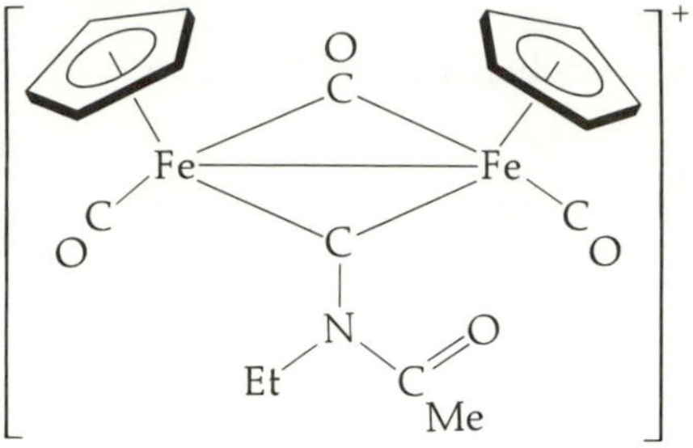

$$Cl^- \quad (11.36)$$

**11.39**

$$trans\text{-}M(CNMe)_2(diphos)_2 + [Me_3O]BF_4 \xrightarrow{C_6H_6}$$

M = Mo or W

$$trans\text{-}[M(CNMe)(CNMe_2)(diphos)_2]^+BF_4^- \quad (11.37)$$

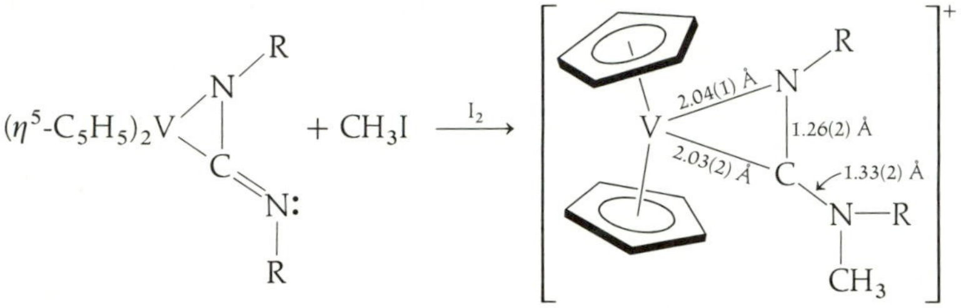

$$+ \ CH_3I \xrightarrow{I_2}$$

$$I_3^- \quad (11.38)$$

R = p-tolyl                                    **11.40**

In equations *11.36* and *11.37* acylation or alkylation of isonitrile ligands affords cationic carbyne complexes. The acyliminium ligand in **11.39** is formally a $\mu_2$-carbyne ligand, and the molybdenum and tungsten complexes contain the terminal carbyne ligand, $M\equiv C\text{-}NMe_2$. Equation *11.38* shows that alkylation of an $\eta^2$-carbodiimide ligand gives the cationic complex **11.40**. The structure of this product reveals a C—N distance of 1.26(2) Å, shorter than the corresponding C—N distance in the neutral reactant complex. This alkylated ligand is formally an $\eta^2$-amidinyl ligand, $R\overset{\cdot\cdot}{N}=\overset{\cdot}{C}\text{-}N(Me)(R)$, which acts as a three-electron donor similar to $\eta^2$-acyl or iminoacyl ligands.

Equations *11.39–11.41* show that alkylation of the S atoms of terminal, $\mu_2$-, or $\mu_3$-CS ligands is possible also. Since sulfur is more basic than oxygen, these alkylation products are more stable than the corresponding products expected from alkylating CO ligands. The products can be classified as mercaptocarbyne complexes because the electronic representation of an alkylated terminal CS ligand can be formulated as $M\equiv C\text{-}\overset{\cdot\cdot}{\underset{\cdot\cdot}{S}}\text{-}R$.

$$R_4N^+[(\eta^5\text{-}C_5H_5)W(CO)_2CS] + CH_3I \xrightarrow{\text{THF}}$$

R = butyl

$$(\eta^5\text{-}C_5H_5)(OC)_2W\equiv C\text{-}SMe + R_4N^+I^- \quad (11.39)$$

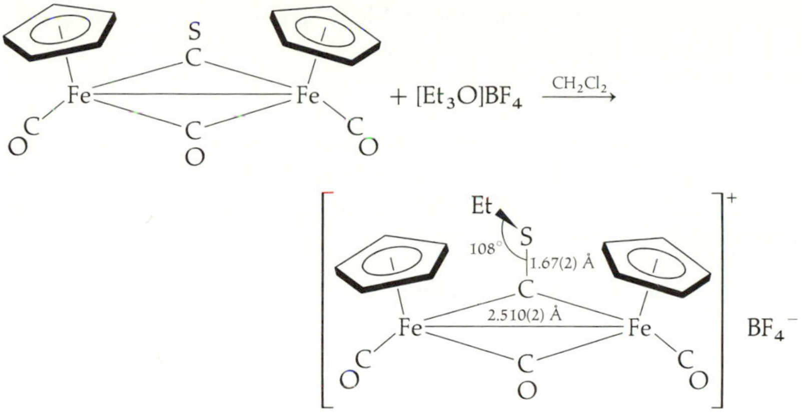

$$+ [Et_3O]BF_4 \xrightarrow{CH_2Cl_2}$$

$$BF_4^- \quad (11.40)$$

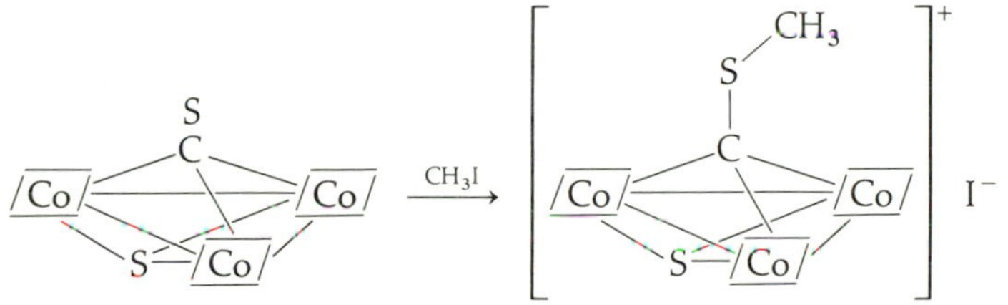

$$\xrightarrow{CH_3I} \qquad I^- \quad (11.41)$$

$$\boxed{Co} = (\eta^5\text{-}C_5H_5)Co$$

## NUCLEOPHILIC ATTACK ON ORGANOMETALLIC COMPLEXES

Organometallic complexes exhibit extensive reactivity toward a large variety of nucleophiles. Commonly used nucleophiles include anionic reagents, such as hydride sources, organolithium or Grignard reagents, alkoxide ions, and even neutral nucleophiles such as amines or phosphines. In general, cationic complexes undergo nucleophilic attack more easily than do neutral or anionic complexes.

Many molecules become more susceptible to nucleophilic attack upon coordination to a metal atom. Since ligand molecules donate electron density to a metal atom, it might be expected that coordinated molecules should be more electrophilic than the uncomplexed molecules. The most electrophilic atom within a ligand is frequently the donor atom, which is therefore usually the site of nucleophilic attack. Although metal to ligand back bonding might reduce the electrophilicity of the donor atom, this electron density usually populates $\pi^*$ molecular orbitals and so is spread over several atoms (such as the $2\pi$ MO in CO). Nucleophilic attack at donor atoms generally affords products that are stabilized by a change in the metal-ligand bonding. In fact, metal stabilization of the products might be the true driving force for these reactions.

The specific site of nucleophilic attack is assumed to be either *charge-controlled* or *frontier-controlled*. Steric effects may be important, also. Charge-control implies that the atom having the greatest positive charge will be the site of nucleophilic attack. Frontier-control means that nucleophilic attack is directed by the LUMO of the complex. Nucleophiles approach a complex as Lewis bases. An electron pair on the incoming nucleophile must populate the LUMO of the complex, and the specific site of attack would be at the atom that contributes most to the LUMO. The extent of this contribution is determined by the size of the atomic orbital coefficient in the LUMO.

Examples of nucleophilic attack at *monohapto* ligands, such as CO, CS, CNR, carbenes, or carbynes, at *dihapto* ligands, such as alkenes, allenes, or alkynes, and at *polyhapto* ligands, such as $\eta^3$-allylic, $\eta^5$-dienylium or $\eta^n$-carbocyclic ligands are presented below.

### Nucleophilic Attack at *Monohapto* Ligands

Nucleophilic attack at CO and related ligands occurs at the carbon donor atom. As shown in equation *11.42*, an M—CO group can be represented as a metallaketene, and nucleophilic attack at the carbon donor atom gives acyl products, which are resonance stabilized by both the oxygen atom (**11.41**) and the metalla group (**11.42**).

$$L_nM{=}C{=}\overset{..}{\underset{.}{O}}. + :\bar{N}uc \longrightarrow L_nM{=}\overset{\overset{:\overset{..}{O}:^-}{|}}{C}{-}Nuc \longleftrightarrow L_n\overset{..}{M}{-}\overset{\overset{:O:}{\|}}{C}{-}Nuc \quad (11.42)$$

**11.41** **11.42**

Organic ketenes react similarly.

Organolithium or Grignard reagents react with a wide variety of metal carbonyls to give acylmetalate anions, as shown for $Cr(CO)_6$ in equation *11.43*. Alkylation of this anion by a trialkyloxonium salt affords neutral carbene complexes such as **11.43,** which is an (alkoxy)(alkyl)carbene complex. E. O. Fischer and his coworkers used this method of adding nucleophiles to terminal carbonyl ligands for the preparation of the first carbene complexes. For this reason, carbene ligands that have a *heteroatom* bonded to the carbene carbon atom are frequently referred to as *Fischer carbenes.*

$$(OC)_5Cr{=}C{=}\overset{..}{O} + \overset{.}{C}H_3Li \xrightarrow{Et_2O} (OC)_5Cr{=}C\overset{O^-Li^+}{\underset{CH_3}{<}}$$

$$\downarrow {[(CH_3)_3O]BF_4}$$

$$(OC)_5Cr{=}C\overset{OCH_3}{\underset{CH_3}{<}} \quad (11.43)$$

**11.43**

Methyllithium adds to a CO ligand in metal carbonyl halide complexes to give *cis*-(halo)(acetyl) metallate complexes (eq. *11.44*). These anions can be protonated to afford *cis*-(halo)[(hydroxy)(methyl)carbene] complexes (**11.44**). Methyllithium, in this case, prefers to attack CO rather than to displace the halide ligand affording $CH_3M(CO)_5$ complexes. Attack at a *cis*-CO ligand is preferred electronically because the halide and acetyl ligands are weaker $\pi$-acids than CO. Weakly $\pi$-acidic ligands usually coordinate *trans* to much stronger $\pi$-acid ligands since *trans* ligands share the same *d*-orbital electron density.

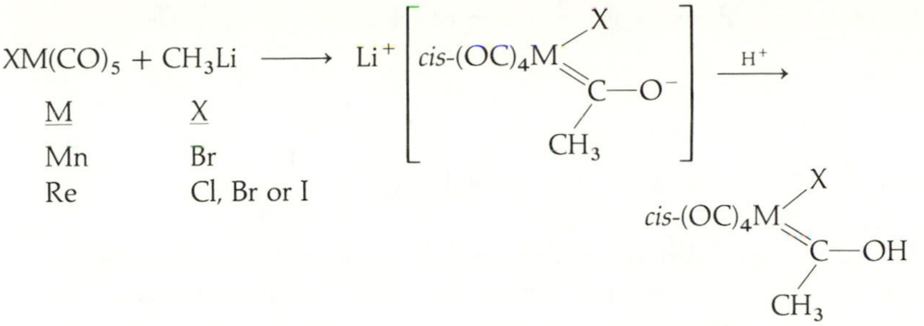

$$XM(CO)_5 + CH_3Li \longrightarrow Li^+ \left[ cis\text{-}(OC)_4M \underset{CH_3}{\overset{X}{\diagup}} C-O^- \right] \xrightarrow{H^+}$$

| M | X |
|---|---|
| Mn | Br |
| Re | Cl, Br or I |

$$cis\text{-}(OC)_4M \underset{CH_3}{\overset{X}{\diagup}} C-OH \qquad (11.44)$$

**11.44**

Acyl metal carbonyls react similarly with organolithium reagents, as shown in equation *11.45*. A charge-controlled nucleophilic attack might be expected to occur at the acyl carbonyl carbon atom to displace $M(CO)_5^-$ anion. However, molecular orbital calculations indicate that frontier-controlled nucleophilic attack should occur at CO. Diacylmetallate anions like **11.45** are metalla analogues to the $\beta$-diketonate anions in which the $sp^2$-CH group is replaced by a metalla moiety.

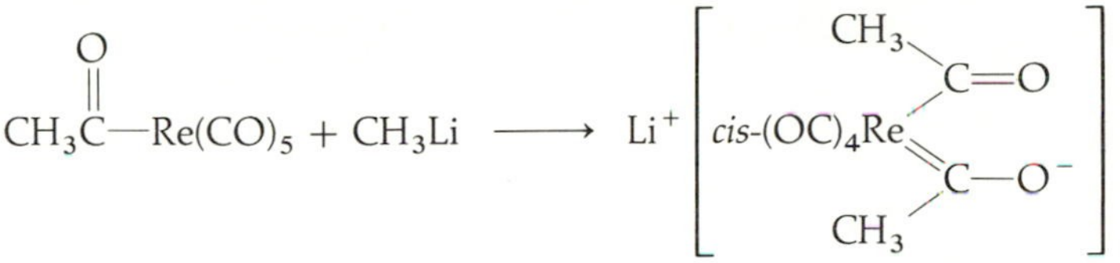

$$CH_3\overset{O}{\overset{\|}{C}}-Re(CO)_5 + CH_3Li \longrightarrow Li^+ \left[ cis\text{-}(OC)_4Re \underset{\underset{CH_3}{}{\diagdown} C-O^-}{\overset{\overset{CH_3}{}{\diagup} C=O}{}} \right]$$

**11.45**

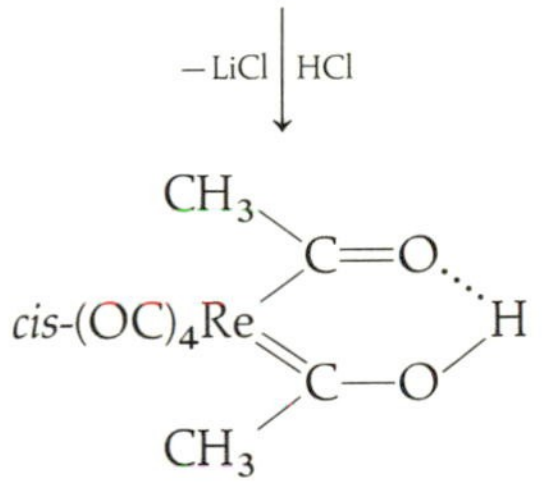

$$\downarrow {-LiCl \,|\, HCl}$$

$$cis\text{-}(OC)_4Re \underset{\underset{CH_3}{}{\diagdown} C-O}{\overset{\overset{CH_3}{}{\diagup} C=O\cdots}{}} H \qquad (11.45)$$

**11.46**

Protonation of **11.45** affords a neutral rhenaacetylacetone molecule, **11.46.**

Equations *8.57* and *8.58* show nucleophilic addition of hydride ion to CO ligands to give formyl complexes. These reactions are analogous to the attack of organolithium reagents at CO ligands.

Nitrogen nucleophiles attack CO ligands at the carbon atom, also, as shown in equations *11.46* and *11.47*. Ammonia and primary or secondary alkyl amines react with cationic metal carbonyls to give carbamoyl complexes (eq. *11.46*). The intermediate ammonium salt is deprotonated by a second equivalent of amine.

$$[(\eta^5\text{-}C_5H_5)Fe(CO)_3]^+PF_6^- + 2HNR_2 \longrightarrow$$

$$(\eta^5\text{-}C_5H_5)(OC)_2Fe-\overset{O}{\overset{\|}{C}}-NR_2 + [R_2NH_2]^+PF_6^- \qquad (11.46)$$

R = H or alkyl

Tertiary amines can add reversibly since deprotonation is not possible:

$$L_nM{=}C{=}O + NR_3 \rightleftharpoons \left\{ L_nM{-}\overset{\overset{\displaystyle O}{\|}}{C}{-}\overset{+}{N}R_3 \right\} \tag{11.47}$$

This acyl ammonium group is a strong *cis*-labilizing ligand and is used occasionally to direct ligand substitution reactions.

Oxygen nucleophiles also attack CO ligands at the carbon atom. Reaction of hydroxide gives metallacarboxylate complexes, which usually eliminate $CO_2$ to afford anionic metal carbonyls or hydrides, as shown in equations *11.48* and *8.28*. Alkoxide ions attack CO to give alkoxycarbonyl complexes (eq. *11.49*).

$$(OC)_4Fe{=}C{=}O + OH^- \longrightarrow (OC)_4Fe{-}\overset{\overset{\displaystyle O}{\|}}{C}{-}OH \tag{11.48}$$

$$\text{with } OH^-$$

$$HFe(CO)_4^- \underset{\xrightarrow{H_2O}}{\rightleftarrows} Fe(CO)_4^{2-} \overset{-CO_2}{\longleftarrow} (OC)_4Fe{-}\overset{\overset{\displaystyle O}{\|}}{C}{-}O^- + H_2O$$

$$[Fe(CO)_2(NO)L_2]^+ + OMe^- \longrightarrow MeO\overset{\overset{\displaystyle O}{\|}}{C}{-}Fe(CO)(NO)L_2 \tag{11.49}$$

$L = Ph_3P$

Water acts as a nucleophile toward $[Re(CO)_6]^+$ and other cationic complexes. When isotopically labeled water (such as $H_2{}^{17}O$ or $H_2{}^{18}O$) is used, an oxygen-exchange reaction is observed:

$$[Re(CO)_6]PF_6 + H_2O^* \overset{fast}{\rightleftharpoons} (OC)_5Re{-}\overset{\overset{\displaystyle O}{\|}}{C}{\underset{O^*H_2}{\diagdown}}$$

**11.47**

$$slow \downarrow H_2O$$

$$(OC)_5Re{-}\overset{\overset{\displaystyle O}{\|}}{C}{-}O^*H + H_3O^+$$

**11.48**

$$fast \updownarrow$$

$$[Re(CO)_5(CO^*)]^+ + 2H_2O \overset{H_3O^+}{\rightleftharpoons} (OC)_5Re{-}\overset{\overset{\displaystyle OH}{|}}{C}{\underset{O^*}{\diagdown}} \tag{11.50}$$

The rate of reaction is first order in $[Re(CO)_6]^+$ and *second order* in $H_2O$. This implies either a stepwise mechanism of nucleophilic attack by $H_2O$ followed by a slower water-initiated deprotonation of **11.47** or else a concerted attack of a dimeric water species on the reactant complex to afford **11.48** directly. The assumption of the presence of the rhenacarboxylic acid intermediate (**11.48**) is supported by the existence of several complexes of this type, such as $(\eta^5\text{-}C_5H_5)Re(CO)(NO)(CO_2H)$, $(\eta^5\text{-}C_5H_5)FeL_2(CO_2H)$ where L is CO or $Ph_3P$, and $IrCl_2(CO_2H)(CO)L_2$ where L is $Me_2PhP$ or $Me_2PhAs$.

The use of amine N-oxides, such as $Me_3N^+\text{—}O^-$, to generate vacant coordination sites under mild thermal conditions has been discussed previously (equation 5.51). A reasonable mechanism for this reaction is shown in equation *11.51*.

$$L_nM\!=\!C\!=\!O + \overset{-}{O}\!-\!\overset{+}{N}R_3 \longrightarrow [L_nM\!-\!\overset{O}{\overset{\|}{C}}\!-\!O\!-\!\overset{+}{N}R_3] \longrightarrow$$

$$L_nM + CO_2 + NR_3 \quad (11.51)$$

Nucleophilic attack by the oxygen atom of the N-oxide on a CO ligand affords an intermediate that can eliminate $CO_2$ and free amine, thus generating a metal fragment that is coordinatively unsaturated. In some instances, the tertiary amine coordinates to this metal fragment; however, this ligand is easily displaced by other molecules.

Examples of nucleophilic attack on neutral and cationic CS complexes are shown in equations *11.52* and *11.53*. Attack occurs at the CS carbon atom, as with CO, and the resulting products can be alkylated to afford neutral dithio- and cationic dithio- or (alkoxy)(thioalkoxy)carbene complexes (**11.49** and **11.50**, respectively).

$$W(CO)_5(CS) + SR^- \longrightarrow (OC)_5W\!=\!C\overset{S^-}{\underset{SR}{\diagdown}} \overset{R'I}{\longrightarrow}$$

$$(OC)_5W\!=\!C\overset{SR'}{\underset{SR}{\diagdown}} \quad (11.52)$$

**11.49**

R and R′ = alkyl

$$[(\eta^5\text{-}C_5H_5)(OC)_2Fe(CS)]^+ + YR^- \longrightarrow (\eta^5\text{-}C_5H_5)(OC)_2Fe\!-\!\overset{S}{\overset{\|}{C}}\!-\!YR$$

$$YR^- = MeO^-, PhO^-, PhS^-, \text{ or } PhSe^-$$

$$\Big\downarrow MeOSO_2F$$

$$\left[(\eta^5\text{-}C_5H_5)(OC)_2Fe\!=\!C\overset{SMe}{\underset{YR}{\diagdown}}\right]^+ FSO_3^- \quad (11.53)$$

**11.50**

Isonitrile ligands react with organolithium reagents to give iminoacyl complexes, as shown in equation *11.54*. Methanol attacks isonitrile ligands to give (alkoxy)(amino)carbene complexes (eq. *11.55*). This O—H addition across the C≡N multiple bond represents a nucleophilic attack by methoxide ion at the carbon atom with concomitant protonation of the nitrogen atom.

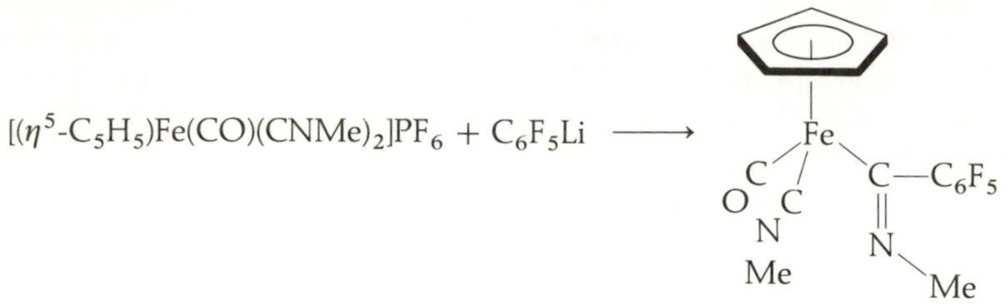

$$[(\eta^5\text{-}C_5H_5)Fe(CO)(CNMe)_2]PF_6 + C_6F_5Li \longrightarrow \qquad (11.54)$$

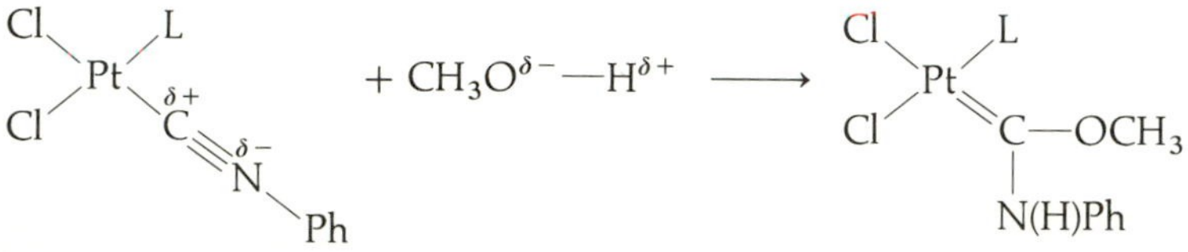

$$L = Et_3P$$

$$\text{Cl} \diagdown \text{Pt} \diagup \text{L} \qquad + CH_3O^{\delta-}\!\!-\!\!H^{\delta+} \longrightarrow \qquad (11.55)$$

In contrast to the electron deficient alkylidene complexes of niobium and tantalum, which possess nucleophilic carbene carbon atoms (eqs. *11.14* and *11.15*), the analogous carbon atoms of (alkoxy)(alkyl)carbene ligands and carbene complexes of low-valent, coordinatively saturated metals are electrophilic because of weaker M → C back bonding. Carbene carbon atoms in these latter complexes are, therefore, susceptible to nucleophilic attack.

An excellent example of this reactivity is the general aminolysis reaction:

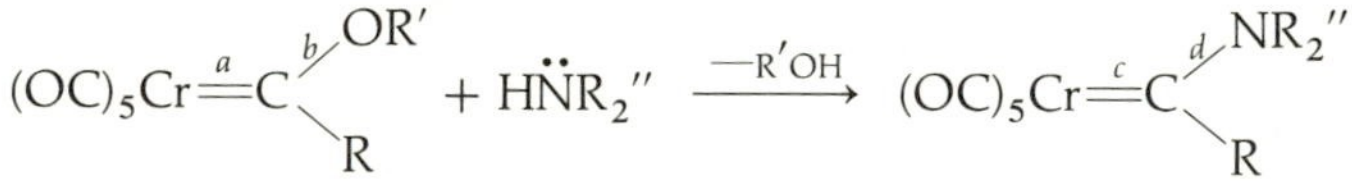

$$(OC)_5Cr\overset{a}{=}C\overset{b}{\underset{R}{\overset{OR'}{\diagup}}} + H\ddot{N}R_2'' \xrightarrow{-R'OH} (OC)_5Cr\overset{c}{=}C\overset{d}{\underset{R}{\overset{NR_2''}{\diagup}}} \qquad (11.56)$$

Nucleophilic attack by the nitrogen lone pair on the carbene carbon atom leads to elimination of alcohol and formation of the (amino)(alkyl)carbene product. This reaction is an organometallic analogue to the conversion of an organic ester into an amide by treatment with a free amine. Apparently, the $Cr(CO)_5$ moiety is electronically similar to a carbonyl oxygen atom.

Structural data indicate that the bonds *a* and *b* have bond orders of about 1.25 and 1.33, respectively, while bonds *c* and *d* have bond orders of about 1.05 and 1.70, respectively. Therefore, electronic structures **11.51** and **11.52** must contribute significantly to the ground-state structures of these molecules. The contribution from the oxonium structure (**11.51**) is less than that of the iminium structure

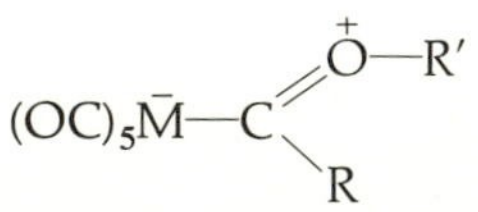

$$(OC)_5\bar{M}\!\!-\!\!C\overset{\overset{+}{O}-R'}{\underset{R}{\diagup}} \qquad\qquad (OC)_5\bar{M}\!\!-\!\!C\overset{\overset{+}{N}R_2''}{\underset{R}{\diagup}}$$

$$\textbf{11.51} \qquad\qquad\qquad \textbf{11.52}$$

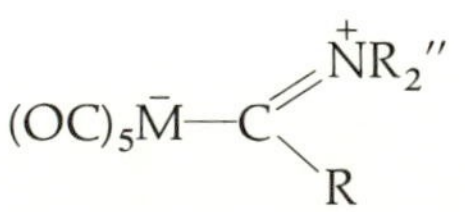

(**11.52**). The C—O and C—N bond orders in these carbene ligands are the same as those determined for organic esters and amides.

The rhenaacetylacetone complex (**11.46**) reacts similarly with a wide variety of primary amines:

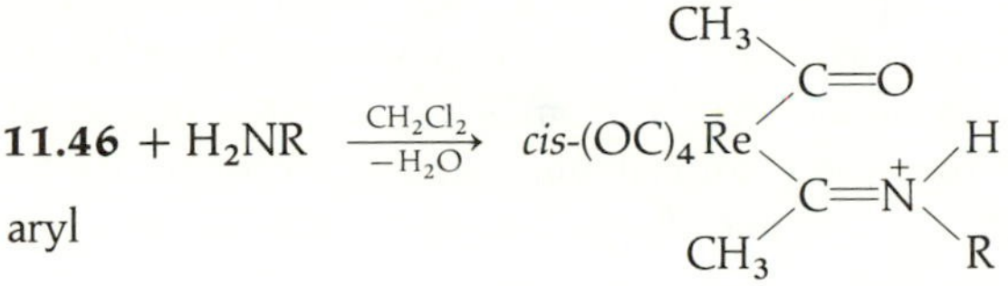

$$\textbf{11.46} + H_2NR \xrightarrow[-H_2O]{CH_2Cl_2} \textit{cis-}(OC)_4\bar{Re} \underset{\underset{CH_3}{|}}{\overset{\overset{CH_3}{|}}{\bigg\langle}} \begin{matrix} C{=}O \\ \\ C{=}\overset{+}{N} \end{matrix} \begin{matrix} H \\ \\ R \end{matrix} \qquad (11.57)$$

R = H, alkyl or aryl

This reaction is an organometallic analogue to a Schiff-base condensation. The x-ray analysis of the rhena-$\beta$-ketoimine products confirms the interesting and unusual zwitterionic electronic structure shown above. Geometrical isomerization about the iminium C—N double bond is slow, and the two geometrical isomers can usually be isolated separately.

Nucleophilic attack by organolithium or hydride reagents at an (alkoxy)(R)carbene carbon atom can be used to prepare carbene ligands that lack heteroatom substituents, as shown in equations *11.58* and *11.59*.

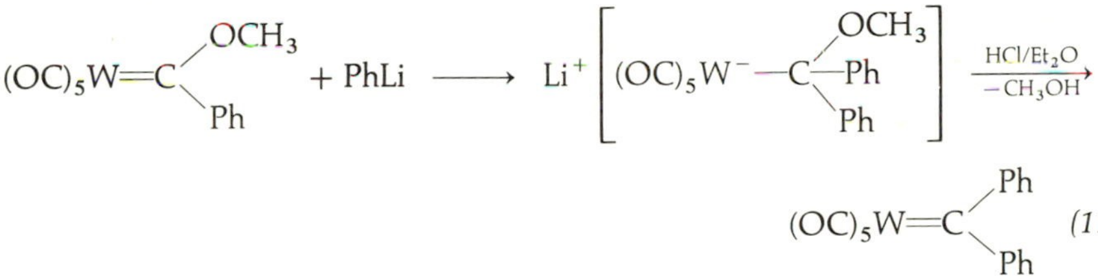

$$(OC)_5W{=}C\underset{Ph}{\overset{OCH_3}{\big\langle}} + PhLi \longrightarrow Li^+ \left[ (OC)_5W^- {-}C\underset{Ph}{\overset{OCH_3}{\big\langle}} Ph \right] \xrightarrow[-CH_3OH]{HCl/Et_2O}$$

$$(OC)_5W{=}C\underset{Ph}{\overset{Ph}{\big\langle}} \qquad (11.58)$$

**11.53**

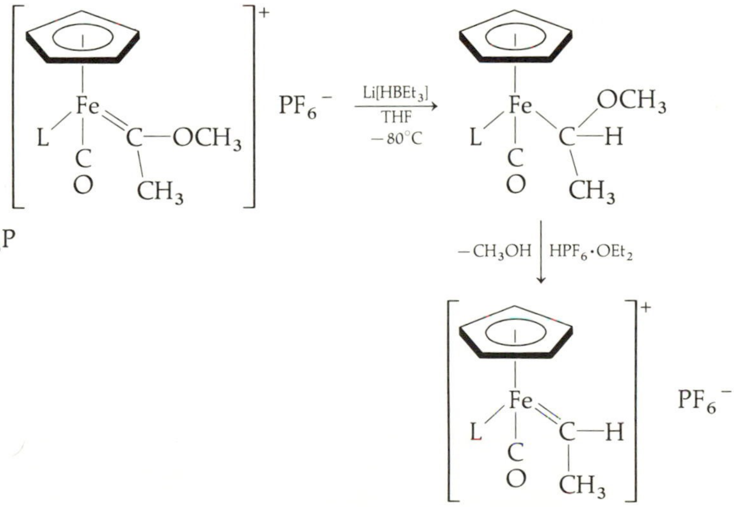

L = Ph₃P

(11.59)

**11.54**

Protonation of the alkoxyalkyl intermediates causes elimination of methanol, forming a diphenylcarbene complex (**11.53**) or an ethylidene complex (**11.54**). When complex **11.54** is treated with a second equivalent of Li[HBEt$_3$], the complex ($\eta^5$-C$_5$H$_5$)Fe(L)(CO)Et is formed by hydride attack at the ethylidene carbon atom.

Some carbene complexes react with tertiary phosphines to give ylide (or phosphorane) complexes (eqs. *11.60–11.62*). In compounds **11.55** and **11.56,** the ylide carbon atom is the donor atom; in complex **11.57** the donor atom is one of the sulfur atoms of the newly formed ylide ligand. Phosphorus-ylide carbon atoms usually bear a formal negative charge because of the zwitterionic character of the phosphorus-carbon bond.

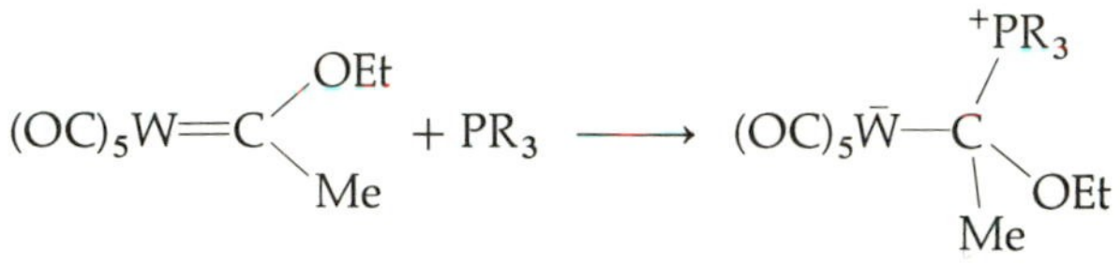

$$(OC)_5W{=}C\underset{Me}{\overset{OEt}{<}} + PR_3 \longrightarrow (OC)_5\overset{+PR_3}{\underset{\underset{Me}{|}}{W{-}C}}\overset{}{\underset{OEt}{<}} \qquad (11.60)$$

R = *n*-butyl

**11.55**

$$\textbf{11.54} + PPh_3 \longrightarrow \qquad (11.61)$$

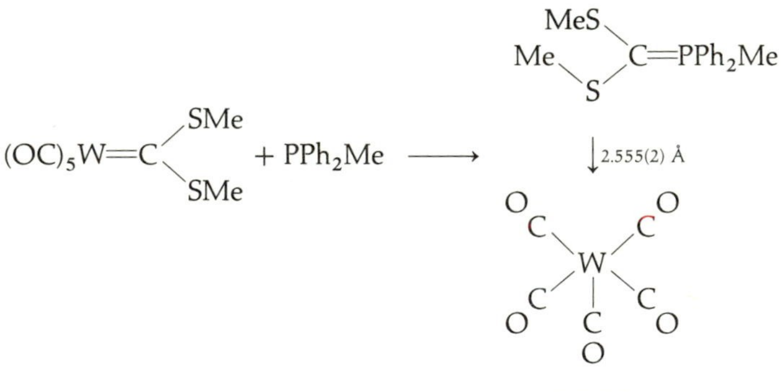

**11.56**

$$(OC)_5W{=}C\underset{SMe}{\overset{SMe}{<}} + PPh_2Me \longrightarrow \qquad (11.62)$$

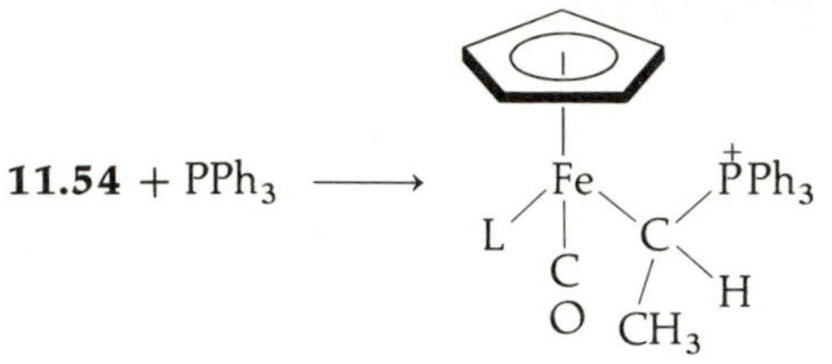

2.555(2) Å

**11.57**

When the benzyl complex (**11.58**) is treated with trityl cation, hydride abstraction gives **11.59** as the kinetically controlled product but **11.60** as the thermodynamically more stable isomer:

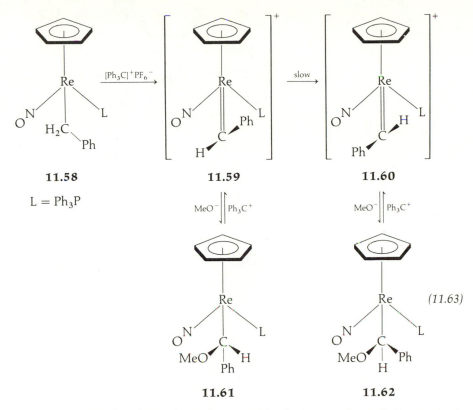

**11.58**                    **11.59**                    **11.60**

L = Ph₃P

(11.63)

**11.61**                              **11.62**

An x-ray structural analysis of complex **11.60** has been carried out. Quite unexpect-edly, isomers **11.59** and **11.60** undergo nucleophilic attack by methoxide ion to give *different* diastereomers, **11.61** and **11.62**, respectively, instead of a diastereo-meric mixture in each reaction. Methoxide attack occurs from the *same side* of the benzylidene ligand in both reactions. This stereospecificity is maintained in the reverse reactions, in which trityl cation removes methoxide ion from the neutral methoxybenzyl complexes (**11.61** and **11.62**). The steric bulk of the metal moiety directs the stereospecificity of this nucleophilic attack.

Some cationic (alkoxy)(alkyl)carbene complexes react like oxonium ions. Pre-sumably, these complexes have a considerable contribution from structure **11.51** in the ground electronic state. Two examples of this type of reaction are shown in equations *11.64* and *11.65*.

Complexes **11.63** and **11.64** are cationic carbene complexes written as oxonium Lewis structures. Reaction with iodide or methoxide ions affords neutral acyl com-plexes by nucleophilic displacement of the acyl carbonyl oxygen atom. In both reactions, a neutral acyl moiety acts as a leaving group as the carbene ligand al-kylates the incoming nucleophile.

Carbyne ligands of 18-electron complexes usually react with nucleophiles at the carbyne carbon atom. Conversion of a carbyne complex into (alkoxy) or (amino) (phenyl)carbene complexes via nucleophilic attack is shown in equation *11.66*.

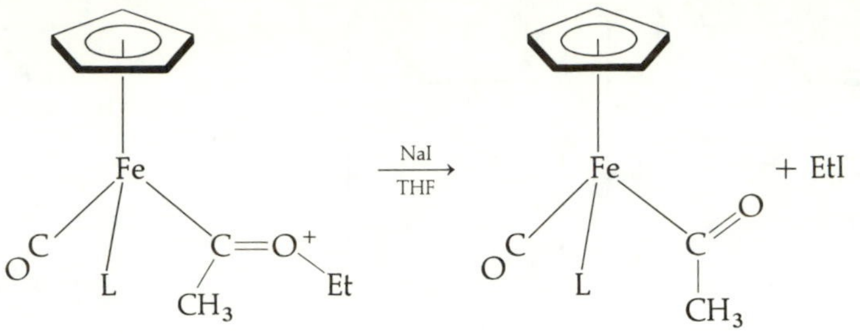

(11.64)

**11.63**

$L = Ph_3P$; $BF_4^-$ salt

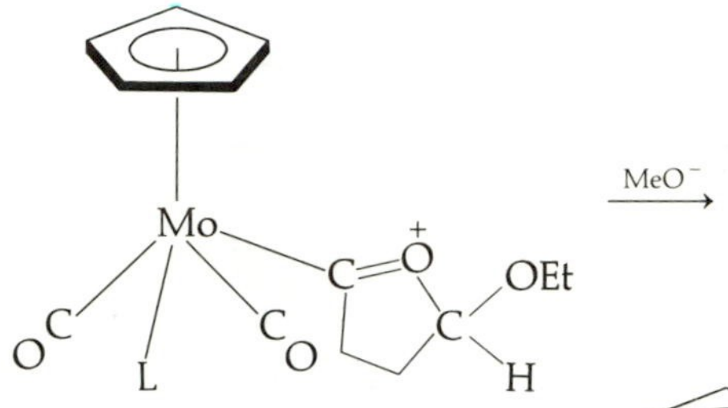

**11.64**

$L = Ph_3P$; $BF_4^-$ salt

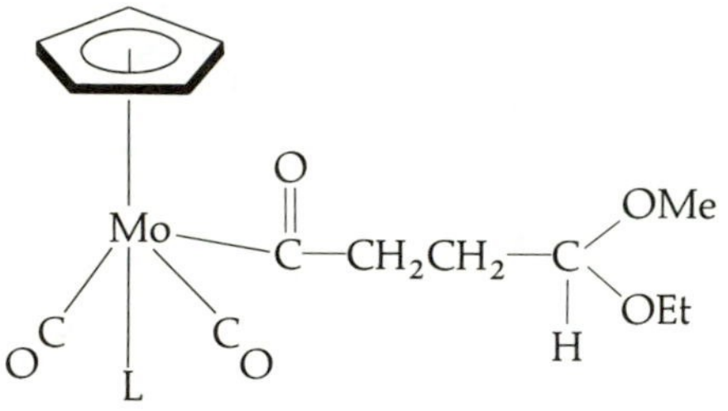

(11.65)

$$[(\eta^5\text{-}C_5H_5)(OC)_2Mn\equiv CPh]^+ \xrightarrow{\text{NaOR}} (\eta^5\text{-}C_5H_5)(OC)_2Mn=C\underset{Ph}{\overset{OR}{<}}$$

(11.66)

$$\xrightarrow[-H_2NR_2^+]{2HNR_2} (\eta^5\text{-}C_5H_5)(OC)_2Mn=C\underset{Ph}{\overset{NR_2}{<}}$$

As with carbene complexes, the carbon donor atom of carbyne ligands can be either electrophilic (eq. *11.66*), or nucleophilic (eq. *11.13*).

## Nucleophilic Attack at *Dihapto* Ligands

Coordinated alkenes, including unactivated olefins like ethylene, are very susceptible to nucleophilic attack. This chemical activation of alkenes upon complexation has been used in stoichiometric organic synthesis. $\eta^2$-Alkene complexes are either presumed or known intermediates in several catalytic processes, also.

Nucleophilic attack on an $\eta^2$-olefin complex usually occurs at either the metal atom or the $\eta^2$-olefin ligand:

$$L_nM\text{---Nuc} + R_2C\text{=}CR_2 \xleftarrow{\ddot{\text{Nuc}}^-} \left[ L_nM \leftarrow \overset{\overset{\displaystyle R_2}{\displaystyle C}}{\underset{\underset{\displaystyle R_2}{\displaystyle C}}{\|}} \right]^+ \xrightarrow{\ddot{\text{Nuc}}^-}$$

$$L_nM\text{---}CR_2\text{---}\overset{\displaystyle \text{Nuc}}{\underset{\displaystyle R}{C}}\text{---}R \qquad (11.67)$$

Attack at M usually displaces the alkene as in a simple ligand substitution reaction. Attack on the coordinated olefin gives an alkyl complex. Alternatively, attack at M to give an intermediate of higher coordination number followed by a 1,2-addition to the $\eta^2$-alkene ligand would also give an alkyl complex. However, this mechanism would proceed through a higher-energy intermediate than direct olefin attack does—particularly if the original olefin complex were coordinatively saturated. As discussed below, nucleophiles attack olefin ligands *trans* (or *exo*) to the metal moiety. This stereospecificity is more consistent with direct attack on the alkene than with attack at M followed by a 1,2-addition. Most 1,2-addition products reflect *cis* (or *syn*) addition.

The next three equations show methoxide attack on $\eta^2$-olefins to give *exo*-products. This stereochemical result implies direct attack on the alkene from the face opposite to the bulky metal moiety.

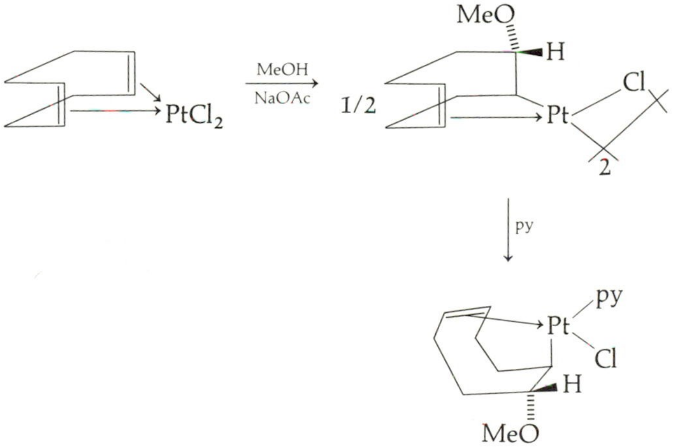

(11.68)

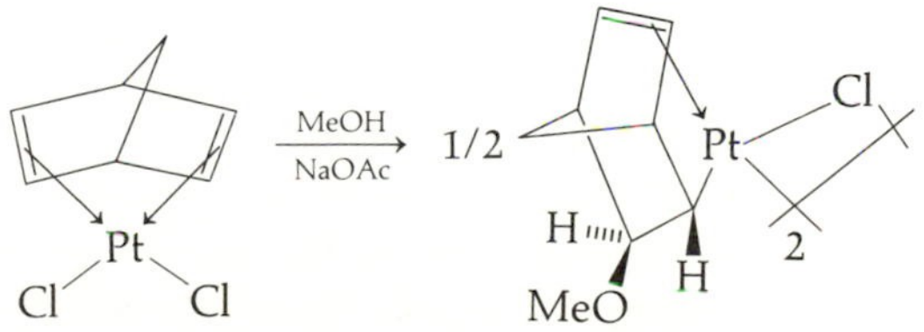

(11.69)

(11.70)

Olefin ligands undergo addition by a variety of nucleophiles including alkoxide ions, amines, phosphines, azide ion, and stabilized carbanions, such as $CN^-$ or $^-CH(CO_2Et)_2$. As expected, nucleophilic attack on $\eta^2$-alkene ligands is more facile when the complex is cationic.

Phosphines and amines attack $\eta^2$-alkene ligands to give alkyl-phosphonium (**11.66**) and alkyl ammonium (**11.68**) complexes.

$$(\eta^2\text{-}C_2H_4)Ru(CO)Cl_2L_2 \xrightarrow[-10°C]{2L} [(PhMe_2\overset{+}{P}CH_2CH_2)Ru(CO)ClL_3]Cl^- \quad (11.71)$$

**11.65**                                               **11.66**

L = PMe₂Ph                                              70%

$$cis\text{-}(\eta^2\text{-}C_2H_4)PtCl_2(NHMe_2) \xrightarrow{NHMe_2} Me_2H\overset{+}{N}\text{—}CH_2CH_2\text{—}\underset{\underset{Cl}{|}}{\overset{\overset{Cl}{|}}{Pt}}{}^-\leftarrow NHMe_2 \quad (11.72)$$

**11.67**

**11.68**

In the formation of **11.66**, one of the original chloride ligands is displaced by Me₂PhP. When complex **11.65** is treated with one equivalent of Me₂PhP at 35°C, the ethylene ligand is displaced by the phosphine. In complexes unrelated to **11.65** or **11.67**, nucleophilic attack on an $\eta^2$-alkene ligand at low temperature but on the metal atom at higher temperature is observed for several complexes. This result might indicate a *kinetic* preference for attack on the olefin ligand.

Anionic organometallic complexes that act as nucleophiles can attack $\eta^2$-alkene ligands, as shown in equation 11.73.

The anions $(\eta^5\text{-}C_5H_5)M(CO)_3{}^-$ where M is Mo or W react similarly. Nucleophilic attack by $[Re(CO)_5]^-$ on the ethylene ligand affords the heterodinuclear complex **11.69**. A second equivalent of $[Re(CO)_5]^-$ displaces the tungsten complex to give the 1,2-dirhenaethane complex **11.70**.

Hydride complexes can attack olefin ligands as nucleophilic hydride reagents (eq. 11.74). The hydride adds regiospecifically to the less substituted olefinic carbon atom. Loss of hydride generates a coordinatively unsaturated cationic complex, which then coordinates a solvent molecule.

The numerous examples of nucleophilic addition to $\eta^2$-alkene ligands have generated much recent interest in understanding *how* metals activate olefins toward nucleophilic attack. Considering the Dewar–Chatt–Duncanson model of metal-

$$[(\eta^5\text{-}C_5H_5)(OC)_3W(\eta^2\text{-}C_2H_4)]^+BF_4^- \xrightarrow[-50°C]{\text{Na[Re(CO)}_5]}$$

$$(\eta^5\text{-}C_5H_5)(OC)_3W\text{—}CH_2CH_2\text{—}Re(CO)_5$$

**11.69**, 78%

$\xrightarrow[-50°C]{\text{Na[Re(CO)}_5]}$

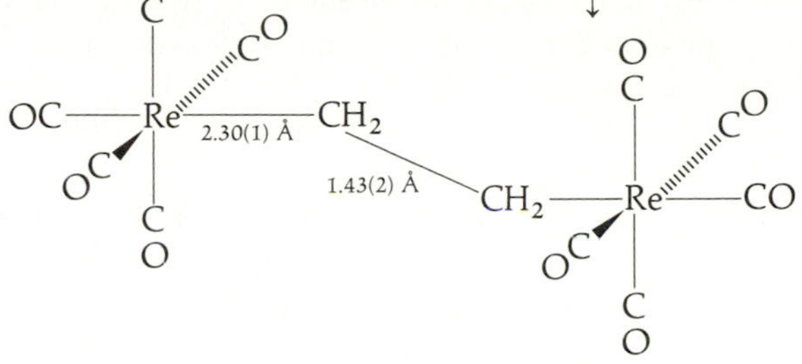

(structure **11.70**)

2.30(1) Å

1.43(2) Å

(11.73)

**11.70**

$$[(\eta^5\text{-}C_5H_5)(OC)_2Fe(\eta^2\text{-}CH_2CRR')]^+PF_6^- + (\eta^5\text{-}C_5H_5)Fe(CO)(L)H \xrightarrow{CH_3CN}$$

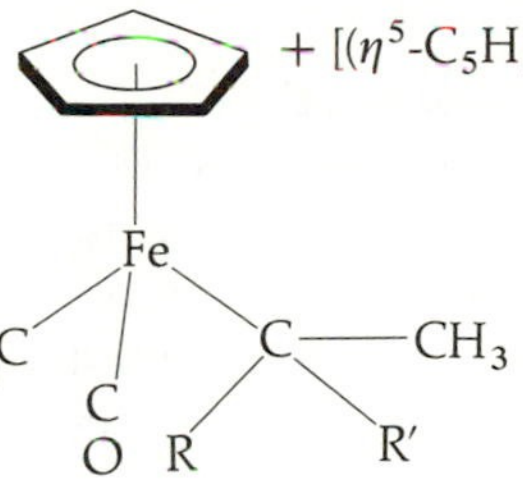

(Fe structure) $+ [(\eta^5\text{-}C_5H_5)Fe(CO)(L)(CH_3CN)]^+PF_6^-$　(11.74)

R, R' = H or Me; L = Ph_3P

olefin bonding (figure 5.2), we know that the alkene donates π-electron density to the metal. However, metal to olefin back bonding transfers negative charge back to the olefin. Even though the net bonding effects may leave the olefin ligand slightly positively charged (and therefore more susceptible to nucleophilic attack), a computational or experimental determination of the *actual* charge on a coordinated ligand is very difficult to obtain.

From a frontier orbital approach, a nucleophile such as a hydride ion has a filled valence orbital that interacts with the LUMO of the $\eta^2$-alkene complex to give the product.

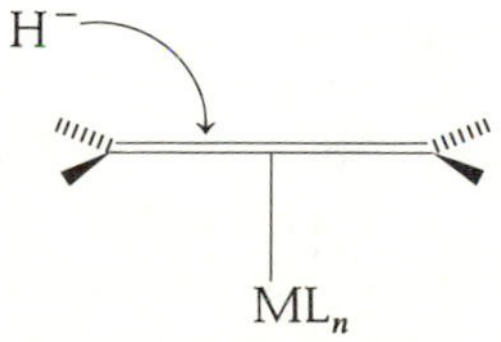

| 11.71 | 11.72 | 11.73 |

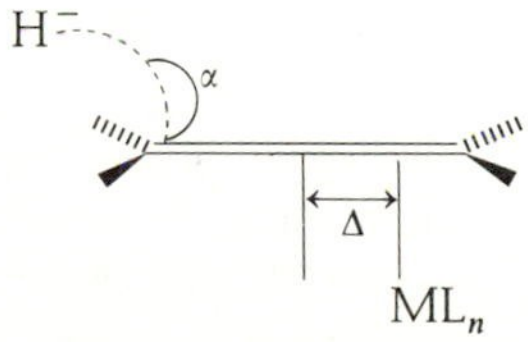

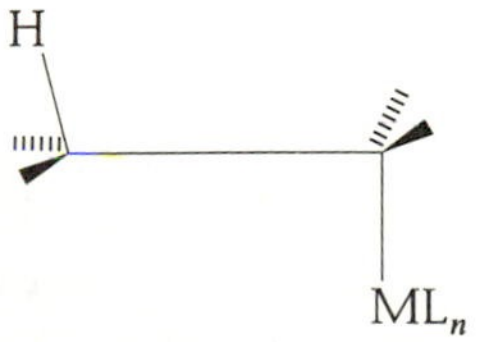

Calculations reveal that if the $\eta^2$-alkene ligand maintains its symmetrical structure through the transition state of the reaction, as in **11.71**, then the olefin is *not* activated toward the potential nucleophilic attack. However, if the metal moiety has slipped from the symmetrical structure by a distance $\Delta$ (**11.72**), then the alkene becomes activated toward nucleophilic addition. When $\Delta$ is about 0.69 Å, the alkene becomes a *monohapto* ligand similar to the alkyl product complex (**11.73**). Calculations confirm this activation of the alkene upon shifting the metal moiety and also indicate that the alkene-nucleophile overlap (when $\Delta = 0.69$ Å) is maximized at an approach angle $\alpha$ of about 110°.

Some degree of $\eta^2$-$\eta^1$ slippage is apparently required for alkene activation via complexation. Presumably, this slippage occurs in the early part of the reaction, although this aspect is difficult to determine. *Exo*-addition to the extended structure (**11.72**) is still preferred electronically because *endo*-addition of the nucleophile generates an antibonding interaction between the valence orbitals of the nucleophile and of the metal atom.

If metal slippage is required in the transition state to activate the alkene ligand, then conformationally constrained alkene complexes that contain slipped metal-olefin bonding in the *ground-state* structure *may show enhanced reactivity* toward nucleophilic attack. The chemistry of complex **11.74** appears to substantiate these predictions.

The x-ray structural analysis of (*endo*-dicyclopentadiene)PdCl$_2$ (**11.74**) reveals two different alkene bonds. The double bond between carbon-5 and carbon-6 is bonded symmetrically to the Pd atom (with distances from palladium to carbons-5 and 6 of 2.21(2) Å and 2.20(2) Å, respectively). The other double bond between carbons-1 and 2, however, is coordinated unsymmetrically to the Pd atom. This alkene-Pd bond is distorted by a lateral slip of 0.34 Å from a symmetrical structure.

**11.74**

Furthermore the C1-C2 bond vector is tilted by 13° away from perpendicularity to the PdCl$_2$ plane. These distortions are a result of the constrained cyclic structure of the ligand and afford unequal Pd-C1 and Pd-C2 distances of 2.28(1) Å and 2.19(1) Å, respectively.

Interestingly, complex **11.74** is known to react with alkoxide ions, $\beta$-diketones, and amines by nucleophilic attack at carbon-1. This site corresponds to the carbon farther removed from the metal (see **11.72**). These results indicate that structural deformation from a symmetrical $\eta^2$-alkene ground state geometry *probably* plays an important role in directing nucleophilic attack, as predicted by theory for the structures of *transition state species.*

As with $\eta^2$-alkene ligands, the coordinated olefinic bonds of $\eta^2$-allenes undergo nucleophilic attack (eqs. *11.75* and *11.76*). Complex **11.75** is a zwitterionic alkenyl-ammonium complex similar to complex **11.68**.

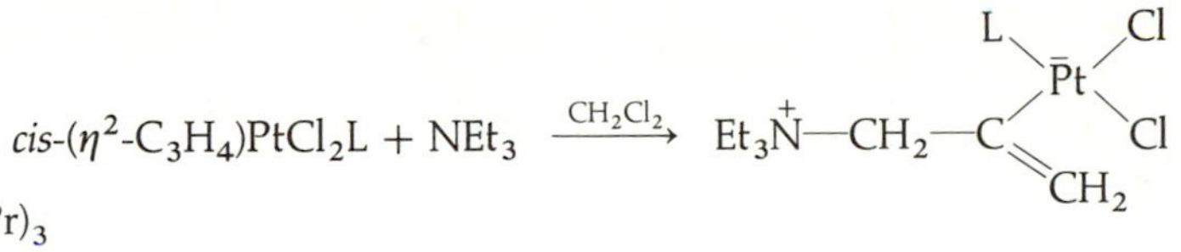

$$cis\text{-}(\eta^2\text{-}C_3H_4)PtCl_2L + NEt_3 \xrightarrow{CH_2Cl_2} \underset{CH_2}{Et_3\overset{+}{N}-CH_2-C}\overset{\displaystyle L\!\diagdown\!\underset{\textstyle Pt}{}\!\diagup\!Cl}{\diagdown Cl} \quad (11.75)$$

L = P($n$-Pr)$_3$

**11.75**

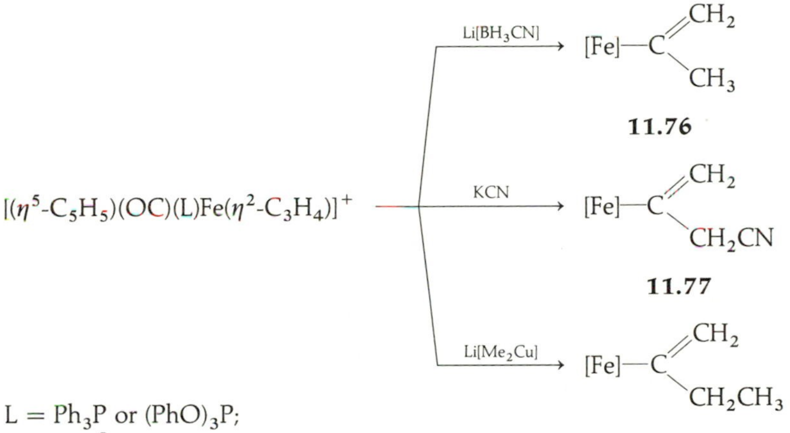

$[(\eta^5\text{-}C_5H_5)(OC)(L)Fe(\eta^2\text{-}C_3H_4)]^+$

$\xrightarrow{Li[BH_3CN]}$ $[Fe]-C\overset{\diagup CH_2}{\diagdown CH_3}$

**11.76**

$\xrightarrow{KCN}$ $[Fe]-C\overset{\diagup CH_2}{\diagdown CH_2CN}$ $(11.76)$

**11.77**

$\xrightarrow{Li[Me_2Cu]}$ $[Fe]-C\overset{\diagup CH_2}{\diagdown CH_2CH_3}$

**11.78**

L = Ph$_3$P or (PhO)$_3$P;
[Fe] = $(\eta^5\text{-}C_5H_5)(OC)(L)Fe$

The cationic ($\eta^2$-allene)Fe complexes in equation *11.76* react with anionic nucleophiles like cyano borohydride, cyanide ion, and dimethylcuprate to give the $\eta^1$-alkenyl complexes **11.76–11.78**. Complexes **11.77** and **11.78** undergo a 1,3 base-catalyzed hydrogen shift to give, respectively, the *cis*-substituted $\eta^1$-alkenyl complexes, **11.79** and **11.80**.

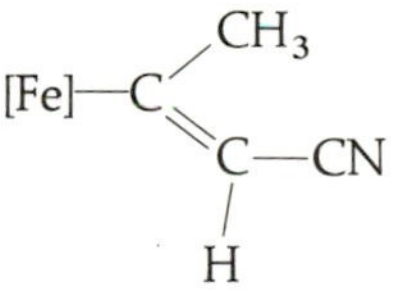

$[Fe]-C\overset{\diagup CH_3}{\diagdown \underset{|}{C}-CN}$ $\quad$ $[Fe]-C\overset{\diagup CH_3}{\diagdown \underset{|}{C}-CH_3}$
$\qquad\qquad\;\; H \qquad\qquad\qquad\qquad\quad H$

**11.79** 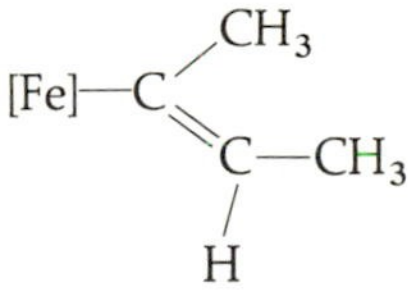 $\qquad\qquad\qquad\quad$ **11.80**

*Dihapto*-alkyne ligands are susceptible to nucleophilic attack to give $\eta^1$-alkenyl complexes (eq. *11.77*). Additions *trans* to the metal moieties are observed except with borohydride addition of hydride ion.

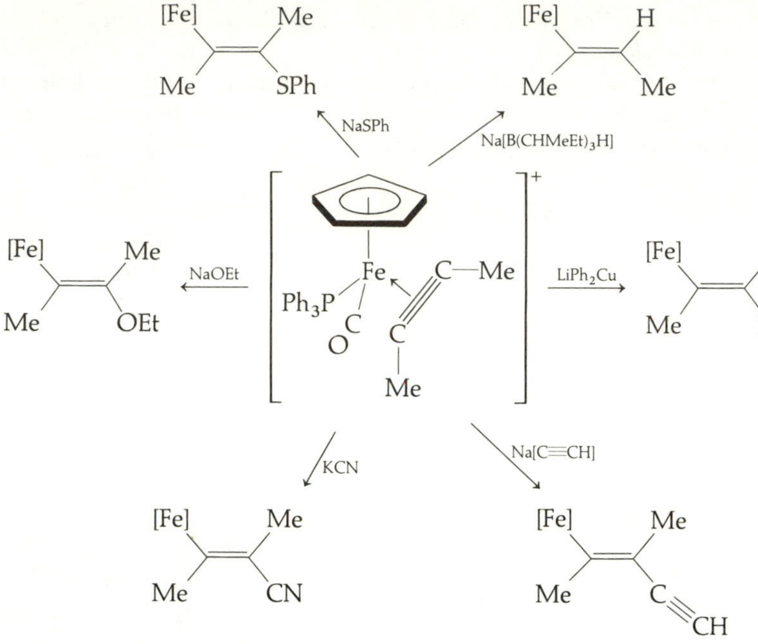

[Fe] Me / Me SPh   ←NaSPh

[Fe] H / Me Me   ←Na[B(CHMeEt)$_3$H]

[Fe] Me / Me OEt   ←NaOEt

$$\left[ \begin{array}{c} \eta^5\text{-}C_5H_5 \\ Ph_3P\text{--}Fe\text{---}C\text{---}Me \\ OC\text{---}C\text{---}C\text{---}Me \end{array} \right]^{+}$$

[Fe] Me / Me Ph   ←LiPh$_2$Cu                                         (11.77)

[Fe] Me / Me CN   ←KCN

[Fe] Me / Me C≡CH   ←Na[C≡CH]

[Fe] = ($\eta^5$-C$_5$H$_5$)Fe(CO)(Ph$_3$P)

With $\eta^2$-PhC≡CMe complexes, *trans*-addition of dimethylcuprate occurs regio-specifically at the phenyl-substituted alkyne carbon atom (eq. *11.78*). An x-ray structural determination of the (PhO)$_3$P complex confirms this regiospecificity. Hydride addition to the $\eta^2$-PhC≡CMe ligands occurs with the same regiospecificity but with *cis*-stereochemistry.

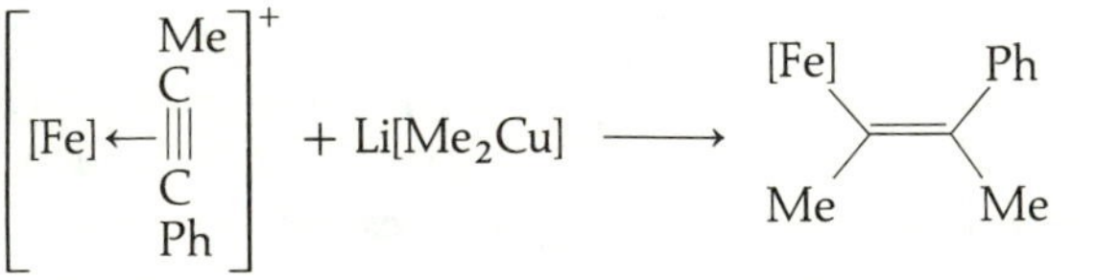

$$\left[ \begin{array}{c} Me \\ C \\ [Fe]\leftarrow \||| \\ C \\ Ph \end{array} \right]^{+} + Li[Me_2Cu] \longrightarrow \begin{array}{c} [Fe] \quad Ph \\ Me \quad Me \end{array}$$   (11.78)

[Fe] = ($\eta^5$-C$_5$H$_5$)(OC)(L)Fe; L = Ph$_3$P or (PhO)$_3$P

The stereochemistry of hydride addition was confirmed by the x-ray analysis of complex **11.81** (eq. *11.79*). Attack at the methyl-substituted alkyne carbon atom to

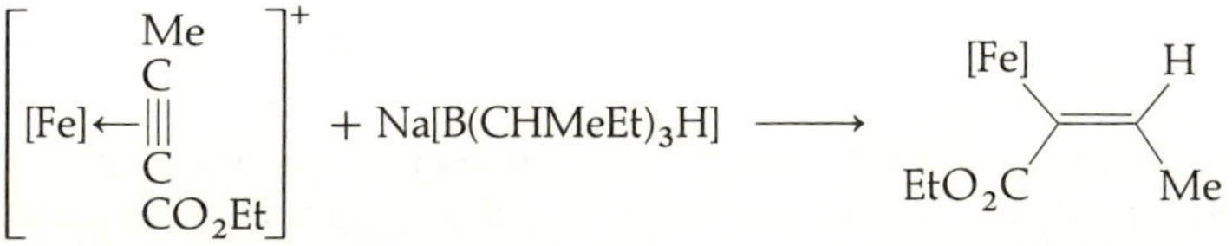

$$\left[ \begin{array}{c} Me \\ C \\ [Fe]\leftarrow \||| \\ C \\ CO_2Et \end{array} \right]^{+} + Na[B(CHMeEt)_3H] \longrightarrow \begin{array}{c} [Fe] \quad H \\ EtO_2C \quad Me \end{array}$$   (11.79)

**11.81**

[Fe] = ($\eta^5$-C$_5$H$_5$)(OC)[(PhO)$_3$P]Fe

form **11.81** is predicted by the same arguments used previously to explain metal activation of $\eta^2$-alkene ligands.

If the metal moieties of alkyne complexes undergo a similar slipping displacement upon nucleophilic attack as shown in **11.72** for olefin complexes, then the iron complex would be displaced toward the more negatively polarized alkyne carbon, the carbon that bears the $CO_2Et$ substituent. Attack by nucleophiles on the now more "exposed" methyl-substituted alkyne carbon atom would afford the observed regiospecificity.

The mechanism of the cis-addition of hydride ion is not known. Perhaps trans-addition occurs first, followed by a rapid geometrical isomerization of the $\eta^1$-alkenyl product. Alternatively, the hydride ion might add to the CO ligand to give a formyl complex. Transfer of hydride from the formyl ligand to the endo-side of the alkyne ligand would then form a cis-addition product.

Oxidative-cleavage of an Fe-$\eta^1$-alkenyl bond can be effected by using $Br_2$, $I_2$, NBS, or $[py \cdot Br]^+Br^-$ (eq. 11.80). The alkene stereochemistry is retained in these cleavage reactions.

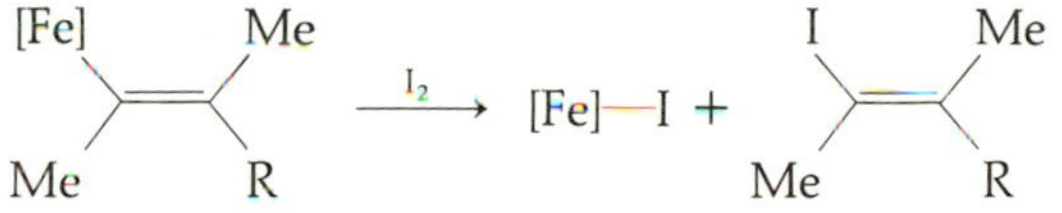

$$[Fe] = (\eta^5\text{-}C_5H_5)(OC)(Ph_3P)Fe; \ R = H \text{ or } Ph$$

**Nucleophilic Attack at Selected Acyclic *Polyhapto* Ligands**

Nucleophilic attack on an $\eta^3$-allyl ligand affords an $\eta^2$-alkene complex:

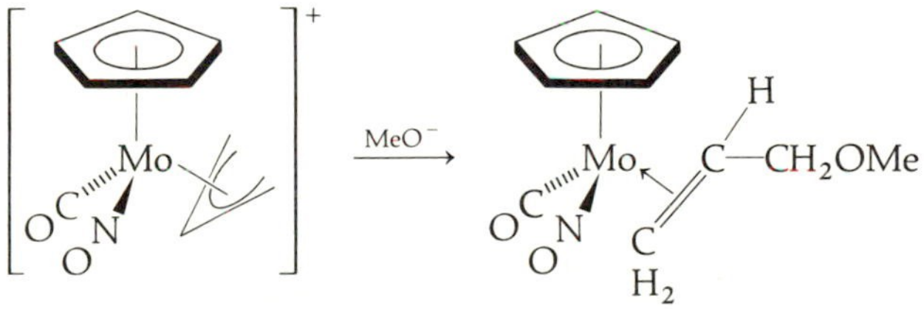

**11.82**

A formal representation of the electronic rearrangement associated with this reaction is as follows:

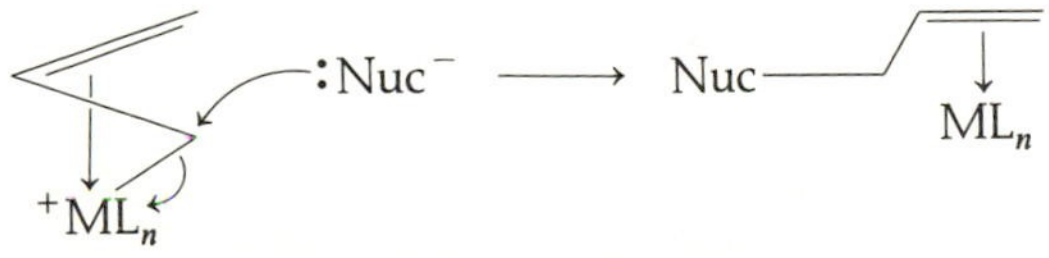

**11.83**            **11.84**

Structure **11.83** depicts an $\eta^3$-allyl ligand as an $(\eta^2$-alkene)(alkyl) Lewis structure. Attack of the incoming nucleophile on the alkyl terminal carbon atom of the $\eta^3$-

allyl ligand displaces the M—C bonding electron pair to form the carbon-nucleophile $\sigma$ bond. The metal atom is formally reduced by one electron (or by two if the alkyl ligand is regarded as an anionic two-electron ligand). However, the electron count on M does not change because the three-electron allyl ligand is converted into a two-electron alkene ligand.

If the other equivalent Lewis structure of **11.83** is used, then attack by a nucleophile occurs at the other terminal carbon atom giving **11.84**, also.

According to the molecular orbital analysis of ($\eta^3$-allyl)M bonding (see figure 5.5), an incoming nucleophile would place an electron pair into the LUMO of the $\eta^3$-allyl ligand. This orbital would be the partially filled, nonbonding $2\pi$ MO. Since this MO has atomic contributions from only the two terminal carbon atoms, a nucleophile would attack, presumably, at only these positions. Soft nucleophiles are most reactive, probably because the lobes of the $2\pi$ orbital of the allyl ligand are relatively diffuse and easily polarized like the HOMO lobes of soft nucleophiles.

An intricate and interesting study of nucleophilic additions to the cationic molybdenum complex **11.82** illustrates subtle regio- and stereospecificity associated with these reactions. Complex **11.82** exists in two rotameric conformations shown as **11.85** and **11.86**. These differ only in the orientation of the $\eta^3$-allyl ligand. In the *endo*-isomer, the central carbon of the allyl ligand points away from the $\eta^5$-$C_5H_5$ ring; in the *exo* isomer it points toward the $\eta^5$-$C_5H_5$ ring.

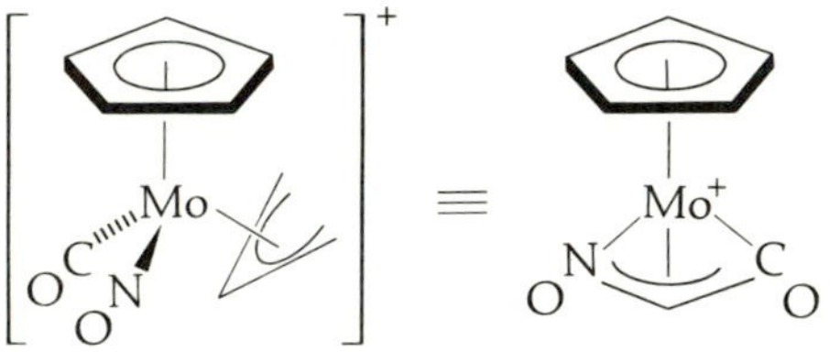

*endo* isomer

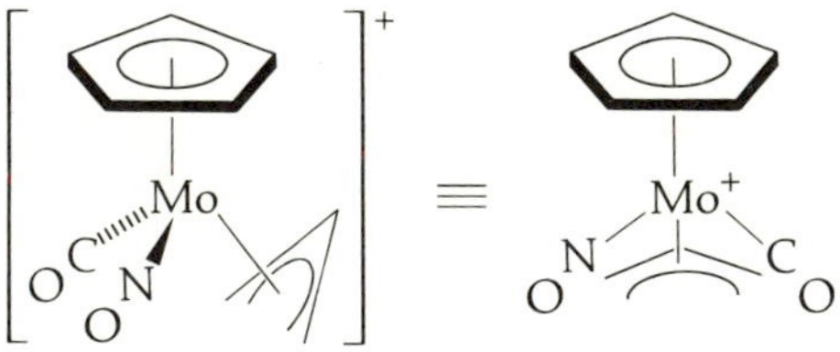

**11.85**

*exo* isomer

**11.86**

Nucleophiles such as $(Me_3C)S^-$, $[HC(O)]Me_2C^-$, and $Na[BH_3CN]$ (a hydride source) add to **11.85** and **11.86** stereospecifically, as shown in equation *11.83*. Since the *endo* and *exo* isomers interconvert very slowly, the chemistry of each rotamer can be studied separately.

Because of the chirality at the pseudo-tetrahedral molybdenum center, the two ends of the $\eta^3$-allyl ligand in both **11.85** and **11.86** are chemically different. No-

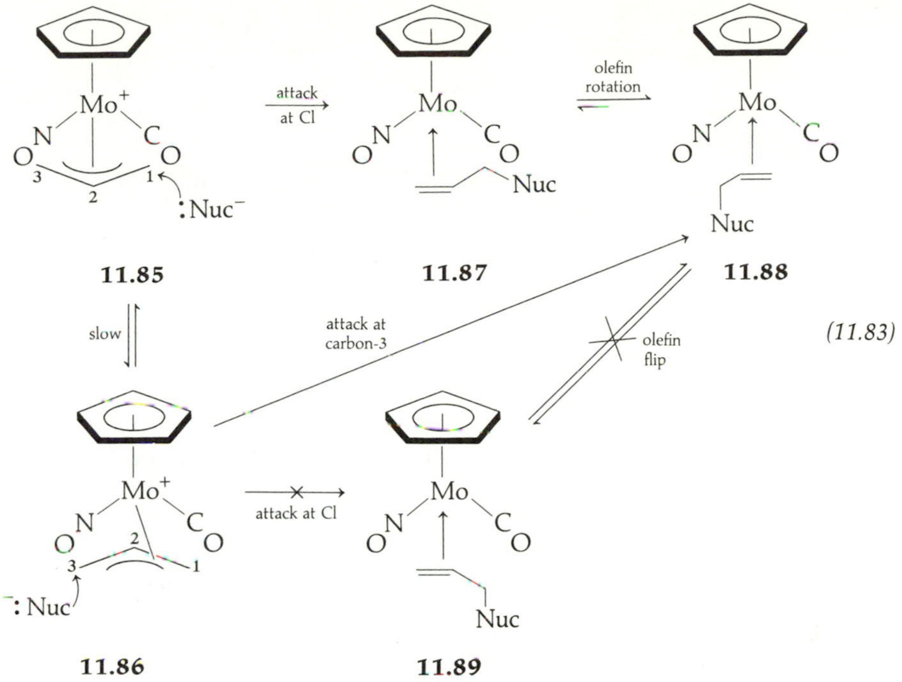

$$(11.83)$$

tice that in each rotamer, the carbon-1 terminus is closer to the CO ligand, and the carbon-3 terminus is closer to the NO ligand.

Nucleophilic attack on **11.85** occurs only at carbon-1 to give alkene complex **11.87**. This complex rearranges by a 180° rotation about the alkene-Mo bond to give **11.88** as the most stable structure. Nucleophilic attack on **11.86** occurs only at carbon-3 to give *again* complex **11.88**. The interesting observation is that diastereomer **11.88** is formed in *both* reactions.

If nucleophilic attack on **11.86** occurred at C1, then complex **11.89** would be formed. Since complex **11.89** is the other diastereomer of complex **11.88**, a mixture of diastereomers should be observed. The possibility of diastereomeric alkene complexes results from the complexation of a monosubstituted alkene to a metal complex, which destroys the mirror symmetry of the free alkene (see chap. 5). Complex **11.89** can be converted to **11.88** by rotating the alkene ligand by 180° about the olefin-Mo bond and then replacing the alkene by its mirror image. Alternatively, the $\eta^2$-alkene ligand of **11.89** can be rotated and then flipped over and recoordinated to the Mo atom.

This study demonstrates high stereospecificity. Nucleophilic addition occurs only at carbon-1 in complex **11.85** and only at carbon-3 in complex **11.86**. This stereospecificity might be frontier-controlled, but the preferred formation of the most stable $\eta^2$-alkene complex (**11.88**) is probably the principal driving force.

Equation *11.84* reveals that nucleophilic addition occurs from the face of the $\eta^3$-allyl ligand opposite the Mo moiety and only at the carbon-3 terminus. The stereochemistry at the tertiary-substituted carbon atom was determined by x-ray diffraction.

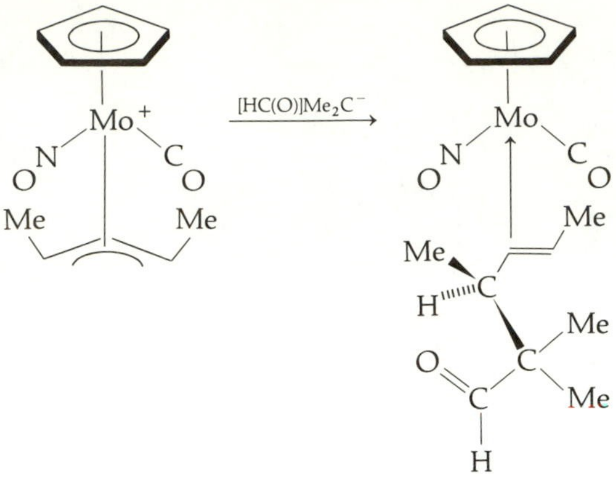

$$(11.84)$$

For the unsymmetrically substituted $\eta^3$-allyl complexes **11.90** and **11.91**, nucleophilic attack gives only a *single* product:

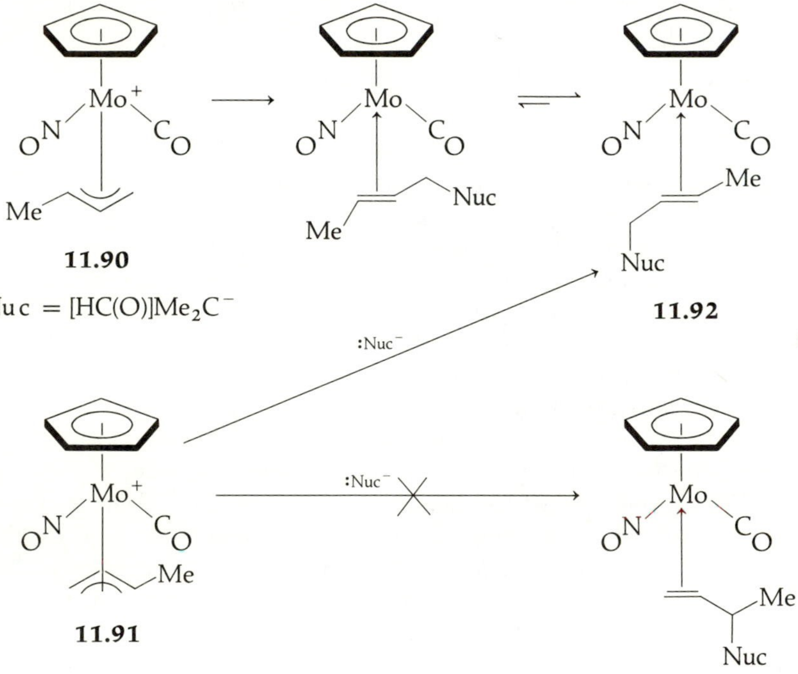

**11.90**

$\vec{:}Nuc = [HC(O)]Me_2C^-$

**11.92**

$$(11.85)$$

**11.91**

**11.93**

Attack occurs regiospecifically at the allylic $CH_2$ terminus to give the most stable alkene complex (**11.92**). Attack at the substituted allylic terminus to give complex **11.93** does not occur. Apparently, an internal olefinic product is more stable than the terminal olefinic structure.

Nucleophilic alkylation at allylic positions of olefins can be effected by using ($\eta^3$-allyl)Pd complexes as reaction intermediates:

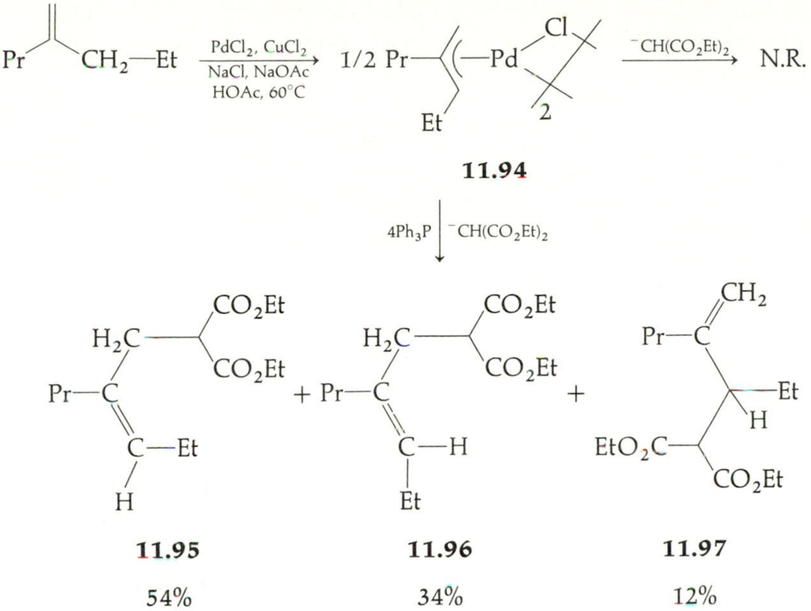

$$\text{Pr} \overset{\text{O}}{\underset{}{\|}} \text{CH}_2-\text{Et} \xrightarrow[\substack{\text{NaCl, NaOAc} \\ \text{HOAc, 60°C}}]{\text{PdCl}_2, \text{CuCl}_2} 1/2 \; \text{Pr} \begin{matrix} \\ \end{matrix} (-\text{Pd} \begin{matrix} \text{Cl} \\ \end{matrix})_2 \xrightarrow{{}^-\text{CH(CO}_2\text{Et)}_2} \text{N.R.}$$

**11.94**

$$\downarrow \; 4\text{Ph}_3\text{P} \quad {}^-\text{CH(CO}_2\text{Et)}_2$$

**11.95** + **11.96** + **11.97**   (11.86)

$$\begin{matrix} \text{11.95} & \text{11.96} & \text{11.97} \\ 54\% & 34\% & 12\% \end{matrix}$$

The reaction conditions shown are those for alkene-alkylations, which are catalytic in palladium. Conversion of the alkene reactant to the $(\eta^3$-allyl)Pd(II) complex (**11.94**) involves the loss of HCl and is one of the well-known methods of preparing $\eta^3$-allyl complexes (see eq. 5.80). In this step, one of the allylic hydrogen atoms of the alkene is lost. The NaOAc/HOAc buffer solution presumably takes up the evolved HCl. Sodium chloride is needed to dissolve the PdCl$_2$ (probably as the PdCl$_4{}^{2-}$ complex).

When complex **11.94** is treated with diethyl malonate anion, no reaction is observed. However, if this reaction is attempted in the presence of Ph$_3$P, then nucleophilic addition to the less substituted allylic terminus generates the alkenes **11.95** and **11.96,** and nucleophilic attack at the more substituted allylic terminus gives the alkene **11.97.** Copper(II) ion oxidizes the reductively eliminated Pd$^0$ to Pd(II), thereby regenerating the Pd(II) reactant.

Other soft nucleophiles, such as $^-$CH(CO$_2$Me)(SO$_2$Me) and $^-$CH(CO$_2$Me)-(SOMe), add similarly to $(\eta^3$-allyl)Pd complexes. Hard nucleophiles, such as MeLi or MeMgI, do not usually add to these allyl complexes.

Nucleophilic attack occurs on the face of the allyl ligand opposite that occupied by the palladium moiety:

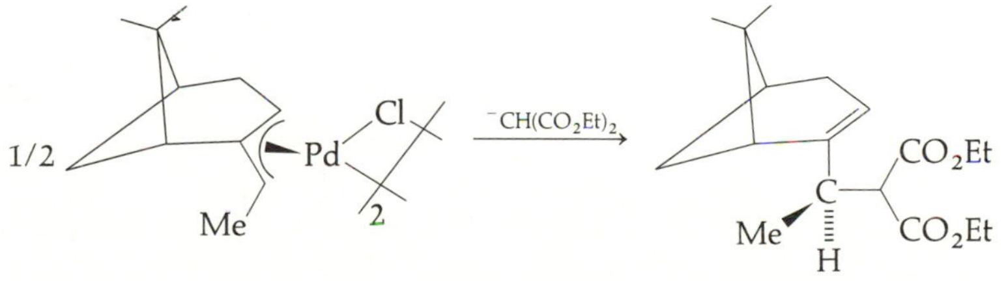

$$1/2 \quad \xrightarrow{{}^-\text{CH(CO}_2\text{Et)}_2} \qquad (11.87)$$

The influence of the phosphine ligand is evident during the step involving nucleophilic attack. When an optically active phosphine L* is used, a slight amount of asymmetric induction is observed, as shown in equation *11.88*.

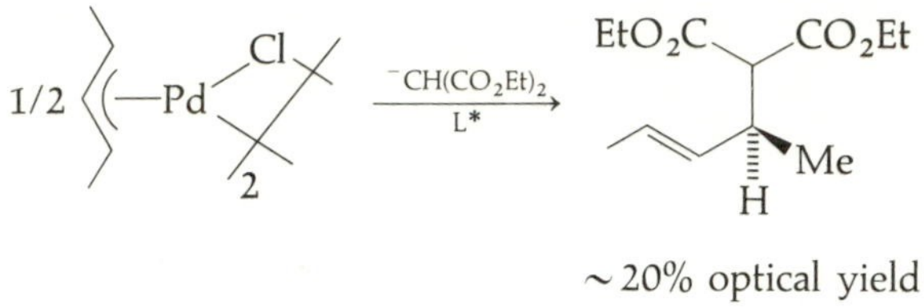

$$1/2 \left\langle \left( -Pd \stackrel{Cl}{\diagup} \right)_2 \right. \xrightarrow[L^*]{^-CH(CO_2Et)_2} \qquad (11.88)$$

~ 20% optical yield

Phosphine ligands presumably convert the ($\eta^3$-allyl)Pd chloride dimers into cationic ($\eta^3$-allyl)PdL$_2$$^+$ complexes. These complexes then undergo facile nucleophilic attack. Palladium(0) complexes, such as Pd(PPh$_3$)$_4$ or Pd(diphos)$_2$, also afford nucleophilic alkylation at the allylic position of olefins. Oxidative-addition of these Pd(0) complexes to olefins would generate the same ($\eta^3$-allyl)PdL$_2$$^+$ intermediates.

Nucleophilic attack on $\eta^5$-dienyl ligands is a well-studied reaction. The product is usually an ($\eta^4$-1,3-diene)complex (eq. *11.89*). This reaction is formally the reverse of a common abstraction route to the formation of $\eta^5$-dienyl complexes (see equations *5.83–5.87*).

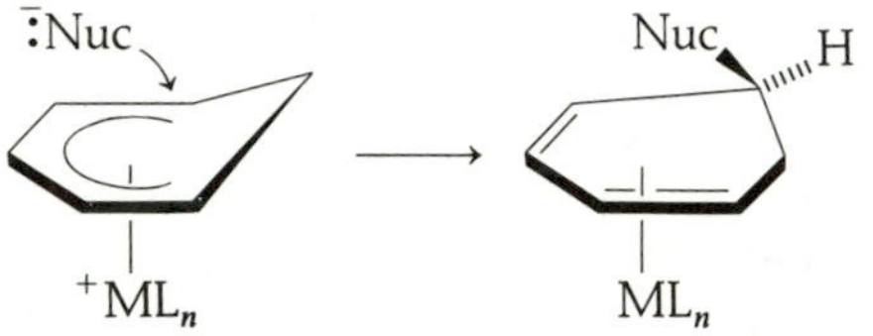

$$(11.89)$$

Nucleophilic attack on $\eta^5$-cyclohexadienyl or cycloheptadienyl ligands in cationic metal complexes takes place at either the $\eta^5$-dienyl ligand, M, or at an ancillary CO ligand. Usually, nucleophilic addition occurs on the $\eta^5$-dienyl ligand. With only a few exceptions, this attack is stereospecifically *exo*, as shown in equation *11.89*.

The best-studied $\eta^5$-dienyl system is the cationic ($\eta^5$-cyclohexadienyl)Fe(CO)$_3$$^+$ complex (**11.98**) and related substituted analogues. Nucleophilic additions occur with hydride reagents such as Li[Et$_3$BH], enolate anions, cyanide, amines, and phosphines. Tertiary amines and phosphines give ammonium and phosphonium salts (eq. *11.90*), and secondary amines give neutral amine complexes (eq. *11.91*).

**11.98**

$$(11.90)$$

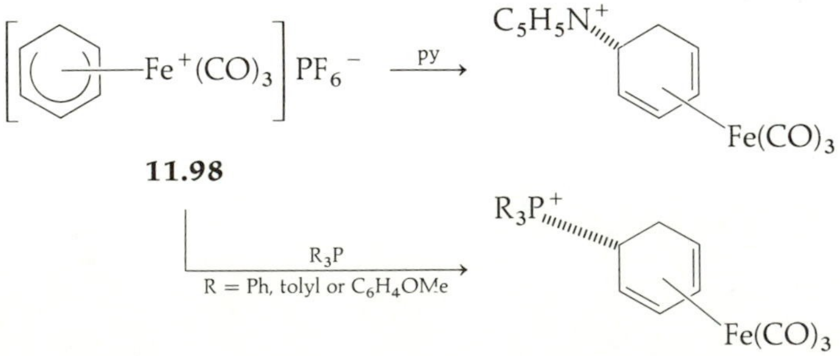

R = Ph, tolyl or C$_6$H$_4$OMe

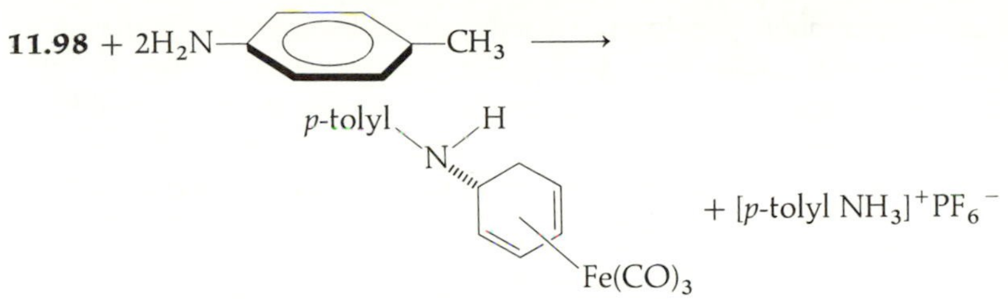

$$11.98 + 2H_2N-\langle\bigcirc\rangle-CH_3 \longrightarrow$$

p-tolyl, H
$$\text{N}$$

$$+ [p\text{-tolyl } NH_3]^+PF_6^-$$

$$\text{Fe(CO)}_3$$

(11.91)

Nucleophilic addition to $\eta^5$-dienyl ligands usually occurs at either the carbon-1 or carbon-5 terminus of the dienyl system, as illustrated by equations *11.92* and *11.93*.

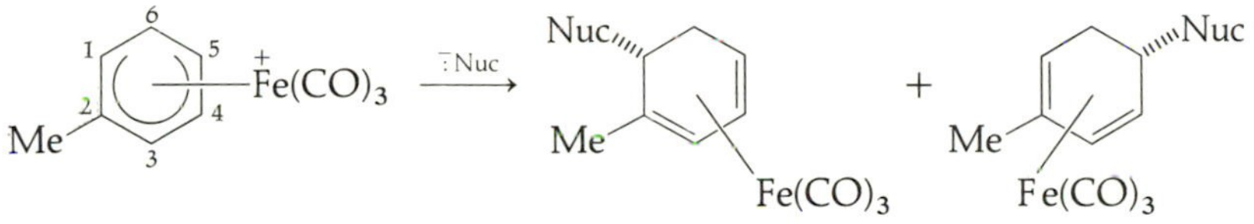

$$\text{11.98} \quad \xrightarrow{\text{:Nuc}} \quad \text{11.99} \quad + \quad \text{11.100}$$

(11.92)

**11.98**             **11.99**             **11.100**

$$\text{11.101} \quad \xrightarrow{\text{:Nuc}} \quad \text{11.102} \quad + \quad \text{11.103}$$

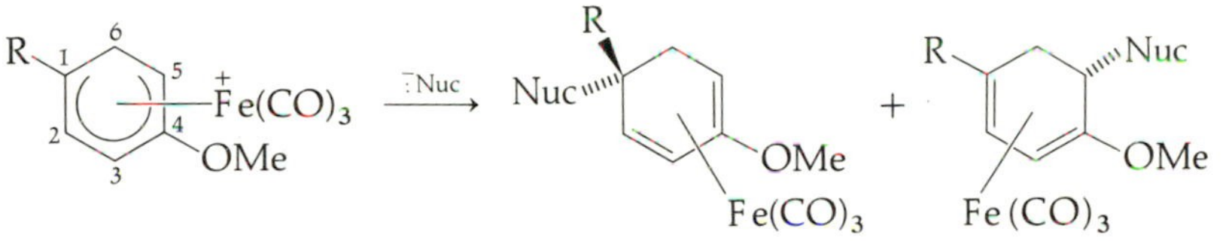

(11.93)

**11.101**             **11.102**             **11.103**

R = Me or Et

With substituted $\eta^5$-dienyl ligands, the two possible products, **11.99** or **11.100** and **11.102** or **11.103,** are different compounds. In this case product mixtures are obtained. Since the Fe(CO)$_3$ moiety can be removed readily by mild oxidation, stereospecific reactions such as those of equations *11.92* and *11.93* are being applied to syntheses of more complex organic molecules. The most desired products are those derived from regiospecific attack at carbon-1. This mode of attack is apparently enhanced when the anionic nucleophile is highly dissociated from its cationic counterion; for example, potassium salts are better than lithium salts for this purpose. The regiospecificity of attack is presumably dictated by frontier control.

### Nucleophilic Attack at Selected Carbocyclic *Polyhapto* Ligands

Phosphines add to $\eta^4$-C$_4$H$_4$ ligands to give phosphonium $\eta^3$-ring adducts (**11.104**). The phosphine adds to the *exo*-face of the cyclobutadiene ligand:

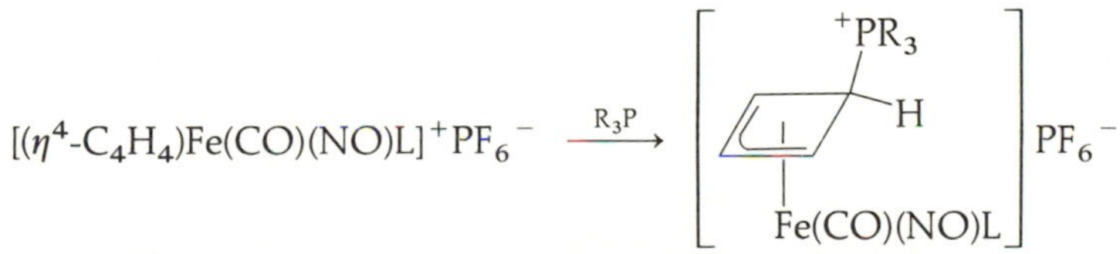

$$[(\eta^4\text{-C}_4H_4)Fe(CO)(NO)L]^+PF_6^- \xrightarrow{R_3P} \left[\begin{array}{c} ^+PR_3 \\ \text{H} \\ \text{Fe(CO)(NO)L} \end{array}\right] PF_6^- \quad (11.94)$$

L = CO or a phosphine; R = alkyl or aryl             **11.104**

For essentially all cationic complexes containing unsaturated hydrocarbon ligands, nucleophilic addition by hydride reagents occurs *exo* with respect to the metal rather than directly on the metal. The site of attack is determined by using deuterated reagents.

Hydride additions to $\eta^5$-$C_5H_5$ ligands give $\eta^4$-1,3-cyclopentadienyl complexes:

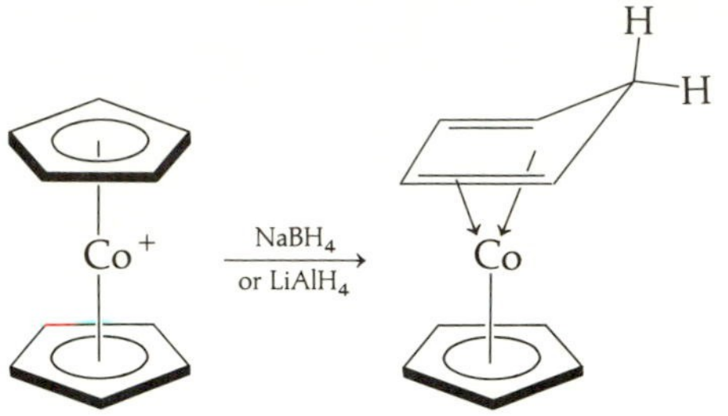

$$(11.95)$$

**11.105**

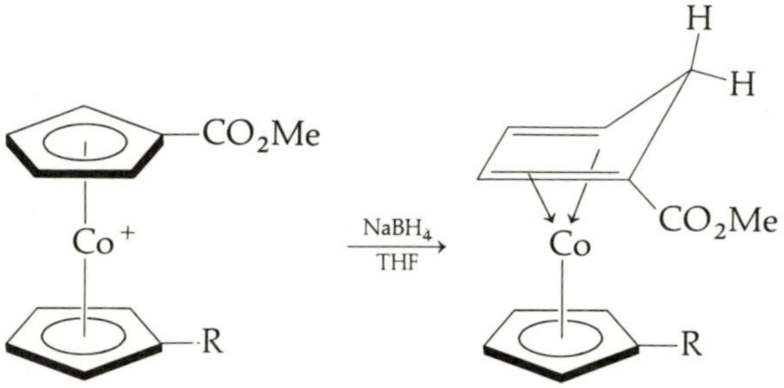

$$(11.96)$$

$R = H$ or $CO_2Me$                                                                      **11.106**

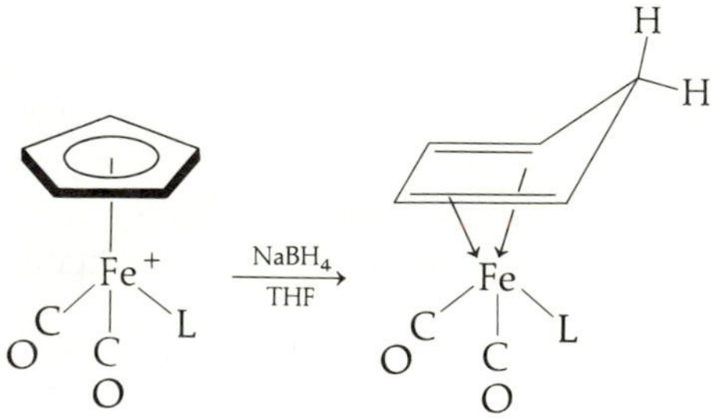

$$(11.97)$$

$L = Ph_3P$                                                                                  **11.107**

Cobalticenium ion adds hydride to form **11.105**. Phenyllithium adds similarly to give the *exo*-phenyl substituted cyclopentadienyl derivative of **11.105**. In equation *11.96*, when R is hydrogen, the hydride adds to the substituted ring at the carbon atom adjacent to the carbomethoxy substituent.

Equation *11.97* illustrates hydride addition to an $\eta^5$-$C_5H_5$ ligand when non-carbocyclic ancillary ligands are present. Hydride addition to a related cationic iron complex (**11.108**) gave the first evidence of direct hydride addition to the metal atom:

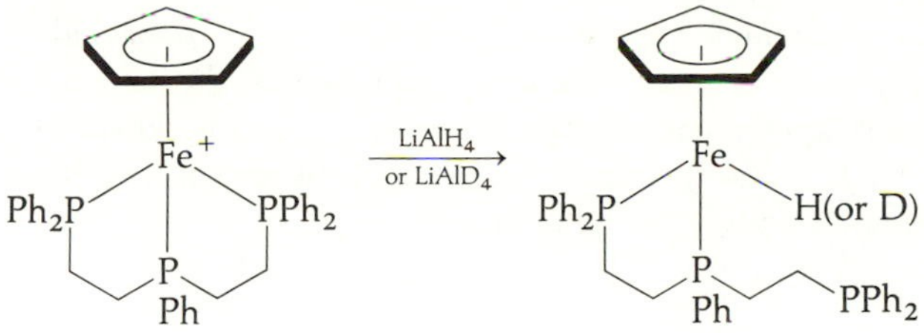

$$\text{(11.98)}$$

**11.108**　　　　　　　　　　　　　　**11.109**

Presumably, complex **11.108** has considerable ring strain within the triphos ligand because of the small covalent radius of the iron atom. As a result of this chelate ring strain, one of the terminal phosphine ligands readily dissociates from iron affording a 16-electron intermediate. This intermediate adds a hydride ligand to give **11.109**. The ruthenium analogue to **11.108** adds hydride or deuteride to the ring to give the normal $\eta^4$-cyclopentadiene *exo*-addition product. Apparently, the triphos ligand remains coordinated to the larger Ru atom.

In cationic $\eta^5$-cyclopentadienyl complexes containing CO ligands, nucleophilic attack can occur at either the $\eta^5$-$C_5H_5$ or a CO ligand. Attack by hydride at CO is illustrated in equation 8.58. However, in equation 11.99 methyllithium adds at both sites to give a mixture of neutral acetyl (**11.110**) and $\eta^4$-cyclopentadienyl (**11.111**) products.

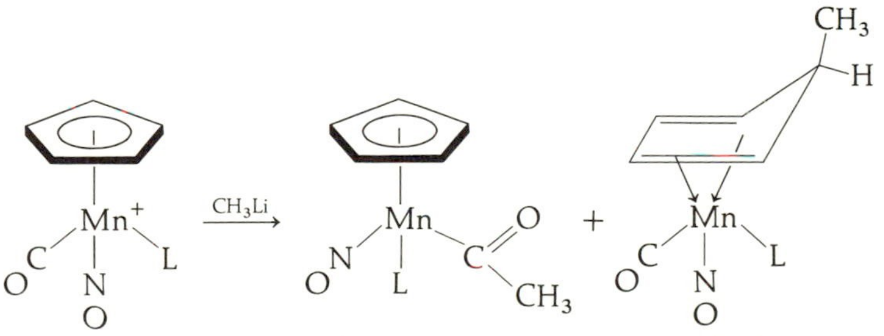

$$\text{(11.99)}$$

$L = Ph_3P$　　　　　　　**11.110**　　　　　　　**11.111**

Coordinated arenes are very susceptible to nucleophilic attack by a wide variety of nucleophiles. The $\eta^6$-chlorobenzene complex (**11.112**) is converted to an $\eta^6$-anisole complex (**11.113**) when treated with methoxide ion as shown in equation

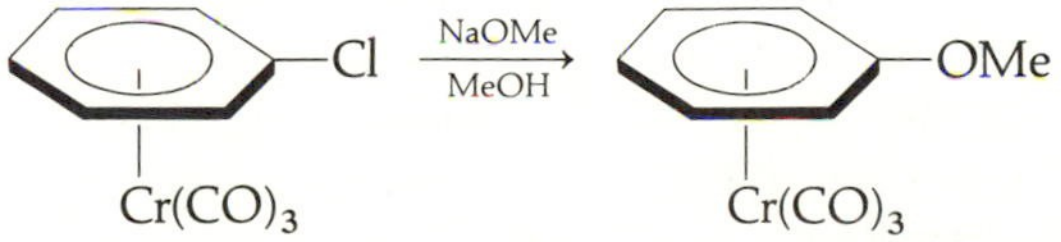

$$\text{(11.100)}$$

**11.112**　　　　　　　　　　　　**11.113**

*11.100.* The rate of this substitution reaction is comparable to that for chloride displacement from *uncomplexed* p-nitrochlorobenzene. The *hexahapto*-fluorobenzene complex reacts similarly.

Soft carbon nucleophiles add to $\eta^6$-arenes to give ($\eta^5$-pentadienyl) metallate intermediates, such as **11.114**. Oxidative-cleavage of the ligand Cr bond by $I_2$ gives a free substituted arene; alkylation by methyl iodide regenerates the starting complex (eq. *11.101*). X-ray analysis of **11.114** (where R = 2-lithio-1,3-dithiane) confirms the stereochemistry of addition.

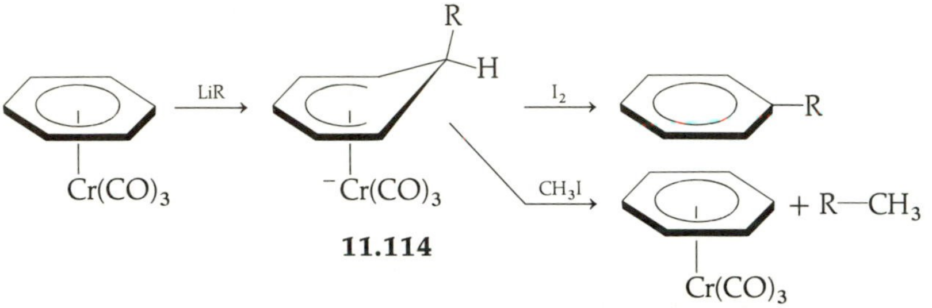

R = enolate carbanion or $^-CH_2CN$, $^-CMe_3$, and others

When a more reactive carbanion, like methyllithium or n-butyllithium, is used, a metallation reaction is observed, as shown in equation *11.102*. Oxidation of the metallation product by $I_2$ affords iodobenzene.

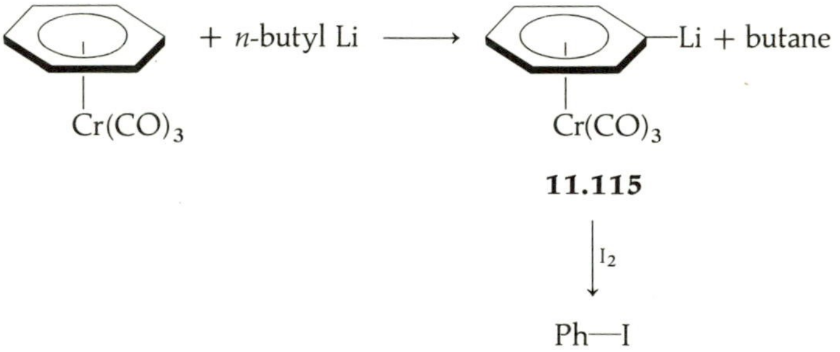

For mono-substituted $\eta^6$-arene ligands, the site of attack by an incoming nucleophile can frequently be rationalized from electronic considerations. Electron-donating substituents usually direct attack at *meta* carbon atoms, and electron-withdrawing substituents direct attack at the *para* carbon atom. These sites have the most positive charge localization and correspond qualitatively to the sites of greatest atomic contribution to the LUMO of the free arene also.

Cationic ($\eta^7$-$C_7H_7$)M complexes undergo *exo*-addition of nucleophiles to the cycloheptatrienyl ring to give neutral $\eta^6$-cycloheptatriene complexes such as **11.117**. Equation *11.103* illustrates typical examples of this reaction. When the cationic complex (**11.116**) is treated with $C_5H_5^-$ anion, a seven-substituted $\eta^6$-cycloheptatriene complex is formed, as expected. However, addition of excess $C_5H_5^-$ anion presumably removes a proton from this product complex and initiates a rearrangement leading to ($\eta^6$-$C_6H_6$)Cr(CO)$_3$. The overall reaction represents an unusual ring contraction (eq. *11.104*).

$$(\eta^7\text{-}C_7H_7)Cr(CO)_3{}^+ + Y^- \longrightarrow$$

**11.116**

$(11.103)$

**11.117**

$Y^- = {}^-OMe, {}^-SH, Me^-, Ph^-, PhC_2{}^-, CN^-,$
${}^-CH_2(CO_2Et), {}^-CH(CO_2Et)_2,$
$NHMe_2, PPhEt_2, PBu_3$

$$\textbf{11.116} + C_5H_5{}^- \longrightarrow \textbf{11.117} \ (Y = C_5H_5)$$

$C_5H_5{}^-$

$(11.104)$

$+ \ C_5H_6 \ +$

An interesting question arises regarding the preferred site of nucleophilic attack when a complex contains two different polyene ligands. For example, mixed-sandwich complexes have two different types of $\eta^n\text{-}C_nH_n$ carbocyclic ligands. Complexes that contain both an *open* conjugated polyene ligand (an $\eta^5$-pentadienyl ligand, for example) and a *closed* conjugated polyene ligand (such as an $\eta^5\text{-}C_5H_5$ ligand) are also known.

A survey of experimental data regarding nucleophilic additions to cationic 18-electron metal-polyene complexes containing various types of polyene ligands or ancillary ligands has led to the formulation of three general rules. These rules are consistent with a theoretical model for nucleophilic addition to polyene ligands. The basis of this model is that kinetically controlled nucleophilic attack is preferred at the particular polyene carbon atoms that bear the greatest positive charge. The degree of positive charge on a particular polyene carbon atom is directly related to the degree of contribution that carbon atom makes to the polyene HOMO. In cationic complexes of this type, electron donation from the polyene HOMO to the metal atom is probably more important than metal-to-LUMO back bonding.

The first rule states that nucleophilic attack occurs preferentially at *even* polyene ligands. Relatively unsophisticated calculations indicate that the total charge on an *even* polyene ligand is between 0 and $+2$ and that the total charge on an *odd* polyene ligand is between $-1$ and $+1$. Therefore, between an even or odd polyene ligand, the site of greatest atomic charge is probably within the even polyene ligand. Equations *11.105* and *11.106* demonstrate this rule for mixed-sandwich complexes. Because of the unique nonaromatic electronic structure of cyclobutadiene, nucleophiles prefer to add to all other even polyenes before adding to $\eta^4\text{-}C_4H_4$ ligands.

The second rule is that, given the choice of addition to *open* or *closed* polyene ligands, nucleophiles prefer to attack an *open* polyene ligand. Theoretically, this preference for attacking open polyenes is expected because the individual atomic

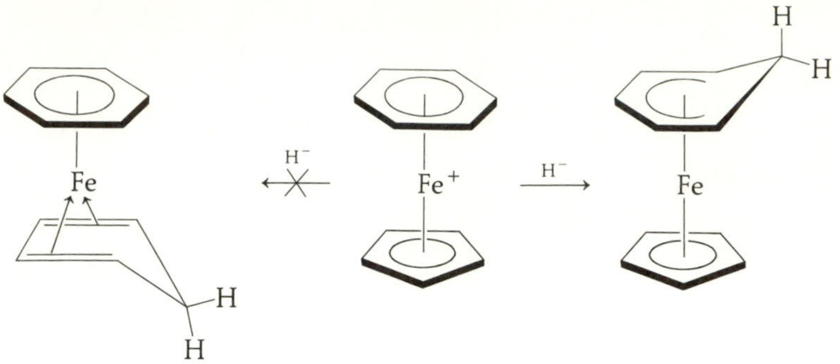

$$(11.105)$$

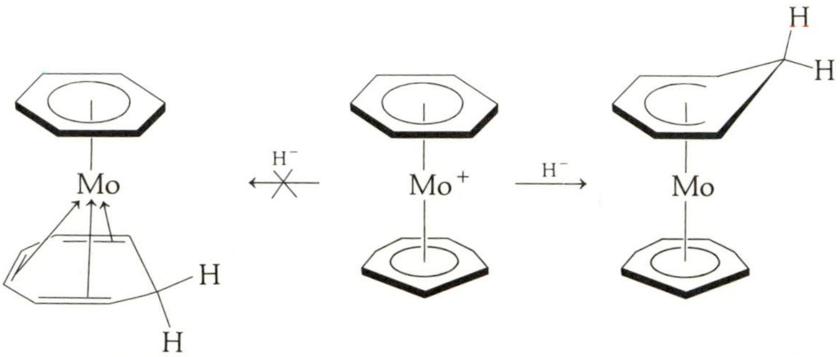

$$(11.106)$$

charges on these ligands tend to be nonuniform, some carbon atoms having a high degree of positive charge. In *closed* polyene ligands, the total charge on the ligand tends to be averaged over all of the carbon sites. Equation *11.107* illustrates this preference.

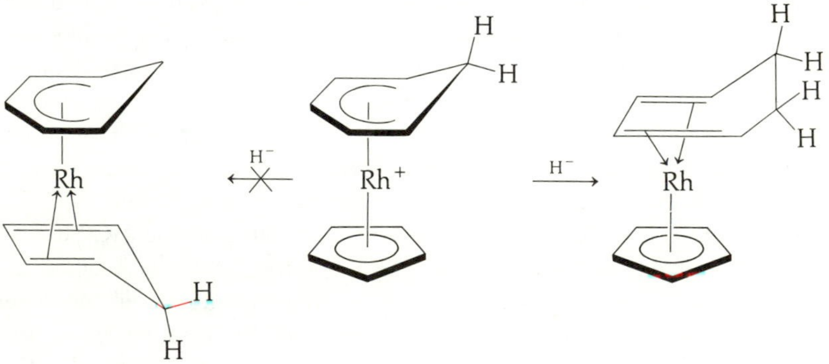

$$(11.107)$$

The third rule is more subtle. It states that nucleophilic attack on *even open* polyene ligands is always preferred at the terminal carbon atoms. For an *odd open* polyenyl ligand (like an $\eta^3$-allyl), nucleophilic attack at the terminal carbon atoms occurs only if the metal moiety is very electron withdrawing. From the examples discussed in this chapter, nucleophilic attack at one of the terminal carbon atoms is generally preferred even for odd open polyene ligands.

As with any set of generalizations, specific examples that violate these rules are known. Nucleophilic attack on ancillary ligands such as CO can occur rather than on a polyene ligand present in the complex. Some additions are known to occur initially on a polyene ligand but to undergo subsequent transfer of the nucleophile to an ancillary ligand. The reverse reaction has also been observed. These results demonstrate the effect of thermodynamic control on the eventual site to which the nucleophile is bonded. Of course, solvent effects and steric or electronic features of the nucleophile itself can greatly influence the type of products formed from any particular reaction involving nucleophilic attack on a coordinated molecule.

## STUDY QUESTIONS

1. When $trans$-$(\eta^5$-$C_5H_5)(OC)_2(Ph_3P)MoC(O)CH_2CH_2C(OEt)_2H$ is treated with $BF_3$, complex **11.64** forms. When complex **11.64** is treated with either $Et_3N$ or DMSO, the ethoxy-substituted carbon atom is converted to an aldehyde group. Write mechanisms to explain these two observed reactions.

2. Predict the organometallic products of the following reactions:

   (a) $Cr(CO)_6 + o$-$Li_2C_6H_4$ Product $A$ $\xrightarrow{2[Et_3O]BF_4}$ Product $B$

   (b) $W(CH)(PMe_3)_4Cl + CF_3SO_3H \longrightarrow$

   (c) $[(Me_2N)(Cl)C]Cr(CO)_5 + AgBF_4 \longrightarrow$

   (d) $[(\eta^5$-$C_5H_5)_2Co]^+ + MeLi \longrightarrow$ Product $A$ $\xrightarrow{[Ph_3C]BF_4}$ Product $B$

   (e) $[(\eta^5$-$C_5H_5)Fe(CO)_3]^+ + H_3\overset{+}{N}CH_2CH_2Br \xrightarrow[CH_3CN]{2Et_3N}$ Product $A$ (ionic)

   $\qquad\qquad\qquad \Big|\xrightarrow[\text{NaH}]{HOCH_2CH_2Br}$ Product $B$ (ionic)

   (f) $CH_3C(O)Re(CO)_5 + 2MeLi \longrightarrow$

3. Write a mechanism to explain the formation of the products of the following reactions:

   (a) $[(\eta^5$-$C_5H_5)Cr(NO)_2(CNMe)]^+ \xrightarrow{KOH}$

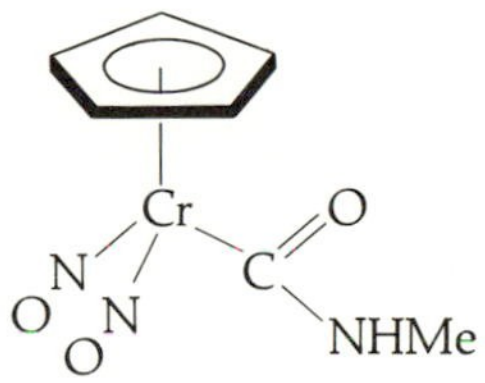

   (b) $Os(Cl)_2(PPh_3)_2(CO)(CCl_2) \xrightarrow{MeNH_2} Os(Cl)_2(PPh_3)_2(CO)(CNMe)$

   (c) $R_1R_2C{=}C(H)CH_2OAc + Li[CH(NO_2)(SO_2Ph)] \xrightarrow{Pd(PPh_3)_4}$

   $\qquad\qquad\qquad\qquad R_1R_2C{=}C(H)CH_2CH(NO_2)(SO_2Ph)$

   (d) $(\eta^5$-$C_5H_5)(OC)_3Mo(CH_2)_3Br + [(\eta^5$-$C_5H_5)Fe(CO)_2]^- \longrightarrow$

   $\qquad\qquad\qquad (\eta^5$-$C_5H_5)(OC)_2Fe(CH_2)_3Fe(CO)_2(\eta^5$-$C_5H_5)$

(e) $Br(CH_2)_3Mn(CO)_5 + [Mn(CO)_5]^-$ $\longrightarrow$ $(OC)_5Mn\!-\!Mn(CO)_4$

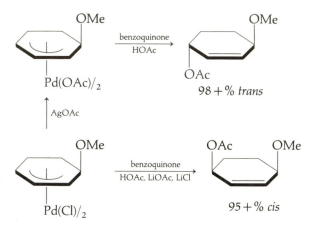

4. Explain the high stereospecificity of the two reactions below. Postulate an intermediate for each reaction. The benzoquinone is used to oxidatively remove the Pd from the olefin product.

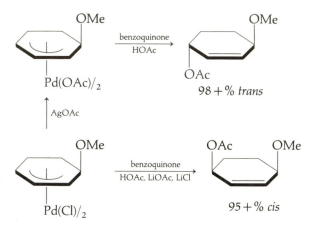

5. For the reaction below sketch the structures of the products $B$ and $C$, and explain how the changes in the IR spectra of figure 11.1 are consistent with your answers.

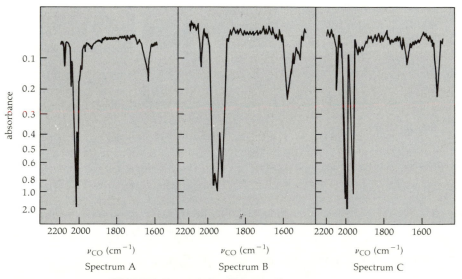

FIGURE 11.1 Infrared spectra for question 5.

$$\text{CH}_3\overset{\displaystyle\text{O}}{\overset{\displaystyle\|}{\text{C}}}\text{Re(CO)}_5 \text{ (spectrum } A) + \text{CH}_3\text{Li} \longrightarrow \text{Product } B \text{ (spectrum } B)$$

$$\Big\downarrow \text{HCl/Et}_2\text{O}$$

Product $C$ (spectrum $C$: Note, the band at $\sim 1670 \text{ cm}^{-1}$ is an $\text{O} \cdots \text{H} \cdots \text{O}$ stretching band)

## SUGGESTED READING

### General References

Johnson, M. D. *Acc. Chem. Res.* **11**, 57 (1978).
White, D. A. *Organometal. Chem. Rev.* **A3**, 497 (1968).

### Nucleophilic Attack on CO: Carbenoid and Carbyne Complexes

Angelici, R. J. *Acc. Chem. Res.* **5**, 335 (1972).
Brown, F. J. *Progr. Inorg. Chem.* **27**, 1 (1980).
Cardin, D. J., Cetinkaya, B., and Lappert, M. F. *Chem. Rev.* **72**, 545 (1972).
Cardin, D. J., Cetinkaya, B., Doyle, M. J., and Lappert, M. F. *Chem. Soc. Rev.* **2**, 99 (1973).
Casey, C. P. *Trans. Metal. Organometal. Org. Syn.* **1**, 190 (1976).
Cotton, F. A., and Lukehart, C. M. *Progr. Inorg. Chem.* **16**, 487 (1972).
Fischer, E. O. *Adv. Organometal. Chem.* **14**, 1 (1976).
Fischer, E. O., and Schubert, U. *J. Organometal. Chem.* **100**, 59 (1975).
Lukehart, C. M. *Acc. Chem. Res.* **14**, 109 (1981).

### Nucleophilic Attack on Olefins and Acetylenes

Birch, A. J., and Jenkins, I. D. *Trans. Metal. Organometal. Org. Syn.* **1**, 2 (1976).
Chisholm, M. H., and Clark, H. C. *Acc. Chem. Res.* **6**, 202 (1973).
Eisenstein, O., and Hoffmann, R. *J. Am. Chem. Soc.* **103**, 4308 (1981).
Rosenblum, M. *Acc. Chem. Res.* **7**, 122 (1974).

### Nucleophilic Attack on Polyenes and Polyenyls

Adams, R. D., Chodosh, D. F., Faller, J. W., and Rosan, A. M. *J. Am. Chem. Soc.* **101**, 2571 (1979).
Baker, R. *Chem. Rev.* **73**, 487 (1973).
Davies, S. G., Green, M. L. H., and Mingos, D. M. *Tetrahedron.* **34**, 3047 (1978).
Jaouen, G. *Trans. Metal. Organometal. Org. Syn.* **2**, 66 (1978).
Pauson, P. L. *J. Organometal. Chem.* **200**, 207 (1980).
Semmelhack, M. F. *J. Organometal. Chem. Library.* **1**, 361 (1976).
Trost, B. M. *Acc. Chem. Res.* **13**, 385 (1980).
Tsuji, J. *Acc. Chem. Res.* **6**, 8 (1973).

# APPLICATIONS OF TRANSITION METAL ORGANOMETALLIC CHEMISTRY

## PART TWO

# Applications
# to Organic Synthesis

## CHAPTER 12

Within the last decade, the boundary between inorganic and organic chemistry has become sufficiently blurred that a formal distinction between these areas can no longer be made. Both main-group and transition metal organometallic compounds are common reagents in organic synthesis.

Organometallic compounds provide new methods of effecting carbon-carbon bond formation while maintaining desired chemoselectivity, regioselectivity, and stereoselectivity. Chemoselectivity refers to functional group selection, regioselectivity, to directed attack at a specific atom within a functional group, and stereoselectivity, to the selection of the relative stereochemistry (diastereoselectivity) or absolute stereochemistry (enantioselectivity) of the product. For example, a metal complex may activate an unsymmetrical alkene toward external nucleophilic attack without affecting the reactivity of other functional groups in the molecule (chemoselectivity), while also directing the nucleophile to only one of the two olefinic carbon atoms (regioselectivity) from only the face of the alkene opposite to that occupied by the metal complex (stereoselectivity).

Particularly stable organometallic complexes can be used as *protecting groups* to prevent chemical reaction at specific functional groups. In addition, isolable complexes of organic compounds that are unstable as free molecules can be used as sources of these molecules (or synthons).

Ideally, relatively inexpensive metals should be used for stoichiometric reactions involving organometallic intermediates. When more expensive metals are used in stoichiometric reactions, the process must either be catalytic in the metal or allow for easy recovery of the metal.

Several applications of transition metal organometallic compounds to organic synthesis are described in this chapter. Emphasis is placed on applications from the recent chemical literature that lead to the synthesis of organic molecules of current interest such as natural products. These reactions involve primarily carbon-carbon bond formation and are performed on a relatively small scale. Both stoichiometric and catalytic applications are discussed. These reactions are classified according to the major function of the metal moiety or the principal organometallic transformation that occurs. Large-scale catalytic processes used to prepare commercial quantities of organic chemicals are discussed in chap. 13.

## ORGANOMETALLIC COMPLEXES AS PROTECTING GROUPS AND ORGANIC SYNTHONS

Organometallic moieties can be used as protecting groups when the resulting metal complex is unreactive to the subsequent required chemical reagents. Ideally such protecting groups are relatively inexpensive and easily removed.

In equation *12.1*, the olefinic bond of an enyne is preferentially coordinated to the cationic $(\eta^5\text{-}C_5H_5)Fe(CO)_2{}^+$ moiety. The uncomplexed alkyne bond is catalytically reduced, and then the Fe moiety is removed by adding NaI.

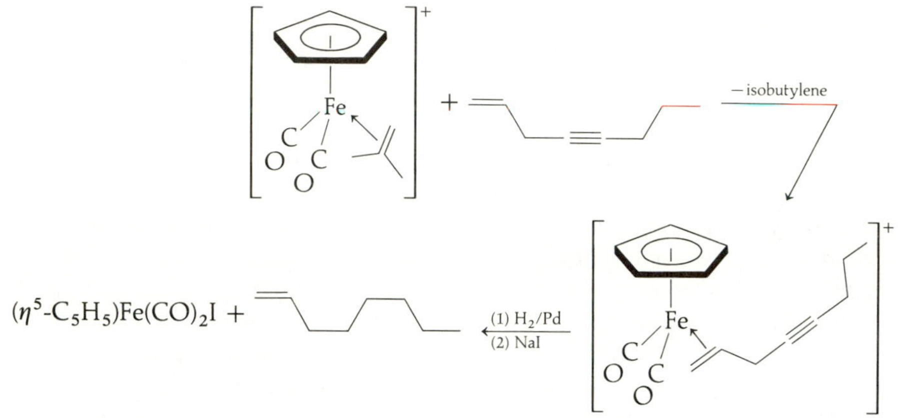

Fe
C O
C O
− isobutylene
$(\eta^5\text{-}C_5H_5)Fe(CO)_2I$ +
(1) $H_2$/Pd
(2) NaI


*(12.1)*

(*Tetrahapto*-diene)$Fe(CO)_3$ complexes undergo neither Diels–Alder additions nor catalytic hydrogenation. The distal alkene bonds of myrcene and ergosteryl acetate are selectively hydrogenated by using $Fe(CO)_3$ protection of the diene (eqs. *12.2* and *12.3*, respectively).

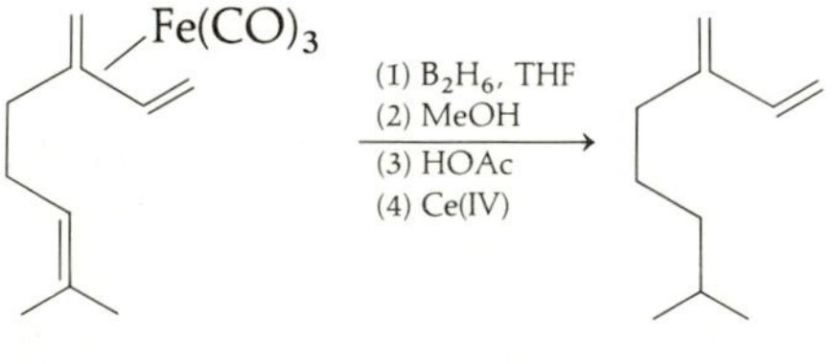

$Fe(CO)_3$
(1) $B_2H_6$, THF
(2) MeOH
(3) HOAc
(4) Ce(IV)


*(12.2)*

myrcene 43%

AcO
$Fe(CO)_3$
$PtO_2/H_2$
$(PhCH_2)Me_2SiH$


*(12.3)*

94%

Enol ethers can also be protected from reduction by coordination to $Fe(CO)_3$. Mild oxidation with $Me_3N{-}O$ or Ce(IV) ion removes the $Fe(CO)_3$ group.

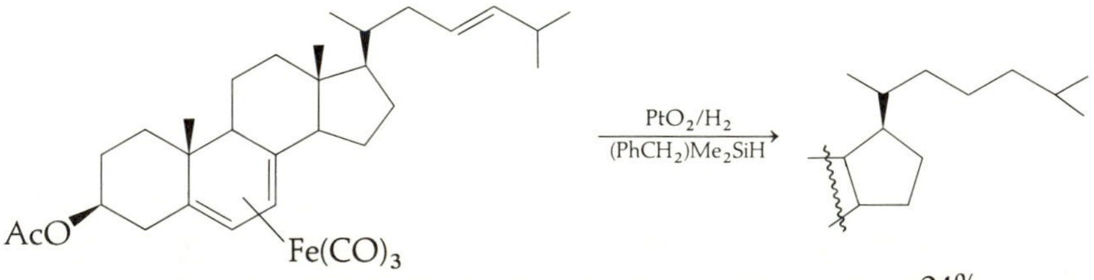

A Rh(I) $\beta$-diketonate moiety has been used to protect the 1,5-diene fragment of a conjugated triene. Catalytic hydrogenation of the uncomplexed central alkene bond followed by deprotection with cyanide ion afforded the reduced organic product in good yield:

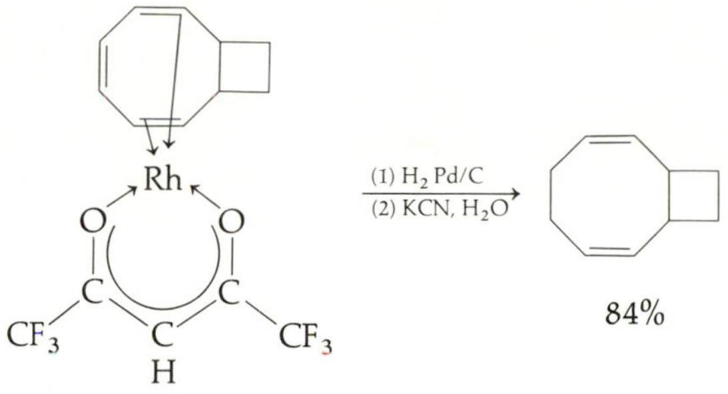

$$(12.4)$$

84%

Group VI (alkoxy)carbene complexes react with amino acid esters or peptide esters to give (amino)carbene complexes via the usual aminolysis reaction (eq. 11.56). As shown in equation 12.5, the products (**12.1**) are stable to normal peptide coupling conditions. It was with this method of N-terminus protection by an organometallic moiety that the 14−17 amino acid sequence of human proinsulin C-peptide was accomplished. Generation of the free amino acid or peptide is effected by treatment with trifluoroacetic acid (TFA).

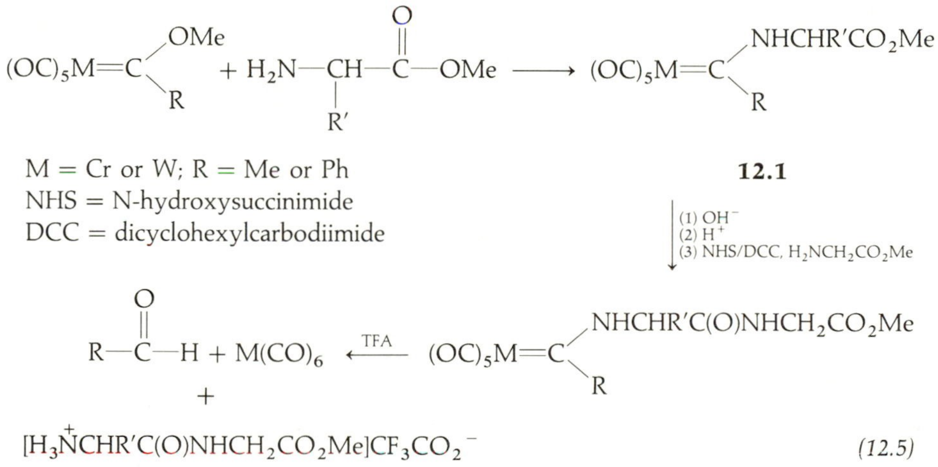

M = Cr or W; R = Me or Ph
NHS = N-hydroxysuccinimide
DCC = dicyclohexylcarbodiimide

**12.1**

(1) OH⁻
(2) H⁺
(3) NHS/DCC, $H_2NCH_2CO_2Me$

$$(12.5)$$

Rhena-$\beta$-ketoimine complexes of amino acid (or peptide) esters (**12.2**) are formed similarly, as shown in equation 12.6.

Saponification of the esters to the carboxylic acid followed by normal peptide coupling to another amino acid ester gives rhena-$\beta$-ketoimine derivatives of dipeptides (**12.3**). The rhena moiety acts as an N-terminus protecting group as well as a heavy-atom label of these systems. Removal of the rhenium with HCl/acetone affords the free amino acid; oxidative removal with iodosobenzene, PhIO, gives the N-acetyl amino acid derivatives.

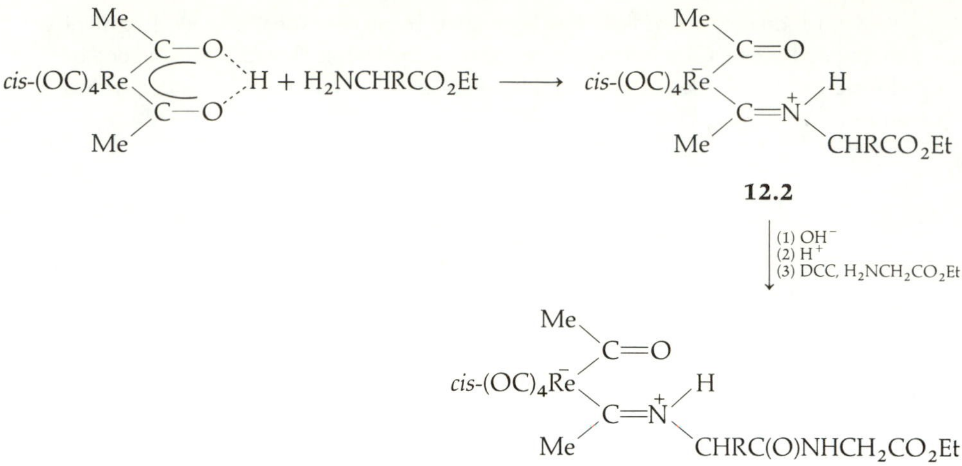

$$\text{cis-}(OC)_4Re \underset{\displaystyle Me}{\overset{\displaystyle Me}{\bigg\langle}} \begin{matrix} C\!-\!O \\ C\!-\!O \end{matrix} \!\!:\!H \; + \; H_2NCHRCO_2Et \; \longrightarrow \; \text{cis-}(OC)_4\bar{Re} \underset{\displaystyle Me}{\overset{\displaystyle Me}{\bigg\langle}} \begin{matrix} C\!=\!O \qquad H \\ \overset{+}{C}\!=\!N \\ \qquad CHRCO_2Et \end{matrix}$$

**12.2**

(1) OH⁻
(2) H⁺
(3) DCC, H₂NCH₂CO₂Et

$$\text{cis-}(OC)_4\bar{Re} \underset{\displaystyle Me}{\overset{\displaystyle Me}{\bigg\langle}} \begin{matrix} C\!=\!O \qquad H \\ \overset{+}{C}\!=\!N \\ \qquad CHRC(O)NHCH_2CO_2Et \end{matrix}$$

**12.3**                    *(12.6)*

Complexes **12.4** through **12.8** give examples of stabilization of reactive organic molecules against thermal rearrangement or polymerization by a metal. The free ligands are, respectively, cyclobutadiene, trimethylenemethane, 2,4-cyclohexadienone (a phenol tautomer), norbornadiene-7-one, and tricyclo(4.4.0.0$^{2,5}$)deca-7,9-diene. Stable complexes of other reactive organic molecules, such as pentalene, benzyne, and Dewar benzene, are also known.

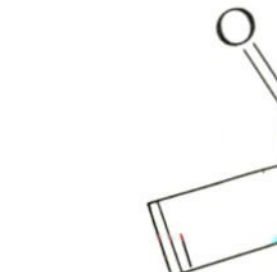

**12.4**                 **12.5** 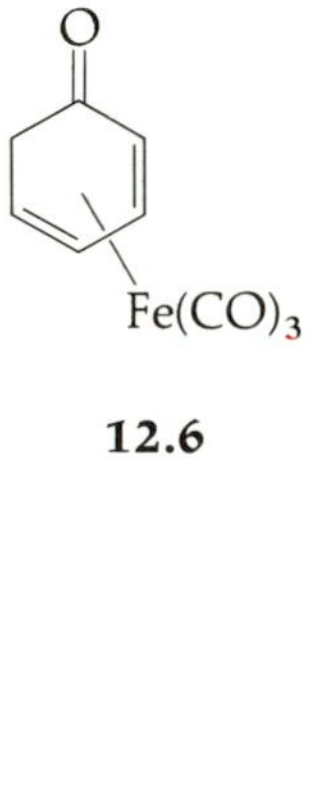                **12.6**

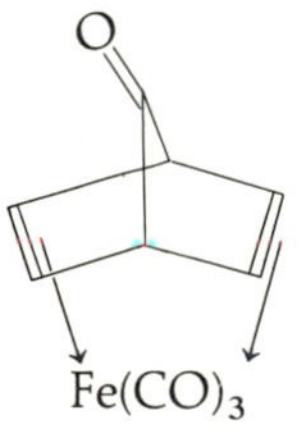

**12.7** 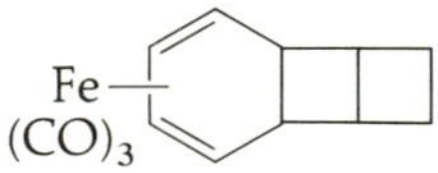                **12.8**

These complexes can be used as *in situ* sources of the free ligand for subsequent chemical synthesis. The cyclobutadiene complex (**12.4**) has been used extensively for this purpose. Removal of the Fe(CO)₃ group by oxidation generates free cyclobutadiene, which then undergoes Diels–Alder and other addition reactions. Complexes of other reactive organic molecules can be expected to become more common reagents in organic synthesis.

Carbene and carbyne complexes act as sources of carbene or carbyne fragments under certain reaction conditions. Thermal decomposition of carbyne complexes affords alkynes (eqs. 12.7 and 12.8). However, the actual intermediate formation of free carbyne radicals has not been demonstrated.

$$trans\text{-}Br(OC)_4Cr\equiv C\text{—}Ph \xrightarrow[30°C]{hexane} PhC\equiv CPh \qquad (12.7)$$

$$PhCCo_3(CO)_9 \xrightarrow[reflux]{diglyme} PhC\equiv CPh \qquad (12.8)$$

The carbene complex **12.9** decomposes thermally to give dimethoxystilbene (eq. 12.9). No cyclopropane derivative is formed during this decomposition when tetramethylethylene is present as a carbene trap.

$$[(Ph)(MeO)C]Cr(CO)_5 \xrightarrow[12\ hrs]{135°C} cis\text{-} \text{ and } trans\text{-}\alpha,\alpha'\text{-}Ph(MeO)C\!=\!C(OMe)Ph \qquad (12.9)$$

$$\textbf{12.9} \qquad\qquad\qquad\qquad\qquad\qquad 60\%$$

When one of the atoms bonded to the carbene carbon bears hydrogen atoms, a hydrogen shift can occur upon thermal decomposition to give a stable unsaturated organic molecule, as shown in the formation of an imine product:

$$[(Ph)(MeHN)C]Cr(CO)_5 \xrightarrow[160°C,\ 100\ atm]{C_6H_6,\ CO} Cr(CO)_6 + PhCH\!=\!NMe \qquad (12.10)$$

Formation of the imine can be formally regarded as proceeding through an intramolecular hydrogen shift within the free carbene species, $(Ph)(MeHN)\ddot{C}$.

Both of these reactions have been used to prepare more complex organic molecules. In equation 12.11, thermal reaction of complex **12.9** with diphenylacetylene gave a good yield of naphthol product. The implication of a free carbene intermediate is only a formalism. Possible assistance by chromium in the cyclization step

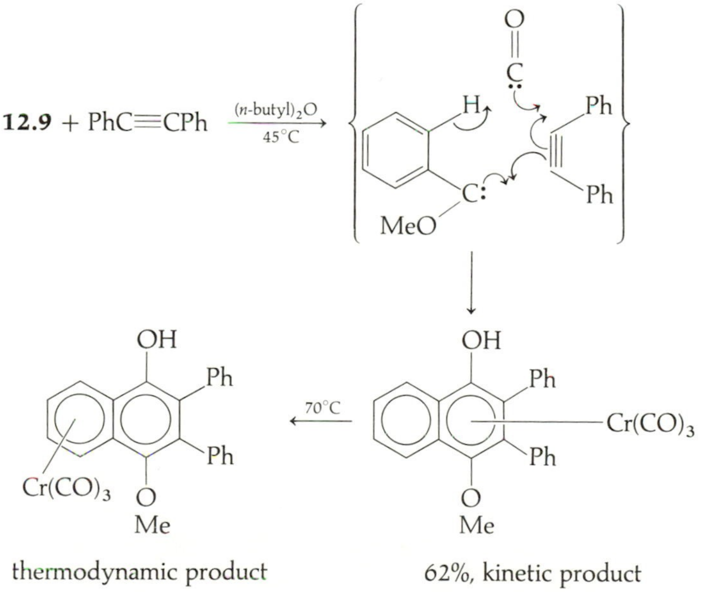

$$\textbf{12.9} + PhC\equiv CPh \xrightarrow[45°C]{(n\text{-}butyl)_2O}$$

thermodynamic product

62%, kinetic product

is supported by the observation of an $(\eta^6\text{-arene})Cr(CO)_3$ complex at the newly formed ring as the kinetic product of the reaction.

Thermal decomposition of the (amino)(alkyl)carbene complex, shown in equation *12.12*, in pyridine presumably generates an imine via a hydrogen atom shift. Subsequent cyclization gives a 17% yield of tetrahydroharman.

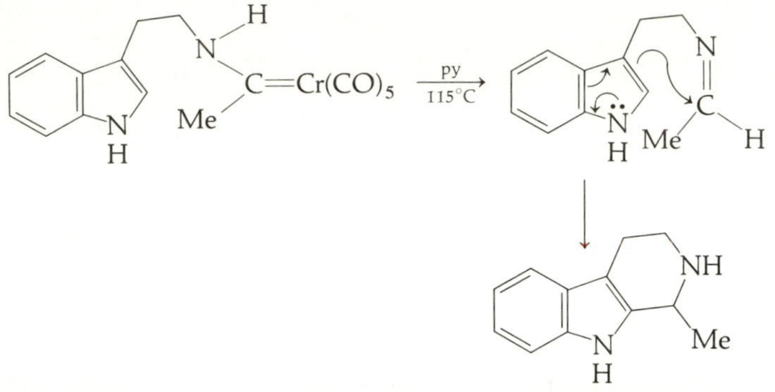

(12.12)

17%, tetrahydroharman

Carbenoid complexes can also be used as *methylene* or *carbene transfer agents*. Several examples of these reactions are shown in equations *12.13–12.16*. Although carbene complexes that have an alkoxy group alpha to the carbene carbon atom (**12.9**, for example) do not react with unactivated alkenes to give cyclopropanes, they do react with *activated* alkenes such as methyl crotonate to give the corresponding product (eq. *12.13*). Carbene transfer to the alkene is highly specific in retaining the original stereochemistry.

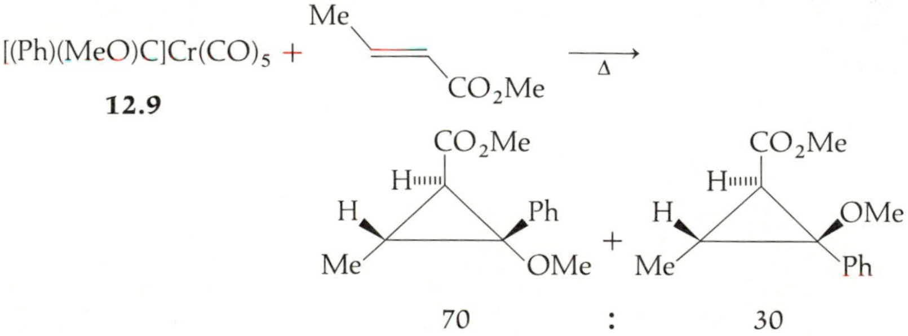

(12.13)

70        :        30

Changing the metal from chromium to molybdenum or tungsten alters the relative abundance of the two cyclopropane isomers. When a CO ligand in **12.9** is replaced by an optically active phosphine, a small amount of asymmetric induction in the cyclopropane products is observed. These results indicate that either the carbene species or the alkene or both remain coordinated to the metal atom during the reaction.

Alkylidene complexes (like **12.10, 12.12** and **12.13**) transfer alkylidene ligands to unactivated alkenes (eqs. *12.14–12.16*). The olefin stereochemistry is retained in the cyclopropane products of these reactions also.

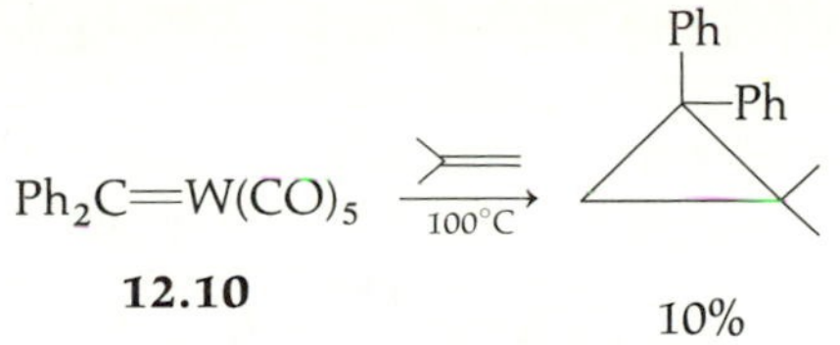
Ph
Ph
$Ph_2C{=}W(CO)_5$
100°C
12.10
10%

(12.14)

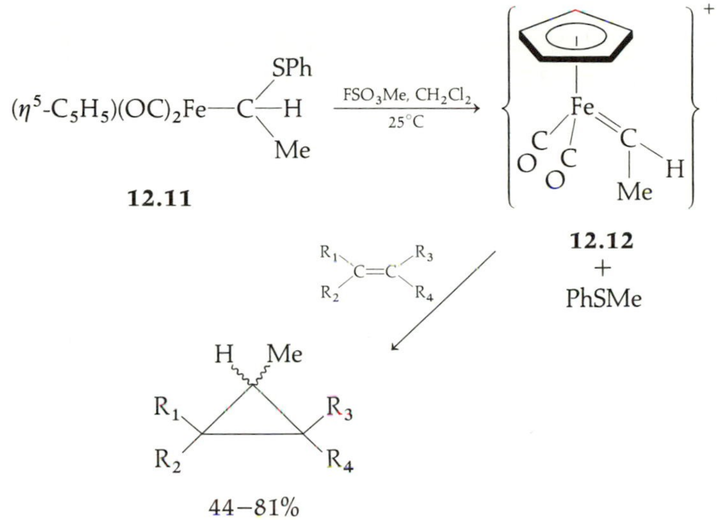
SPh
$(\eta^5\text{-}C_5H_5)(OC)_2Fe$—C—H
Me
12.11
$FSO_3Me$, $CH_2Cl_2$
25°C
Fe
C
H
Me
C
O
C
O
+
$FSO_3^-$
12.12
+
PhSMe
$R_1$ $R_2$ C=C $R_3$ $R_4$
H Me
$R_1$ $R_2$ $R_3$ $R_4$
44–81%

(12.15)

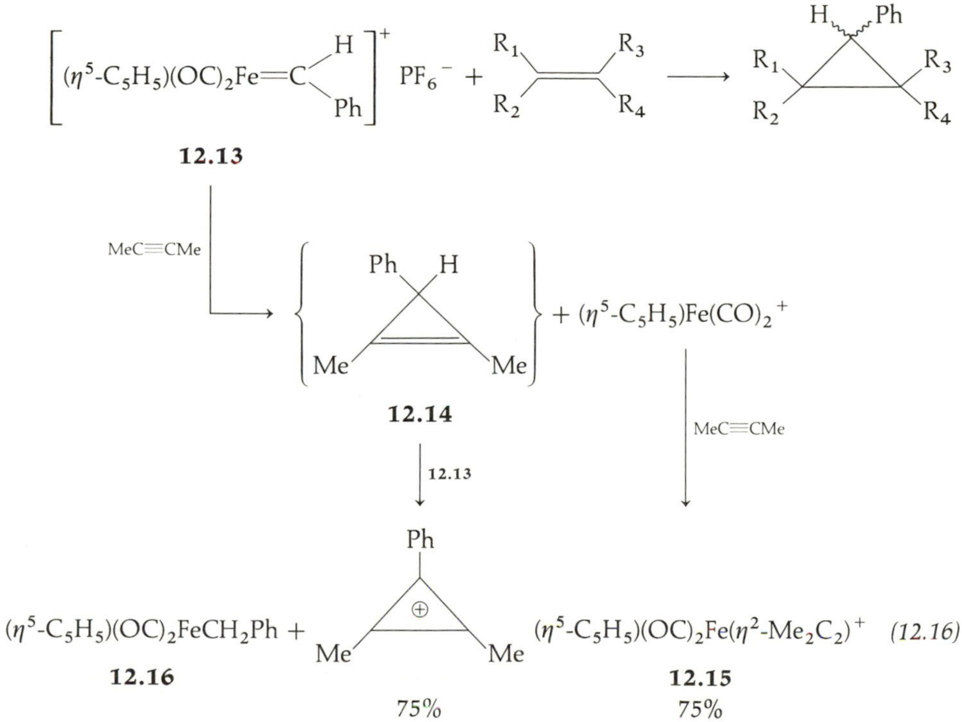
$[(\eta^5\text{-}C_5H_5)(OC)_2Fe{=}C(H)(Ph)]^+$ $PF_6^-$ +
$R_1$ $R_2$ $R_3$ $R_4$
H Ph
12.13
MeC≡CMe
Ph H
Me Me
12.14
+ $(\eta^5\text{-}C_5H_5)Fe(CO)_2^+$
12.13
MeC≡CMe
$(\eta^5\text{-}C_5H_5)(OC)_2FeCH_2Ph$ +
Ph
Me Me
$(\eta^5\text{-}C_5H_5)(OC)_2Fe(\eta^2\text{-}Me_2C_2)^+$
12.16
75%
12.15
75%

(12.16)

The ethylidene complex (**12.12**) is prepared by electrophilic alkylation of the SPh substituent of complex **12.11**. Benzylidene transfer by **12.13** to dimethylacetylene gives, presumably, the expected cyclopropene product (**12.14**) and a 16-electron iron cationic complex (which coordinates a second alkyne to give complex **12.15**). A second molecule of the benzylidene reactant **12.13** apparently abstracts a hydride ion from **12.14** to give the benzyl complex **12.16** and a dimethylphenylcyclopropenium salt.

Other types of methylene transfer complexes include metallacyclic compounds like *Tebbe's reagent*, **12.17**, and **12.18**.

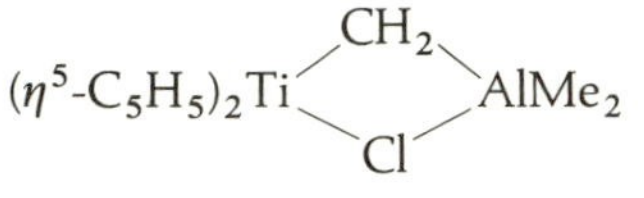

$$(\eta^5\text{-}C_5H_5)_2Ti\underset{Cl}{\overset{CH_2}{<}}AlMe_2 \qquad (\eta^5\text{-}C_5H_5)_2Ti\underset{CH_2}{\overset{CH_2}{<}}C(H)(CMe_3)$$

**12.17** 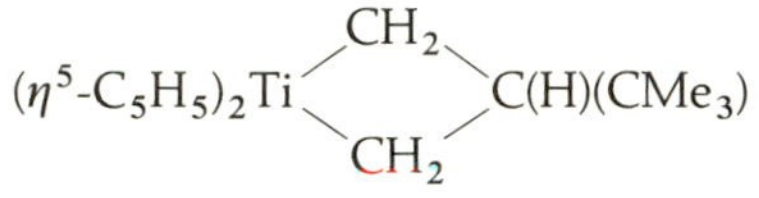  **12.18**

Both of these complexes exhibit Wittig-type chemistry in converting carbonyl groups into alkenes, as shown in equation *12.17* for complex **12.18**. These reagents extend the range of Wittig chemistry to include carbonyl groups of thioesters, amides, and silyl esters, which do not undergo the normal Wittig reaction.

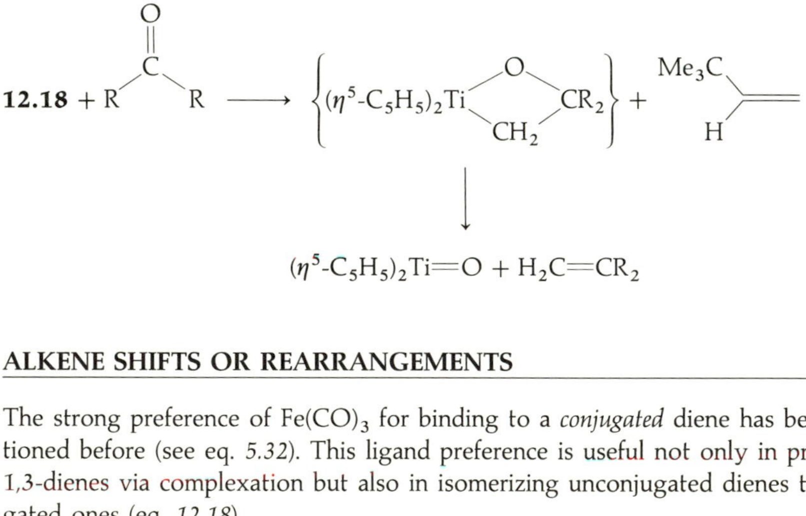

$$\textbf{12.18} + R\overset{O}{\overset{\|}{\underset{}{C}}}R \longrightarrow \left\{(\eta^5\text{-}C_5H_5)_2Ti\underset{CH_2}{\overset{O}{<}}CR_2\right\} + \underset{H}{\overset{Me_3C}{>}}$$

$$\downarrow$$

$$(\eta^5\text{-}C_5H_5)_2Ti{=}O + H_2C{=}CR_2 \qquad (12.17)$$

## ALKENE SHIFTS OR REARRANGEMENTS

The strong preference of $Fe(CO)_3$ for binding to a *conjugated* diene has been mentioned before (see eq. *5.32*). This ligand preference is useful not only in protecting 1,3-dienes via complexation but also in isomerizing unconjugated dienes to conjugated ones (eq. *12.18*).

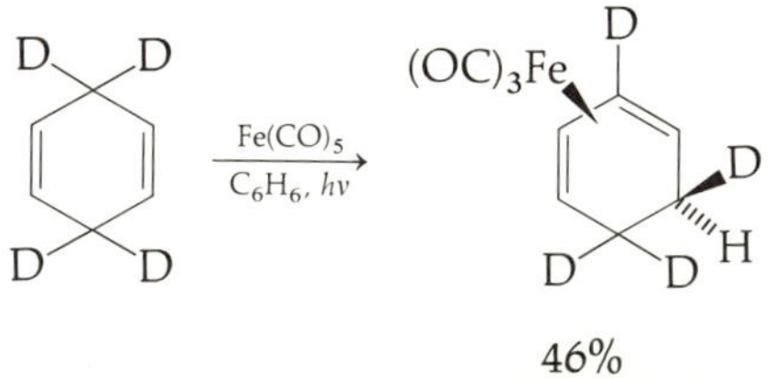

$$(12.18)$$

46%

The transfer of the deuterium atom to an *endo* site in the conjugated diene product complex probably occurs through an $(\eta^3\text{-allylic})Fe(CO)_3D$ intermediate.

Applications of this reaction to steroid and prostaglandin C syntheses are shown in equations *12.19* and *12.20*, respectively. Oxidative removal of the $Fe(CO)_3$ group is accomplished with $FeCl_3$ or $CrO_3$.

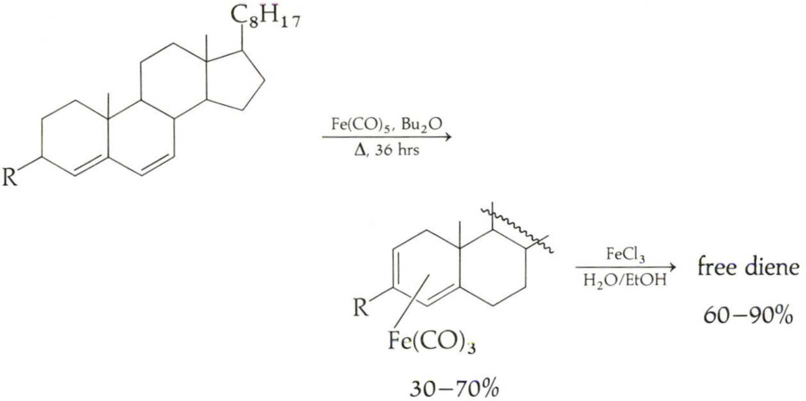

THP = 2-tetrahydropyranyl

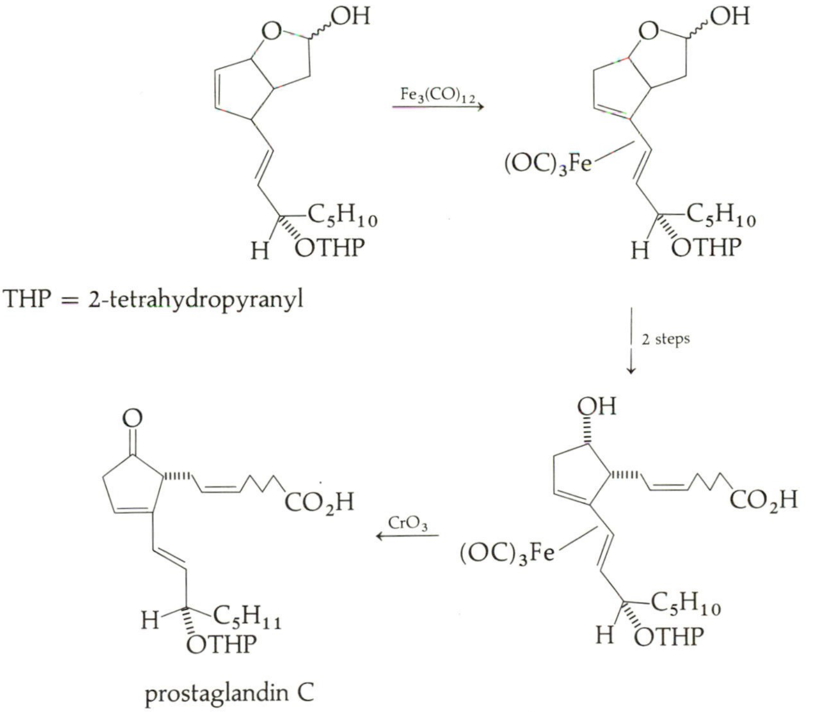

prostaglandin C

An extended alkene reorganization of calciferol is attained through an $\eta^3$-allylic intermediate. Subsequent hydride reduction gives isotachysterol (eq. *12.21*). The $\eta^3$-allylic ligand is formed by the abstraction of a hydrogen from calciferol. Nucleophilic attack on the allyl ligand by hydride ion forms a methyl group and a new olefin skeleton upon elimination of palladium(0).

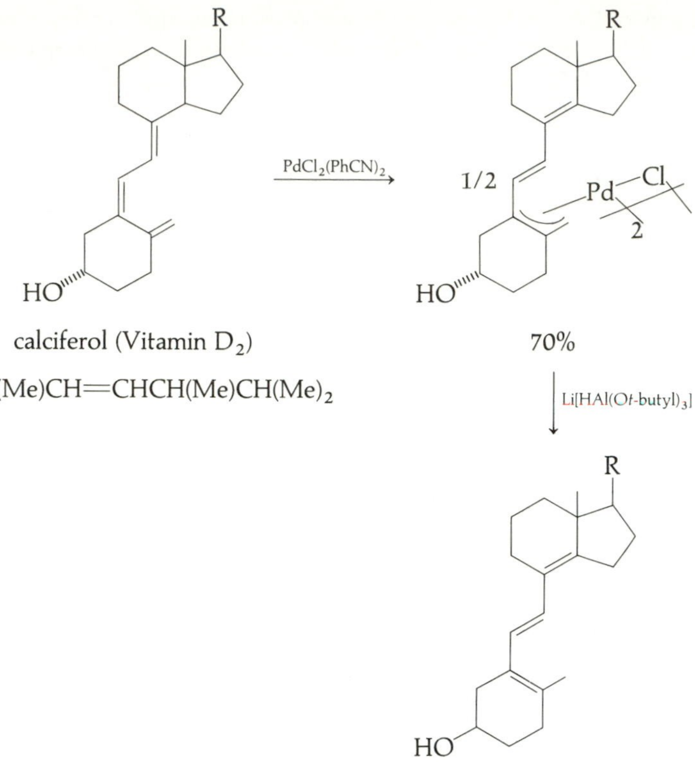

calciferol (Vitamin D₂) [calciferol (Vitamin D$_2$)]

70%

$R = CH(Me)CH{=}CHCH(Me)CH(Me)_2$

PdCl$_2$(PhCN)$_2$

1/2

Li[HAl(O$t$-butyl)$_3$]

(12.21)

isotachysterol

## ORGANOMETALLIC ANIONS AS NUCLEOPHILES

A vast number of reactions are known to involve organometallic complexes as nucleophiles. Most of these give alkyl or acyl complexes (chap. 8). The dianion $Fe(CO)_4{}^{2-}$ is a relatively inexpensive reagent that converts alkyl or acyl halides into alkanes, aldehydes, ketones, carboxylic acids, or acid halides, as shown in scheme 12.22. Initial Fe-C(alkyl) bond formation occurs by nucleophilic attack of $Fe(CO)_4{}^{2-}$ on an alkyl halide. Alkyl migration can insert a carbonyl group if desired, and oxidative or electrophilic cleavage reactions generate the organic product. Therefore, $Fe(CO)_4{}^{2-}$ acts as a Grignard analogue in converting a carbon-halide bond into a variety of other functional groups.

Unsymmetrical ketones are prepared in good yields by this method. The iron dianion is specific for halide displacement. If an organic substrate contains mixed halides, the iron complex displaces only the more reactive halogen. Tertiary and secondary halides undergo a competing elimination reaction because of the strong basicity of the $Fe(CO)_4{}^{2-}$ complex, which has about the same $pK_b$ as that of $OH^-$. Allylic halides that have alkyl groups delta to the halide form stable 1,3-diene iron complexes and therefore should be avoided as reagents in these reactions.

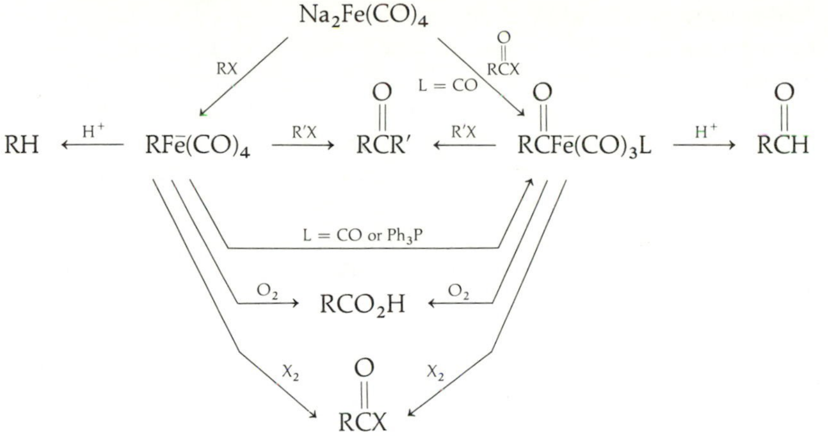

$$\text{RH} \xleftarrow{H^+} \text{RFe(CO)}_4 \xrightarrow{R'X} \text{RCR'} \xleftarrow{R'X} \text{RCFe(CO)}_3\text{L} \xrightarrow{H^+} \text{RCH} \quad (12.22)$$

(reaction scheme 12.22: Na$_2$Fe(CO)$_4$ reacting with RX and RCX; L = CO; products include RCR', RCFe(CO)$_3$L with L = CO or Ph$_3$P; $O_2 \rightarrow RCO_2H$; $X_2 \rightarrow RCX$)

## NUCLEOPHILIC ATTACK ON COORDINATED MOLECULES

Nucleophilic attack by a carbon nucleophile on a coordinated molecule couples two organic fragments to give a larger molecule by forming a new carbon-carbon bond. The regioselectivity and stereoselectivity of the coupling reaction make it especially interesting. Intramolecular nucleophilic attack by heteroatomic nucleophiles generates cyclic molecules. Several examples of applications of these reactions to organic synthesis are provided below.

### Nucleophilic Attack on Olefin Ligands

When the cationic $\eta^2$-alkene complex **12.19** undergoes nucleophilic attack by the $\eta^1$-allyl complex **12.20**, a five-carbon (alkyl)($\eta^2$-alkene) product complex (**12.21**) forms. This reaction also occurs when the alkene contains methyl, phenyl, aldehyde, or ester substituents.

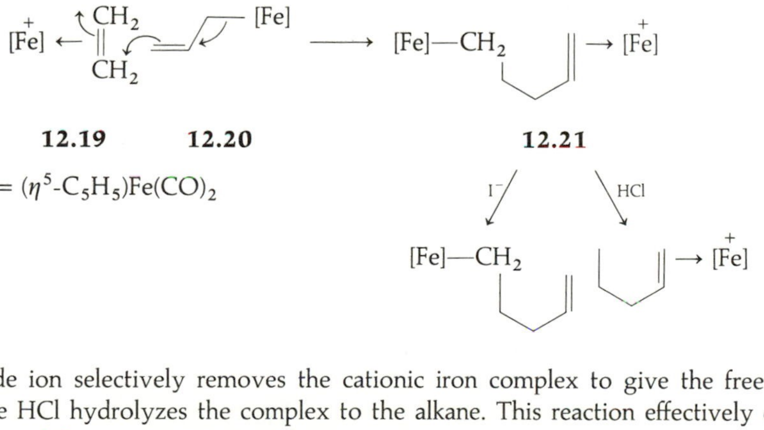

(reaction scheme 12.23)

**12.19**          **12.20**                              **12.21**

$$[\text{Fe}] = (\eta^5\text{-C}_5\text{H}_5)\text{Fe(CO)}_2$$

$$\text{[Fe]—CH}_2 \qquad \rightarrow \text{[Fe]}^+ \quad (12.23)$$

Iodide ion selectively removes the cationic iron complex to give the free alkene while HCl hydrolyzes the complex to the alkane. This reaction effectively couples two- and three-carbon skeletal fragments.

Bicyclic β-lactams are formed by a sequence of reactions shown as scheme *12.24.* Schiff-base condensation of the cationic ketonic complex (**12.22**) gives a neutral pyrroline complex (**12.23**) by the intramolecular nucleophilic attack of the imine nitrogen atom on the $\eta^2$-alkene ligand. Reduction of **12.23** with NaBH$_4$ gives stereoisomeric pyrrolidine complexes **12.24** and **12.25**. Heating complex **12.25** causes CO-insertion to give the chelate complex **12.26**. Complex **12.24** does not undergo CO-insertion to give an analogous chelate complex because of unfavorable steric interactions between the methyl substituent and the rings of the expected bicyclic product complex. Silver oxide oxidation of **12.26** initiates reductive-elimination of the bicyclic β-lactam in good yield (see eq. *10.54*).

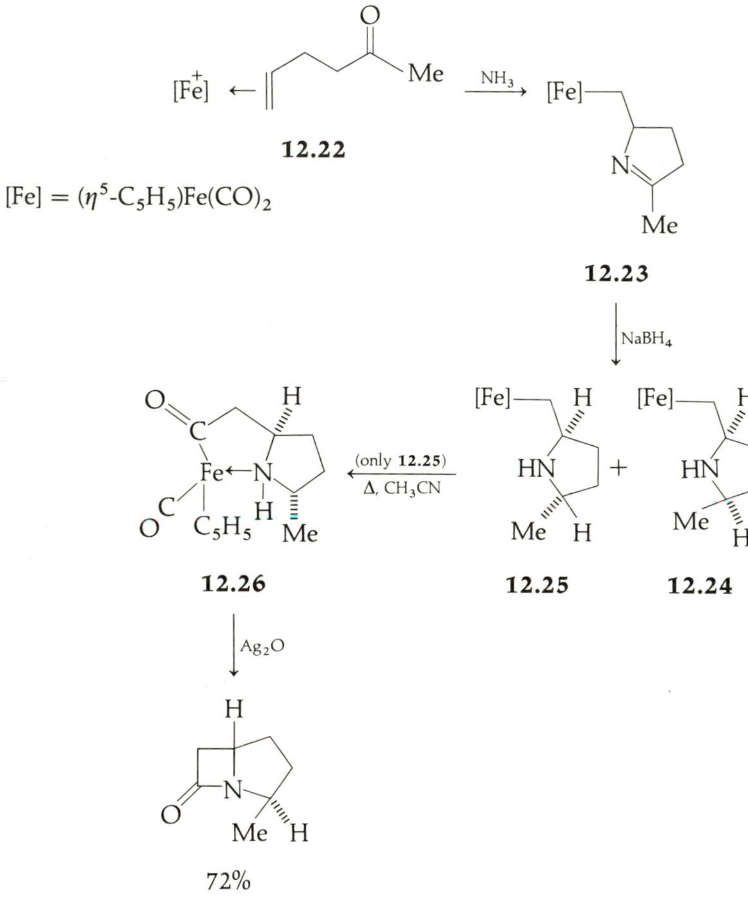

$$[Fe] = (\eta^5\text{-}C_5H_5)Fe(CO)_2$$

(12.24)

Indole ring systems have been prepared similarly both stoichiometrically and catalytically by using palladium intermediates and intramolecular nucleophilic attack on an olefin ligand. A reaction mechanism is shown in equation *12.25*. The ($\eta^2$-alkene)(amine)Pd compound (**12.27**) is formed by ligand displacement of two CH$_3$CN ligands from the reactant complex.

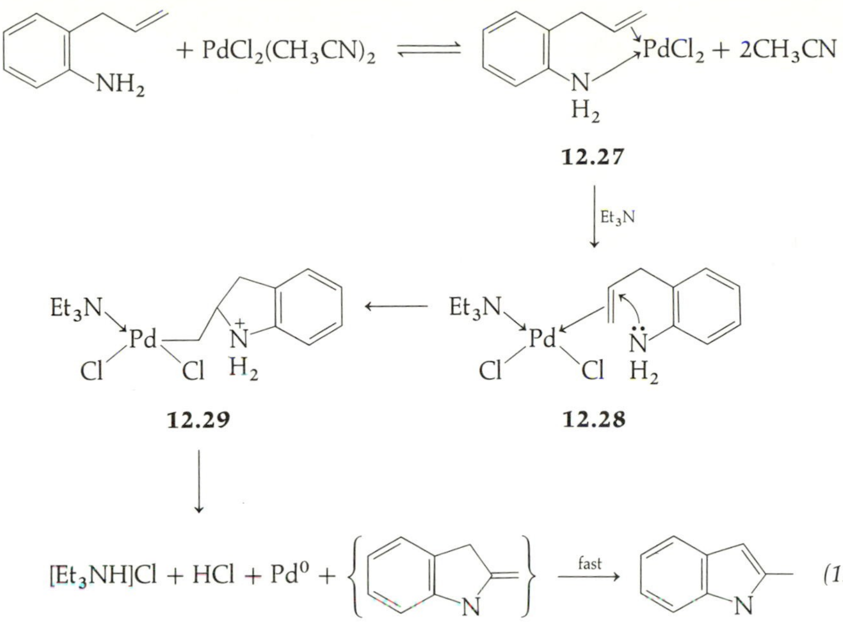

**12.27**

$\downarrow$ Et₃N

**12.29**           **12.28**

$\downarrow$

$[Et_3NH]Cl + HCl + Pd^0 +$ $\xrightarrow{\text{fast}}$    *(12.25)*

Addition of $Et_3N$ presumably displaces the more weakly bound aniline ligand. This amino group adds to the alkene ligand to give the primary alkyl complex **12.29**. The intermediate formation of a complex such as **12.29** was confirmed by trapping with CO gas in methanol solution to give indoline acetic acid ester, a "CO-insertion" product. Frequently alkyl intermediates can be trapped as acyl complexes upon treatment with carbon monoxide. Deprotonation of the ammonium center and reductive-elimination of HPdCl (which decomposes to HCl and $Pd^0$) gives a bicyclic product that spontaneously rearranges to the observed indole product. Substituents on the aromatic ring, the amino nitrogen atom, and the olefin of the allyl aniline reactant lead to substituted indole products.

### Nucleophilic Attack on Allylic Ligands

*Trihapto*-allylic complexes have been used extensively as intermediates in natural product syntheses. To date the metal of choice is palladium since these reactions can usually be made catalytic with respect to the metal. Nucleophilic attack on $\eta^3$-allylic ligands (eq. *11.82*) is important for organic synthesis because it permits considerable control over the regioselectivity and stereoselectivity of alkylations at the allylic positions of olefins. Allylic functionalization under normal organic reaction conditions is less useful because of the low selectivity of these processes.

    Equation sequence *(12.26)* shows the conversion of a monoterpene to a sesquiterpene via an ($\eta^3$-allylic)Pd intermediate. The allyl complex (**12.30**) is formed in the first step followed by phosphine-assisted nucleophilic attack by a sulfonyl ester enolate anion. This reaction gives a product containing a sesquiterpene carbon skeleton, which is converted to farnesol in three additional steps.

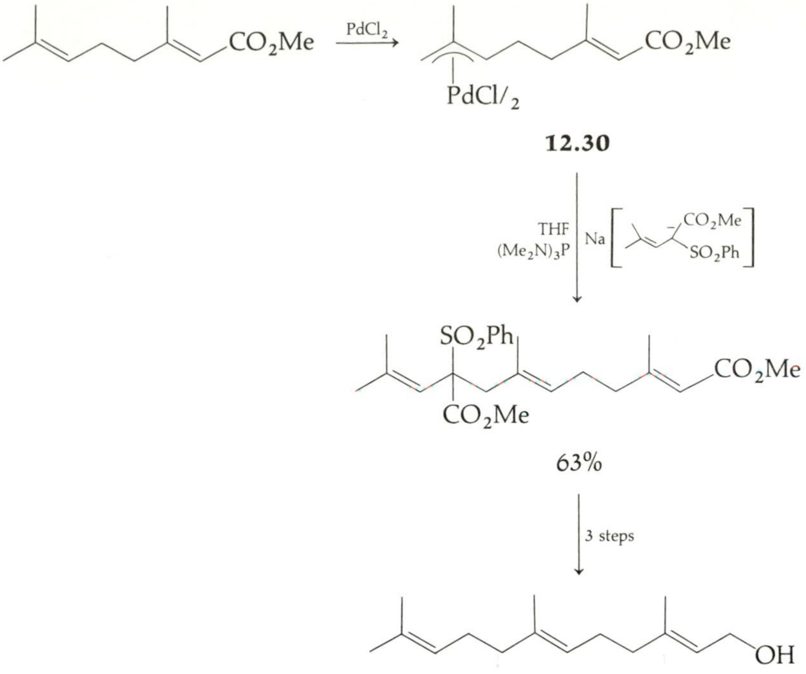

**12.30**

63%

3 steps

OH    *(12.26)*

farnesol

A similar reaction sequence, equation *12.27*, converts a farnesoate sesquiterpene into a diterpene, geranylgeraniol. Again, $(\eta^3$-allyl)Pd complex formation occurs at the terminal olefin giving complex **12.31**. Nucleophilic attack by the sulfonyl ester enolate anion at the least hindered allylic terminus adds a five-carbon unit to the sesquiterpene skeleton. Cleavage of the carboxy and sulfonyl ester groups and reduction of the ester terminus by DIBAL (diisobutylaluminum hydride) gives the diterpene.

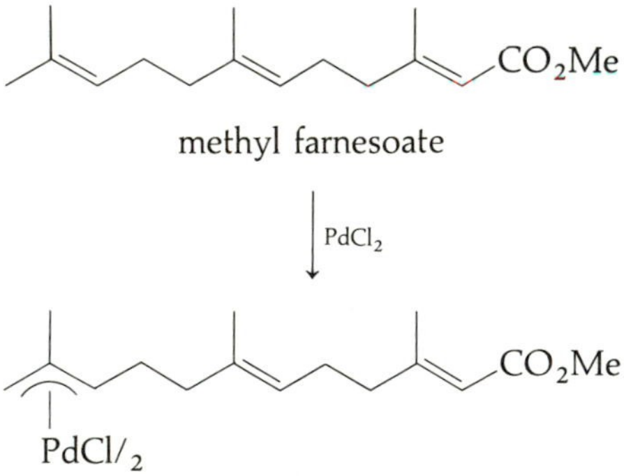

methyl farnesoate

PdCl$_2$

**12.31**

40%

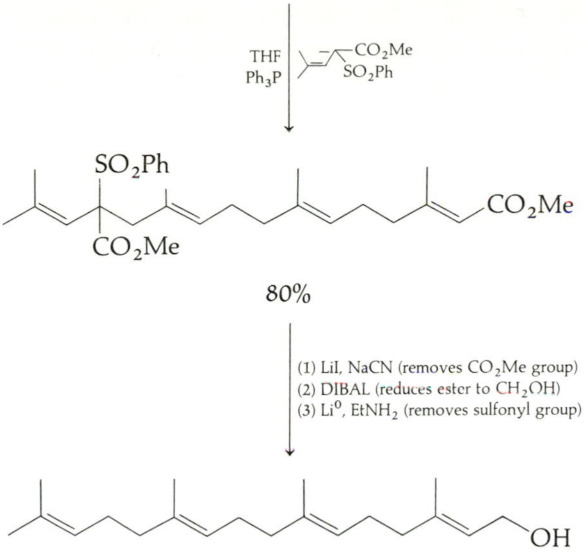

$$(12.27)$$

(E, E, E)-geranylgeraniol

Vitamin A alcohol is prepared from a simpler ($\eta^3$-allylic)Pd complex and a more complex enolate anion (eq. *12.28*). In this sequence, the $\eta^3$-allylic ligand provides a five-carbon fragment to the final carbon skeleton.

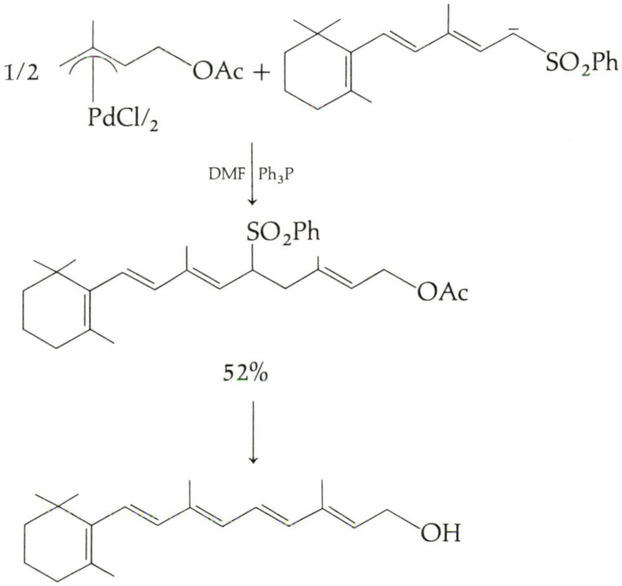

$$(12.28)$$

vitamin A

An example of a Pd(0)-catalyzed allylic nucleophilic alkylation, is shown in equation 12.29 (see also chap. 11). The overall transformation gives a precursor to the sex pheromone of the Monarch butterfly (X = H). An ($\eta^3$-allylic)Pd intermediate is again present in this synthesis.

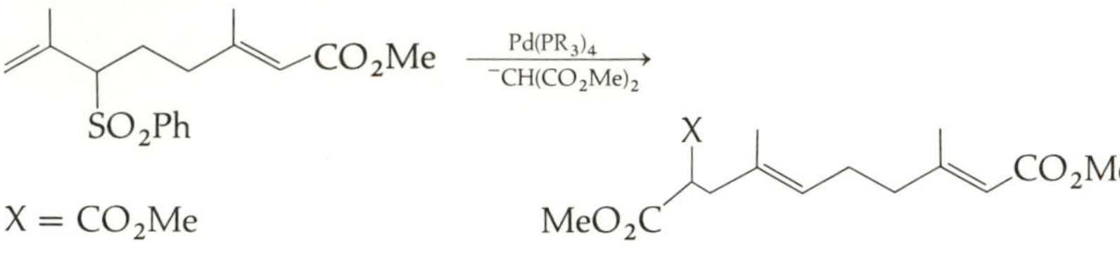

X = CO₂Me (12.29)

X = CO₂Me

Equations 12.30 and 12.31 show macrolide formation via an intramolecular nucleophilic attack on ($\eta^3$-allylic)Pd intermediates. *In situ* formation of an enolate

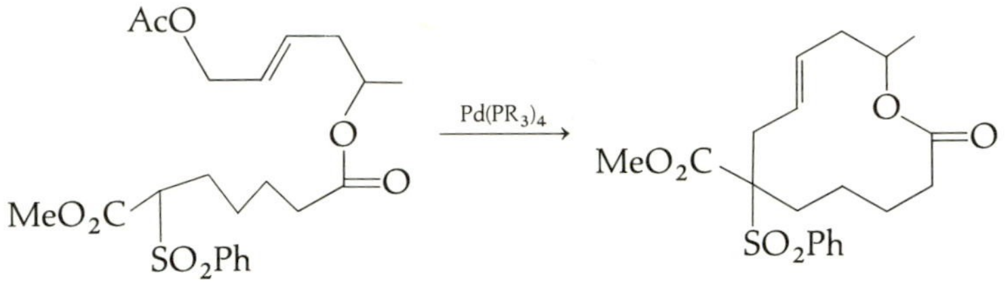

**12.32**, 78%                                    (12.30)

(±)-recifeiolide

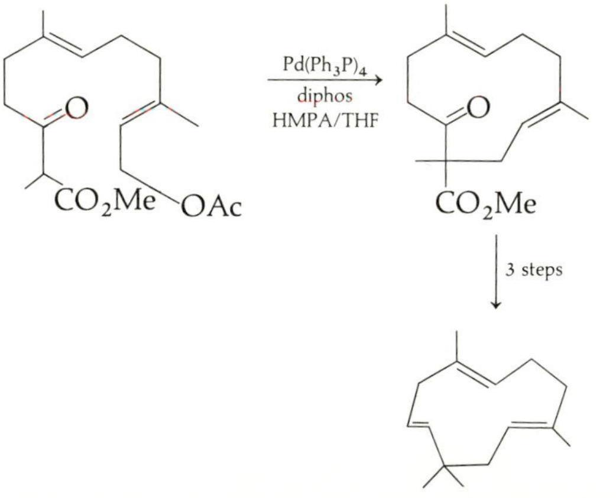

(12.31)

humulene (a sesquiterpene)

anion leads to the nucleophilic attack on the allylic ligands. The 12-membered ring product **12.32** is formed exclusively because the enolate anion attacks the less substituted terminus of the $\eta^3$-allyl intermediate regioselectively. This selectivity represents a significant advance in macrolide synthesis.

The applications of ($\eta^3$-allylic)Pd complexes as intermediates in organic syntheses shown in equations *12.26–12.31* demonstrate the usefulness of the chemical selectivity and regioselectivity of nucleophilic attack on these complexes. Equations *12.32* through *12.34* are applications that take advantage of the stereoselectivity of this reaction.

The steroid synthesis shown in equation *12.32* gives unusual stereochemistry at the carbon-20 center because of the inversion of configuration at the allylic carbon position upon *exo*-nucleophilic attack by the incoming nucleophile. This particular steroidal stereochemistry is found in some marine organisms. Complex **12.33** is formed regardless of the initial olefin stereochemistry. Phosphine-assisted nucleophilic attack by enolate anions selectively gives products with the C20 stereochemistry shown.

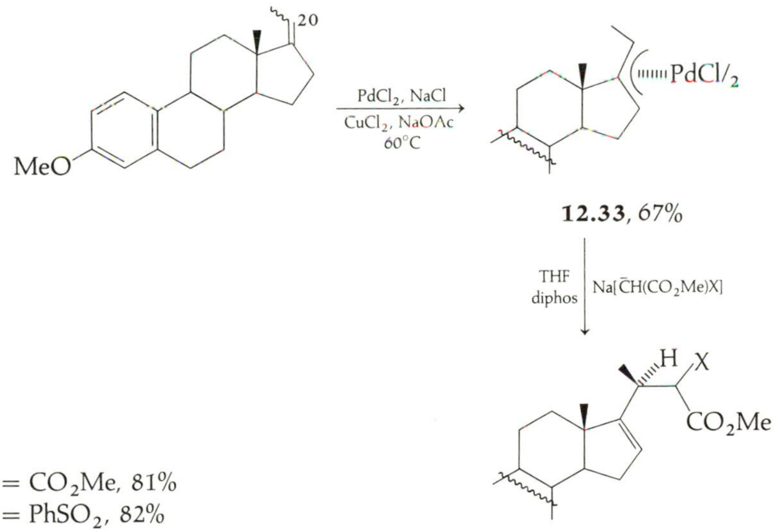

PdCl₂, NaCl
CuCl₂, NaOAc
60°C

**12.33**, 67%

THF
diphos

Na[C̄H(CO₂Me)X]     (12.32)

X = CO₂Me, 81%
X = PhSO₂, 82%

Equation *12.33* shows a general reaction sequence for effecting a 1,5-chirality transfer in acyclic systems by using an ($\eta^3$-allylic)Pd intermediate. Diastereomerically pure **12.36** is obtained if the following conditions are met: (1) Ionization from **12.34** to give **12.35** must occur preferentially from a single conformation; (2) carbon-carbon bond formation from nucleophilic attack on allylic intermediate **12.35** must occur stereoselectively at a rate faster than loss of stereochemistry due to other mechanisms; and (3) carbon-carbon bond formation must occur regioselectively at the allylic terminus farthest from the carboxylate group. Conversion of **12.34** to **12.36** constitutes an allylic rearrangement with chirality transfer.

A successful application of this method to the synthesis of the vitamin E side chain is shown in equation *12.34*. Compound **12.37** is obtained from D-glucose.

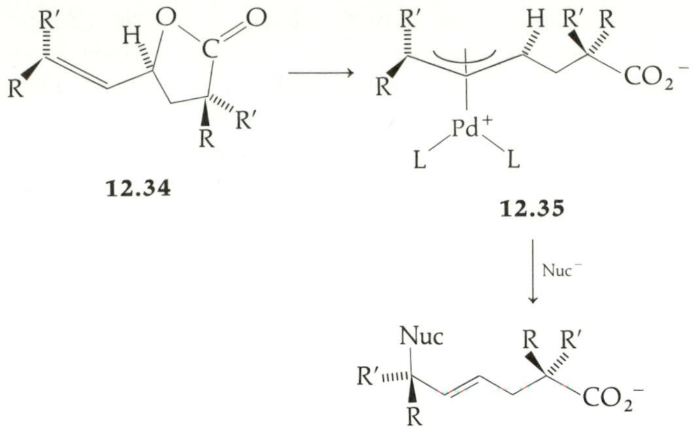

**12.34**

**12.35**

$$\text{(12.33)}$$

**12.36**

Formation of an ($\eta^3$-allylic)Pd intermediate induces conformational rigidity at the olefinic end of the acyclic intermediate. Rapid nucleophilic addition gives **12.38**, which was found to be 95% diastereomerically pure. After several more steps, the proper functionality is obtained for the vitamin E side chain. This reaction apparently proceeds as expected according to the general conditions required for equation *12.33*.

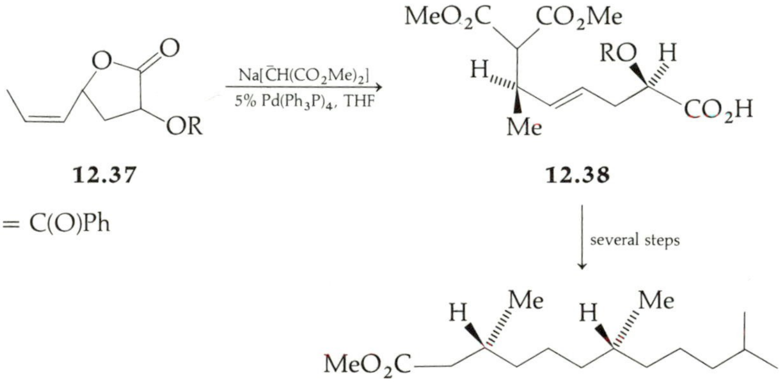

**12.37**

R = C(O)Ph

**12.38**

several steps

$$\text{(12.34)}$$

Recently, alkenyl zirconium and aluminum compounds have been found to react with ($\eta^3$-allylic)Pd complexes to give allylic alkylation products of stereochemistry *opposite* to that obtained by direct alkylation of ($\eta^3$-allylic)Pd complexes. These reagents provide a complementary route to allylic alkylations.

Direct nucleophilic attack on ($\eta^3$-allylic)Pd complexes occurs *exo* to the palladium atom to give inversion of configuration at the allylic carbon atom when the C—C bond forms. These zirconium and aluminum alkenyl reagents, on the contrary, transfer the alkenyl group to the allylic carbon from the *endo* face, affording retention of configuration at the allylic carbon atom. A mechanism to explain these results is shown in equation *12.35*.

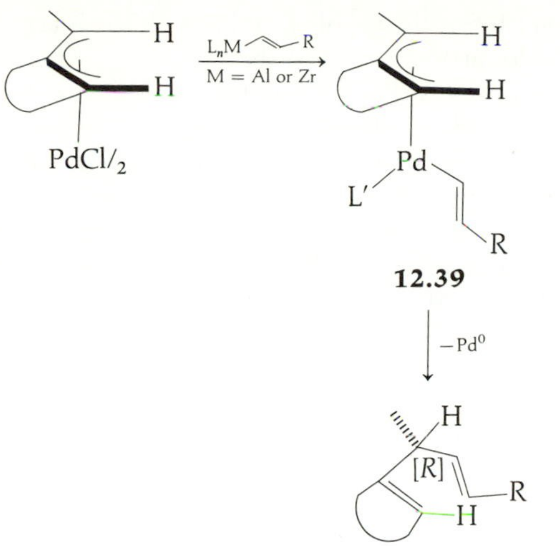

$$(12.35)$$

The ($\eta^3$-allylic)Pd complex reacts with the Al or Zr alkenyl reagents to give a transmetallated product **12.39** in which the alkenyl group has been transferred to the palladium atom. Subsequent addition of the alkenyl ligand to the allylic carbon atom from the *same face* as the Pd gives retention of configuration at the carbon. The Pd atom is removed by the usual reductive-elimination.

Equation *12.36* depicts preparation of 20(R)-cholestan-3-one using a Zr-alkenyl reagent. Alkenyl transfer to the steroidal nucleus occurs at both C20 and C16, which

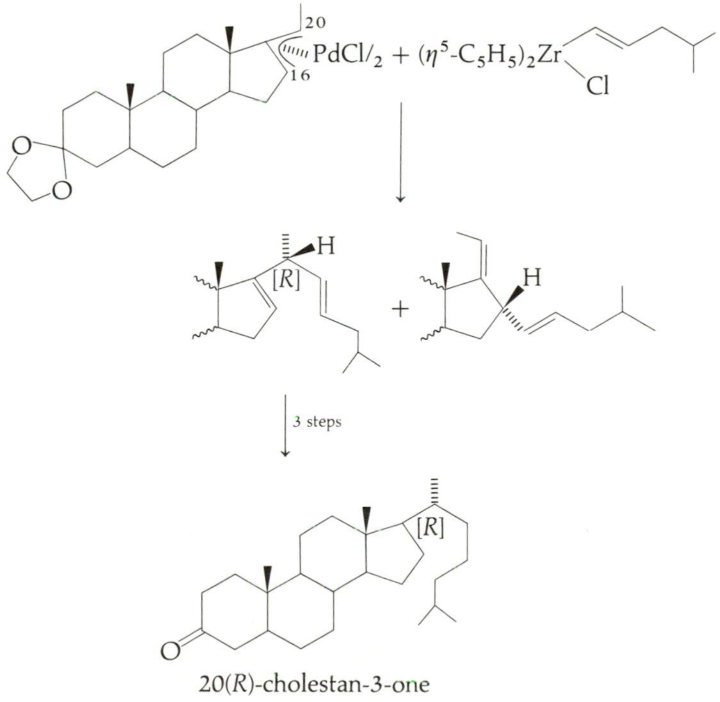

$$(12.36)$$

20(R)-cholestan-3-one

are terminal carbon sites of the ($\eta^3$-allylic)Pd ligand. However, the stereochemistry of the C—C bond formation is *opposite* that found from direct nucleophilic attack on the intermediate ($\eta^3$-allylic)Pd complex (i.e., from the *exo* face).

The Pd(0)-catalyzed addition shown in equation *12.37* occurs similarly. The allylic acetate is converted into an ($\eta^3$-allylic)Pd complex (**12.40**). Transfer of the alkenyl group onto the palladium atom precedes C—C bond formation. This alkenyl ligand attacks the allylic ligand from the *endo* face (relative to Pd) to form the C—C bond affording **12.41** with concomitant reductive-elimination of Pd(0). Compound **12.41** forms in good yield as a 90:10 mixture of stereoisomers.

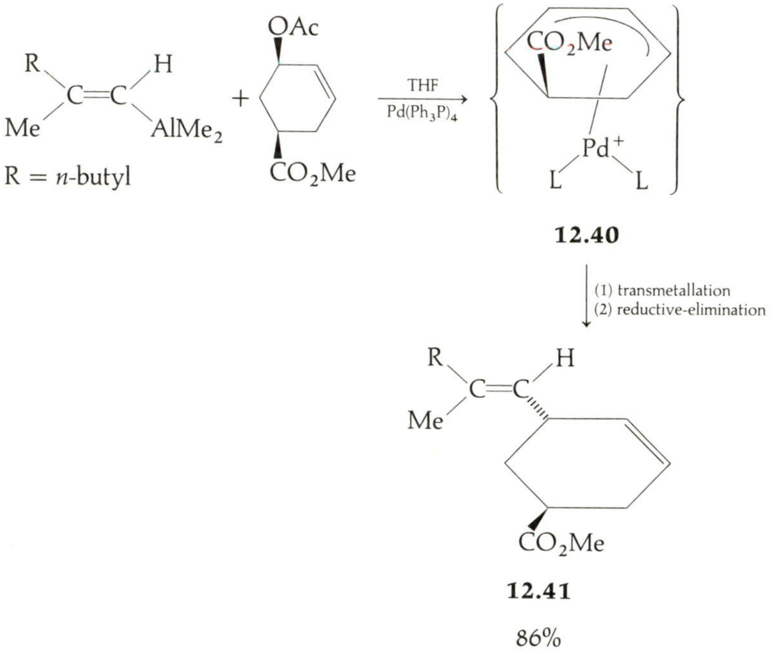

**12.40**

(1) transmetallation
(2) reductive-elimination

(12.37)

**12.41**

86%

## Nucleophilic Attack on $\eta^5$-Dienyl Ligands

Iron tricarbonyl complexes of $\eta^5$-dienyl ligands are also susceptible to nucleophilic attack. As with direct nucleophilic attack on $\eta^3$-allylic ligands, the external nucleophiles add to the $\eta^5$-dienyl ligands from the face opposite, or *exo*, to the Fe(CO)$_3$ moiety.

An application of this organometallic reagent is shown in equation *12.38* for the synthesis of zingiberene. Organocadmium reagents act as external nucleophiles. Removal of the Fe(CO)$_3$ moiety from the 1,3-diene is accomplished by Cu(II) oxidation.

Methoxy-substituted $\eta^5$-cyclohexyldienyl complexes like **12.42** and **12.43** undergo nucleophilic attack such that the incoming nucleophile adds preferentially to the dienyl terminus *para* to the methoxy group. This regiospecificity has extended the applications of these complexes to organic synthesis.

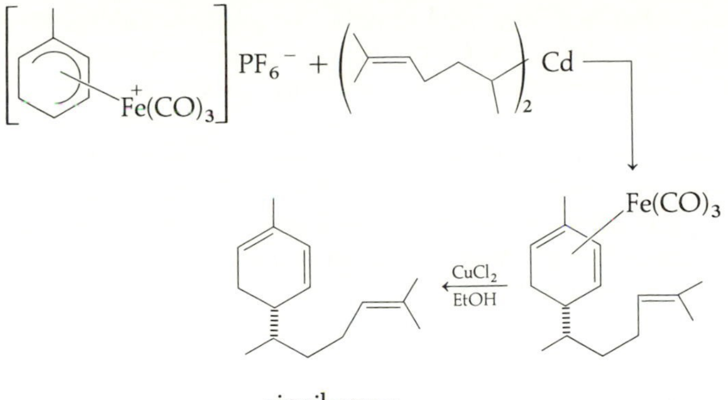

(12.38)

zingiberene

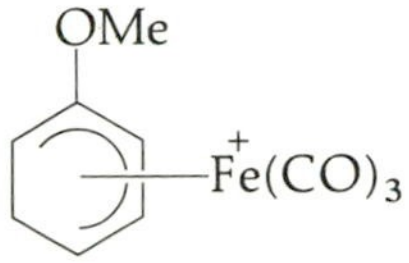

**12.42** 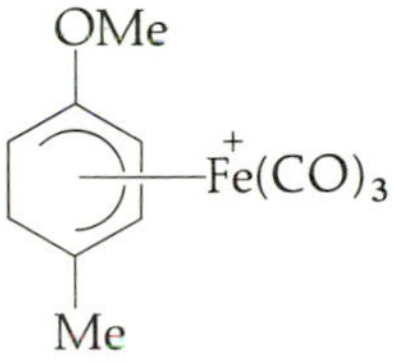 **12.43** 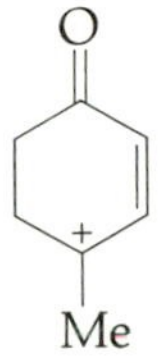 **12.44**

One application of interest is shown in equation *12.39*. When complex **12.43** is treated with an enolate anion R⁻ (or other stabilized carbanion, such as C̄N), nucleophilic addition occurs to give the diene complex **12.45.** Oxidative removal of the $Fe(CO)_3$ group with $Me_3NO$ followed by acid hydrolysis gives a cyclohexenone product. For this reason, complex **12.43** is considered to be a synthon equivalent to the cyclohexenone cation **12.44.**

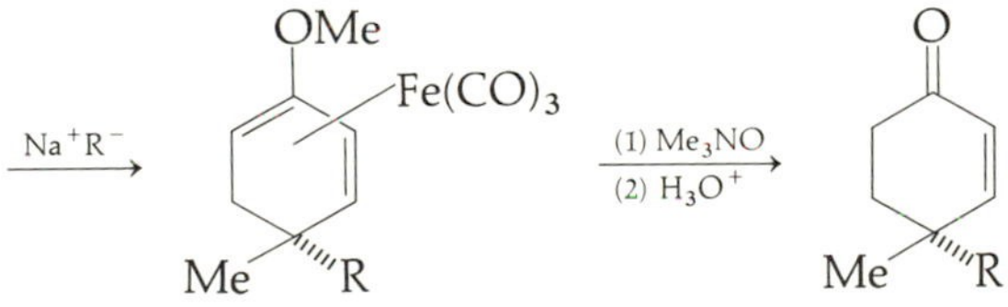

(12.39)

**12.45**

With selected enolate anions used in the initial reaction of equation *12.39*, this reaction has been applied to natural product synthesis. When R⁻ is **12.46**, the corresponding diene complex (**12.45**) is a ring A precursor to trichothecene derivatives.

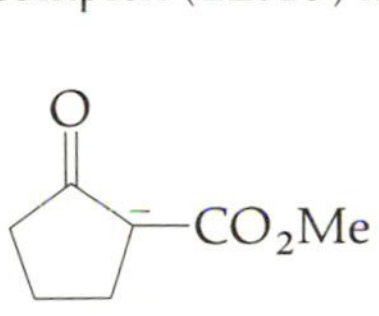

**12.46**

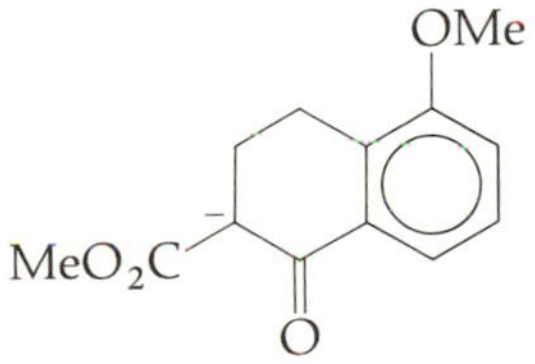

**12.47**

After several subsequent steps, 12,13-epoxy-14-methoxytrichothecene can be synthesized. Similarly, when R⁻ is **12.47**, steroid derivatives can be synthesized. However, several organic transformations are usually required after the organometallic step.

Spirocyclic compounds can also be synthesized by this method (eq. *12.40*). Complex **12.48** reacts with dimethyl malonate anion to give **12.49** with 80% selectivity for the dienyl terminus opposite to the methoxy-substituted position. Removal of the $Fe(CO)_3$ group and generation of an enolate anion gives the spirocyclic products **12.50** and **12.51** via Dieckmann condensations.

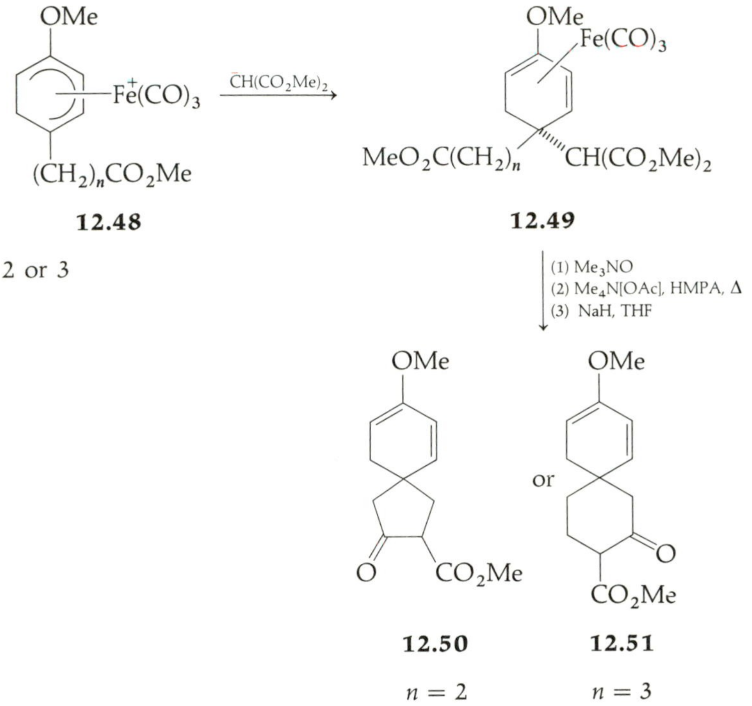

OMe

$\xrightarrow{CH(CO_2Me)_2}$ —Fe⁺(CO)₃

(CH₂)ₙCO₂Me

**12.48**

$n = 2 \text{ or } 3$

OMe / Fe(CO)₃

$MeO_2C(CH_2)_n \quad CH(CO_2Me)_2$

**12.49**

(1) Me₃NO
(2) Me₄N[OAc], HMPA, Δ
(3) NaH, THF

OMe

O    CO₂Me

**12.50**

$n = 2$

or

OMe

O

CO₂Me

**12.51**

$n = 3$

(12.40)

Formation of quaternary carbon centers is effected relatively easily compared to the normal "organic" methods, which are less effective when remote functionality is present. It appears that these $(\eta^5\text{-dienyl})Fe(CO)_3$ complexes offer a new route for the synthesis of precursors to highly functionalized spirocyclic compounds.

## Nucleophilic Attack on $\eta^6$-Arene Ligands

Nucleophilic attack on a *hexahapto*-arene complex can be used to effect halide or hydrogen exchange by the incoming nucleophile (see eqs. *11.100* and *11.101*). Examples of intramolecular carbanion attack on arene ligands to give fused bicyclic or spirocyclic compounds are shown in equations *12.41* and *12.42*.

The $(\eta^6\text{-arene})Cr(CO)_3$ complexes are treated with lithium diisopropylamide (LDA) to generate a carbanion alpha to the cyano substituent. Intramolecular nucleophilic

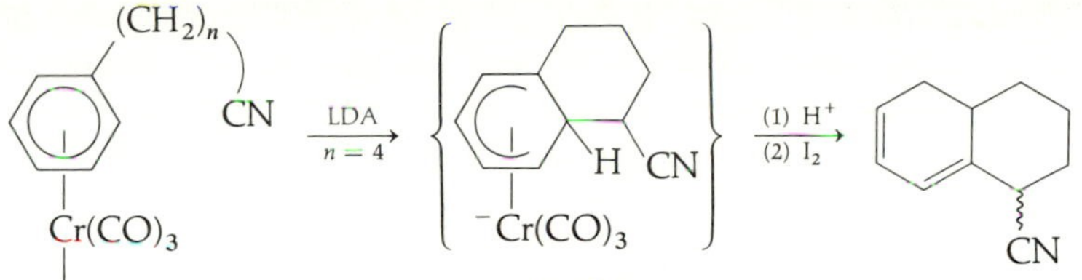

(12.41)

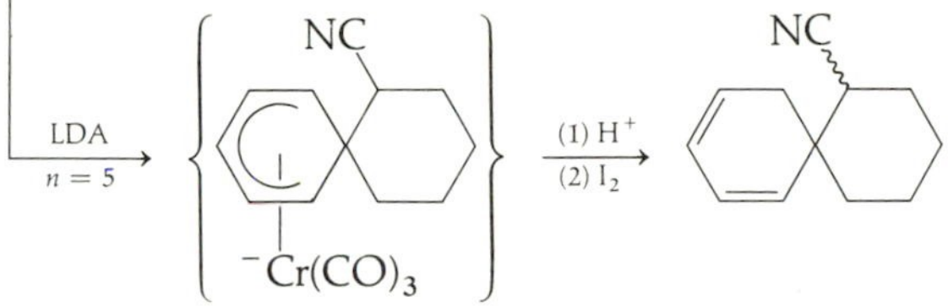

**12.52**

(12.42)

**12.53**

attack on the arene ligand presumably generates intermediates such as **12.52** and **12.53**. Protonation and then oxidative removal of the $Cr(CO)_3$ with iodine gives the organic products as isomeric mixtures.

## METAL STABILIZATION OF CARBANIONIC LIGANDS

Some complexes react with anionic reagents as Brønsted acids rather than as electrophiles. Frequently, the resulting anion is readily formed because the metal moiety can stabilize the negative charge.

For example, the methyl substituents of Group VI (alkoxy)(methyl)carbene complexes are relatively acidic because of metal stabilization of the resulting anion, as shown in equation *12.43*.

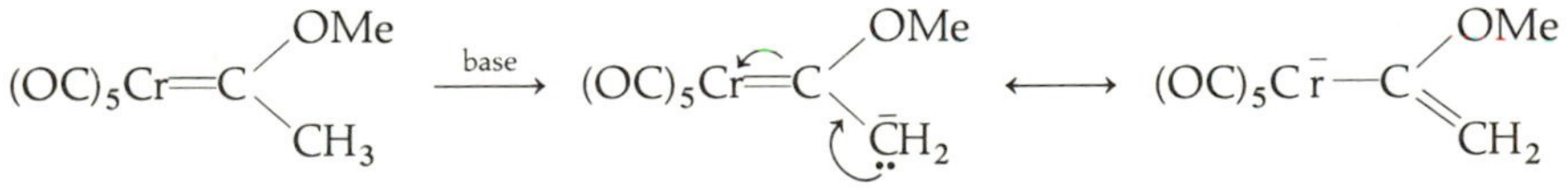

**12.54**                          **12.55**  *(12.43)*

Deprotonation of the methyl group forms the carbanion **12.54**, which is stabilized by resonance structure **12.55**. The carbene ligand is formally converted into an $\eta^1$-alkenyl ligand. The carbene methyl group is about as acidic as the OH group of *p*-cyanophenol. Since the $Cr(CO)_5$ moiety in these complexes is electronically equivalent to a carbonyl oxygen atom (see eq. *11.56*), **12.54** can be regarded as a metalla-enolate anion of an ester.

This facile enolate formation has been applied to the synthesis of α-methylene-γ-butyrolactone compounds, which are of interest in cancer research. In equation *12.44*, enolization of the carbenoid methyl group is effected by *n*-butyllithium. Alkylation of the carbanion with ethylene oxide displaces methoxide ion to give the new carbene complex, **12.56**. By repeating the $\alpha$-enolization, alkylation with chloromethyl methyl ether gives the carbene complex **12.57**. Alumina-assisted extrusion of methanol from **12.57** affords complex **12.58**, and oxidative removal of the Cr complex with Ce(IV) ion gives $\alpha$-methylene-$\gamma$-butyrolactone in good yield.

$$(OC)_5Cr=C\overset{OMe}{\underset{CH_3}{\diagdown}} \quad \xrightarrow[\text{(2) } CH_2-CH_2]{\text{(1) } n\text{-butyl Li}} \quad (OC)_5Cr=C$$

**12.56**

(1) $n$-butyl Li
(2) $ClCH_2OCH_3$

$$O=C \quad \xleftarrow{Ce^{4+}} \quad (OC)_5Cr=C \quad \xleftarrow{Al_2O_3} \quad (OC)_5Cr=C \qquad (12.44)$$

$$CH_2 \qquad\qquad\qquad CH_2 \qquad\qquad\qquad CH_2OMe$$

80%                                    **12.58**                                    **12.57**

                                             64%                                        40%

In a similar reaction, (metalla-$\beta$-diketonato)$BF_2$ complexes such as **12.59** (which can be prepared from metalla-$\beta$-diketones, **11.46**, and $BF_3$) react with Brønsted bases. Equation 12.45 depicts the deprotonation of the methyl substituent of complex **12.60** (a single valence bond representation of the delocalized complex **12.59**) by KH to give the $\alpha$-enolate anion **12.61**. This carbanion can also be represented as complex **12.62** where the rhenium atom has formal carbene and $\eta^1$-alkenyl ligands.

The observed product is an $\eta^2$-allylic complex, which is shown as **12.63** in an all-sigma bonding representation. Apparently, complex **12.62** spontaneously rearranges to **12.63** as shown. Interestingly, this rearrangement is symmetry allowed if complex **12.62** is regarded as a 2-metalla-*transoid*-1,3-butadiene and if complex **12.63** is regarded as a 1-metalla-bicyclo[1.1.0] butane.

Conversion of complex **12.60** to **12.63** represents an interligand C—C bond formation reaction without either a change in the formal oxidation state of the metal or a reductive-elimination reaction. The acyl carbons of **12.60** undergo a transannular coupling across the six-membered chelate ring. The methyl substituent of **12.60**, or related molecules, is sufficiently acidic that even pyridine causes deprotonation.

Carbanions alpha to the aromatic ring of an $\eta^6$-arene ligand are also metal stabilized. Enolate anions of this type are stable enough to be produced in aqueous media and then alkylated via phase-transfer conditions, as shown in equation 12.46. The relatively high electronegativity of the $Cr(CO)_3$ and ester groups of complex **12.64** stabilize enolate anion formation in the basic aqueous layer. Tetraalkyl-ammonium ion phase transfer of these enolate anions into the benzene layer permits

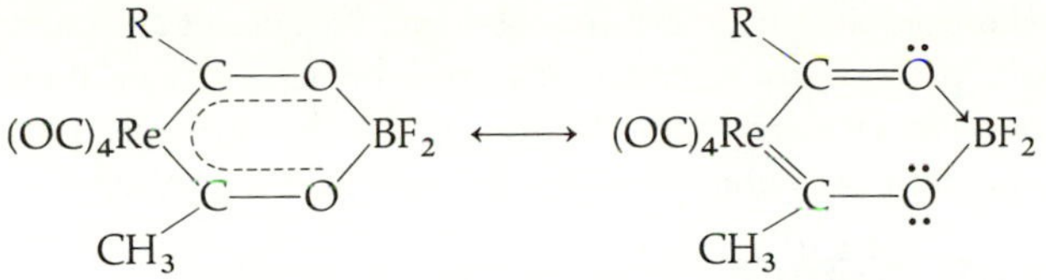

**12.59**

R = CH₃ or *i*-Pr

**12.60**

(one Lewis structure)

KH | −H₂↑

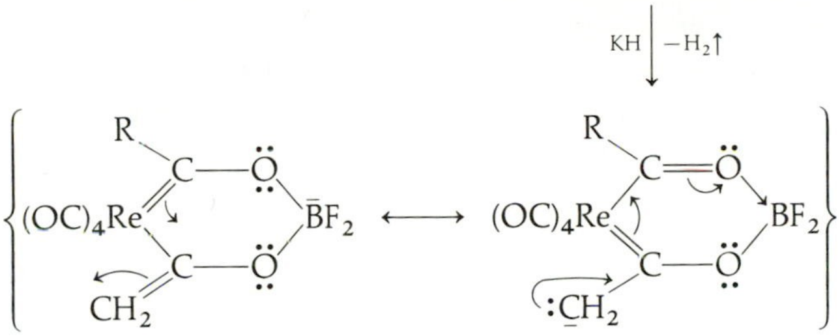

**12.62**

**12.61**

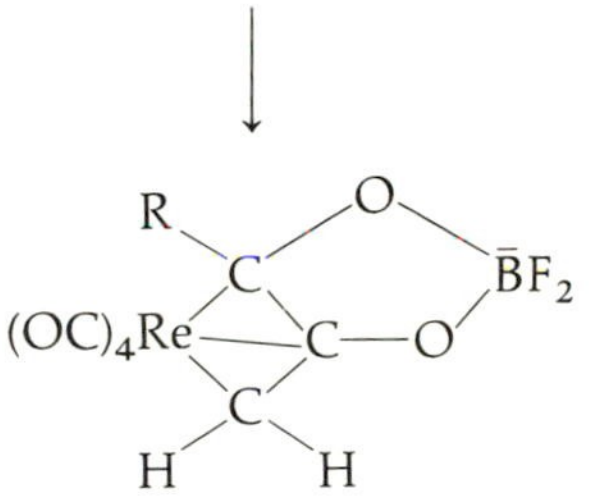

**12.63**

(12.45)

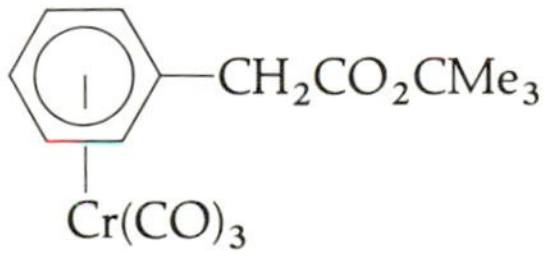

**12.64**

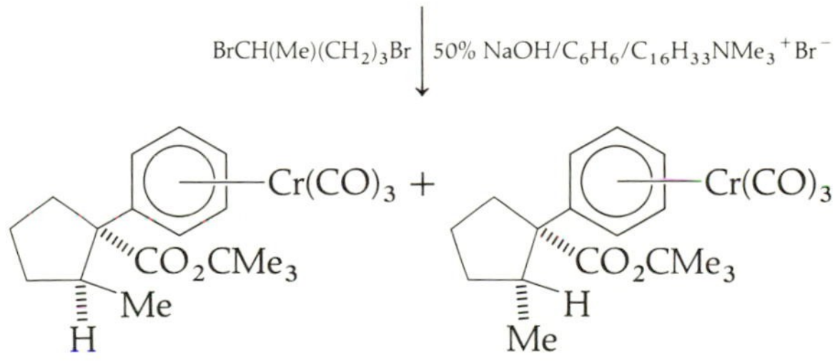

BrCH(Me)(CH₂)₃Br | 50% NaOH/C₆H₆/C₁₆H₃₃NMe₃⁺Br⁻

**12.65** + **12.66** (12.46)

72 : 28

45% total yield

alkylation of these anions by 1,4-dibromopentane. Products **12.65** and **12.66** reveal that both enolic protons are removed and that the second alkylation occurs intra-molecularly to give cyclopentane substituents on the aromatic ring. An unequal diastereomeric mixture is obtained from the ring-closure reaction.

## ELECTROPHILIC ATTACK ON $\eta^3$-ALLYL LIGANDS

Like nucleophilic attack on *trihapto*-allyl ligands, *electrophilic alkylation* of these ligands (eq. *11.20*) has also found application in natural product synthesis. Examples of the introduction of five-carbon isoprenyl units into substrate complexes are shown in equations *12.47* and *12.48*. Electrophilic alkylation of complex **12.67** at the less sterically hindered allylic terminus by using bromoisoprenyl compounds affords geranyl ethyl ether and geranyl acetate directly:

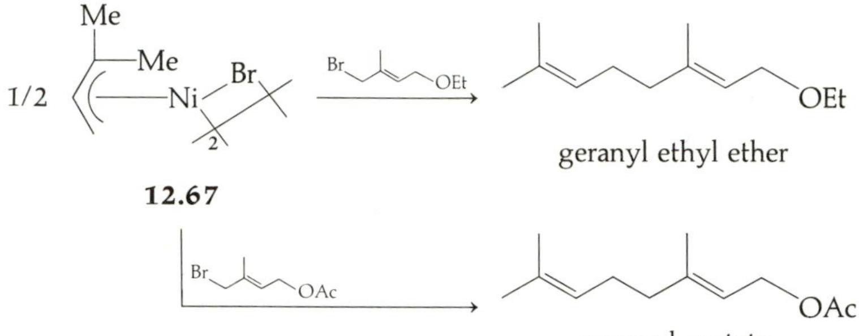

geranyl ethyl ether

**12.67**

geranyl acetate          (12.47)

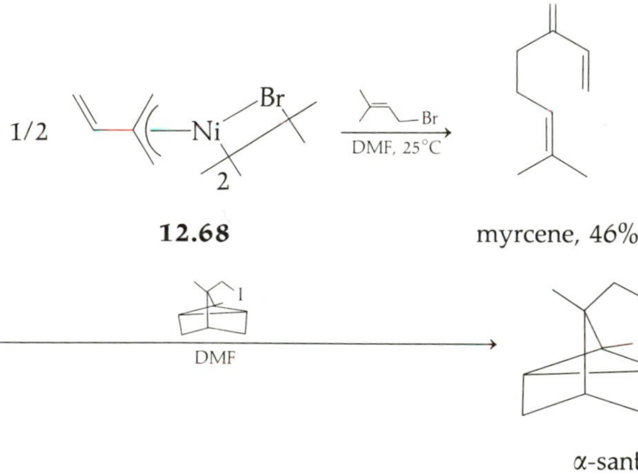

**12.68**          myrcene, 46%          (12.48)

**12.67**

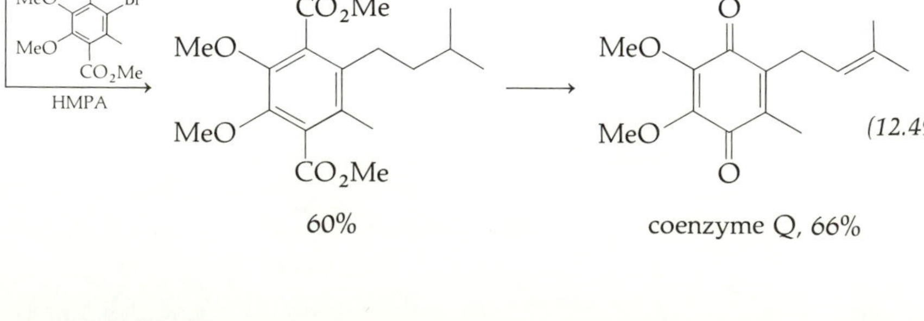

α-santalene

60%          coenzyme Q, 66%          (12.49)

Myrcene is formed in good yield by alkylation of the $\eta^3$-allyl complex **12.68** (eq. *12.48*). These reactions demonstrate diene syntheses of monoterpene products.

Two other natural products synthesized by electrophilic alkylation of the allyl ligand of complex **12.67** are α-santalene and coenzyme Q (eq. *12.49*). These reactions show alkyl iodide and aryl bromide alkylation of the $\eta^3$-allylic ligand at the less sterically hindered allylic terminus. Santalene, which is a sesquiterpene, is prepared directly; coenzyme Q is formed after conversion of the terephthalate ester to the quinone.

## ALKYNE INSERTION REACTIONS

Insertion of unsaturated molecules, particularly alkenes and alkynes, is a very important method of incorporating small substrate molecules into an organometallic complex. Alkene insertion, or the equivalent 1,2-addition of metal-ligand bonds across an olefinic bond, is a frequently proposed step in several large-scale processes that use organometallic compounds as catalysts. Recently, alkyne insertion reactions have also been applied to the synthesis of several types of organic molecules.

Naphthoquinones can be prepared from alkynes as shown in equation *12.50*. The diacyl complexes (**12.69**) are prepared by oxidative-addition of the appropriate metal complexes to benzocyclobutenedione.

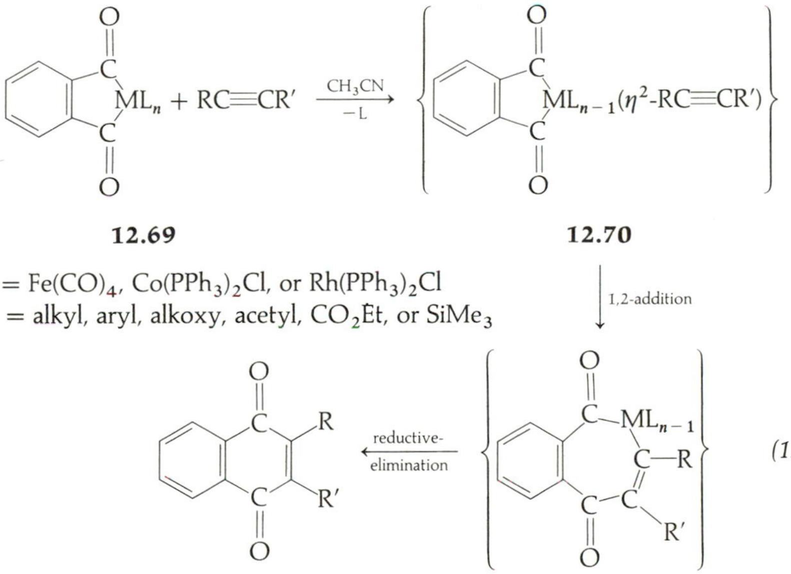

**12.69**                    **12.70**

$ML_n$ = Fe(CO)$_4$, Co(PPh$_3$)$_2$Cl, or Rh(PPh$_3$)$_2$Cl
R, R′ = alkyl, aryl, alkoxy, acetyl, CO$_2$Et, or SiMe$_3$

1,2-addition

naphthoquinones                    **12.71**

(12.50)

Upon treatment with alkynes, these complexes presumably coordinate to an alkyne molecule affording an intermediate like **12.70**. A 1,2-addition to the alkyne ligand gives a metallacyclic intermediate (**12.71**). Subsequent reductive-elimination of the metal moiety generates the naphthoquinone products.

In this series, the iron complex reacts with alkynes directly to form naphtho-quinones; the cobalt and rhodium reactants require silver-ion assistance. Presumably, $AgBF_4$ initially removes the chloride ligand to generate a vacant coordination site for alkyne coordination. The intermediates **12.70** and **12.71** are probably cationic complexes in these cases. Yields of the napthoquinones range from 22% to 100% depending on the choice of metal and of substituents R and R'.

Indan or tetralin ring systems are prepared from 1,6-heptadiyne or 1,7-octadiyne and substituted monoacetylenes (eq. 12.51). This cooligomerization reaction is cata-lyzed by $(\eta^5\text{-}C_5H_5)Co(CO)_2$.

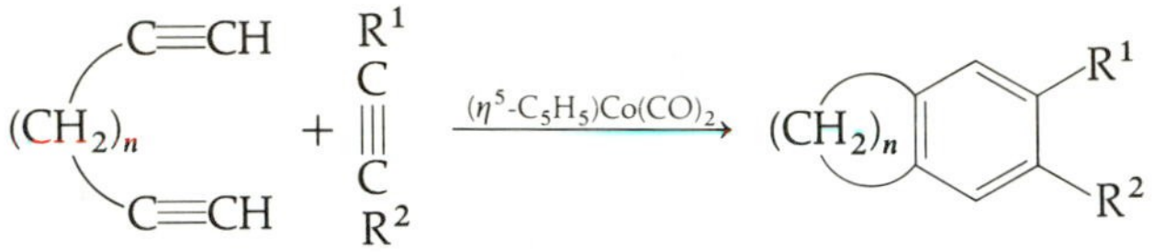

$$(12.51)$$

$n = 3$ or $4$; $R^1, R^2 = CO_2Me$, Ph, $Me_3Si$, and others

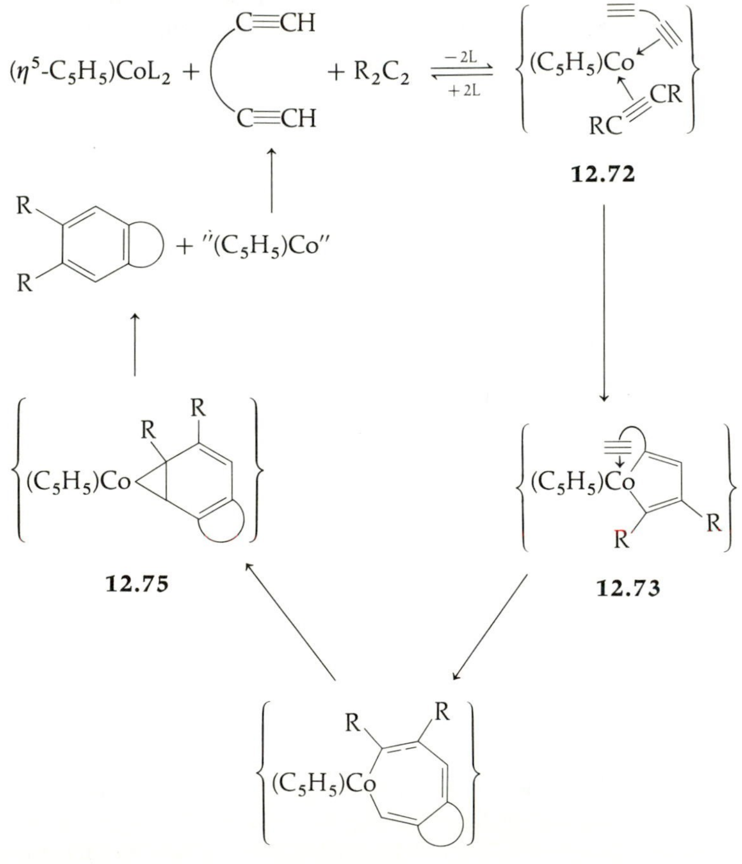

$$(12.52)$$

**12.74**

One mechanism proposed for this catalytic reaction is shown in equation *12.52*. The cobalt catalyst sequentially dissociates its two CO ligands and coordinates to the monoalkyne bond and one of the alkyne bonds of the diyne giving intermediate **12.72.**

This 18-electron intermediate undergoes alkyne coupling to give a cobaltacyclo-pentadiene intermediate (**12.73**). The coupling requires a formal loss of a two-electron donor, and subsequent (or concomitant) coordination to the distal alkyne bond may take place. Insertion of the latter gives the metallacyclic intermediate **12.74.** Coupling of the two vinyl fragments results in the last intermediate (**12.75**).

Since the cobalt atoms in intermediates **12.73−12.75** have a formal oxidation state two units higher than the Co atom of **12.72** (or the $C_5H_5CoL_2$ reactant), the final step of this sequence is a formal reductive-elimination. The bicyclic product is produced along with a coordinatively unsaturated "$(\eta^5\text{-}C_5H_5)Co$" moiety. This cobalt species acts as the catalytically active complex. It can coordinate with a second equivalent of alkyne reagents to give **12.72** and thereby repeat the reaction sequence.

When the cooligomerization shown in equation *12.51* is attempted with 1,5-hexadiynes, benzocyclobutenes are formed (eq. *12.53*).

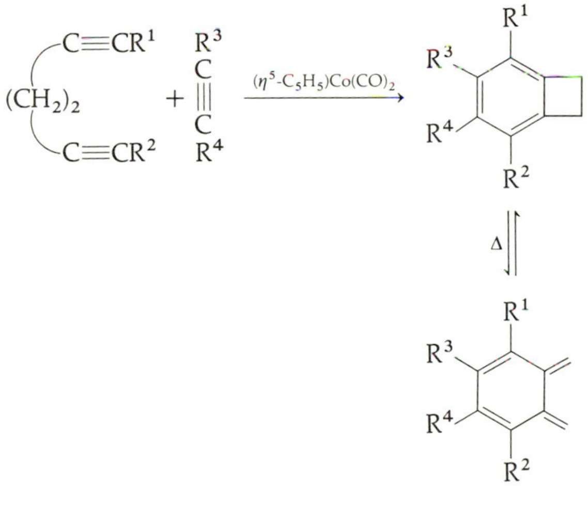

$$o\text{-xylylenes} \tag{12.53}$$

Benzocyclobutenes are known to open thermally to give *o*-xylylenes, which are reactive enophiles in Diels−Alder reactions. This possibility extends the applications of the cobalt-catalyzed cyclization reaction.

An example of coupling the cobalt-catalyzed cooligomerization with a subse-quent intramolecular Diels−Alder addition is shown in equation *12.54*. Reaction of a substituted 1,5-hexadiyne with bistrimethylsilylacetylene (BTMSA) in the presence of $(\eta^5\text{-}C_5H_5)Co(CO)_2$ gives the benzocyclobutene **12.76**. Thermal opening to the *o*-xylylene initiates an intramolecular Diels−Alder addition to give an estratrienone derivative (**12.77**).

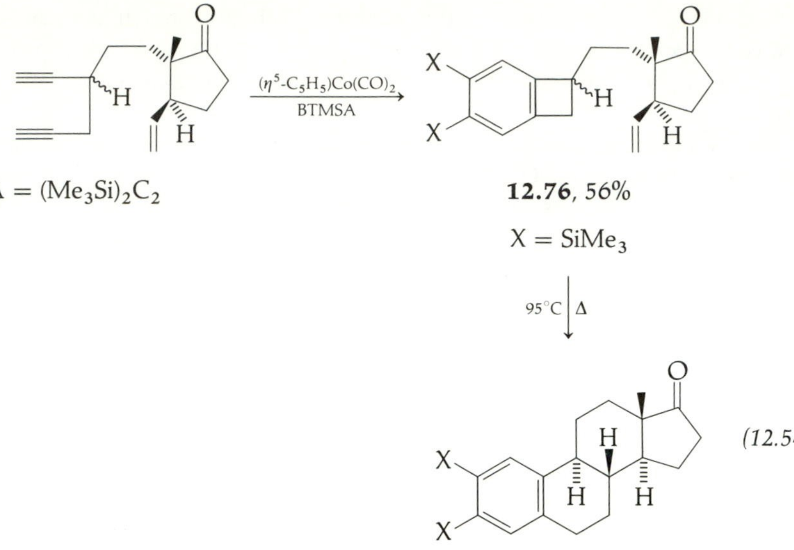

BTMSA = $(Me_3Si)_2C_2$

**12.76**, 56%

X = SiMe$_3$

(12.54)

**12.77**, 18%

Similarly, a synthesis of d,l-estrone using this procedure has been reported. The reaction sequence is shown in equation 12.55.

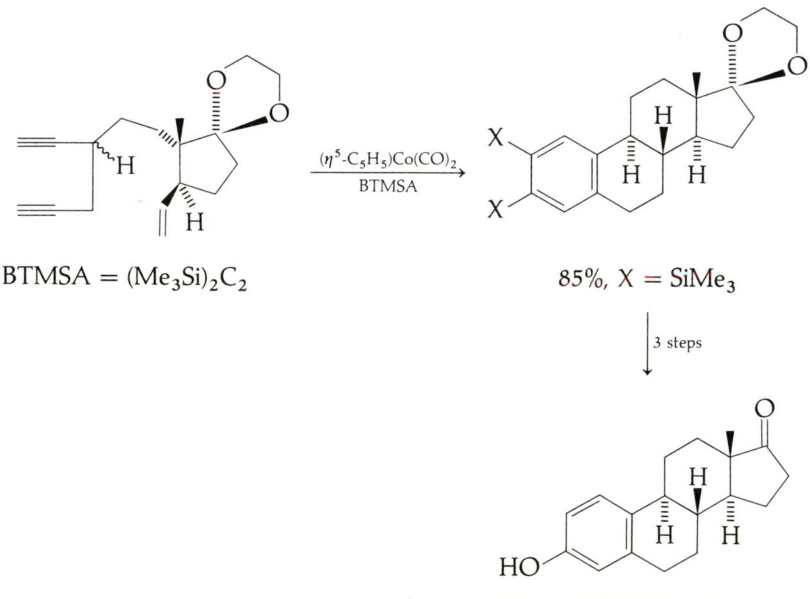

BTMSA = $(Me_3Si)_2C_2$

85%, X = SiMe$_3$

3 steps

(12.55)

d,l-estrone

## REACTIONS INVOLVING INITIAL OXIDATIVE-ADDITION

Several applications of organometallic chemistry to organic synthesis require a formal oxidative-addition of an organic reactant to a metal complex as a crucial first step. Reactions that are either stoichiometric or catalytic in the metal complex are known.

Palladium catalyzes the arylation, vinylation, or benzylation of alkenes. An example of arylation is shown in equation *12.56*. The reactants include an aryl halide, an alkene, and an amine base.

$$p\text{-BrC}_6\text{H}_4\text{CO}_2\text{Me} + \text{CH}_2{=}\text{CHCO}_2\text{Me} + \text{R}_3\text{N} \xrightarrow[\substack{\text{PR}'_3\ (2\ \text{mol-}\%)\\100°\text{C, 7 hrs}}]{\text{Pd(OAc)}_2(1\ \text{mol-}\%)}$$

R = Et or *n*-butyl
R′ = Ph or *p*-tolyl

$$\begin{array}{c}\text{CO}_2\text{Me}\\ \diagup\!\!\diagup\\ \text{C}_6\text{H}_4\text{CO}_2\text{Me-}p\end{array} + [\text{R}_3\text{NH}]\text{Br} \quad (12.56)$$

81%

The products formed represent an overall extrusion of HX from the aryl halide and the olefin to give an ammonium salt and a styrene derivative.

A general reaction mechanism for this catalytic process is shown in equation scheme *12.57*. The initial Pd(II) is reduced to a Pd(0)-phosphine complex by excess olefin in the presence of a phosphine ligand. Oxidative-addition of the aryl bromide to the Pd atom gives a four-coordinate Pd(II) complex (**12.78**). Coordination to the olefin and subsequent 1,2-addition of the Pd-aryl bond across the olefin affords the alkyl complex (**12.79**). Beta-hydrogen elimination gives the arylated olefin product and a palladium-hydride complex. Reductive-elimination of HBr in the presence of base affords the ammonium halide product and regenerates the Pd(0)-phosphine catalyst.

$$\text{PdL}_n + \text{Ar—Br} \xrightarrow{-(n-2)\text{L}} \begin{array}{c}\text{L}\diagdown\quad\diagup\text{Ar}\\ \quad\text{Pd}\\ \text{Br}\diagup\quad\diagdown\text{L}\end{array}$$

L = PR₃

**12.78**

$$\Big\downarrow \text{CH}_2{=}\text{CH(CO}_2\text{Me)}$$

$$\begin{array}{c}\text{CO}_2\text{Me}\\ \diagup\!\!\diagup\\ \text{Ar}\diagup\end{array} \longleftarrow \begin{array}{c}\text{L}\diagdown\quad\diagup\text{CH(CO}_2\text{Me)}\\ \quad\text{Pd}\\ \text{Br}\diagup\quad\diagdown\text{L}\quad\diagdown\text{CH}_2\text{Ar}\end{array}$$

+ **12.79**

$$\{\text{HPdL}_2\text{Br}\} \xrightarrow{\text{R}_3'\text{N}} \text{PdL}_2(\text{recycled}) + [\text{R}_3'\text{NH}]\text{Br} \quad (12.57)$$

Stereochemical results are consistent with a *syn*-addition of the Pd-aryl bond to the alkene and with a *syn*-elimination of the Pd-hydride complex. The aryl moiety adds to the less substituted olefinic carbon atom. This addition is primarily sterically controlled. The organohalide that oxidatively adds to the palladium complex should not contain any $\beta$-hydrogen substituents on $sp^3$-carbon centers; otherwise, alkenes are produced by $\beta$-hydrogen elimination before alkene arylation can occur.

Transition metal complexes can be used as stoichiometric reagents to effect aryl-aryl coupling. These complexes form carbon-carbon bonds in a reaction that may be considered as the low-temperature analogue to the Ullman reaction. Their use in the synthesis of alnusone is shown in equation *12.58*.

R = Me or CH$_2$OMe

$$\xrightarrow{\text{Ni(PPh}_3)_4 \text{ or} \atop (\eta^4\text{-1,5-COD)}_2\text{Ni}}$$

$$\xrightarrow[R = CH_2OMe]{H^+} \quad \text{alnusone} \quad (12.58)$$
$$72\%$$

52%, R = Me
46%, R = CH$_2$OMe

One postulated mechanism involves oxidative-addition of an aryl-iodine bond to a Ni(0) complex to give a Ni(II)-aryl complex as an intermediate. Displacement of the second iodine atom occurs upon aryl coupling with the concomitant formation of NiI$_2$. Alnusone is obtained by removing the alcoholic protecting groups.

Two applications of allylic ligand coupling to natural product syntheses are shown in equations *12.59* and *12.60*. In both cases, a *bis*-($\eta^3$-allyl)Ni complex, prepared by oxidative-addition, is thermally decomposed to form a new C—C bond. A 1,5-diene fragment is generated from the reductive-elimination step in which the allylic ligands are coupled. In equation *12.59*, the initial organic product is treated with Ph$_2$S and light to isomerize one of the olefinic bonds to the *trans* isomer. The final product is humulene. In equation *12.60*, two isomeric products are obtained from allylic coupling; a third product also obtained results apparently from "CO-insertion" into one of the $\eta^1$-allyl to metal bonds before the reductive-elimination step. Hydrogenation of this product gives the macrolide fragrance, muscone.

There are many examples of metal-promoted rearrangements of highly strained organic molecules involving oxidative-addition to a highly strained C—C single bond. Subsequent rearrangement gives a more stable structure. The most common highly strained organic molecules are those with strained ring systems. A recent application of this type of reaction in a synthesis relevant to natural product compounds is shown in equation *12.61*.

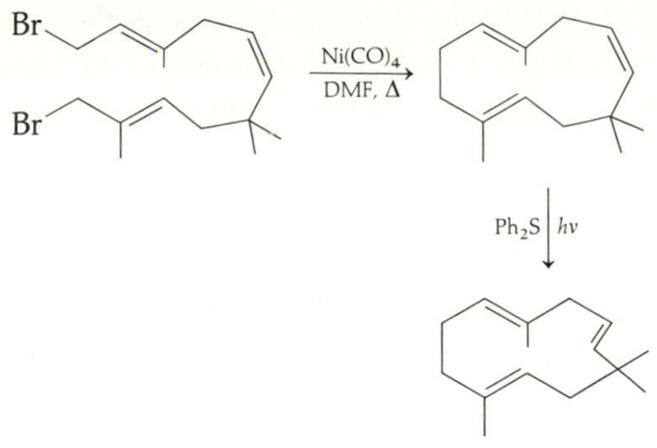

humulene

(12.59)

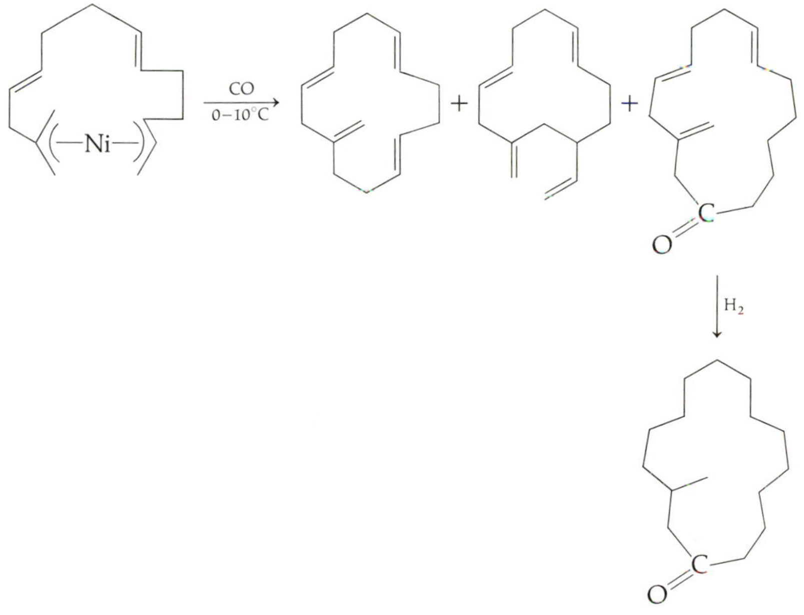

muscone

(12.60)

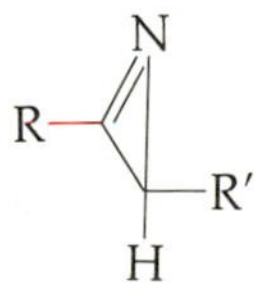

+ CO $\xrightarrow{Pd^0(PPh_3)_4}$ bicyclic β-lactams (see **12.86**)

(12.61)

**12.80** $\xrightarrow{Pd^0(DBA)}$ vinyl isocyanates (see **12.83**)

DBA = dibenzylideneacetone, PhCH=CHC(O)CH=CHPh

Palladium(0) complexes act as catalysts in the carbonylation of azirines like **12.80** to give bicyclic β-lactams when the ancillary ligands are Ph₃P. However, when the ancillary ligand is dibenzylideneacetone (DBA), then vinyl isocyanates are produced.

The reason for this remarkable product dependence on the identity of the ancillary ligands of these catalysts is not obvious. Presumably, a common intermediate is involved in both reaction cycles. A proposed mechanism is shown in equation scheme 12.62 for phenyl azirine.

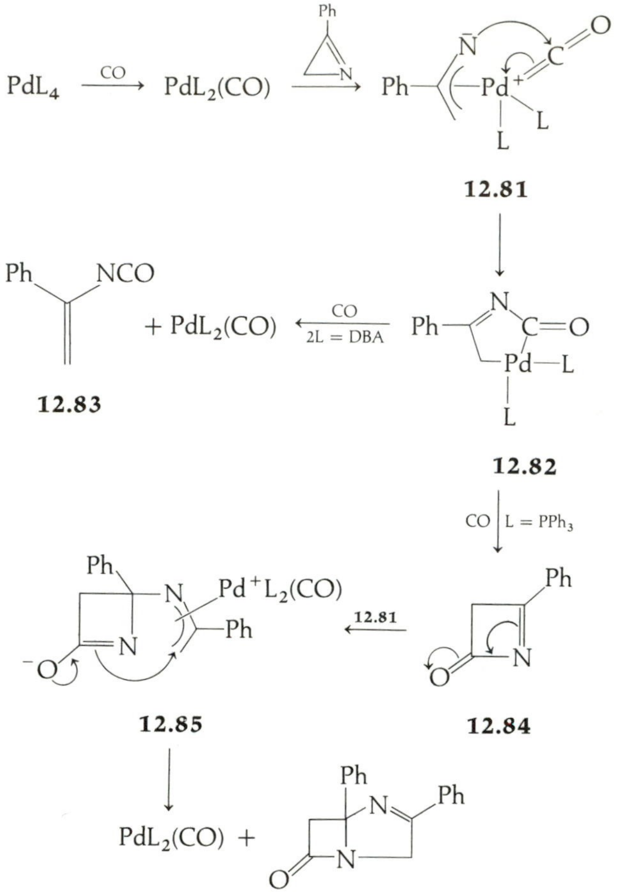

$$PdL_4 \xrightarrow{CO} PdL_2(CO) \longrightarrow \textbf{12.81}$$

**12.81**

$$\textbf{12.83} + PdL_2(CO) \xleftarrow[2L = DBA]{CO} \textbf{12.82}$$

**12.83**

**12.82**

$$CO \mid L = PPh_3$$

**12.85** $\xleftarrow{\textbf{12.81}}$ **12.84**

**12.85**

**12.84**

$$PdL_2(CO) + \textbf{12.86}$$

(12.62)

**12.86**

　　In this mechanism, the PdL$_4$ complexes react with CO to give the catalytically-active three-coordinate 16-electron PdL$_2$(CO) species. These add oxidatively to the C—N single bond of an azirine to give **12.81**, which is an aza analogue of an $\eta^3$-allyl complex. Intramolecular nucleophilic attack by the nitrogen atom at the CO ligand affords the common intermediate **12.82.**

　　Reductive-elimination of the catalyst from complex **12.82** can occur in one of two ways. One way gives an acyclic vinyl isocyanate (**12.83**), which is the observed product when DBA is the ancillary ligand. With Ph$_3$P as ancillary ligands, a cyclic reductive-elimination product (**12.84**), is produced. This $\beta$-lactam undergoes nucleophilic attack by a *second* molecule of **12.81** to give **12.85**. The ring nitrogen atom of complex **12.85** attacks the azaallylic terminus as a nucleophile, thereby initiating the reductive-elimination of the catalyst and generating the observed bicyclic β-lactam product.


## STEREOCHEMICAL INFLUENCE OF
## A METAL MOIETY AT A DISTAL SITE

Complexation of a metal moiety to a planar ligand, such as an olefin or $C_nH_n$ carbocyclic ligand, not only activates such ligands to nucleophilic attack but also permits a stereochemical discrimination between the original two faces (or sides) of the free ligand molecule. A direct result of this stereochemical influence of metal complexation is the predominantly *exo* attack by external nucleophiles (see chap. 11). A metal complex effectively blocks attack at one side of the coordinated molecule.

In addition, metal complexation can affect the stereochemistry at sites *distal* to the actual donor atoms. Coordination to appropriately substituted molecules can generate geometrical isomers, which are usually separable by fractional crystallization or chromatography. A subsequent reaction (stereochemically controlled by the metal complex) at a distal site affords products with specific stereochemistry at that site relative to the positions of the *original* substituents of the ligand.

One example of this type of distal stereochemical influence is shown in figure 12.1 for a synthesis of racemic *cis*- or *trans*-alkyl indanols. Reaction of the racemic alkyl indanone (**12.87**) with $Cr(CO)_6$ gives two racemic sets of geometrical isomers. Complexes **12.88** and **12.89** are enantiomeric ($\eta^6$-arene)$Cr(CO)_3$ complexes in which the chromium moiety and the R group are located on the same face of the ring (i.e., R is *endo* to Cr). Complexes **12.90** and **12.91** are also enantiomers. However, the alkyl substituent and the chromium atom lie on opposite faces of the ring (R is *exo* to Cr). Since the *endo* and *exo* isomers are chemically different, they can be separated by chromatography.

The key step that gives the desired stereospecificity is reduction of the indanone complexes (**12.88–12.91**) to the indanol complexes (**12.92–12.95**) by $KBH_4$. Interestingly, this reduction occurs such that the alcohol substituent is positioned only *endo* to the $Cr(CO)_3$ moiety. Presumably there are steric interactions that create a kinetic preference for attack at this site. Notice that the chromium-controlled stereospecific reduction of the ketone functionality at this distant ligand site determines the relative stereochemistry of the hydroxyl and alkyl substituents. Cleavage of the $Cr(CO)_3$ moieties affords stereospecifically either *cis*- or *trans*-alkyl indanols as racemic mixtures.

Metal complexation of planar prochiral free molecules affords dissymmetric complexes. If a subsequent reaction at a distal ligand site in such complexes is controlled stereochemically by the metal complex, then optically pure products are formed. An example of this type of application is shown in figure 12.2 for the synthesis of optically pure 2-methylindanone.

Resolution of the enantiomeric ($\eta^6$-1-*endo*-indanol)$Cr(CO)_3$ complexes (**12.96** and **12.97**) by conversion to the acid succinate is accomplished by normal methods. The separate optical isomers can be oxidized to the ($\eta^6$-indanone)$Cr(CO)_3$ complexes (**12.98** and **12.99**). Notice that 1-indanone as a free molecule is not chiral,

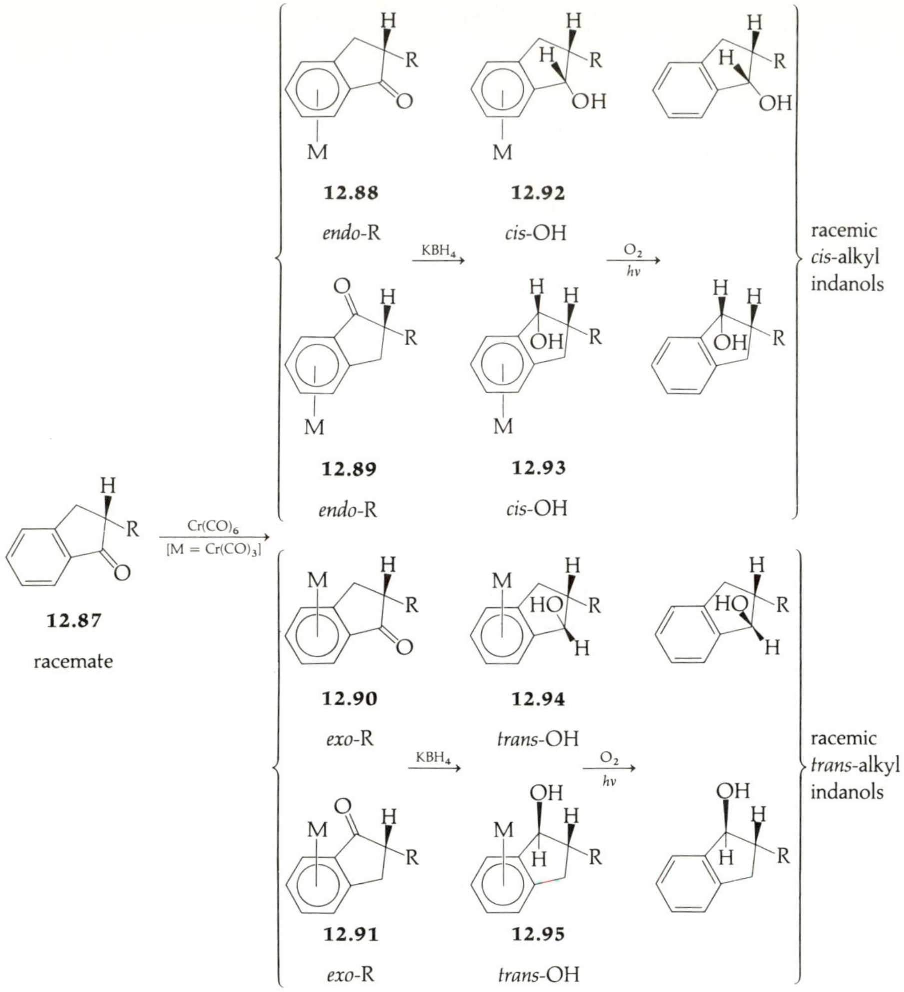

**FIGURE 12.1**   Stereospecific synthesis of racemic *cis-* or *trans*-alkyl indanols.

yet the Cr(CO)$_3$ complex is dissymmetric. Free 1-indanone can be regarded as a prochiral molecule. It possesses only one mirror plane of symmetry, the molecular plane. Complexation of a Cr(CO)$_3$ moiety to 1-indanone removes this plane of symmetry, thereby generating the enantiomers **12.98** and **12.99**. In theory, these enantiomers could be resolved and separated after direct complexation of Cr(CO)$_3$ to indanone. However, it is easier in practice to resolve the enantiomeric indanol complexes.

When enantiomer **12.98** is treated with NaH, an enolate anion forms. Although alkylation of this enolate could occur from either side of the ring, methyl iodide mono-alkylation occurs solely from the face *exo* to the Cr(CO)$_3$ group. This preference is probably determined by steric factors. When the metal moiety of the *exo*-methyl complex is removed, the R(−)-2-methylindanone product (**12.102**) is

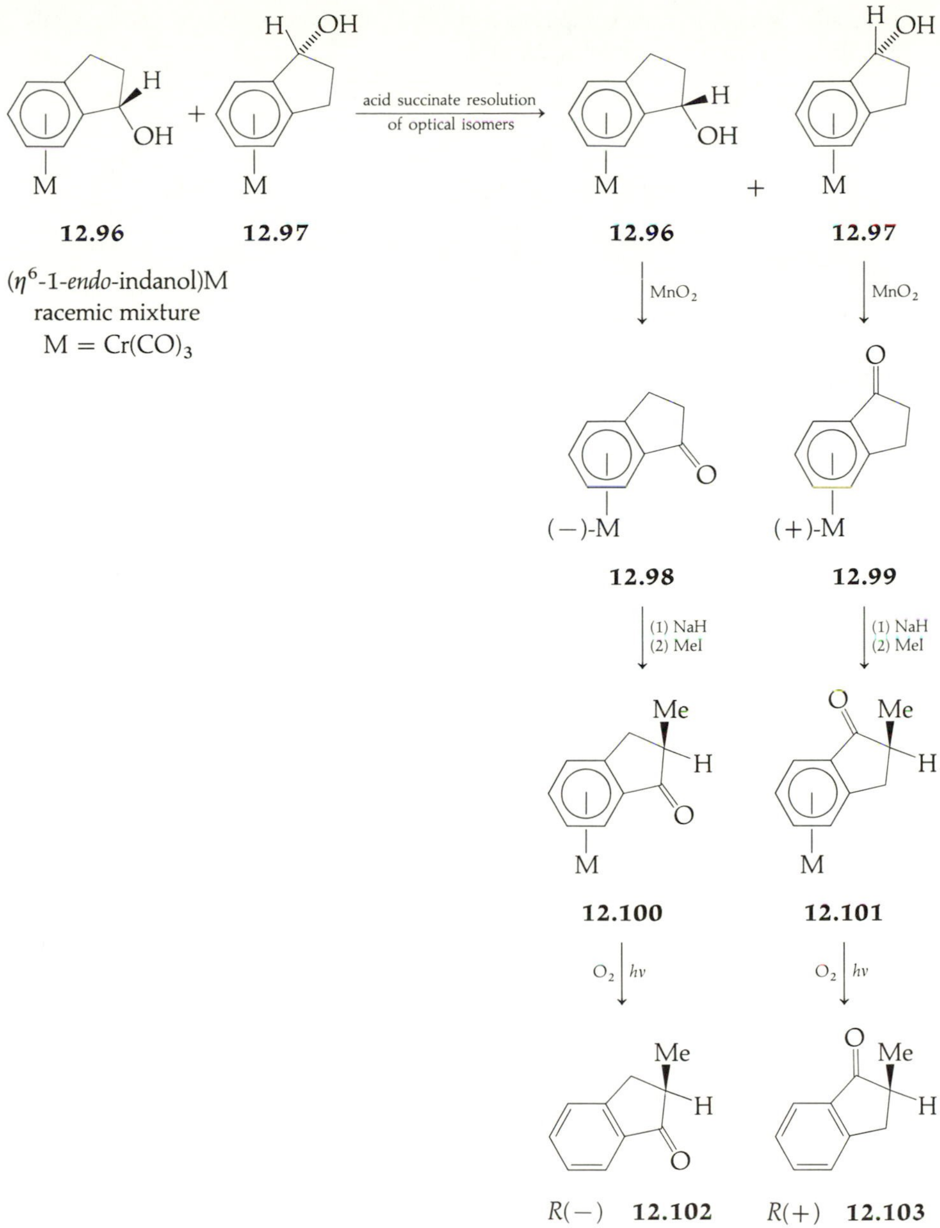

**FIGURE 12.2** Synthesis of optically pure 2-methylindanone.

obtained with an optical purity of 100%. The $R(+)$-2-methylindanone (**12.103**) could presumably be prepared similarly.

Another application of metal-induced dissymmetry in prochiral ligands is shown in equation scheme *12.63* for the synthesis of optically pure secondary alcohols. *Ortho*-methylacetophenone is a prochiral molecule. Complexation to a $Cr(CO)_3$

group affords enantiomeric complexes that can be resolved as separate compounds. One such enantiomer is complex **12.104.**

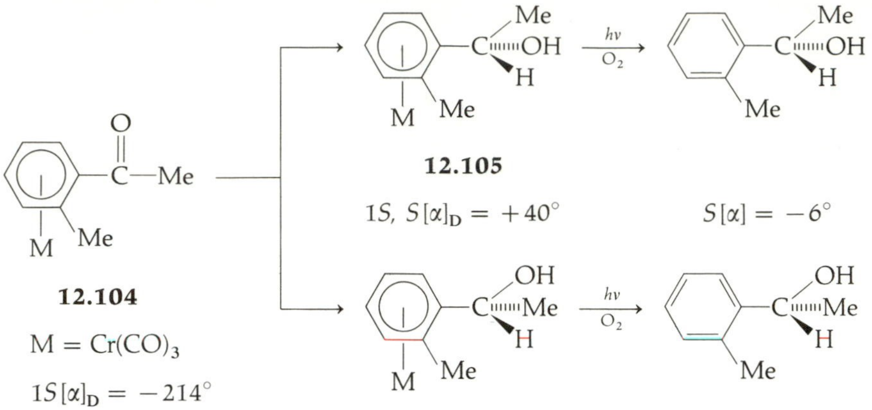

**12.104**

M = Cr(CO)$_3$

1S[α]$_D$ = −214°

**12.105**

1S, S[α]$_D$ = +40°          S[α] = −6°

**12.106**

1S, R[α]$_D$ = +27°          R[α] = +6°          (12.63)

Reduction of the ketonic group to the secondary alcohol generates a new optical center at the alcoholic carbon atom. Although both enantiomers (having either S or R configurations at this carbon center) are formed, the presence of only one optical configuration generated by Cr(CO)$_3$ complexation gives diastereomeric complexes **12.105** and **12.106.** These diastereomers are separated by chromatography, and removal of the Cr(CO)$_3$ moieties gives the optically active secondary alcohols with an optical purity of 100%.

This example differs from the one shown in figure 12.2 in that the reaction at the distal ligand site of an optically active complex is not stereochemically controlled by the presence of the metal complex. However, the formation of a diastereomeric mixture of product complexes permits a chemical separation that ultimately gives optically pure organic products. In figure 12.2, the reaction sequence also involves one enantiomeric resolution, but a subsequent separation of diastereomers is *not* required because of the metal-controlled stereospecificity of the alkylation reaction.

Other examples of the stereochemical influence of metal complexes involving catalytic reactions are discussed in chap. 13.

## STUDY QUESTIONS

1. Complex **12.3** is oxidized by PhIO to give N-acetyl dipeptide derivatives. Explain how this reaction is analogous to the oxidative removal of carbene ligands shown below.

$$[(Me)(PhCH_2NH)C]Cr(CO)_5 \xrightarrow[CH_2Cl_2]{PhIO} MeC(O)N(H)CH_2Ph$$

2. Using the available literature, give examples of the organic chemistry exhibited by complex **12.7.**

3. Write a mechanism for the following reaction. This reaction is catalyzed by $[(\eta^5\text{-}C_5H_5)Fe(CO)_2(THF)]^+BF_4^-$. [See M. Rosenblum and D. Scheck, *Organometallics* **1**, 397 (1982)].

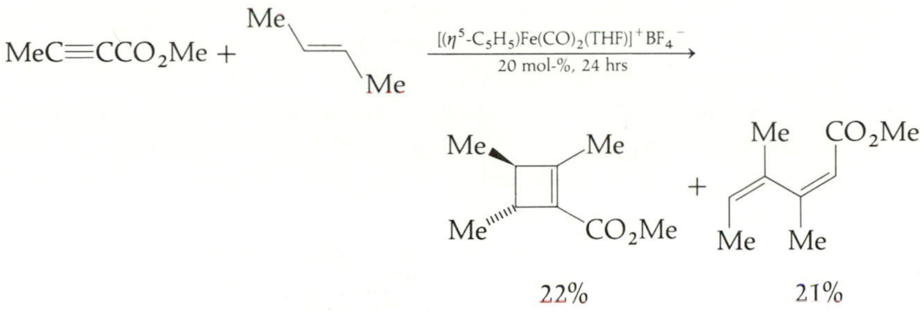
MeC≡CCO2Me +
Me
Me
[(η5-C5H5)Fe(CO)2(THF)]+BF4−
20 mol-%, 24 hrs
Me
Me
Me
CO2Me
+
Me
CO2Me
Me
Me
22%
21%

4. Write reaction mechanisms for equations *12.29–12.31.*
5. Write reaction mechanisms for the equations below. The products are formed with a stereo- and regioselectivity of ≥98%. [See H. Matsushita and E. Negishi, *J. Am. Chem. Soc.* **103**, 2882 (1981).]

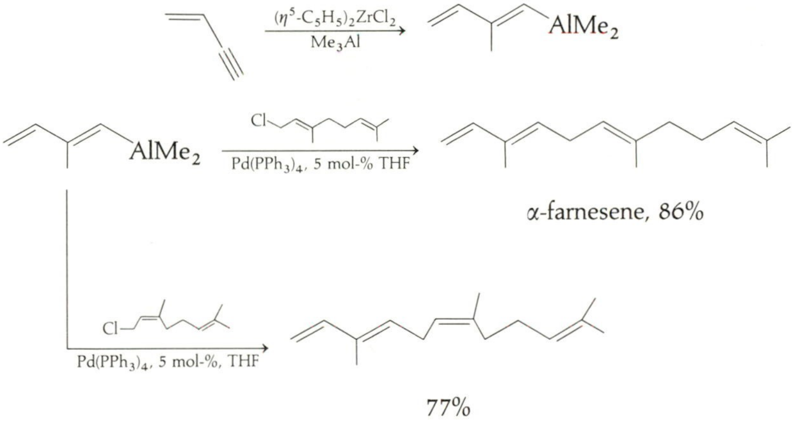
(η5-C5H5)2ZrCl2
Me3Al
AlMe2
AlMe2
Cl
Pd(PPh3)4, 5 mol-% THF
α-farnesene, 86%
Cl
Pd(PPh3)4, 5 mol-%, THF
77%

6. Write a mechanism for the following reaction [see C. M. Lukehart and K. Srinivasan, *Organometallics* **2**, 1640 (1983)]:

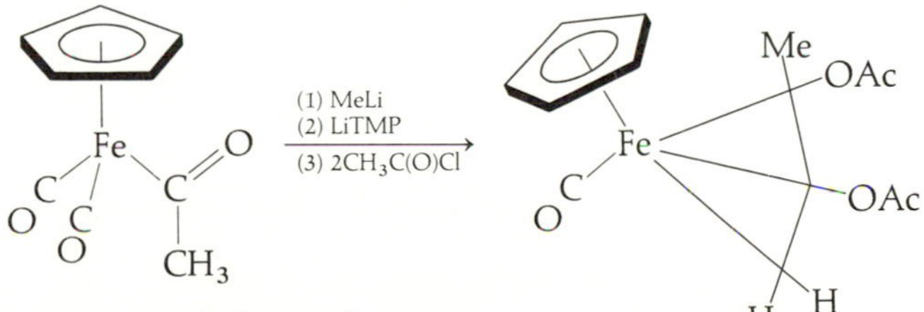
Fe
C
O
C
O
C
O
CH3
(1) MeLi
(2) LiTMP
(3) 2CH3C(O)Cl
Fe
C
O
Me
OAc
OAc
H
H

TMP = 2,2,6,6-tetramethylpiperidine

7. Show how the rearrangement of complex **12.62** to **12.63** is a metalla analogue to the known reverse reaction which is the [$_\sigma2_a + _\sigma2_s$] pericyclic ring opening of bicyclo [1.1.0] butanes to *transoid*-1,3-dienes. [See C. M. Lukehart and K. Srinivasan, *Organometallics* **1**, 1247 (1982).]

8. Show how the coupling of two $\eta^2$-alkyne ligands to give a metalla-cyclopentadiene complex results in a formal loss of two electrons at the metal atom (compare the conversion of complex **12.72** to **12.73**).

9. In the first step of equation *12.55*, two diastereomeric benzocyclobutene isomers are isolated in yields of 27.5% and 29.1%. Show the structures of these compounds. [See R. L. Funk and K. P. C. Vollhardt, *J. Am. Chem. Soc.* **102**, 5253 (1980).]

10. Sketch structures of the isolable intermediate complexes **A–D**, which generate separate diastereomers of 1,2-*bis*(α-hydroxyethyl)benzene, as shown below. [See J. Besançon, et al., *J. Organometal. Chem.* **94**, 35 (1975).]

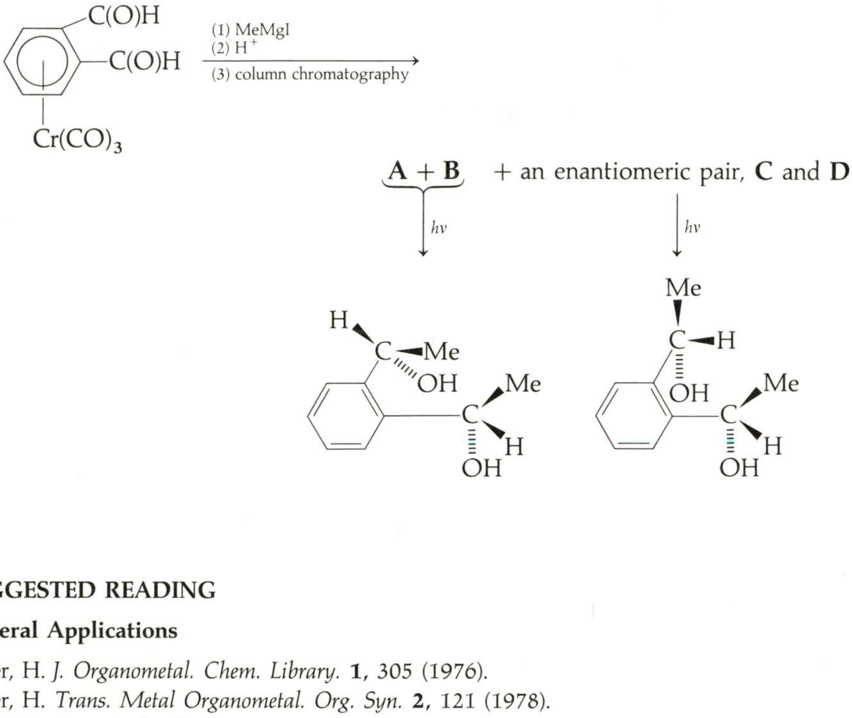

## SUGGESTED READING

### General Applications

Alper, H. *J. Organometal. Chem. Library.* **1**, 305 (1976).

Alper, H. *Trans. Metal Organometal. Org. Syn.* **2**, 121 (1978).

Alper, H. *Pure Appl. Chem.* **52**, 607 (1980).

Heck, R. F. *Organotransition Metal Chemistry: A Mechanistic Approach.* New York: Academic Press, 1974.

Noyori, R. *Trans. Metal Organometal. Org. Syn.* **1**, 83 (1976).

Davies, S. G. *Organotransition Metal Chemistry Applications to Organic Synthesis.* Oxford: Pergamon Press, 1982.

Tsuji, J. *Organic Synthesis by Means of Transition Metal Complexes.* New York: Springer-Verlag, 1975.

Wender, I., and Pino, P., eds. *Organic Syntheses via Metal Carbonyls.* Vol. 1 and Vol. 2. New York: Interscience, 1968 and 1977, respectively.

## Applications Involving Olefin, Alkyne, Allyl, and Arene Complexes

Baker, R. *Chem. Rev.* **73**, 487 (1973).
Birch, A. J., and Jenkins, I. D. *Trans. Metal Organometal. Org. Syn.* **1**, 1 (1976).
Hegedus, L. S. *J. Organometal. Chem. Library.* **1**, 329 (1976).
Heimbach, P., Jolly, P. W., and Wilke, G. *Adv. Organometal. Chem.* **8**, 29 (1970).
Jaouen, G. *Trans. Metal Organometal. Org. Syn.* **2**, 65 (1978).
Nicholas, K. M., Nestle, M. O., and Seyferth, D. *Trans. Metal Organometal. Org. Syn.* **2**, 1 (1978).
Semmelhack, M. F. *J. Organometal. Chem. Library.* **1**, 361 (1976).
Semmelhack, M. F., *et al. Tetrahedron.* **37**, 3957 (1981).
Semmelhack, M. F. *NY Acad. Sci.* **295**, 36 (1977).
Seyferth, D. *Adv. Organometal. Chem.* **14**, 97 (1976).
Vollhardt, K. P. C. *Acc. Chem. Res.* **10**, 1 (1977).

## Applications Involving Carbene Complexes

Brown, F. J. *Progr. Inorg. Chem.* **27**, 1 (1980).
Casey, C. P. *J. Organometal. Chem. Library.* **1**, 397 (1976).
Casey, C. P. *Trans. Metal Organometal. Org. Syn.* **1**, 189 (1976).

## Applications Involving Iron and Zirconium Complexes

Collman, J. P. *Acc. Chem. Res.* **8**, 342 (1975).
Pearson, A. J. *Acc. Chem. Res.* **13**, 463 (1980).
Rosenblum, M. *Acc. Chem. Res.* **7**, 122 (1974).
Schwartz, J. *J. Organometal. Chem. Library.* **1**, 461 (1976).
Schwartz, J. *Pure Appl. Chem.* **52**, 733 (1980).
Weissberger, E., and Laszlo, P. *Acc. Chem. Res.* **9**, 209 (1976).

## Applications Involving Palladium Complexes

Heck, R. F. *Acc. Chem. Res.* **12**, 146 (1979).
Henry, P. M. *Adv. Organometal. Chem.* **13**, 363 (1975).
Maitlis, P. M. *Acc. Chem. Res.* **9**, 93 (1976).
Trost, B. M. *Acc. Chem. Res.* **13**, 385 (1980).
Tsuji, J. *Organic Synthesis with Palladium Compounds.* New York: Springer-Verlag, 1980.
Tsuji, J. *Acc. Chem. Res.* **6**, 8 (1973).
Tsuji, J. *Acc. Chem. Res.* **2**, 144 (1969).

## Applications Involving Phase-Transfer Catalysis

Alper, H. *Adv. Organometal. Chem.* **19**, 183 (1981).

## Applications Involving Chiral Metal Atoms

Brunner, H. *Adv. Organometal. Chem.* **18**, 151 (1980).

# Applications to Industrial Homogeneous Catalysis

## CHAPTER 13

Transition metal organometallic chemistry has played an important role in developing our understanding of the chemistry of several industrial processes. Realization of the potential relevance of organometallic reaction chemistry to large-scale catalytic processes has generated considerable interest in elucidating the mechanisms involved in known systems, developing better catalysts for desired reactions, and discovering organometallic catalysts for other processes of industrial importance. Research in these areas ranges from the very basic to the very applied and developmental. In industrial, academic, and government laboratories research groups are exploring this chemistry.

In this chapter, chemical catalysis is introduced. Because of economic considerations, large-scale chemistry must utilize reagents that can be obtained from natural resources. For this reason, a short discussion of the composition of natural resources and of how chemical reagents are obtained from them is included. In the latter part of the chapter, four industrial processes that utilize homogeneous organometallic catalysts are presented. The fundamental relationship between the known types of organometallic complexes and reactions discussed earlier and our understanding of catalytic reaction cycles is clear. It is highly likely that future industrial processes will be developed from reactions discovered during basic research investigations.

## CHEMICAL CATALYSIS

In the general reaction shown as equation *13.1*, reactants $A$ and $B$ combine to give products $C$ and $D$. Since all reactions can be regarded as equilibria, the relative amounts of reactants and products at equilibrium is defined by the value of the equilibrium constant $K$.

$$A + B \; \overset{K}{\rightleftharpoons} \; C + D \tag{13.1}$$

The change in standard free energy in going from reactants to products is related to $K$ as follows:

$$\Delta G^0 = -RT \ln K \qquad (13.2)$$

When $\Delta G^0$ for a reaction is negative, the reaction is thermodynamically feasible. When $\Delta G^0$ is positive, the yield of products at equilibrium could be quite low. In practice, reactions suitable for large-scale chemistry should have either negative free energy change or possibly slightly positive $\Delta G^0$ values (but no larger than $+ 10$ kcal/mole). For positive values, it is important to drive the reaction toward the product side by alternative means, such as continuous removal of products from the reaction vessel.

Chemical catalysis is *solely a kinetic phenomenon*. Through either a single step or a sequence of steps, a catalyst provides faster product formation than would occur in the uncatalyzed reaction. The rate-determining step in a catalyzed reaction has a lower free energy of *activation* than that of the rate-determining step of the uncatalyzed reaction. A catalytically active complex coordinates the reactants and accelerates the chemical conversion to products. The products eliminate or dissociate from the catalyst thereby regenerating the active catalyst to repeat the cycle.

For example, hydrogenation of ethylene to ethane is very favorable thermodynamically at low temperature (eq. *13.3*), but it is so slow kinetically that mixtures of ethylene and hydrogen are stable nearly indefinitely.

$$CH_2{=}CH_2 + H_2 \xrightarrow{\overset{298\,K\,=\,5.16\,\times\,10^{17}\,atm^{-1}}{\rule{4cm}{0pt}}} C_2H_6 \quad \Delta G^0 = -24.1 \text{ kcal/mole } (13.3)$$

However, when a palladium catalyst is added, hydrogenation proceeds at a rapid rate.

Since a catalyst only accelerates a chemical reaction and does not affect the final position of the equilibrium, it is important to develop catalysts for thermodynamically favorable reactions. Equation *13.3* is a good illustration of these considerations. This reaction has a large negative $\Delta G^0$, and thus lies far to the product side at equilibrium. The palladium metal catalyst only accelerates the hydrogenation reaction. Less thermodynamically favorable reactions might be possible candidates for catalyst development, also. However, reactions that have large positive free energy changes are not practical even when catalyzed because of the low yield of product.

Reactions that have a positive entropy change $\Delta S^0$ can be made more thermodynamically favorable by increasing the temperature according to the equation

$$\Delta G^0 = \Delta H^0 - T \Delta S^0.$$

Catalysis at this higher temperature might become more practical.

Although a catalyst does not affect the final equilibrium, it *can* affect the *initial* product distribution. If products are formed via competitive reactions, then a given catalyst might affect their relative rates differently. This effect is evidenced by a high, nonequilibrium concentration of selected products during the early part of the reaction. However, since all reactions are equilibria, prolonged reaction times will eventually give the theoretical equilibrium mixture of reactants and products.

Metallic catalytic species are classified as either *homogeneous* or *heterogeneous*. The distinction is based on the solubility and type of active site of the catalyst. *Homogeneous organometallic catalysts* are soluble in the reaction medium, and their active site is a metal atom in a molecular complex. This complex might be mononuclear, polynuclear, or a cluster. In polynuclear complexes, more than one metal atom could be involved in the catalytic reaction, making determination of an active site more difficult. When cluster complexes are present in the catalytically active reaction medium, two questions arise. One question is whether the cluster *is* the active catalyst or whether some degree of cluster fragmentation occurs at high temperature and pressure to give minute concentrations of very reactive mononuclear catalysts. The second question is one of particle size. Is a cluster complex that contains 15 to 25 metal atoms plus ancillary ligands a large soluble molecule or a very small metallic particle? If such a large molecule were a particle (as indicated by light scattering or other measurements), then the catalyst would be best classified as heterogeneous. Usually, turbidity and light scattering measurements are used to determine if small particles are actually present in the reaction medium. Soluble complexes have a maximum dimension of 100 Å or less.

*Heterogeneous catalysts* are insoluble in the reaction medium, and the active site is located on a solid surface. Metallic heterogeneous catalysts usually consist of a deposited layer of metal or a metallic compound on a solid support such as alumina, silica, zeolites, or other refractory materials. If the metal atoms are in a reduced oxidation state, then the active site might be a small crystallite of the pure metal.

A general comparison of some major features of homogeneous and heterogeneous catalysts is shown in table 13.1. Although the industrial output of chemicals synthesized via heterogeneous catalysis vastly outweighs chemical production via homogeneous catalysis, there has been a significant shift toward developing and instituting homogeneous molecular catalysis in industrial processes. This shift is

**TABLE 13.1   Comparison of homogeneous and heterogeneous catalysts.**

| Property | Homogeneous catalyst | Heterogeneous catalyst |
|---|---|---|
| Solubility | soluble | insoluble |
| Composition | discrete molecular complex | metal deposition on a solid support |
| Thermal stability | usually decompose at high temperatures | usually stable to high temperatures |
| Ease of product separation | usually a difficult problem | usually easily recovered |
| Active site | usually a single metal atom | a surface site, such as on a metal crystallite |
| Reaction conditions | lower temperature and pressures | usually higher temperature and pressure |
| Reaction selectivity | generally good | generally good to fair |
| Ease of modification | relatively easy | relatively difficult |
| Ease of study | relatively easy | relatively difficult |

dictated by several factors. Due to rising costs in energy and natural resources, the higher selectivity and lower energy requirements of homogeneous catalysts are becoming more important economically. It is also easier to modify a molecular catalyst to effect a specific reaction than it is to modify a surface catalyst of variable composition. Soluble species can be examined by IR and NMR during a reaction, whereas surface species are difficult to study even under ideal laboratory conditions.

As we shall discuss later, the chemistry of homogeneous mononuclear catalysts can be understood in terms of the known reaction chemistry of organometallic complexes. Catalyst selectivity and chemical reactivity can be understood at a molecular level. Appropriate modification of a homogeneous catalyst might be effected by simply changing the type of ancillary ligands.

Such a fundamental understanding of the chemistry occurring at surface catalytic sites of a heterogeneous catalysis is not yet possible. Model studies of the chemical activation of hydrocarbons on a surface of crystalline platinum reveal several different types of metal atom environments each with different chemical reactivity. As shown in figure 13.1, platinum atoms occupy terrace, step, and kink environments on a crystalline platinum surface. The coordination number of surface Pt atoms decreases in going from terrace, to step and finally to kink sites. Hydrocarbon reactivity *increases* in the same sequence: terrace atoms < step atoms < kink atoms. Step atoms are active for H—H and C—H bond cleavage, but kink atoms are more reactive and can also cleave C—C bonds. Presumably the metal atom reactivity increases with decreasing metallic coordination number of the site environment. Greater coordination unsaturation yields higher chemical reactivity. The observation of different chemical reactivity for each type of metal surface environment might explain the observed low selectivity of some heterogeneous catalysts.

Organometallic cluster chemistry provides a link between our understanding of the chemistry of homogeneous mononuclear catalysts and that of bulk-solid surface heterogeneous catalysts. Large clusters have a relatively large metal core that might approximate a small particle of metal. Cluster fragmentation reactions (eq. *14.20*) can

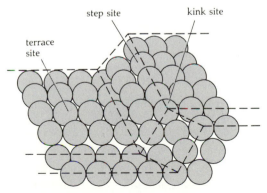

**FIGURE 13.1** Types of metal atom environments on a crystalline platinum surface. *Adapted from* Strem Chemiker IX, No. 1 (1981).

"expose" selected metal atoms of the cluster to the solution environment by generating "holes" or vacancies in the original metal core, as, for example, in the fragmentation of a *closo*-cluster to an *arachno*-fragmented cluster. These cluster fragments
might possess "kink-type" metal atom environments that would show considerable
chemical reactivity. Chemical reactions of these clusters could be studied more easily
than those of an active site on a surface.

Clusters containing peripheral monatomic ligands such as carbido ligands (chap. 6)
might model chemical species known to form at surface metal atoms. For example,
carbon monoxide is believed to react with a step-kink surface of platinum to undergo
complete C—O bond dissociation giving metal-carbido, $M{\equiv}C$, and metal-oxo,
$M{=}O$, species. Organometallic cluster species having exposed carbido ligands
might imitate some of the chemical reactivity of these surface species obtained from
the dissociative chemisorption of carbon monoxide.

The search for a solution to the practical problem of separating a homogeneous
catalyst from the reaction medium has led to the development of *supported catalysts.*
These catalysts are hybrids of conventional homogeneous and heterogeneous catalysts. Attaching a complex known to be active as a homogeneous catalyst to a
solid support immobilizes the complex and simplifies separation of the catalyst.
Complexes are covalently attached to solid supports through remote functional
groups on ancillary ligands.

Among the materials used to support metal complexes are silica, zeolites, glass,
clay, alumina, silica gel, polystyrene, polyamines, polyvinyls, polyallyls, polybutadiene, polyamino acids, urethanes, acrylic polymers, cellulose, cross-linked dextrans,
and agraose. Solid supports that contain surface hydroxyl groups, such as silica,
react directly with metal alkyl or $\eta^3$-allyl compounds to bind metal complexes
(eq. *13.4*). Reaction with two surface hydroxyl groups gives a chelate coordination.

$M = $ Ti, Zr, Nb, Cr, Mo, W
$R = CH_2Ph$, $\eta^3$-allyl

A very common, albeit less direct, method of immobilizing a metal complex on
a solid support is to covalently attach a desired ancillary ligand to a support. Examples of this approach are shown in equations *13.5*–*13.10*. Several variations of
the reactions shown are also used. For example, the functionalized ligand shown in
equation *13.5* might be coordinated to a metal complex first and then attached to
the polymer support. Polystyrene supports have been used extensively (eqs. *13.7*
and *13.8*).

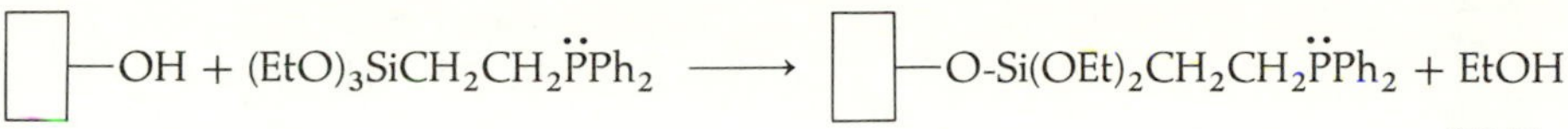

$$\boxed{\ }-OH + (EtO)_3SiCH_2CH_2\ddot{P}Ph_2 \longrightarrow \boxed{\ }-O\text{-}Si(OEt)_2CH_2CH_2\ddot{P}Ph_2 + EtOH \qquad (13.5)$$

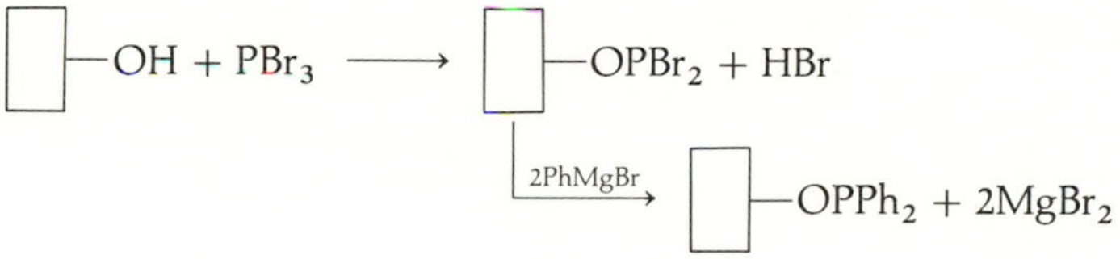

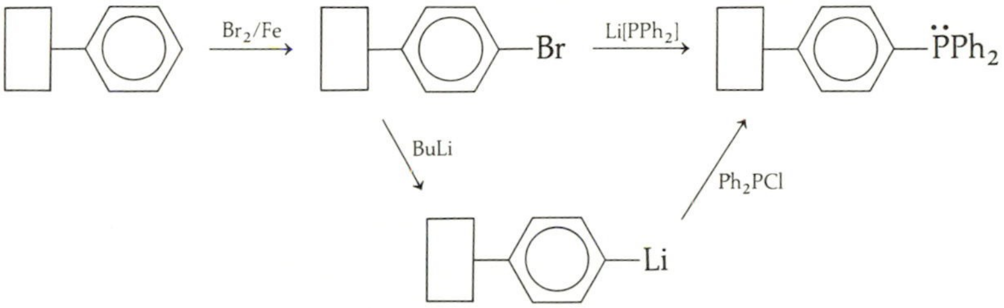

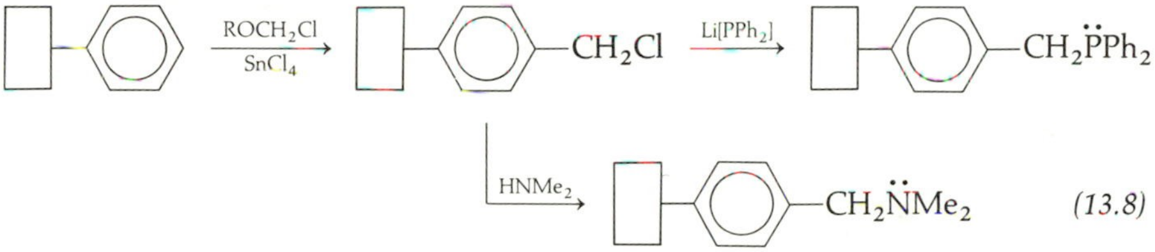

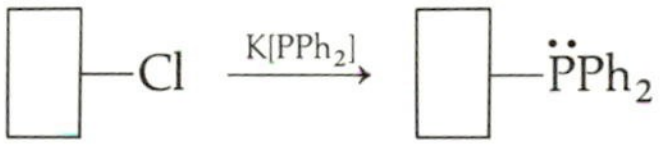

$$\boxed{\ }-OH + PBr_3 \longrightarrow \boxed{\ }-OPBr_2 + HBr$$

$$\xrightarrow{2PhMgBr} \boxed{\ }-OPPh_2 + 2MgBr_2 \qquad (13.6)$$

$$\boxed{\ }\!\!-\!\!\bigcirc \xrightarrow{Br_2/Fe} \boxed{\ }\!\!-\!\!\bigcirc\!\!-Br \xrightarrow{Li[PPh_2]} \boxed{\ }\!\!-\!\!\bigcirc\!\!-\ddot{P}Ph_2$$

$$\searrow^{BuLi} \qquad \nearrow^{Ph_2PCl}$$

$$\boxed{\ }\!\!-\!\!\bigcirc\!\!-Li \qquad (13.7)$$

$$\boxed{\ }\!\!-\!\!\bigcirc \xrightarrow[SnCl_4]{ROCH_2Cl} \boxed{\ }\!\!-\!\!\bigcirc\!\!-CH_2Cl \xrightarrow{Li[PPh_2]} \boxed{\ }\!\!-\!\!\bigcirc\!\!-CH_2\ddot{P}Ph_2$$

$$\searrow^{HNMe_2} \qquad \boxed{\ }\!\!-\!\!\bigcirc\!\!-CH_2\ddot{N}Me_2 \qquad (13.8)$$

$$\boxed{\ }-Cl \xrightarrow{K[PPh_2]} \boxed{\ }-\ddot{P}Ph_2 \qquad (13.9)$$

$$\xrightarrow[\substack{(1)\ HBr \\ (2)\ K[PPh_2]}]{} \qquad (13.10)$$

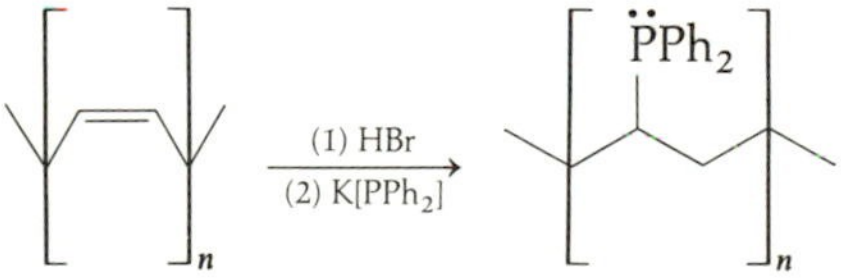

Another way of preparing poly-p-bromostyrene than that shown in equation *13.7* is to polymerize p-bromostyrene directly. Complexes of styrene such as ($\eta^6$-styrene) Cr(CO)$_3$ can be prepared and then copolymerized with either styrene or methylacrylate. Ligand attachment to polyvinyl chloride and polybutadiene is shown in equations *13.9* and *13.10*. The carbon backbone of these ligand-functionalized polymers is saturated. Cyclopentadienyl groups can be attached to polymers as shown in equation *13.11* for a polystyrene support.

Metal complexation to these functionalized solid supports can occur by four routes: (1) ligand displacement, (2) ligand-induced fragmentation of a dimeric complex, (3) reductive-ligation, and (4) oxidative-addition to halo-functionalized supports. These reactions are analogous to the normal methods of forming organometallic complexes.

$$(13.11)$$

Essentially all of the known homogeneous metal-catalyzed reactions have been studied using supported catalysts. Several commercial processes involve silica-supported metal complexes. Phosphinated polystyrene supports have not been utilized in large-scale commercial processes yet, but they have considerable potential in practical applications if polymer instability and loss of metal through leaching processes can be controlled. These supported catalysts may provide a relatively convenient method of coupling the synthetic chemical advantages of a homogeneous catalyst with the physical advantages of a heterogeneous one.

## NATURAL RESOURCES

A basic problem confronting the industrial chemist is: How can we make the chemicals we need from our natural resources? Two approaches are (1) to extract the needed compounds directly from a natural resource, and (2) to convert chemical compounds obtained from our natural resources into the desired compounds. Methane is an example of a desired chemical that can be obtained directly from a natural resource, natural gas. Vinyl acetate or butanal are examples of desired compounds that must be synthesized from resource-derived chemicals. When large-scale chemical synthesis is required, economic considerations demand that syntheses be efficient and consist of as few steps as possible. Products should be obtained selectively in high yield and at a low energy consumption. A continuous, or flow, process is usually preferable to a batch process. Expensive reagents, such as metals, should be used as catalytic (rather than as stoichiometric) reagents.

Our natural resources consist of mineral deposits, hydrocarbon sources, air, and water. The nearly alchemical task of converting "earth," air, fire, and water into feedstock organic chemicals is interesting and challenging! Minerals provide a source of inorganic elements, acids, bases, and salts. Air contains mainly nitrogen, oxygen, noble gases, and some carbon dioxide. Water is both a resource itself and a source of hydrogen and oxygen. Carbon is obtained from hydrocarbon resources. The most convenient oxidizing agent is molecular oxygen, and the most convenient reducing agent is molecular hydrogen.

Hydrocarbon resources consist of natural gas, petroleum-containing sources, coal, and general biomass. Natural gas is 60–95% methane, the remainder consisting primarily of ethane, propane, and butane. Separation and liquefaction of the propane

and butane components yields liquefied petroleum gas (LPG). Methane is also produced by the bacterial degradation of biomass.

Petroleum contains a variety of saturated hydrocarbons, including a small amount
of natural gas. The heavier components are characterized by the number of carbon
atoms per molecule within a given boiling point range as shown in table 13.2.
Only small amounts of aromatic hydrocarbon compounds are present in crude
petroleum.

TABLE 13.2   Approximate composition of the heavier fractions
of crude petroleum.

| Fraction | Approximate composition | Approximate boiling range (°C) |
|---|---|---|
| Petroleum ether (naphtha) | $C_5-C_7$ | 20–100 |
| Gasoline | $C_6-C_{12}$ | 50–200 |
| Kerosene | $C_{10}-C_{16}$ | 200–300 |
| Fuel oil, diesel oil, gas oil | $C_{16}-C_{20}$ | 300–500 |
| Lubricating oils, greases, petroleum jelly, paraffin wax | $C_{18}-C_{20}$ | 400–higher |
| Asphalt, petroleum coke | | residue |

Coal is a heterogeneous material of complex chemical composition. It appears to
consist of a hydrocarbon matrix with an amorphous phase that can be extracted
with pyridine and carbon disulfide. The structure of the hydrocarbon matrix is
probably different for each type of coal, but apparently it consists of fused aromatic
rings with hydroxyl substituents. Fused aromatic units are connected by aryl ether
and benzylic methylene bridges. Some saturated rings may also be present. A proposed structure of a coal macromolecule is shown in figure 13.2.

Table 13.3 shows the overall elemental composition of coal taken from data
supplied by an electric power generating station. In most coals, the C/H atomic
ratio of the carbonaceous material is 0.8–0.9 ($CH_{0.8-0.9}$). In order to produce more
hydrogen-rich materials like gasoline (composition about $CH_{2.2}$) from coal, hydrogen must be added to coal by *hydroliquefaction processes.* An alternative approach
to utilizing the hydrocarbon and energy content of coal is to effect reactions of coal
with steam and oxygen that produce a gasification product mixture of CO and $H_2$.
This mixture of gases is called *synthesis gas* or *syn gas.* The coal acts primarily as a
carbon source; the oxygen atom and molecular hydrogen come from water. Synthesis gas can be used to make a variety of organic materials including gasoline,
methane, and methanol.

Direct pyrolysis of soft coal produces coke (94%), coal tar (~6%), and some
volatile compounds. Coke, a purer carbon source than coal, is used in large quantities by steel producers. Coal tar contains about 200 different compounds. From
one ton of coal the following compounds are obtained as major components: naphthalene (5 lbs.), pyridine (2 lbs.), cresols (2 lbs.), benzene (2 lbs.), aniline (1 lb.),
phenathroline (1 lb.), phenol (0.5 lb.), toluene (0.5 lb.) and xylenes (0.1 lb.).

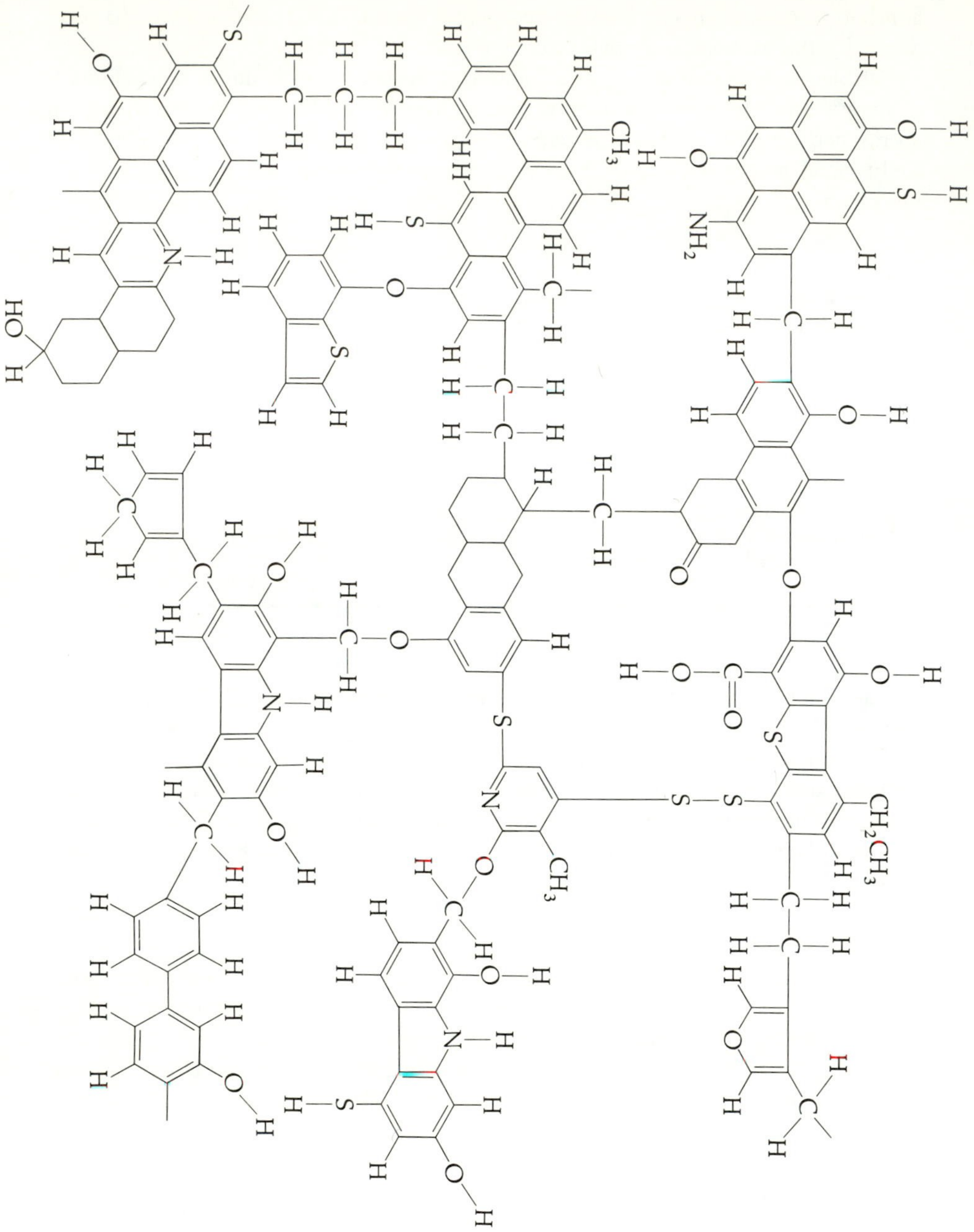

**FIGURE 13.2**  Proposed structure of a coal macromolecule. *Adapted from J. Haggin*, Chem. Eng. News, *Aug. 9, 21 (1982).*

**TABLE 13.3   Elemental analysis of a coal sample after drying.**

| Elemental analysis (%) | | Coal ash composition (%) | |
|---|---|---|---|
| C | 68.52 | $SiO_2$ | 48.5 |
| H | 4.41 | $Fe_2O_3$ | 19.6 |
| N | 1.50 | $Al_2O_3$ | 21.0 |
| S | 3.36 | $TiO_2$ | 1.1 |
| coal ash | 13.36 | $Mn_3O_4$ | 0.1 |
| O | 8.85 | CaO | 0.7 |
| Cl | 0.016 | MgO | 1.3 |
| | | $Na_2O$ | 0.2 |
| | | $K_2O$ | 3.1 |
| | | $SO_3{}^a$ | 1.9 |

[a] Sulfur present as pyritic sulfur.
*Source*: Oak Ridge National Laboratory (1973).

## CHEMICAL REAGENTS FROM NATURAL RESOURCES

In order to be suitable for large-scale syntheses of organic compounds, reagent chemicals should be easily derivable from natural resources. For the catalytic processes to be discussed below, some of the more important chemicals include water, oxygen, carbon from natural gas (mostly $CH_4$), coal, or coke, and alkenes from catalytic and thermal cracking of hydrocarbons. The commercial value of ethylene alone produced by cracking processes in 1981 was $7 billion. Aliphatic hydrocarbons are used directly for a variety of purposes, especially as combustion fuels.

Commercial hydrogen production in 1981 was about 105 billion cubic feet. Molecular hydrogen is produced by *steam reforming* of natural gas as shown in equation *13.12*. Only about 1% of the hydrogen sold is produced by electrolysis of water. In steam reforming, methane reacts with water in the vapor phase over a heterogeneous catalyst to produce a 1:3 $CO/H_2$ synthesis gas mixture.

$$CH_4 + H_2O \xrightarrow[\text{650–1000°C, 10 atm}]{\text{NiO cat.}} CO + 3H_2 \qquad (13.12)$$

More hydrogen can be obtained via the *shift reaction* shown in equation *13.13*. In this step, the CO produced reacts with more water to give $CO_2$ and $H_2$. Removal of the $CO_2$ by scrubbing the product gas with solutions of KOH or monoethanolamine and removal of traces of CO by reduction over a nickel catalyst provides very pure $H_2$ (99+%).

$$CO + H_2O \xrightarrow[\text{375°C}]{\text{metal oxide cat.}} CO_2 + H_2 \qquad (13.13)$$

Synthesis gas (syn gas) is a very useful reagent itself. It can be prepared from methane (eq. *13.12*) or from other carbon sources such as coal or coke:

$$C + H_2O \xrightarrow[\text{1–30 atm}]{\text{925–1375°C}} CO + H_2 \qquad (13.14)$$

This reaction, which is called the *water gas reaction,* yields a $1:1$ $CO/H_2$ syn gas product. If more hydrogen is needed in the syn gas product for a subsequent synthesis, then the shift reaction, equation *13.13,* converts part of the CO into $CO_2$ and $H_2$.

Synthesis gas reacts directly with metallic heterogeneous catalysts under various conditions of temperature and pressure to give important industrial reagents. As shown in equation *13.15,* these products include methane, methanol, and gasoline.

$$CO + H_2 \xrightarrow{250-400°C} \begin{cases} \xrightarrow{\text{Ni cat.}} CH_4 + H_2O \\ \xrightarrow{\text{Zn/Cu cat.}} CH_3OH \\ \xrightarrow{\text{Fe cat.}} \text{alkanes (gasoline fuels)} \end{cases} \qquad (13.15)$$

Synthetic fuel production from syn gas is known as the *Fischer–Tropsch process.* This process was developed around 1925 and was used for fuel production during World War II. South Africa now uses this process to produce fuels in the Sasol II and Sasol III synthetic fuels plants.

## ORGANIC CHEMICALS FROM INDUSTRIAL PROCESSES

We have discussed how natural resources can be used to generate important industrial feedstock reagents such as syn gas and $H_2$. Alkenes are obtained from our natural resources by hydrocarbon cracking. This section describes several industrial processes, each of which is believed to involve a homogeneous organometallic complex as the catalytically active species. Although a comprehensive discussion of each process is not intended, the known or postulated mechanisms for these processes can be described as a cyclical series of fundamental organometallic reactions. These mechanisms underscore the importance of detailed model studies of the reactions discussed in earlier chapters and provide an analytical concept useful both in understanding known catalytic processes and in developing new ones.

### Hydroformylation or the Oxo Process

Hydroformylation is the most important industrial reaction that uses metal carbonyl catalysts. By definition, it refers to the addition of hydrogen and a formyl group, $HC=O$, to an alkene. The term *Oxo process* refers to the "oxonation reaction," the addition of oxygen to a double bond. Although both names refer to the same reaction, the term *hydroformylation* describes the overall reaction more accurately.

The main application of hydroformylation is the synthesis of butanal from propene as shown in equation *13.16.* As in all hydroformylation reactions, a $C_n$ alkene is converted to a $C_{n+1}$ aldehyde.

$$CH_3CH{=}CH_2 + CO + H_2 \xrightarrow{\text{catalyst}} CH_3CH_2CH_2\overset{\displaystyle O}{\overset{\displaystyle \|}{C}}H \qquad (13.16)$$

Approximately six billion pounds of butanal are produced annually by this process. Butanal is subsequently hydrogenated to butanol, an important industrial solvent, or converted to 2-ethylhexanol via aldol condensation and hydrogenation reactions. 2-Ethylhexanol is esterified with phthalic anhydride to give dioctyl phthalate, which is a plasticizer for polyvinyl chloride resins.

An additional two to five billion pounds of other aldehydes are produced annually by hydroformylation. Longer chain olefins give higher molecular weight aldehydes. Hydrogenation of these aldehydes affords fatty alcohols used in the synthesis of biodegradable detergents, surfactants, lubricants, and other plasticizers.

Hydroformylation was discovered in 1938 by O. Roelen while working for Ruhrchemie AG in Germany. He observed a marked increase in aldehyde products when an olefinic Fischer–Tropsch fraction was recycled over a cobalt heterogeneous catalyst in the presence of synthesis gas. Subsequent study revealed that this reaction is promoted by homogeneous catalysts. The three types of catalyst used are (1) simple hydrido cobalt carbonyl complexes, (2) hydrido cobalt carbonyl complexes modified by addition of tertiary phosphine ligands, and (3) tertiary phosphine hydrido rhodium carbonyl species. Other metals such as Ru, Mn, Fe, Cu, Ag, and Pt are also known catalysts. A platinum-tin coupled catalyst has received considerable attention recently. Several polymer-supported and other heterogeneous catalysts have been developed also.

The principal hydroformylation systems are based on cobalt or rhodium. Although most industrial processes still use cobalt, commercial introduction of a Rh-based catalyst by Union Carbide in 1976 may initiate a rapid conversion to rhodium catalysts.

Two aspects of the hydroformylation reaction of great importance to industrial chemists are the minimization of undesired by-products and the avoidance of expensive operations in providing high pressures and high temperatures during the reaction.

A complete description of the hydroformylation of propene is shown in equation *13.17*. The desired product is the normal or unbranched aldehyde.

$$CH_3CH{=}CH_2 + CO + H_2 \xrightarrow{\text{cat.}} \underset{n\text{-butanal}}{CH_3CH_2CH_2\overset{\overset{\displaystyle O}{\|}}{C}H} + \underset{\text{isobutanal}}{(CH_3)_2CH\overset{\overset{\displaystyle O}{\|}}{C}H} +$$

$$C_4 \text{ alcohols} + \text{dipropyl ketones} + \text{alkanes} + \text{aldol products} \quad (13.17)$$

Branched aldehydes are generally not desired because they or their alcohol derivatives are less useful. Isobutanal for example, cannot be used to prepare 2-ethylhexanol. Selectivity toward *n*-aldehydes is therefore required of the catalyst. High temperature and pressure conditions tend to increase the yield of by-products.

A simple cobalt-catalyzed reaction mixture is prepared by adding finely divided cobalt metal or a cobalt(II) salt, 1:1 synthesis gas, solvent, and propene to a reactor. Reaction temperature and pressure are usually in the ranges $110°{-}180°C$ and $200{-}350$ atm. Under these conditions, $Co_2(CO)_8$ is formed by reductive-ligation, and this dimer is cleaved reductively by $H_2$ to give $HCo(CO)_4$, the catalyst precursor. Aldehyde products are obtained in a total yield of 78–82% with a linear to branched ratio of 2.5–4/1. Alcohols are formed in 10–12% yield; propane in 2% yield; and other by-products in 8–10% yield.

As shown in equation *13.18*, the reaction rate has a direct first-order dependence on the concentration of alkene, cobalt complex, and the partial pressure of hydrogen, and an inverse first-order dependence on the partial pressure of carbon monoxide.

$$d[\text{aldehyde}]/dt = k_{\text{obs}}[\text{alkene}][\text{Co}]P_{H_2}/P_{CO} \tag{13.18}$$

A reaction mechanism of the simple cobalt-catalyzed hydroformylation process is shown in figure 13.3.

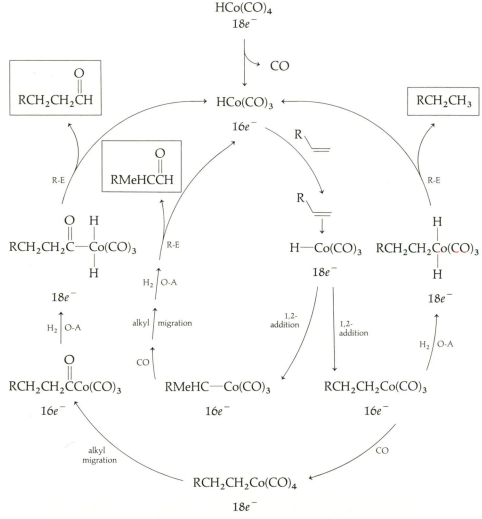

**FIGURE 13.3** Mechanism of hydroformylation by an unmodified cobalt carbonyl catalyst.

In this mechanism, loss of CO from $HCo(CO)_4$ generates the 16-electron, catalytically active species $HCo(CO)_3$. This complex coordinates alkene to give an $(\eta^2$-olefin)(hydride)cobalt intermediate. Anti-Markownikoff addition of the Co—H bond to the olefinic double bond gives an $n$-alkyl complex. Alkyl migration affords an $n$-acyl ligand. Oxidative-addition of $H_2$ followed by reductive-elimination gives the $n$-aldehyde product and regenerates $HCo(CO)_3$. Markownikoff addition of the Co—H bond to the alkene double bond forms a branched alkyl complex. Repeating the same sequence of reactions forms the branched aldehyde.

A reasonable mechanism for alkane formation is also shown. Oxidative-addition of $H_2$ to alkyl complex intermediates rather than to acyl complex intermediates gives alkane products rather than aldehyde products upon reductive-elimination.

The rate-determining step seems to be the reductive-elimination of aldehyde. Inhibition of the reaction rate by high CO pressures is caused by the necessary requirement of having 16-electron species as intermediates. High CO pressure also favors $Co_2(CO)_8$ formation via the following reaction sequence:

$$HCo(CO)_3 + HCo(CO)_4 \rightleftharpoons Co_2(CO)_7 + H_2$$

$$\big\updownarrow \text{CO}$$

$$Co_2(CO)_8 \qquad\qquad (13.19)$$

However, some CO pressure is needed to prevent the decomposition of $HCo(CO)_4$ to metallic cobalt (which precipitates from solution). At 120°C, this $P_{CO}$ value is about 10 atm; at 200°C, it rises to about 100 atm.

Tertiary phosphine-modified hydrido cobalt carbonyl catalysts usually give complete reduction to alcohols under hydroformylation conditions. Tributylphosphine is commonly used. The catalyst precursor is $HCo(CO)_3L$, and the catalytically-active species is the 16-electron derivative, $HCo(CO)_2L$. Because these catalysts are more stable, synthesis gas pressures as low as 5−10 atm at 100−180°C can be used. The linear to branched product ratios are as high as 8:1. However, these reactions are considerably slower than reactions catalyzed by unmodified cobalt catalysts. The increased steric crowding by the bulky phosphine ligand is responsible for the greater proportion of linear product.

An important recent development in hydroformylation chemistry is the discovery that tertiary phosphine hydrido rhodium carbonyl complexes act as very selective catalysts under very mild reaction conditions. Union Carbide began using these catalysts in a commercial hydroformylation process in 1976. Concurrent research in Wilkinson's laboratory on the mechanism of hydroformylation with $HRh(CO)(PPh_3)_3$ as a catalyst led to a good understanding of the role of these complexes in the commercial catalytic cycle. Because of this work, the reagent complex $ClRh(PPh_3)_3$ is known as *Wilkinson's catalyst.*

The rhodium catalyst is usually prepared by reacting finely dispersed rhodium metal or rhodium complexes, such as $Rh_4(CO)_{12}$, $[Rh(CO)_2Cl]_2$, or $RhCl(CO)L_2$, where L is a tertiary phosphine ligand, with synthesis gas in the presence of $Ph_3P$ or a phosphite. The immediate catalyst precursor is $HRh(CO)(PPh_3)_3$ but other complexes, including $HRh(CO)(PPh_3)_2$, $HRh(CO)_2(PPh_3)_2$, $HRh(CO)_2(PPh_3)$,

HRh(CO)(PPh$_3$)$_2$, and HRh(CO)$_3$(PPh$_3$), are probably present through ligand dissociation and ligand complexation equilibria. Reaction temperatures and pressures fall in the ranges of 80–120° and 10–25 atm. The presence of excess phosphine ligand increases the selectivity of linear aldehyde formation and stabilizes the tertiary phosphine hydrido rhodium carbonyl complexes. Aldehyde products can be distilled directly from the reaction solution. Excess phosphine ligand also depresses rhodium-catalyzed hydrogenation of the alkene.

For example, with HRh(CO)(PPh$_3$)$_3$ as the catalyst precursor, hydroformylation of propene at 100°C in 35 atm of 1:1 CO/H$_2$ gives a linear to branched aldehyde ratio of slightly over 1:1. With a ten-fold excess of Ph$_3$P, the selectivity of linear aldehyde is about doubled. In molten Ph$_3$P (mp 79°C) as solvent, the linear to branched aldehyde ratio increases to 15.3:1.

Two mechanisms proposed for the modified rhodium catalyzed hydroformylation reaction are shown in figures 13.4 and 13.5. A dissociative mechanism (figure 13.4) involves the initial loss of a PPh$_3$ ligand to generate a 16-electron four-coordinate complex that coordinates the alkene molecule. Subsequent 1,2-addition of the

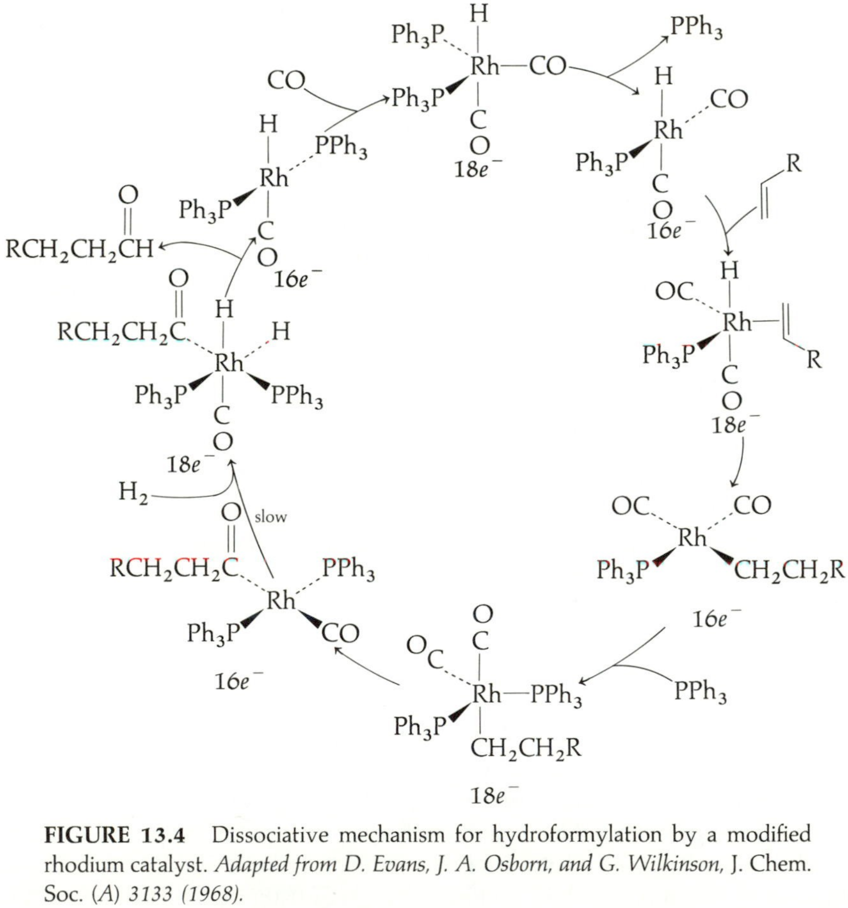

**FIGURE 13.4**  Dissociative mechanism for hydroformylation by a modified rhodium catalyst. *Adapted from D. Evans, J. A. Osborn, and G. Wilkinson, J. Chem. Soc. (A) 3133 (1968).*

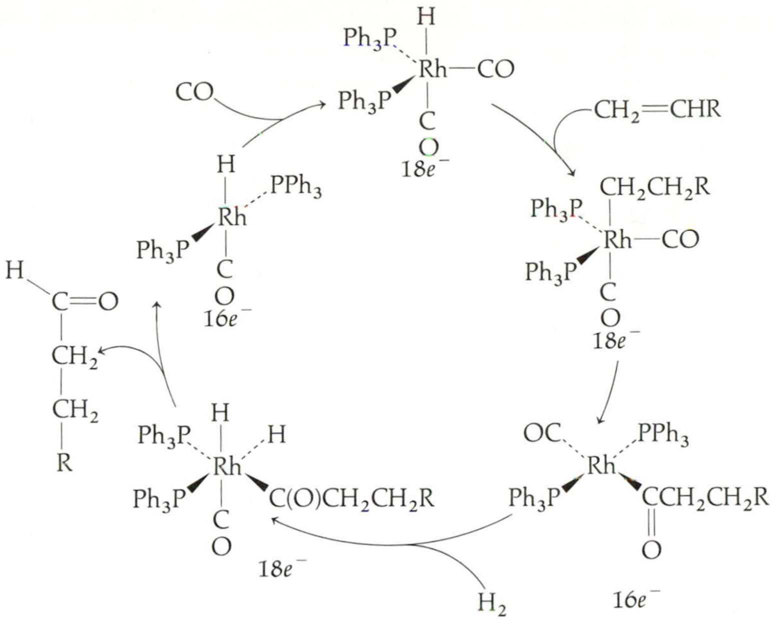

H
Ph₃P.
Rh—CO
Ph₃P
C
O
18e⁻

CO
PPh₃

H
Rh
Ph₃P
C
O
16e⁻

H
C=O
CH₂
CH₂
R

Ph₃P
H
H
Rh
Ph₃P
C
O
18e⁻
C(O)CH₂CH₂R

—CH₂=CHR

CH₂CH₂R
Ph₃P
Rh—CO
Ph₃P
C
O
18e⁻

OC
PPh₃
Rh
Ph₃P
CCH₂CH₂R
O
16e⁻

H₂

**FIGURE 13.5** Associative mechanism for hydroformylation by a modified rhodium catalyst. *Adapted from D. Evans, J. A. Osborn, and G. Wilkinson, J. Chem. Soc. (A) 3133 (1968).*

Rh—H bond to the $\eta^2$-alkene ligand gives an alkyl complex that can then undergo alkyl migration to a CO ligand. Oxidative-addition of $H_2$ to form a *cis*-dihydrido acyl intermediate is the slow step of the cycle. Reductive-elimination of aldehyde affords a four-coordinate complex, which adds another CO ligand to complete the cycle.

At very high concentrations of excess tertiary phosphine an associative mechanism (figure 13.5) is postulated. In this mechanism, the alkene coordinates to the 18-electron five-coordinate complex, $HRh(CO)_2(PPh_3)_2$, to eventually afford an 18-electron five-coordinate alkyl complex. Direct coordination of alkene to $HRh(CO)_2(PPh_3)_2$ would give a high-energy 20-electron intermediate. However, a concerted associative mechanism for this step involving Rh—H bond cleavage concomitant with Rh-alkene bond formation might be possible. A dissociative route involving initial loss of CO, addition of alkene, then 1,2-addition and recombination with CO, has been ruled out by NMR spectroscopy. The remaining steps of the associative mechanism are as expected. Oxidative-addition of $H_2$ is again the slowest step in the cycle.

The high selectivity for linear aldehyde shown by these modified rhodium catalysts in the presence of excess phosphine ligand is presumably determined at the alkene insertion step to give the alkyl complex. At high phosphine concentration, two bulky phosphine ligands are present in equatorial sites. As the alkyl ligand is being formed, the large cone angles of these phosphine ligands sterically favor the formation of the least bulky alkyl ligand, which is linear rather than branched.

Spectroscopic examination of reaction solutions under catalytic reaction conditions is being used to investigate the species present in these systems. High-pressure infrared data is particularly informative. For example, the more stable and less reactive catalyst, $HIr(CO)_3(PR_3)$, where R is isopropyl, can be treated sequentially with 200 psi of ethylene at 50°C followed by 200 psi of $H_2$ at 50°C. Infrared examination of these reactions reveals a reaction sequence of $HIr(CO)_3(PR_3) \rightarrow EtIr(CO)_3(PR_3) \rightarrow EtC(O)Ir(CO)_3(PR_3) \rightarrow EtCHO + HIr(CO)_3(PR_3)$. Although the dihydride complex resulting from oxidative-addition of $H_2$ is not observed spectroscopically, related iridium complexes, like $HIr(Cl)(acyl)(CO)L_2$, are known compounds.

As table 13.4 reveals, the modified rhodium catalysts are superior to the unmodified or modified cobalt catalysts for hydroformylation of terminal alkene feedstocks to linear aldehydes. These rhodium catalysts are very selective for linear aldehydes. Linear aldehydes are produced in high yield with very little by-product formation. Reaction temperature and pressure are also lower for the rhodium processes. Although rhodium is about 3500 times more expensive than cobalt, these rhodium catalysts are 100 to 1000 times more reactive than the cobalt catalysts, and their use means cost savings on plant construction and reduced energy consumption because of the milder reaction conditions of the rhodium process. Efficient catalyst recovery will make rhodium-based processes attractive for commercial utilization.

An interesting application of hydroformylation chemistry to organic synthesis is the possible preparation of optically active aldehydes. As shown in equations 13.20 and 13.21, addition of H and CHO groups across a carbon-carbon double bond in either terminal or internal prochiral alkenes can afford optically active aldehydes.

$$(R)(R')C{=}CH_2 \xrightarrow[+CHO]{+H} (R)(R')\overset{*}{C}HCH_2CHO + (R)(R')\overset{*}{C}(CHO)CH_3 \quad (13.20)$$

$$(R)(H)C{=}C(H)(R') \xrightarrow[+CHO]{+H} RCH_2\overset{*}{C}H(CHO)R' + R\overset{*}{C}H(CHO)CH_2R' \quad (13.21)$$

Notice that the optical center is formed when the alkene inserts into the Rh—H bond. Ancillary phosphine ligands influence the *regioselectivity* of this insertion, and opti-

**TABLE 13.4   Comparison of cobalt and rhodium catalysts for hydroformylation of alkenes.**

|  | Unmodified cobalt | Ligand modified cobalt | Ligand modified rhodium |
|---|---|---|---|
| Temperature (°C) | 140–180 | 160–200 | 80–120 |
| Pressure (atm) | 250–350 | 50–100 | 15–25 |
| Cobalt concentration (% metal/alkene) | 0.1–1.0 | 0.5–1.0 | $10^{-2}$–$10^{-3}$ |
| Linear: branched ratio | 3–4:1 | 6–8:1 | 10–14:1[a] |
| Aldehydes (%) | ~80 | — | ~96 |
| Alcohols (%) | ~10 | ~80 | — |
| Alkanes (%) | ~1 | ~15 | ~2 |
| Other products (%) | ~9 | ~5 | ~2 |

[a] With terminal alkenes as feedstock.
*Source*: C. Masters, *Homogeneous Transition-Metal Catalysis.* (New York: Chapman and Hall, 1981).

cally active phosphine ligands might also influence its *stereoselectivity*, which is fixed on complexation of the alkene. Such stereoselectivity has been observed when using optically active phosphine ligands, such as DIOP (**13.1**), or **13.2**. Optical yields in excess of 44% have been reported, indicating at least partial asymmetric induction.

**13.1**

DIOP

**13.2**

## The Wacker Process

Before about 1960, acetaldehyde was produced commercially from ethanol, LPG, or acetylene. In the late 1950s, a research group at Wacker Chemie developed a homogeneous palladium-catalyzed process for oxidizing ethylene to acetaldehyde. Although newer processes are now being adopted, the Wacker processes for synthesizing acetaldehyde from ethylene or acetone from propene are still used.

Conversion of ethylene to acetaldehyde differs from hydroformylation in that a $C_n$ olefin is converted to a $C_n$ aldehyde, *not* a $C_{n+1}$ aldehyde. The commercial feasibility of the Wacker synthesis of acetaldehyde is demonstrated by equations *13.22–13.25.*

$$PdCl_4^{2-} + C_2H_4 + H_2O \longrightarrow CH_3CHO + Pd^0 + 2HCl + 2Cl^- \quad (13.22)$$

$$Pd^0 + 2CuCl_2 + 2Cl^- \longrightarrow PdCl_4^{2-} + 2CuCl \quad (13.23)$$

$$2CuCl + \tfrac{1}{2}O_2 + 2HCl \longrightarrow 2CuCl_2 + H_2O \quad (13.24)$$

$$\overline{\qquad C_2H_4 + \tfrac{1}{2}O_2 \longrightarrow CH_3CHO \qquad} \quad (13.25)$$

In equation *13.22,* ethylene is oxidized in aqueous solution to acetaldehyde by Pd(II). This reaction was reported in 1894. Since a stoichiometric amount of palladium is reduced for each mole of ethylene oxidized, this reaction is not attractive economically. However, the Wacker chemists realized that cupric chloride oxidizes palladium metal back to Pd(II) (eq. *13.23*) so that a mole of acetaldehyde could be formed at the cost of converting two moles of Cu(II) to Cu(I). This was a key observation because classical redox chemistry reveals that cuprous ion can be oxidized *by air* back to cupric ion in acidic media, as shown in equation *13.24.* The net reaction, equation *13.25,* shows ethylene oxidation to acetaldehyde by molecular oxygen from air. The overall process is catalytic in both palladium and copper.

The pedagogic importance of coupling three known reactions to give an economically important catalytic process cannot be over emphasized. Certainly new processes will be developed by using a similar approach.

Figure 13.6 shows a proposed mechanism for the palladium-catalyzed oxidation of ethylene to acetaldehyde in a homogeneous process. Although some questions still exist on details of the mechanism, much work supports the catalytic cycle given. Kinetic data are consistent with the rate law of equation 13.26. Notice that the reaction is inhibited by high concentrations of acid or chloride ion.

$$d[CH_3CHO]/dt = k[PdCl_4^{2-}][C_2H_4]/[Cl^-]^2[H^+] \qquad (13.26)$$

The first two steps of the catalytic cycle involve ligand substitution of chloride by ethylene and water, which is the solvent, to give the neutral *dihapto*-ethylene complex, $[PdCl_2(H_2O)(\eta^2\text{-}C_2H_4)]$. Both of these steps are equilibria; thus, high chloride concentration will be inhibitory. Formation of the neutral $(\eta^2\text{-alkene})Pd$ complex presumably renders the coordinated ethylene more susceptible to nucleophilic attack by water.

There is general agreement that hydroxypalladation to afford a $\beta$-hydroxyethyl complex is the next step of the cycle. By using *cis*-CHD=CHD instead of ethylene, the stereochemistry of the addition was established as *anti*. This result precludes the intermediate formation of a hydroxy ligand, as in $[cis\text{-}PdCl_2(OH)(\eta^2\text{-}C_2H_4)]^-$, which could then undergo a *cis*- or *syn*-1,2-addition of the Pd—OH bond across the olefinic C=C bond to give $[PdCl_2(CH_2CH_2OH)(H_2O)]^-$. Kinetic arguments rule out external nucleophilic attack by hydroxide ion under usual reaction conditions. At present, formation of the $\beta$-hydroxyethyl complex is believed to occur by *trans* attack by an external water molecule on the *dihapto*-ethylene ligand.

Decomposition of the $\beta$-hydroxyethyl complex to acetaldehyde and metallic palladium is obviously an important step in the catalytic cycle. When the reaction is conducted in $D_2O$, *no* deuterium is incorporated into the acetaldehyde product (eq. 13.27).

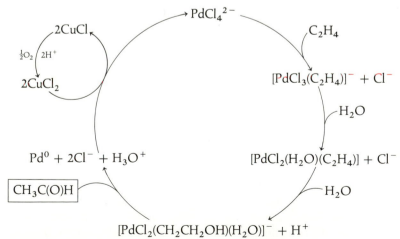

**FIGURE 13.6**   The oxidation of ethylene to acetaldehyde by palladium(II).

$$[PdCl_2(CH_2CH_2OD)(D_2O)]^- \longrightarrow CH_3CHO + Cl^- + Pd^0 \quad (13.27)$$

Clearly, a 1,2-hydrogen shift to form acetaldehyde from the $\beta$-hydroxyethyl complex must occur faster than any H—D exchange or ligand-exchange process. The hydroxyl proton must also be lost.

A sequence of steps is proposed to explain the mechanism of acetaldehyde elimination. As shown in reaction scheme *13.28*, loss of chloride ion from the $\beta$-hydroxyethyl complex gives a coordinatively unsaturated intermediate which rapidly undergoes $\beta$-hydrogen elimination affording a (hydrido)($\eta^2$-vinyl alcohol)Pd complex.

$$[(H_2O)Cl_2PdCH_2CH_2OH]^- \xrightarrow[\text{slow}]{-Cl^-} [(H_2O)ClPd\overset{\displaystyle H_2COH}{\underset{\displaystyle |}{-CH_2}}]$$

$$\left[ (H_2O)Cl_2Pd\overset{\displaystyle CH_3}{\underset{\displaystyle \underset{O}{\overset{|}{-}}C-H}{|}}_{\displaystyle \diagup H} \right]^- \xleftarrow[\text{Cl}^- \text{ (fast)}]{\text{1,2-addition}} \left[ (H_2O)ClPd\overset{\displaystyle H\;\;HCOH}{\underset{\displaystyle \underset{CH_2}{\|}}{\underset{|}{\leftarrow}}} \right]$$

$\beta$-hydrogen elimination (fast) $\downarrow$

$$\xrightarrow[\text{elimination}]{\beta\text{-hydrogen}} CH_3CHO + \{[HPdCl_2(H_2O)]^-\} \longrightarrow Pd^0 + 2Cl^- + H_3O^+$$

$$(13.28)$$

Rapid 1,2-addition in the opposite direction gives an $\alpha$-hydroxyethyl intermediate. *Beta*-hydrogen elimination of the hydroxyl proton forms acetaldehyde and the hydrido complex, $[HPdCl_2(H_2O)]^-$. In the absence of more stabilizing ligands, this hydride complex undergoes reductive-elimination of HCl to give metallic palladium. The Pd(II) catalyst is regenerated according to equations *13.23* and *13.24*.

## Monsanto Acetic Acid Process

Most of the acetaldehyde produced commercially is used to prepare acetic acid. Acetic acid and acetic anhydride are major industrial chemicals involved in the manufacture of vinyl acetate, cellulose acetate, pharmaceuticals, dyes, and pesticides. Acetic acid is also the solvent for the oxidation of *p*-xylene to terephthalic acid, which is used in synthetic fibers, and acetate esters are widely used solvents.

The three major industrial routes to acetic acid or its derivatives all rely on homogeneous catalysis. These methods are (1) oxidation of ethylene *via* acetaldehyde (Wacker Process), (2) free-radical oxidation of butane or naphtha (a mixture of alkanes that boils between 80 and 110°C) by manganese or cobalt salts, and (3) the carbonylation of methanol. Carbonylation of methanol to acetic acid is effected by cobalt or rhodium catalysts:

$$CH_3OH + CO \xrightarrow{\text{cat.}} CH_3CO_2H \quad (13.29)$$

In 1970, the Monsanto process introduced the use of rhodium catalysts. World-wide annual production of acetic acid in 1977 was about 2.5 million tons. By the mid-1980s, about 1 million tons of acetic acid will be produced by the rhodium-based Monsanto process.

As with hydroformylation, rhodium catalysts are proving superior for carbonylation to the older cobalt catalysts because they require milder reaction conditions and show greater selectivity for acetic acid production (see table 13.5). Since hydrogen is a common impurity in industrial carbon monoxide streams, the insensitivity of the rhodium catalyst to hydrogen eliminates the formation of hydrogenation by-products as observed with cobalt catalysts.

Rhodium is added to the reaction vessel as $RhX_3 \cdot 3H_2O$ salts or as Rh(I)-phosphine complexes such as $RhX(CO)(PPh_3)_2$. Iodide ion is introduced as $CH_3I$, aqueous HI, or $I_2$. When water is the solvent, acetic acid forms. If methanol is the solvent, then methyl acetate is produced. Regardless of the source of rhodium, the catalyst is the rhodium-iodide complex $[RhI_2(CO)_2]^-$. Equation *13.30* describes the formation of this complex in aqueous media by reductive-carbonylation.

$$RhI_3 + 3CO + H_2O \longrightarrow [RhI_2(CO)_2]^- + CO_2 + 2H^+ + I^- \quad (13.30)$$

This reaction is known to proceed under the carbonylation conditions of the Monsanto process conducted in hydroxylated solvents. If rhodium-phosphine species are used as sources of rhodium, then the phosphine ligands dissociate from the metal and are quaternized by methyl iodide, which is generated in the system.

Figure 13.7 shows a mechanism for the rhodium-catalyzed carbonylation of methanol to acetic acid. Notice that the reaction is catalytic in both rhodium and iodide ion. Under the reaction conditions used, a molecule of methanol is converted into methyl iodide by reaction with HI. Essentially all of the HI formed during the cycle is consumed by converting methanol to methyl iodide.

Methyl iodide undergoes oxidative-addition to the $cis$-$[RhI_2(CO)_2]^-$ complex to give a six-coordinate methyl complex. Methyl migration to a $cis$-CO ligand gives a five-coordinate acetyl complex. Complexation of a CO molecule affords the six-coordinate acetyl complex, $fac$-$\{RhI_3(CO)_2[C(O)CH_3]\}^-$. Reductive-elimination of acetyl iodide regenerates the catalyst, $cis$-$[RhI_2(CO)_2]^-$. Acetyl iodide reacts with the aqueous solvent to form acetic acid and HI. The HI formed then converts another molecule of methanol to methyl iodide, and the cycle repeats.

**TABLE 13.5   Comparison of cobalt and rhodium catalysts for the carbonylation of methanol to acetic acid.**

|  | *Cobalt catalyst* | *Rhodium catalyst* |
|---|---|---|
| Metal concentration ($M$) | $\sim 10^{-1}$ | $\sim 10^{-3}$ |
| Temperature (°C) | $\sim 230$ | $\sim 180$ |
| Pressure (atm) | 500–700 | 30–40 |
| Selectivity (on MeOH) | 90% | >99% |
| Hydrogen effect | $CH_4$, $CH_3CHO$, EtOH as by-products | no effect |

*Source*: C. Masters, *Homogeneous Transition-Metal Catalysis*. (New York: Chapman and Hall, 1981).

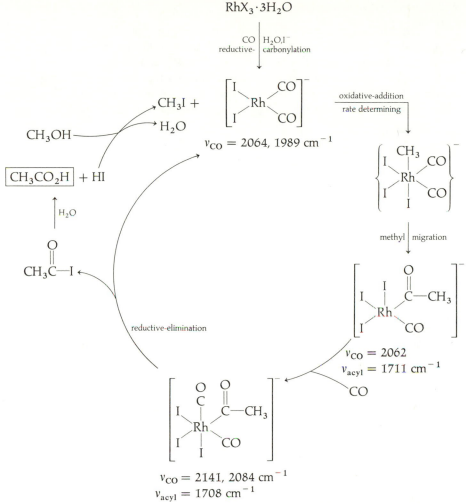

**FIGURE 13.7**   Carbonylation of methanol to acetic acid by a rhodium iodide catalyst. *Adapted from D. Forster, Adv. Organometal. Chem.* **17**, 255 (1979).

As indicated by the listed $\nu_{CO}$ stretching frequencies, all of the rhodium complexes except for the methyl intermediate have been characterized. Methyl migration of this methyl intermediate occurs without the need of external CO. An x-ray structure of the $Me_3PhN^+$ salt of the five-coordinate acetyl complex reveals a weakly bonded dimeric structure with two Rh—I—Rh bridging bonds. Addition of CO to this complex to give the six-coordinate acetyl complex occurs at only about one atmosphere of CO pressure, so the reaction rate shows no dependence on CO pressure. Kinetic data indicate that the oxidative-addition of methyl iodide to $[RhI_2(CO)_2]^-$ is the rate-determining step.

Acetyl iodide reacts with the solvent to give acetic acid in aqueous media or methyl acetate when methanol is the solvent. Higher alcohols can be converted to the corresponding carboxylic acids in aqueous media via this same process.

### Monsanto L-DOPA Synthesis

Asymmetric hydrogenation of prochiral olefins to give optically active saturated compounds by using homogeneous optically active transition metal catalysts is being studied extensively. In 1973 Monsanto chemists, under the direction of W. S. Knowles, developed optically active rhodium catalysts that selectively hydrogenate prochiral N-acylaminocinnamic acids to either D- or L-N-acylamino acids. This process is used to prepare L-DOPA, a drug used in the treatment of Parkinson's disease:

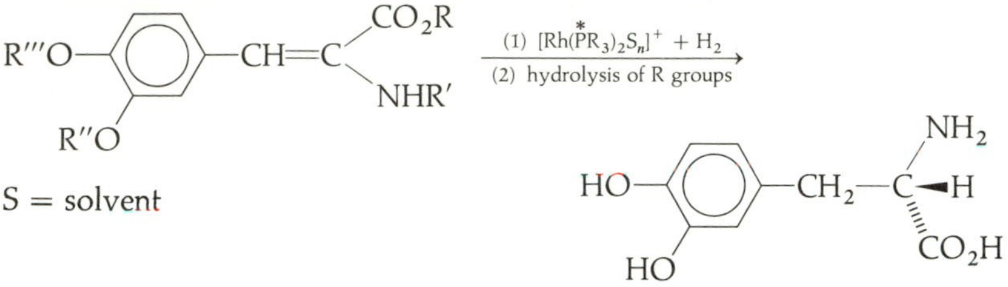

$$R'''O-\underset{R''O}{\bigcirc}-CH=C\underset{NHR'}{\overset{CO_2R}{}} \xrightarrow[\text{(2) hydrolysis of R groups}]{\text{(1) }[Rh(\overset{*}{P}R_3)_2S_n]^+ + H_2}$$

$$HO-\underset{HO}{\bigcirc}-CH_2-\underset{CO_2H}{\overset{NH_2}{C}}\!\!\!-\!\!\!H \qquad (13.31)$$

S = solvent

S-3,4-dihydroxyphenylalanine
(L-DOPA)

The rhodium catalyst is prepared by treating cationic complexes of 1,5-COD, such as $[(\eta^4\text{-}1,5\text{-COD})Rh(\overset{*}{P}R_3)_2]BF_4$, with molecular hydrogen. The COD ligand is hydrogenated to cyclooctane, and solvent molecules coordinate to the cationic rhodium atom. Several optically active phosphine ligands, $\overset{*}{P}R_3$, have been used, including bisphosphine ligands like **13.1** and **13.2**. An optically active monophosphine ligand used in this process is the $R_P$-CAMP phosphine (**13.3**).

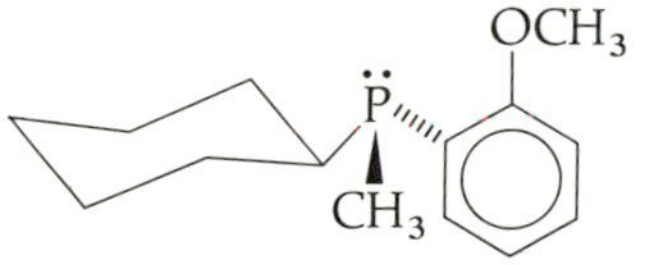

**13.3**

$R_P$-CAMP

One enantiomer of the rhodium catalyst leads to one enantiomer of the amino acid. The opposite enantiomer affords the amino acid of opposite configuration. Optical yields of up to 95% are obtained. The presumed mechanism entails coordination of the prochiral alkene to the optically-active rhodium catalyst. This $\eta^2$-alkene complex is diastereomeric. Oxidative-addition of hydrogen to rhodium gives a dihydrido complex, which then reduces the alkene ligand with high stereoselectivity (see chap. 14).

### STUDY QUESTIONS

1. During hydroformylation of terminal alkenes higher than propene isomerization of the olefinic bond (i.e., a shift in position along the carbon chain) is sometimes observed. Write a mechanism to explain this shift.

2. Write a mechanism for the catalytic hydrogenation of aldehydes to alcohols under hydroformylation conditions.

3. When the Wacker process is run with ethylene in acetic acid solvent, vinyl acetate forms. Write a mechanism for this catalytic process.

4. Hydrocyanation and hydrosilylation of alkenes, both of which are useful reactions for preparing compounds of interest to polymer chemists, are catalyzed by low-valent metals complexes such as those of nickel and platinum. Write a proposed mechanism for each of these reactions:

(a) $R_2C{=}CR_2 + HCN \xrightarrow{\text{cat.}} R_2HC{-}CR_2CN$

(b) $R_2C{=}CR_2 + HSiR_3 \xrightarrow{\text{cat.}} R_2HC{-}CR_2SiR_3$

5. Ethylene glycol can be synthesized from syn gas using metal carbonyl catalysts. Propose a mechanism for this reaction.

6. Acrylic acid can be prepared catalytically from acetylene, water, carbon monoxide, and $NiBr_2$ as follows:

$$HC{\equiv}CH + H_2O \xrightarrow[\substack{30 \text{ atm CO} \\ 150°C, \text{ THF}}]{\text{Ni cat.}} CH_2{=}CHCO_2H$$

The catalyst is believed to be $Ni(H)(X)(CO)_2$. Show how the catalyst can be formed from the reactants, and propose a reasonable mechanism for this process.

7. Methanol homologation to ethanol and higher alcohols can be accomplished with syn gas (160−180°C, 300 atm pressure) in the presence of $Co_2(CO)_8$ and an iodide promoter. Suggest a mechanism for this catalytic reaction. [See G. W. Parshall, *Homogeneous Catalysis*. (New York: John Wiley & Sons, 1980), p. 97.]

8. Hydroquinone is formed in iron- and ruthenium-catalyzed reactions of acetylene:

$$2HC{\equiv}CH + 3CO + H_2O \longrightarrow HO{-}\langle\bigcirc\rangle{-}OH + CO_2$$

Propose a mechanism for this process utilizing an iron carbonyl catalyst. [See P. Pino and G. Braca, *Organic Syntheses via Metal Carbonyls*, Vol. 2, I. Wender and P. Pino, eds. (New York: John Wiley & Sons, 1977), p. 419.]

9. Write a mechanism for the oligomerization of ethylene to butene using $HNiL_3X$ as a catalyst.

10. Consult the available literature and propose a mechanism for the polymerization of ethylene to polyethylene by a Ziegler−Natta catalyst.

## SUGGESTED READING

### General References

Collman, J. P. *Acc. Chem. Res.* **1**, 136 (1968).

Khan, M. M. T., and Martell, A. E. *Homogeneous Catalysis by Metal Complexes.* Vol. 1. New York: Academic Press, 1974.

Masters, C. *Homogeneous Transition-Metal Catalysis.* New York: Chapman and Hall, 1981.

Parshall, G. W. *Homogeneous Catalysis.* New York: John Wiley & Sons, 1980.

Parshall, G. W. *Science.* **208**, 1221 (1980).
Ugo, R., ed. *Aspects of Homogeneous Catalysis.* Vol. 1–4, Boston: D. Reidel Publishing.

### Surface Catalysis

Buchholz, J. C., and Somorjai, G. A. *Acc. Chem. Res.* **9**, 333 (1976).
Castner, D. G., and Somorjai, G. A. *Chem. Rev.* **79**, 233 (1979).
Kesmodel, L. L., and Somorjai, G. A. *Acc. Chem. Res.* **9**, 392 (1976).
Mason, R., and Roberts, M. W. *Inorg. Chim. Acta.* **50**, 53 (1981).
Muetterties, E. L. *Science.* **196**, 839 (1977).
Muetterties, E. L., Rodin, T. N., Band, E., Brucker, C. F., and Pretzer, W. R. *Chem. Rev.* **79**, 91 (1979).
Ozin, G. A. *Acc. Chem. Res.* **10**, 21 (1977).
Schaefer, H. F. *Acc. Chem. Res.* **10**, 287 (1977).
Somorjai, G. A. *Acc. Chem. Res.* **9**, 248 (1976).
Tully, J. C. *Chem. Res.* **14**, 188 (1981).

### Supported Catalysts

Hartley, F. R., and Vezey, P. N. *Adv. Organometal. Chem.* **15**, 189 (1977).
Manecke, G., and Storck, W. *Angew. Chem. International Ed.* **17**, 657 (1978).
Yermakov, Y. I. *Catal. Rev.* **13**, 77 (1976).

### Hydroformylation

Chalk, A. J., and Harrod, J. F. *Adv. Organometal. Chem.* **6**, 119 (1968).
Jardine, F. H. *Progr. Inorg. Chem.* **28**, 63 (1981).
Orchin, M. *Acc. Chem. Res.* **14**, 259 (1981).
Pino, P. *J. Organometal. Chem.* **200**, 223 (1980).
Pruett, R. L. *Adv. Organometal. Chem.* **17**, 1 (1979).
Siegel, H., and Himmele, W. *Angew. Chem. International Ed.* **19**, 178 (1980).

### Wacker Process

Aguiló, A. *Adv. Organometal. Chem.* **5**, 321 (1967).
Lyons, J. E. *Adv. Chem. Series.* **132**, 64 (1974).
Maitlis, P. M. *The Organic Chemistry of Palladium.* Vol. II, *Catalytic Reactions.* New York: Academic Press, 1971.

### Monsanto Acetic Acid Process

Forster, D. *Adv. Organometal. Chem.* **17**, 255 (1979).
Forster, D. *N. Y. Acad. Sci.* **295**, 79 (1977).

### Monsanto L-DOPA Synthesis

Knowles, W. S. *Acc. Chem. Res.* **16**, 106 (1983).
Knowles, W. S., Sabacky, M. J., and Vineyard, B. D. *Homogeneous Catalysis-11.* D. Forster and J. F. Roth, eds. p. 274. Washington: American Chemical Society, 1974.
Knowles, W. S., and Sabacky, M. J. U. S. Patent (1974) 3, 849, 480.

# Research in Catalysis

## CHAPTER 14

Because of possible applications to chemical catalysis, several areas of organometallic chemistry are the subject of intensive basic research. Cluster chemistry is one such topic that has been included in many of the preceding chapters. Although it is difficult to assess which basic research results will prove directly relevant to the development of catalytic processes, it is possible to summarize briefly a few areas of research that are currently being investigated for potential applications. The research results may also increase our understanding of the likely intermediates in catalytic systems.

### HOMOGENEOUS ANALOGUES TO THE SHIFT REACTION

The water gas shift reaction (eq. 13.13) has a $\Delta G^0_{298}$ of $-4.76$ kcal/mole using liquid water, and $-6.82$ kcal/mole with water vapor. The development of homogeneous metal catalysts for this reaction appears to be favorable thermodynamically, and a homogeneous process would require much less energy than is now expended in the heterogeneous one. Furthermore, in homogeneous systems that obtain hydrogen directly from water in the presence of carbon monoxide, the utilization of free $H_2$ might not be required.

Iron pentacarbonyl forms $H_2$ when treated with basic water ($pH \approx 10.7$) according to the water gas shift reaction (eq. 14.1). Nucleophilic attack on a carbonyl ligand gives a metallacarboxylic acid (eqs. 11.48 and 11.50), which decomposes to $CO_2$ and the basic anionic complex $HFe(CO)_4^-$. Protonation by the protic solvent gives $H_2Fe(CO)_4$, which generates $H_2$ by reductive-elimination, and the eventual regeneration of $Fe(CO)_5$ in the presence of carbon monoxide.

$$Fe(CO)_5 + OH^- \longrightarrow HFe(CO)_4^- + CO_2$$

$$\uparrow CO \qquad\qquad \downarrow H_2O$$

$$Fe(CO)_4 + H_2 \longleftarrow H_2Fe(CO)_4 + OH^- \qquad\qquad (14.1)$$

Under more basic conditions, $pH \approx 12$, $Fe(CO)_5$ is converted to $HFe(CO)_4^-$ only, and the dihydride complex is not formed in significant amounts. Nitrobenzenes can

413

be reduced to anilines under these conditions, as shown for nitrobenzene in equation 14.2. A postulated mechanism for the generation of the reducing agent for this catalytic process is shown in equation 14.3.

$$PhNO_2 + 3CO + H_2O \longrightarrow PhNH_2 + 3CO_2 \tag{14.2}$$

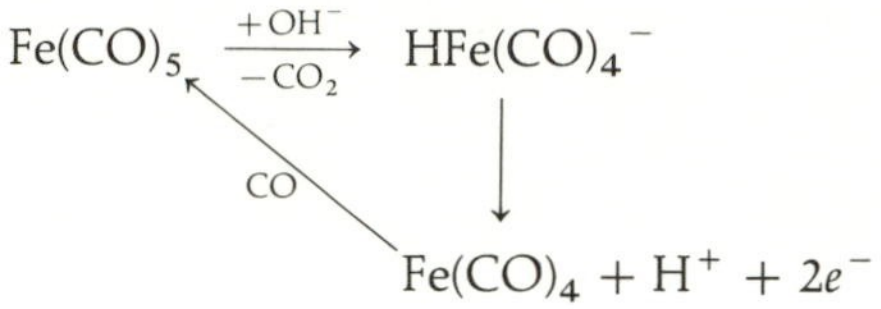

$$Fe(CO)_5 \xrightarrow[-CO_2]{+OH^-} HFe(CO)_4^-$$

$$\downarrow$$

$$Fe(CO)_4 + H^+ + 2e^- \tag{14.3}$$

When $D_2O$ is the solvent, the *minimum* incorporation of deuterium in the aniline product is 88%, indicating that water is indeed the source of the reducing agent.

Many research groups have discovered that a wide variety of binary metal carbonyl complexes catalyze the water gas shift reaction. A general reaction scheme for this catalysis is shown in equation 14.4. Hydroxide attack at a CO ligand causes loss of $CO_2$ and formation of an anionic hydride complex. Protonation by water gives a dihydride intermediate, which reductively eliminates $H_2$, and a coordinatively unsaturated metal carbonyl fragment, which picks up another molecule of CO. The process can then be repeated.

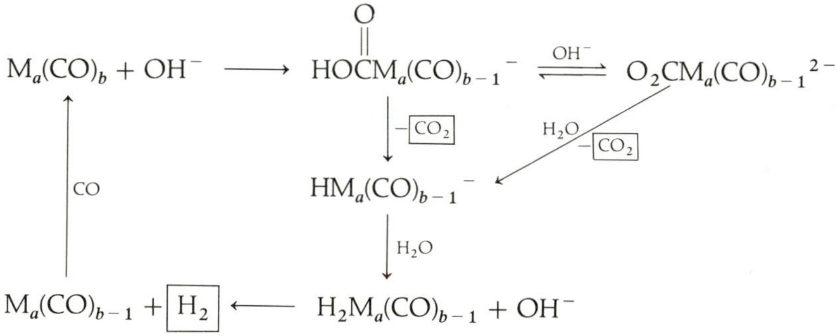

$$M_a(CO)_b + OH^- \longrightarrow HOCM_a(CO)_{b-1}^- \overset{OH^-}{\rightleftharpoons} O_2CM_a(CO)_{b-1}^{2-}$$

$$\uparrow CO \qquad \downarrow -\boxed{CO_2} \qquad \overset{H_2O}{\underset{-\boxed{CO_2}}{\diagup}}$$

$$HM_a(CO)_{b-1}^-$$

$$\downarrow H_2O$$

$$M_a(CO)_{b-1} + \boxed{H_2} \longleftarrow H_2M_a(CO)_{b-1} + OH^- \tag{14.4}$$

One of the more active shift catalysts is the platinum A-frame complex **14.1.** Although the mechanism of reaction is not known, the catalysis presumably pro-

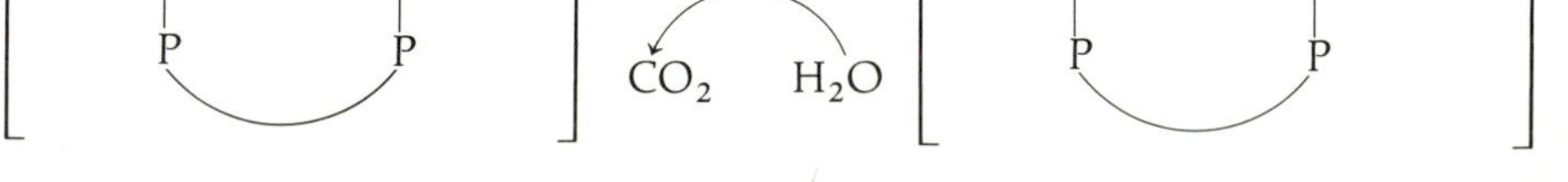

**14.1** (14.5)

$$P \frown P = Ph_2PCH_2PPh_2$$

ceeds according to equation *14.5*. Either the $PF_6^-$ or $BPh_4^-$ salts of **14.1** can be used. The reaction is carried out at 100°C in a solvent consisting of two parts methanol to one part water by volume with a carbon monoxide partial pressure of 110 lb/in². Under these conditions, the turnover rate, which is defined as the number of moles of reactant consumed per mole of catalyst in unit time, is 3.0 ± 0.2 moles of $H_2$ or CO per mole of **14.1** per hour.

A platinum-catalyzed process for the shift reaction is shown in figure 14.1. This process involves coupling Pt(0)-Pt(II) and Pt(II)-Pt(IV) redox couples with a Sn(II)-Sn(IV) couple. Tin(IV) oxidizes Pt(0) to Pt(II) in the top $CO_2$ cycle, and Sn(II) reduces Pt(IV) to Pt(II) in the lower $H_2$ cycle. Still, $CO_2$ is produced by water attack on a carbonyl complex, and $H_2$ is produced by proton addition to a hydride complex.

The rationale for examining a Pt-Sn system arose from the observations (1) that CO reduces Pt(II) to Pt(0) forming $CO_2$; (2) that Pt(II), HCl, and $SnCl_2$ in aqueous HOAc give $H_2$ evolution; and (3) that $SnCl_3^-$ ligands *inhibit* reduction of Pt(II) to Pt(0) *by hydrogen*. Specific reaction conditions for this catalytic system are 0.23 mmol $K_2PtCl_4$, 7.1 mmol $SnCl_4 \cdot 5H_2O$, 40 ml glacial HOAc, 10 ml concentrated HCl, and 10 ml $H_2O$. At 88°C, about 25 turnovers of $CO_2$ or $H_2$ per day per platinum ion are observed with no loss of activity. The approach of coupling redox couples of main group elements with those of transition metals seems very promising. However, very few specific redox potentials have been measured.

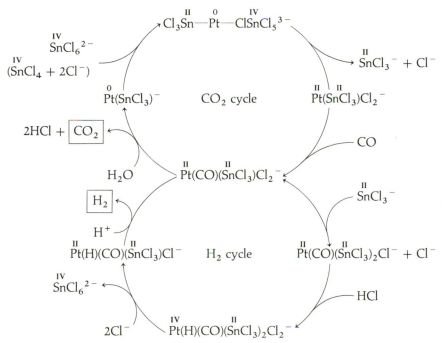

**FIGURE 14.1**   A postulated mechanism of a platinum-tin coupled catalyst for the water gas shift reaction. *Adapted from C.-H. Cheng and R. Eisenberg J. Am. Chem. Soc.* **100**, *5968 (1978).*

## RESEARCH ON FISCHER–TROPSCH AND
## RELATED SYN GAS CHEMISTRY

Conversion of CO and $H_2$ into alkenes or alkanes via the Fischer–Tropsch process requires high temperatures even though these reactions are exothermic. The mechanisms by which the products are formed are not clearly understood. Surface species such as carbide (**14.2**), methylene (**14.3**), $\mu_2$-methylene (**14.4**), hydroxymethylene (**14.5**), hydroxyalkyl (**14.6**), formyl (**14.7**), and $\eta^2$-formaldehyde (**14.8**) ligands have been postulated as important intermediates.

**14.2**          **14.3**          **14.4**          **14.5**

**14.6**          **14.7**          **14.8**

While some organometallic chemists are working to develop homogeneous catalysts to effect Fischer–Tropsch chemistry, others are attempting to understand the observed chemistry via model studies. The isolation and characterization of complexes that contain ligand types such as **14.2–14.8** helps in the search for the true intermediates.

Two homogeneous chemical conversions of paramount importance to syn gas chemistry are reduction of a CO ligand to an $sp^2$- or $sp^3$-C center, and the coupling of a one-carbon ligand, such as a CO reduction product, to another one-carbon ligand to give a two-carbon ligand. These reactions have general importance to all syn gas chemistry regardless of whether totally deoxygenated products (for example, olefins or alkanes from the Fischer–Tropsch process) or oxygen-containing organic products (such as acetic acid, which is discussed later) are desired. In all cases, CO must be reduced, and usually the desired organic product contains two or more carbon atoms.

In previous chapters, several examples of CO reduction and C—C bond forming reactions have been presented. Facile CO reduction occurs via nucleophilic attack on CO ligands (see chap. 11). Carbon-carbon bond formation can occur via usual insertion or reductive-elimination reactions (chapters 9 and 10) or by unusual interligand coupling reactions (eq. 12.45). Selected examples of other model studies of these reactions are included here.

Direct formation of alkanes by reacting metal carbonyl complexes with reducing agents is shown in equations 14.6–14.9. In all cases, the yield of hydrocarbons is quite low. In equation 14.9, the presence of the strong Lewis acid $AlCl_3$ in the solvent melt greatly increases the yield and rate of alkane formation over the con-

$[(\eta^5\text{-}C_5H_5)Fe(CO)_2]_2 + LiAlH_4 \xrightarrow[\text{CO, 36 hrs}]{\text{toluene}}$

$$CH_4 + C_2H_4 + C_2H_6 + C_3H_6 + C_3H_8 + C_4H_8 + C_4H_{10} \quad (14.6)$$

$$1 \quad : \quad 1.6 \quad : \quad 0.9 \quad \text{trace} \quad \text{trace} \quad \text{trace} \quad \text{trace}$$

$$Ru_3(CO)_{12} \xrightarrow[\text{THF}]{\text{AlH}_3} CH_4 + C_2H_4 + C_2H_6 + C_3H_6 + C_3H_8 \quad (14.7)$$

$$1 \quad : \quad 1.7 \quad : \quad 0.5 \quad : \quad 0.2 \quad : \quad 0.1$$

$$Os_3(CO)_{12} \text{ or } Ir_4(CO)_{12} \xrightarrow[\substack{180°C \\ 5 \text{ days}}]{CO + H_2} CH_4 \quad (14.8)$$

$$10\text{--}15\%$$

$$Ir_4(CO)_{12} \xrightarrow[\substack{NaCl/AlCl_3 \text{ melt,} \\ 170\text{--}180°C}]{CO/H_2, \text{ 1 atm}} C_1 \text{ to } C_8 \text{ saturated hydrocarbons} \quad (14.9)$$

ditions shown in equation *14.8*. Presumably, Lewis acid coordination to the oxygen atom of a CO ligand activates CO reduction. Several examples of *isocarbonyl inter-actions*, M—C—O $\cdots$ M', are known (eq. *11.23*).

An example of *four* isocarbonyl-metal interactions within one compound occurs in the complex $[(\eta^5\text{-}C_5Me_5)_2Yb]_2Fe_3(CO)_{11}$. This complex is prepared by reacting $(\eta^5\text{-}C_5Me_5)_2Yb(OEt_2)$ with $Fe_3(CO)_{12}$. Two Lewis structures representing the delocalized electronic structure are shown as **14.9** and **14.10**. Each chelating pair of isocarbonyl ligands represent formal 1,3,5-trimetalla-2,4-pentandionato ligands, similar to the metalla-$\beta$-diketonate ligands (see eq. *11.45*). The isocarbonyl C—O distances and the Fe—C distances to the isocarbonyl ligands are 1.21(1) Å and 1.943(2) Å, respectively.

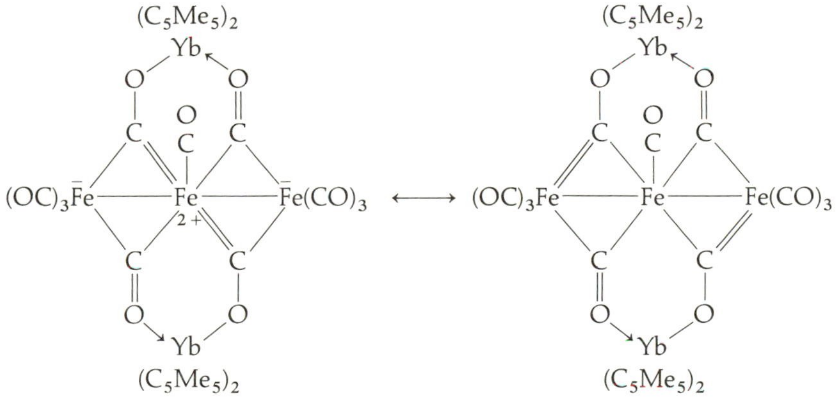

**14.9** **14.10**

Metal carbonyl anions can be protonated to give methane directly as shown in equation *14.9a*. When a deuterated acid is used, $CD_4$ is obtained.

$$Fe_4(CO)_{13}{}^{2-} + HOSO_2CF_3 \longrightarrow CH_4 + \text{other products} \quad (14.9a)$$

Although the detailed mechanism is not known, the reaction probably proceeds by protonation of a CO ligand with elimination of $H_2O$ to give a peripheral iron carbido cluster like $Fe_4C(CO)_{12}{}^{2-}$. Reduction probably occurs through oxidation of the iron core of the cluster.

A recent example of alkane formation via a reductive cleavage of an alkyl ligand is shown in equation *14.10*. Mechanistic studies indicate that a dinuclear metal dihydride and metal dimethyl complex of unknown composition might be involved in the reductive-elimination of methane. Such reactions model the termination step in alkane formation in Fischer–Tropsch chemistry.

$$(\eta^5\text{-}C_5H_5)CoMe_2L + H_2 + L \xrightarrow[45°C]{C_6D_6} 2CH_4 + (\eta^5\text{-}C_5H_5)CoL_2 \quad (14.10)$$

L = $PPh_3$

Several examples of CO-ligand reduction to give reduced one-carbon ligands are shown in equations *14.11–14.15*. In equation *14.11*, a zirconium hydride complex adds to a CO ligand to give a zirconoxymethylidene complex (**14.11**).

$$(\eta^5\text{-}C_5H_5)_2W(CO) + (\eta^5\text{-}C_5Me_5)_2ZrH_2 \longrightarrow$$

$$(\eta^5\text{-}C_5H_5)_2W \overset{2.005(4)\ \text{Å}}{=\!=\!=\!=} C \overset{H}{\underset{\underset{\underset{Zr(\eta^5\text{-}C_5Me_5)_2}{H}}{\overset{1.973(4)\ \text{Å}}{|}}}{\overset{1.354(3)\ \text{Å}}{\underset{166.4(7)°}{\diagdown}}} O} \quad (14.11)$$

**14.11**

Cyanoborohydride reduction of a CO ligand in a cationic complex gives an alkyl ether ligand (**14.12**). In this reaction the CO carbon atom has been reduced to an $sp^3$ center.

$$[(\eta^5\text{-}C_5H_5)Fe(CO)_3]^+BF_4{}^- \xrightarrow[ROH]{Na[B(CN)H_3]} (\eta^5\text{-}C_5H_5)Fe(CO)_2(CH_2OR) \quad (14.12)$$

R = Me or Et                                                                    **14.12**

Several interesting reduction reactions are shown in equation scheme *14.13*. Attack at a CO ligand in the cationic rhenium complex by $HB(Oi\text{-}Pr)_3{}^-$ gives the neutral formyl complex **14.13**. Reduction of the $sp^2$-formyl carbonyl carbon atom by $BH_3$ converts the formyl ligand into an $sp^3$-methyl ligand in complex **14.14**. Reduction of a CO ligand in the original complex by diethylaluminum hydride affords the hydroxymethyl complex **14.15**. This ligand is deoxygenated by $BH_3$ to form the methyl complex **14.14**.

Another example of the sequential reduction of a CO ligand is shown in equation scheme *14.14*. Hydride attack on $Os_3(CO)_{12}$ forms an anionic formyl complex, **14.16**, which can be protonated to give the hydroxymethylene **14.17**. Compound **14.17** is reduced further to a hydroxymethyl complex **14.18** by one equivalent of **14.16**. Formyl ligands are known to be excellent hydride donors. Protonation of **14.18** and loss of water gives the methylene osmium cluster **14.19**. This complex reacts with $H_2$ at one atmosphere and 75°C to give $CH_4$ in about 20% yield.

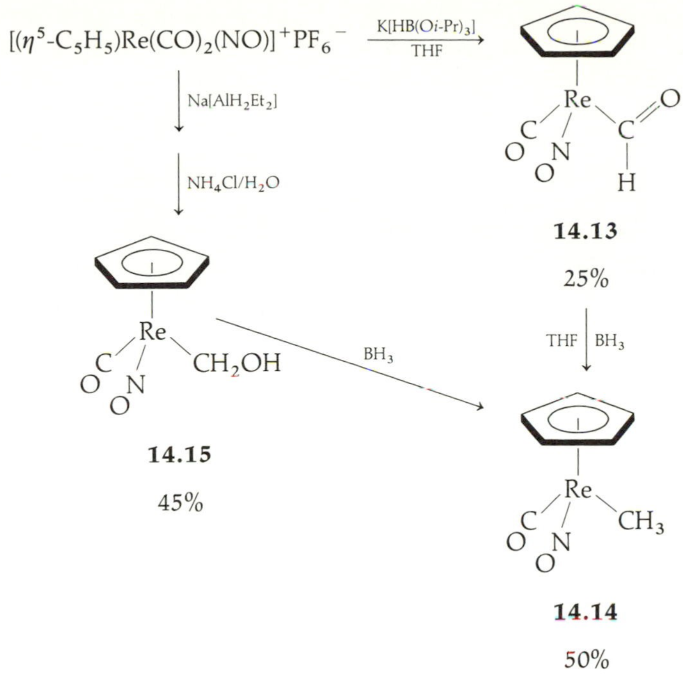

$$[(\eta^5\text{-}C_5H_5)Re(CO)_2(NO)]^+PF_6^- \xrightarrow[\text{THF}]{K[HB(Oi\text{-}Pr)_3]}$$

**14.13**

25%

$\Big\downarrow$ Na[AlH$_2$Et$_2$]

$\Big\downarrow$ NH$_4$Cl/H$_2$O

**14.15**

45%

$\xrightarrow{\text{BH}_3}$

THF $\Big|$ BH$_3$

(14.13)

**14.14**

50%

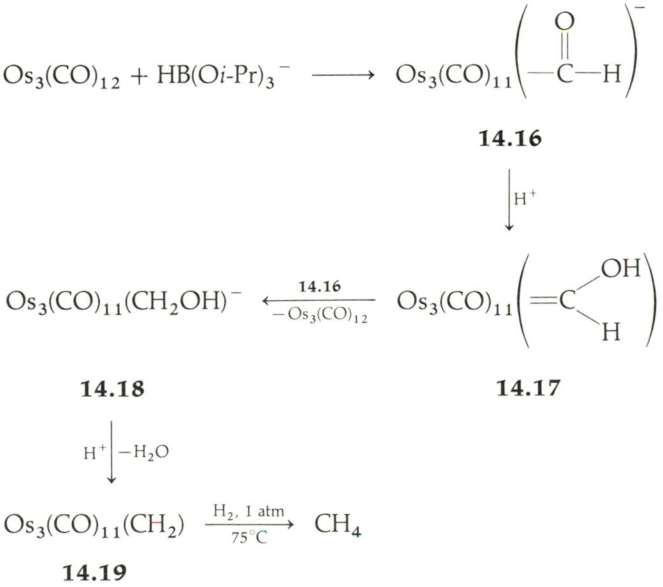

$$Os_3(CO)_{12} + HB(Oi\text{-}Pr)_3^- \longrightarrow Os_3(CO)_{11}\left(-\overset{\overset{\textstyle O}{\|}}{C}-H\right)^-$$

**14.16**

$\Big\downarrow$ H$^+$

$$Os_3(CO)_{11}(CH_2OH)^- \xleftarrow[-Os_3(CO)_{12}]{\textbf{14.16}} Os_3(CO)_{11}\left(=C\overset{\textstyle OH}{\underset{\textstyle H}{}}\right)$$

**14.18**                    **14.17**

$\Big\downarrow$ H$^+$ $-$H$_2$O

$$Os_3(CO)_{11}(CH_2) \xrightarrow[75°C]{H_2,\ 1\ atm} CH_4 \qquad (14.14)$$

**14.19**

An example of reduction of a CO ligand directly to a methyl ligand is shown in equation *14.15* for cationic molybdenum or tungsten complexes.

$$[(\eta^5\text{-}C_5H_5)M(CO)_3(PPh_3)]^+ \xrightarrow[\text{THF}]{\text{NaBH}_4} (\eta^5\text{-}C_5H_5)M(CH_3)(CO)_2(PPh_3) \qquad (14.15)$$

M = Mo or W                                  M = Mo, 69%

                                                    M = W, 27%

The reducing agent here is $NaBH_4$. Acetyl ligands are not readily reduced with $NaBH_4$, but $BH_3 \cdot THF$ reduces acetyl to ethyl ligands very smoothly (eq. *14.16*). Acetyl complexes of other metals also undergo this reduction.

$$(\eta^5\text{-}C_5H_5)(OC)(L)M\overset{O}{\overset{\|}{-}}C\,CH_3 \xrightarrow{H_3B\cdot THF} (\eta^5\text{-}C_5H_5)(OC)(L)M\text{-}CH_2CH_3 \quad (14.16)$$

M = Fe or Ru; L = a phosphine ligand

Coordination of the acyl oxygen atom to $BH_3$ presumably precedes the reduction. Reduction of an acyl ligand in the organoactinide complexes **14.20** to an alkoxide ligand in **14.21** occurs with hydrogen:

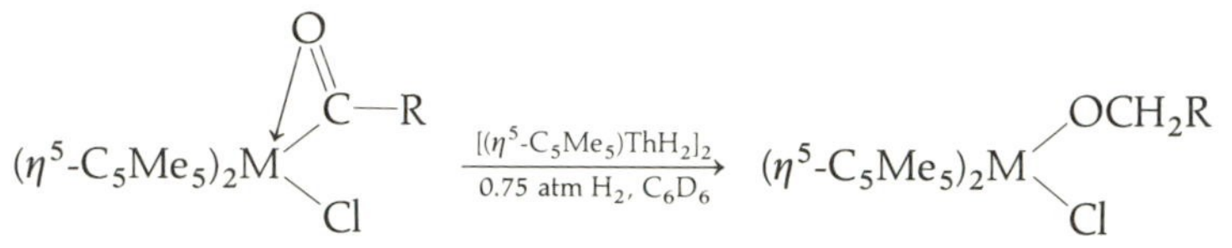

$$(\eta^5\text{-}C_5Me_5)_2M\overset{\overset{O}{\overset{\triangle}{\diagup}}C\text{—}R}{\diagdown Cl} \xrightarrow[0.75\ atm\ H_2,\ C_6D_6]{[(\eta^5\text{-}C_5Me_5)ThH_2]_2} (\eta^5\text{-}C_5Me_5)_2M\overset{\diagup OCH_2R}{\diagdown Cl} \quad (14.17)$$

$$\textbf{14.20} \qquad\qquad\qquad\qquad\qquad\qquad \textbf{14.21}$$

M = U; R = Ph
M = Th; R = $CH_2CMe_3$

A thorium hydride dimer acts as catalyst for this reaction. In contrast to equation *14.16*, here the oxygen is retained in the reduction products (**14.21**) because the acyl reactants are actually oxycarbenoid complexes (see **8.49**) due to the high oxo-philicity of uranium and thorium.

A model reaction for the metal-mediated activation of formaldehyde is shown in equation *14.18*. The substituted osmium carbonyl complex (**14.22**) reacts directly with formaldehyde to give the $\eta^2$-formaldehyde complex (**14.23**). When the latter complex is heated at 75°C, C—H oxidative-addition gives the formyl complex (**14.24**).

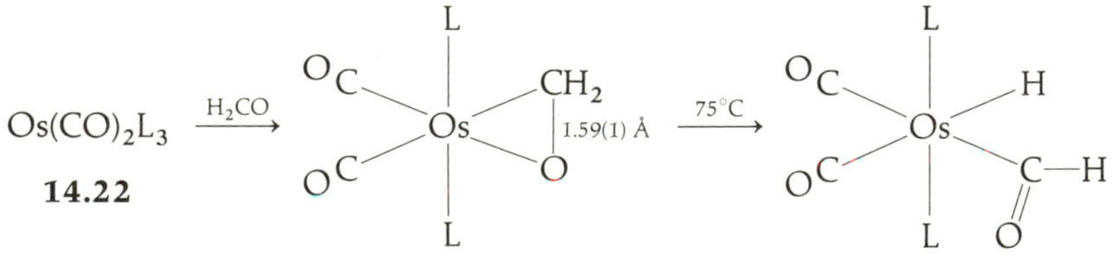

$$Os(CO)_2L_3 \xrightarrow{H_2CO} \quad\quad\quad \xrightarrow{75°C} \quad\quad\quad (14.18)$$

$$\textbf{14.22} \qquad\qquad\qquad \textbf{14.23} \qquad\qquad\qquad \textbf{14.24}$$

L = $Ph_3P$

## METAL-ACTIVATION OF C—H BONDS

The development of homogeneous catalysts that react directly with alkanes to give functionalized organic products would be a significant chemical achievement. Al-kanes obtained directly from natural resources or through Fischer–Tropsch pro-cesses could then be used *directly* as chemical reagents in organic synthesis. Alkane functionalization should, in theory, be very specific and should occur by low-temperature thermal reactions rather than by photochemically induced radical processes. The early designation of alkanes as "paraffins" denotes their resistances to chemical reagents and portends the difficulty in attaining this goal.

Studies of the homogeneous metal-activation of C—H bonds have developed in three areas: (1) observing oxidative-addition of C—H bonds (primarily within ligand molecules) to metal atoms; (2) obtaining direct spectroscopic or structural data indicating an M · · · H—C bonding interaction; and (3) effecting a direct chemical reaction between an alkane and a metal to give isolable and well-characterized complexes.

Oxidative-addition of C—H bonds within ligand molecules to metal atoms and other reactions of ligand C—H bonds in organometallic complexes have been discussed earlier as *intermolecular abstraction* reactions (eqs. *9.70* and *9.75*), *intramolecular abstraction* reactions (eqs. *9.77–9.82*), including *cyclometallation reactions* (eqs. *9.83, 9.84*, and *10.41–10.44*). This third class of reactions includes *ortho*-metallation as well as more distal C—H activation as shown in equation *14.19*.

$$IrL_4Cl + LiCH_2CMe_3 \xrightarrow{-L} \underset{\textbf{14.25}}{\begin{array}{c} H \\ L \,|\, CH_2 \\ {}^{\diagdown}Ir{\diagup}\phantom{x}CMe_2 \\ L \,|\, CH_2 \\ L \end{array}} \qquad (14.19)$$

$$L = Me_3P$$

In this reaction, complex **14.25** is presumably formed by halide displacement to give a neopentyl alkyl complex as an intermediate. Loss of an L ligand would give a four-coordinate Ir(I) alkyl complex, which then undergoes oxidative-addition of a distal C—H bond to afford the six-coordinate Ir(III) complex (**14.25**).

There is considerable continued interest in obtaining direct spectroscopic or structural evidence for an M · · · H—C interaction in ground-state structures. These interactions are believed to be representative of the activation step necessary for metal-mediated C—H bond cleavage.

Two geometrical extremes for M · · · H—C bonding interactions are structures **14.26** and **14.27**. Structure **14.26** represents an *open* three-center two-electron bond;

$$\underset{\textbf{14.26}}{-\overset{|}{\underset{\diagup}{C}}-H \cdots ML_n} \qquad \underset{\textbf{14.27}}{\overset{|}{\underset{\diagup}{-C}}\underset{\cdot M \cdot}{\overset{\diagdown}{\cdots}}\underset{L_n}{\cdots}H}$$

structure **14.27** represents a corresponding *closed* three-center two-electron bond. The metal atom within the complex is formally electron deficient in the *absence* of a C—H interaction. Donation of electron density from the two electron C—H σ bond to M reduces this electron deficiency. In structure **14.26,** complete electronic saturation at M might be accomplished by *hydride* transfer to M. However, heterolytic cleavage of the C—H bond with formation of a carbenium ion should require a high activation energy. Structure **14.27** could possibly afford facile oxidative-addition of the C—H bond to M. In this geometry, the two electrons of the C—H bond could be donated to M via a homolytic cleavage of the C—H bond possibly by a concerted oxidative-addition since $d$-electron density on M could be back bonded into the $\sigma^*$ molecular orbital of the C—H bond (compare the analogous concerted oxidative-addition of $H_2$ in chap. 10). As might be expected, the observed M $\cdots$ H—C interactions fall between these two structural extremes.

Spectroscopic evidence for an M $\cdots$ H—C bonding interaction includes (1) a shift of the C—H stretching vibrational frequency to lower energy; (2) an up-field shift in the PMR resonance of the C—H hydrogen atom; and (3) a reduction in the $^{13}$C—H coupling constant within the bonded C—H moiety.

Structural evidence for an M $\cdots$ H—C bonding interaction includes (1) a metal to carbon contact distance that is less than the sum of a normal C—H distance plus the van der Waals radii of H and M; (2) an M $\cdots$ H distance that is only slightly longer than the known M—H covalent bond distance, or that is less than the sum of the van der Waals radii of M and H; (3) a greater C—H distance than usual; and (4) angular distortions of H—C—R angles away from the angles expected from unperturbed carbon hybridization. Unfortunately, the position of the hydrogen atom and the determination of a precise van der Waals radius for M *in a specific complex* is difficult to obtain. Location of a hydrogen atom position by x-ray diffraction gives artificially short C—H distances. Accurate hydrogen positions can be obtained from neutron diffraction however. For the complexes shown below, many or all of these factors indicate some degree of M $\cdots$ H—C bonding interaction.

Complexes **14.28** through **14.31** are examples of compounds with M $\cdots$ H—C interactions involving alkyl carbon centers. In the copper complex (**14.28**) the bridgehead C—C—C angle of 91(2)° indicates a distorted $sp^3$-C hybridization. The Cu $\cdots$ H distance of 2.01(2) Å and a Cu $\cdots$ H—C angle of 158(17)° indicate a possible Cu $\cdots$ H—C interaction also. The 16-electron molybdenum atom in **14.29** interacts with one C—H $\sigma$ bond as shown to attain an 18-electron configuration. The observed Mo $\cdots$ H distance of 2.27(8) Å is probably actually closer to 2.15 Å after correcting for the misplacement of the H atom as refined from x-ray diffraction data. Normal Mo $\cdots$ H van der Waals distances are found in the range of 2.6–2.9 Å. Proton NMR data also indicate an Mo $\cdots$ H—C interaction.

Complexes **14.30** and **14.31** have $\eta^3$-allylic ligands in which a homoallylic $C(sp^3)$-H bond is donating electron density to the metal atom. In both complexes, the iron and manganese atoms would be 16-electron metals in the absence of the M $\cdots$ H—C interaction.

A neutron diffraction structure of complex **14.30** with L = $P(OMe)_3$ reveals an Fe $\cdots$ H distance of 1.874(3) Å. This distance is only about 0.24 Å longer than an Fe—H *covalent* bond distance. The C—H bond length of 1.164(3) Å in this complex is one of the longest C—H distances observed in a crystalline complex. Spectroscopic evidence supports M $\cdots$ H—C interactions in other complexes of type **14.30** and in complex **14.31**. Interestingly, complex **14.31**, which is prepared by protonation of the anionic complex [($\eta^4$-1,3-cyclohexadiene)Mn(CO)$_3$ $^-$], is fluxional on the PMR timescale. The Mn $\cdots$ H—C interaction exchanges between the two endo homoallylic sites. The iron complexes like **14.30** are prepared similarly by protonation of the appropriate neutral $\eta^4$-1,3-diene complexes. An x-ray structural

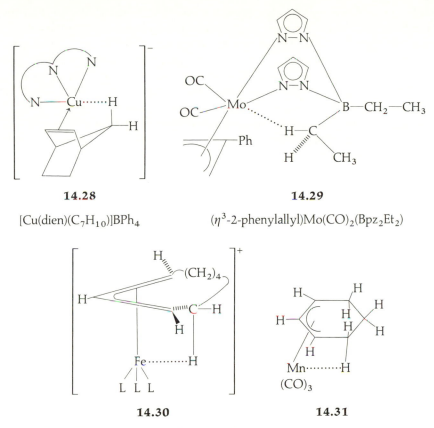

**14.28**

[Cu(dien)(C$_7$H$_{10}$)]BPh$_4$

**14.29**

($\eta^3$-2-phenylallyl)Mo(CO)$_2$(Bpz$_2$Et$_2$)

**14.30**

**14.31**

L = CO, phosphine, or phosphite ligand

analysis of the 6-*endo*-methyl derivative of complex **14.31** confirms a nearly closed Mn $\cdots$ H—C interaction.

Complex **14.32** is obtained by protonation of the $\mu_2$-methylene complex ($\eta^5$-C$_5$H$_5$)$_2$Fe$_2$($\mu_2$-CH$_2$)($\mu_2$-CO)(PPh$_2$CH$_2$PPh$_2$). The x-ray analysis constitutes the first structural characterization of an asymmetrical doubly bridging CH$_3$ ligand. The Fe $\cdots$ H distance and Fe $\cdots$ H—C angle are 1.71(4) Å and 102(3)°, respectively.

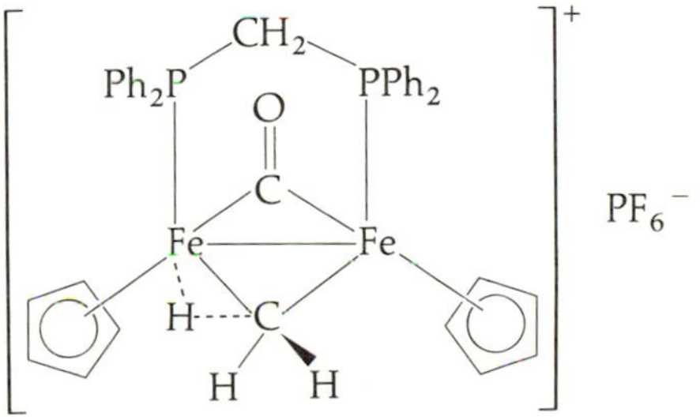

**14.32**

[($\eta^5$-C$_5$H$_5$)$_2$Fe$_2$($\mu_2$-CH$_3$)($\mu_2$-CO)(PPh$_2$CH$_2$PPh$_2$)]$^+$PF$_6^-$

This Fe $\cdots$ H distance is only slightly greater than that observed for Fe—H *covalent* bond distances. The Fe—C($\mu_2$-methyl) distance on the side involving the Fe $\cdots$ H interaction is about 0.08 Å longer than the other Fe—C($\mu_2$-methyl) distance. Presumably, the incoming proton protonates one of the Fe—C($\mu_2$-CH$_2$) bonds in the precursor complex to give **14.32**. Complex **14.32** substantiates the existence of asymmetrical $\mu_2$-CH$_3$ ligands as suggested earlier for Os$_3$(CO)$_{10}$(CH$_3$)($\mu_2$-H) (**9.71**).

Compounds **14.33** and **14.34** are tantalum-neopentylidene complexes that have a strong M $\cdots$ H—C($sp^2$) bonding interaction between the Ta atoms and a C$_\alpha$-hydrogen atom. Both molecular structures have been determined by neutron diffraction; thus the hydrogen atom positions are known accurately.

**14.33**                                                    **14.34**

[Ta(CHCMe$_3$)Cl$_3$(PMe$_3$)]$_2$          ($\eta^5$-C$_5$Me$_5$)Ta($\eta^2$-C$_2$H$_4$)(CHCMe$_3$)(PMe$_3$)

These structures indicate a highly distorted $sp^2$-hybridization at C$_\alpha$ of the neopentylidene ligands. Both Ta—C$_\alpha$ bond distances are *slightly shorter* than a normal Ta—C double bond length. Both Ta $\cdots$ H$_\alpha$ distances are only about 0.35 Å longer than a covalent Ta—H distance of 1.774(3) Å, and the C$_\alpha$—H$_\alpha$ distances are significantly *longer* than the average C—H distance of 1.084(2) Å for all of the other C—H bonds in each complex. Both Ta—C$_\alpha$—H$_\alpha$ angles are *less* than 90°.

Such structures most certainly result from the donation of C$_\alpha$—H$_\alpha$ $\sigma$-bond electron density to the very electron deficient 14- or 16-electron Ta atoms. The best structural representation is shown as **14.35**.

**14.35**

Both complexes have low C—H$_\alpha$ stretching frequencies and low $^{13}$C$_\alpha$—H$_\alpha$ coupling constants. Theoretical studies reveal that complete transfer of H$_\alpha$ to the tantalum atom to give a Ta-(alkylidyne)(hydride) complex is a symmetry-forbidden process. Perhaps this explains why strong Ta $\cdots$ H—C bonding interactions such as these are observable. Such metal-activated C—H bonds are quite susceptible to intramolecular $\alpha$-hydrogen abstraction (compare eq. *9.81*).

Two other complexes that have M $\cdots$ H—C($sp^2$) bonding interactions are **14.36** and **14.37**.

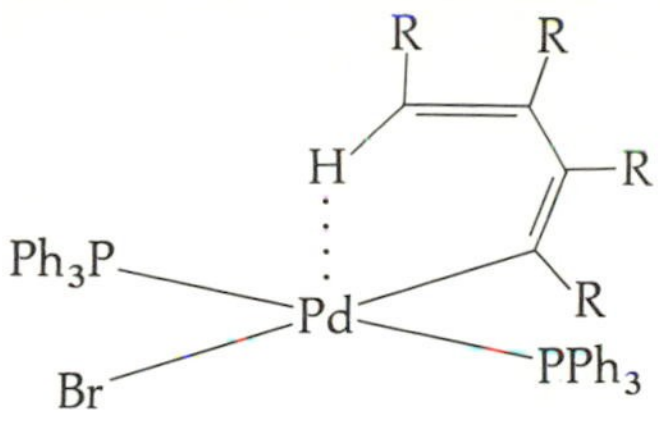

**14.36** 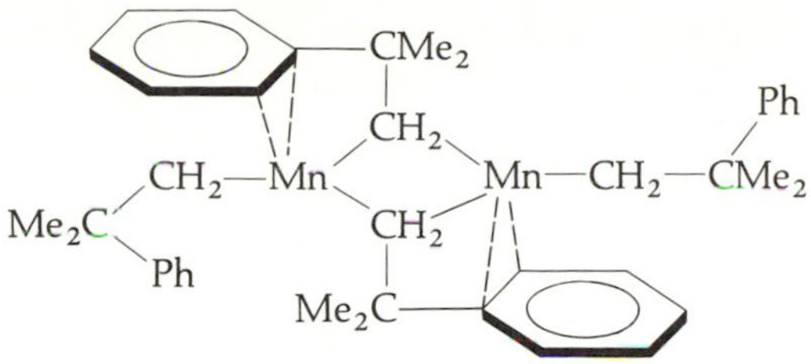 **14.37**

$R = CO_2Me$

The x-ray diffraction structures of both complexes have been determined. In complex **14.36**, the estimated Pd $\cdots$ H distance of about 2.3 Å is 0.8 Å shorter than the estimated Pd $\cdots$ H van der Waals contact distance of 3.1 Å. Examination of the interatomic distances of complex **14.37** reveals a close interaction between the manganese atoms and an *ortho*-hydrogen atom of two phenyl rings, as shown. This geometry resembles that of a possible metal-activated transition state in the well-known *ortho*-metallation reaction. In fact, when complex **14.37** is heated, *ortho*-metallation of the phenyl rings occurs with elimination of *t*-butylbenzene. Several other Mn(II) complexes of the type $[Mn(CH_2Z)_2]_n$, where Z is $CMe_2Ph$, $SiMe_3$, or $CMe_3$, are known. When Z is $SiMe_3$ or $CMe_3$, a polymeric or tetrameric structure, respectively, is observed. These complexes also have C—H stretching vibrations at low frequency, and an Mn $\cdots$ H—C interaction analogous to that found in **14.37** has been postulated.

The tetranuclear complex **14.38** is prepared by two routes as shown in equation *14.20*.

$$[Et_4N]_2[Fe_5C(CO)_{14}] \xrightarrow{3HCl} HFe_4(\eta^2\text{-}CH)(CO)_{12}$$

**14.38**

$$Fe_4C(CO)_{12}{}^{2-} \xrightarrow{2AgBF_4} \{Fe_4C(CO)_{12}\} \underset{}{\overset{H_2}{\longrightarrow}} \qquad (14.20)$$

Neutron and x-ray diffraction structural determinations of complex **14.38** reveal an $Fe_4$ butterfly cluster in which the methine group lies between the wings of the cluster as shown in **14.39**.

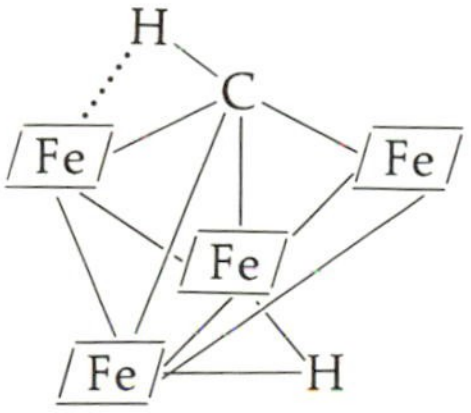

**14.39** 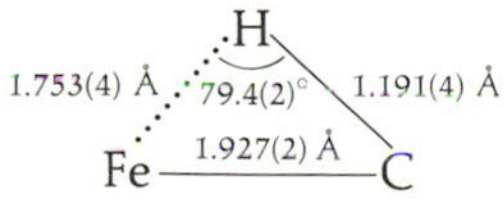 **14.40**

1.753(4) Å    79.4(2)°    1.191(4) Å

1.927(2) Å

$\boxed{Fe} = Fe(CO)_3$

The detailed geometry about the CH group is shown in **14.40**. A very short Fe $\cdots$ H distance and a long C—H distance indicates a strong Fe $\cdots$ H—C bonding interaction. In fact, the CH group can be regarded as a *dihapto* ligand. This C—H distance of 1.191(4) Å is the longest C—H distance yet confirmed structurally. As shown in **14.40**, the Fe $\cdots$ H—C interaction is best represented as a closed three-center two-electron bond like that of **14.27**.

Metal-activation of the methine group of **14.39** is strong enough that the methine hydrogen atom is easily removed by weak bases such as methanol. NMR experiments indicate that the methine hydrogen atom and the $\mu_2$-hydride ligand exchange sites at a detectable rate. Cluster species like **14.39** might be good models for related surface species on heterogeneous catalysts. If so, metal surface bound methine groups might react very differently from C—H bonds in molecular organic compounds. Tetranuclear clusters such as **14.39** might be involved in methane formation from CO ligands, as shown in equation *14.9a*.

Three recent reports of a *direct reaction* between alkanes or saturated C—H bonds and metal atoms should be mentioned here. One study reports an unusual dehydrogenation reaction effected by iridium complexes, as shown in equation *14.21*.

$$[IrH_2S_2L_2]BF_4 + 3\ t\text{-BuC(H)}=CH_2 + \ \bigcirc \ \xrightarrow[80°C]{ClCH_2CH_2Cl}$$

**14.41**

S = acetone or $H_2O$
L = $Ph_3P$

$$BF_4^- + 3\ t\text{-BuCH}_2CH_3 + 2S \quad (14.21)$$

**14.42**

40%

Complexes **14.41** were prepared by treating $[(\eta^4\text{-}1,5\text{-COD})IrL_2]BF_4$ with $H_2$ in the appropriate solvent. Reaction of these complexes with *cyclopentane* in the presence of 3,3-dimethyl-1-butene gives complex **14.42** and 2,2-dimethylbutane. In this reaction, complex **14.41** dehydrogenates cyclopentane to afford a cyclopentadienyl ligand. A similar, though less complete, dehydrogenation is observed with cyclooctane where $[(\eta^4\text{-}1,5\text{-COD})IrL_2]BF_4$ is isolated. Cycloheptane is dehydrogenated to give $[(\eta^5\text{-}C_7H_7)Ir(H)L_2]BF_4$.

These reactions do not appear to proceed via radical, carbonium ion, or heterogeneous mechanisms. Rather, a dehydrogenation mechanism has been proposed in which the metal directly cleaves the C—H bonds of the alkane. Several successive dehydrogenations on these cyclic alkanes eventually give an unsaturated ligand that can coordinate to the iridium atom as a very thermodynamically stable $\pi$-ligand, as does $\eta^5\text{-}C_5H_5$ in **14.42**.

Several key features of equation *14.21* should be noted:

1. Complex **14.41** contains two very weakly coordinating solvent ligands. Loss of these ligands would generate a species such as $IrH_2L_2^+$, which is a 14-electron $d^6$-Ir(III) complex. The Ir atom in this complex is very electron deficient and has considerable electrophilic character.

2. The electrophilic character of the Ir atom encourages *reductive-addition* (chap. 10) of the C—H bonds of alkanes rather than oxidative-addition of solvent C—H or C—Cl bonds or of the aryl C—H bonds of the $Ph_3P$ ligands.

3. The 3,3-dimethyl-1-butene increases the yield of **14.42** by acting as a hydrogen acceptor. An olefinic hydrogen acceptor must be bulky enough that it does not coordinate too strongly to the iridium atom thus competing with the alkane for a metal coordination site. This olefin should not undergo dehydrogenation under the reaction conditions if it is to act as an efficient hydrogen acceptor.

Attempts to convert the reaction of equation *14.21* into a catalytic process are underway.

Perhaps the best example of direct reaction of a metal complex with C—H bonds is shown in equation *14.22*.

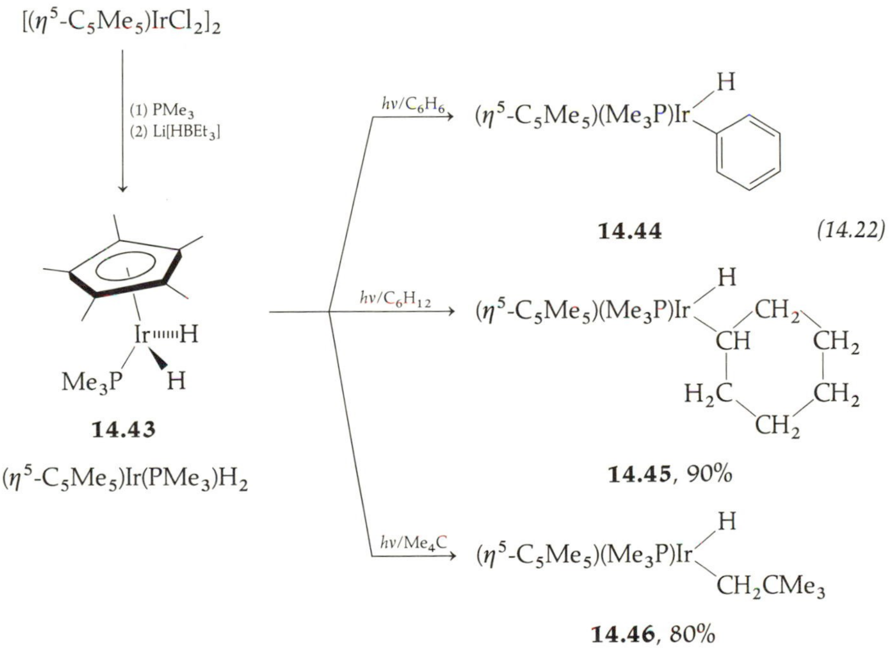

The Ir(III) complex **14.43** is prepared as shown from the dimeric complex $[(\eta^5-C_5Me_5)IrCl_2]_2$. Photolyzing **14.43** in benzene, cyclohexane, and neopentane gives the product complexes **14.44**–**14.46**, respectively. When the photolysis is conducted in $C_6D_{12}$, only $H_2$ and $(\eta^5-C_5Me_5)(Me_3P)Ir(D)(C_6D_{11})$ are obtained.

This result supports the postulated mechanism of photolytically initiated reductive-elimination of $H_2$ from **14.43** to give the neutral 16-electron Ir(I) complex ($\eta^5$-$C_5Me_5$)Ir($PMe_3$) as an intermediate. Oxidative-addition of a C—H bond of a solvent molecule gives the observed products **14.44–14.46**. When a 1:1 mixture of cyclohexane and neopentane is used as the solvent, complexes **14.45** and **14.46** are formed in a ratio of 0.88 to 1.0. The oxidative-addition reaction favors a primary over a secondary C—H bond (the opposite selectivity from that expected for a radical mechanism).

In a related system, photolysis of the Ir(I) complex **14.47** in neopentane solution at room temperature gives the C—H oxidative-addition product **14.48**.

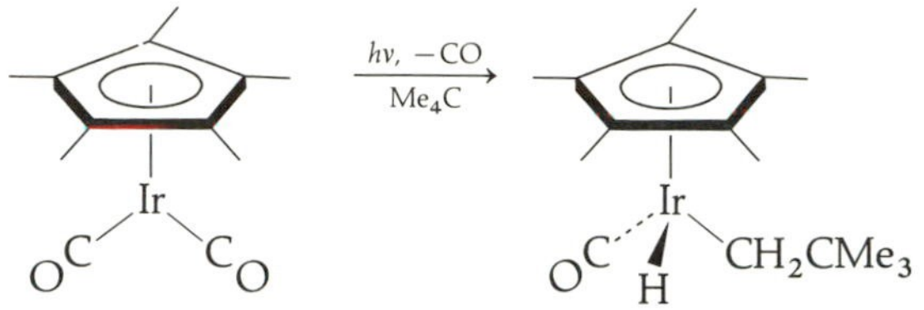

$$\text{(14.23)}$$

**14.47**

($\eta^5$-$C_5Me_5$)Ir($CO$)$_2$

**14.48**

Presumably, photolytic ligand dissociation of a CO ligand generates ($\eta^5$-$C_5Me_5$)Ir($CO$) as the intermediate, which then reacts with one of the C—H bonds of a neopentane molecule.

As an organometallic tour de force, two of the above iridium complexes have been shown to generate intermediates that react with methane.

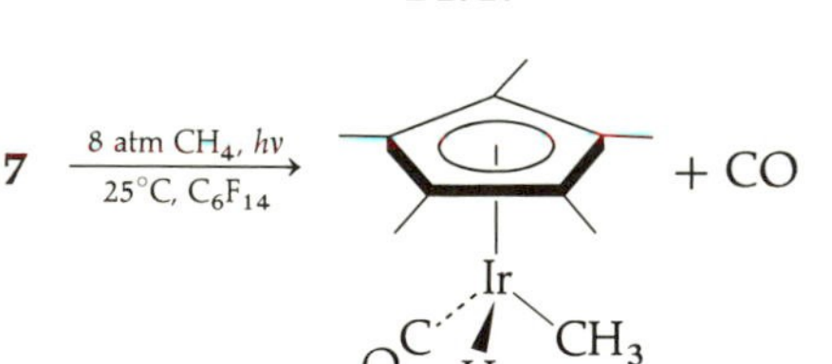

$$\text{(14.24)}$$

**14.49**

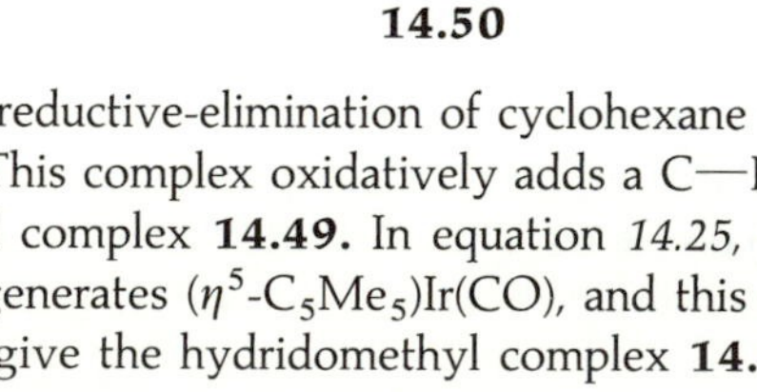

$$\text{(14.25)}$$

**14.50**

In equation *14.24*, thermal reductive-elimination of cyclohexane from **14.45** generates ($\eta^5$-$C_5Me_5$)Ir($PMe_3$). This complex oxidatively adds a C—H bond of methane to give the hydridomethyl complex **14.49**. In equation *14.25*, photochemical dissociation of a CO ligand generates ($\eta^5$-$C_5Me_5$)Ir($CO$), and this intermediate reacts similarly with methane to give the hydridomethyl complex **14.50**.

As equations *14.21* and *14.25* demonstrate, the propensity of Ir(I) complexes to undergo oxidative-addition (or reductive-addition) reactions and the thermodynamic stability of Ir(III) bonds to hydrogen and carbon have been very useful in developing model reactions of *direct* reaction between transition metal complexes and saturated alkanes. Future work may yield catalytic processes of considerable industrial importance.

## ASYMMETRIC HYDROGENATION

The mechanistic details of a metal-catalyzed asymmetric reduction related to that involved in the L-DOPA synthesis (chap. 13) have been elucidated. Hydrogenation of ethyl (Z)-α-acetamidocinnamate to the ethyl ester of N-acetyl-D-phenylalanine is accomplished stereospecifically by an optically active rhodium(I) catalyst in methanol solution. The specific catalyst is [Rh(chiraphos)(MeOH)$_2$]$^+$, where *chiraphos* is the asymmetric biphosphine ligand **14.51**.

**14.51**

chiraphos
2S, 3S-*bis*(diphenylphosphino)butane

Spectroscopic studies reveal the reaction mechanism shown in figure 14.2. Complexation of the prochiral olefin to the chiral rhodium catalyst occurs by dissociation of a solvent ligand. This resulting $\eta^2$-alkene complex exists as two possible diastereomers. Since diastereomers are not required to have the same free energy, they are present in unequal amounts. The major diastereomer can be isolated, and its x-ray structure has been determined. The minor diastereomer could not be detected spectroscopically, so it must be present in a yield of less than 5%.

Oxidative-addition of hydrogen to the ($\eta^2$-alkene)Rh(I) complex gives an ($\eta^2$-alkene)Rh dihydride intermediate. Addition of one Rh—H bond to the olefin occurs regio- and stereoselectively to give an (alkyl)(hydrido)Rh(III) intermediate as only one diastereomer. Reductive-elimination of alkane affords ethyl N-acetyl-D-phenylalaninate and regenerates the Rh(I) catalyst.

Asymmetric reactions are usually believed to occur because of preferential formation of a particularly stable diastereomer. In this study, however, oxidative-addition of hydrogen to the *observed* ($\eta^2$-alkene)Rh(I) diastereomer would afford the amino acid derivative of *opposite* configuration from that of the isolated product. This result implies that the unobserved *minor* ($\eta^2$-alkene)Rh(I) diastereomer is the *actual intermediate in the catalytic process.* Stereoselectivity in this system apparently results from a great difference between the *rates* of oxidative-addition of hydrogen to each

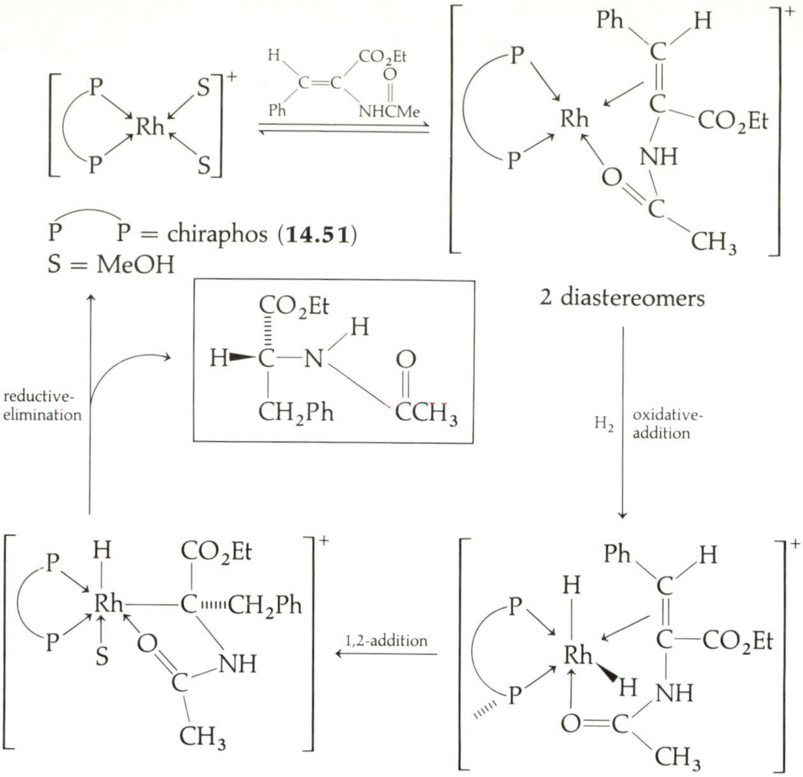

P   P = chiraphos (**14.51**)
S = MeOH

reductive-
elimination

2 diastereomers

H₂ oxidative-addition

1,2-addition

**FIGURE 14.2**   Asymmetric hydrogenation of ethyl (Z)-α-acetamidocinnamate to ethyl N-acetyl-D-phenylalaninate by a rhodium(I) catalyst. *Adapted from J. Halpern,* Inorg. Chim. Acta. **50**, *11 (1981).*

($\eta^2$-alkene)Rh(I) diastereomer rather than from a great difference between the thermodynamic stabilities of the ($\eta^2$-alkene)Rh(I) diastereomers.

## OLEFIN METATHESIS

Olefin metathesis (eq. *9.34*) is defined by the interchange of methylene fragments between two alkenes. The reaction was discovered in the mid-1960s. Both homogeneous and heterogeneous transition metal catalysts have been developed. Most commercial processes produce α-alkenes from internal alkenes by a metathesis exchange with ethylene:

$$R'HC=CHR + H_2C=CH_2 \rightleftharpoons RHC=CH_2 + R'HC=CH_2 \quad (14.26)$$

Terminal alkenes are feedstock chemicals for the manufacture of detergents via fatty alcohols. Phillips uses a metathesis process to produce commercial quantities of neohexene from 2,4,4-trimethyl-2-pentene (eq. *14.27*). This alkene product is used as a perfume intermediate.

$$Me_2C{=}CHCMe_3 + H_2C{=}CH_2 \longrightarrow CH_2{=}CHCMe_3 + Me_2C{=}CH_2 \quad (14.27)$$
$$\text{neohexene}$$

Much of the organometallic research with metal carbene complexes (see chapters 8, 11, and 12) and metallacyclic compounds, such as **12.17** and **12.18,** were directed toward understanding the mechanism of olefin metathesis. Two general mechanisms have been considered for this process.

The first type of mechanism proposed is shown in figure 14.3. These early mechanisms, which are now disfavored, require a pairwise exchange of the alkylidene fragments between two alkenes as indicated for the conversion of **14.52** into **14.53.** Three specific mechanisms proposed for this metathesis are shown. These include a

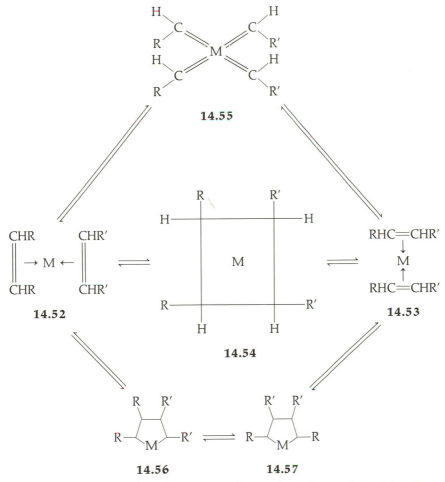

**FIGURE 14.3**  Disfavored pairwise mechanisms for olefin metathesis. *Adapted from R. H. Grubbs,* Comprehensive Organometallic Chemistry. *Vol. 8, Chap. 54 (Oxford: Pergamon Press, 1982).*

concerted route involving a quasi-cyclobutane complex (**14.54**), a route involving a tetracarbene intermediate (**14.55**), and a route with metallacyclopentane intermediates (**14.56** and **14.57**). Although these mechanisms were consistent with preliminary experimental results, they were not adequate for rationalizing the data obtained from labeling experiments.

Considerable evidence from mechanistic and model studies now strongly favors an exchange of alkylidene fragments by a non-pairwise mechanism. It appears that only mononuclear catalysts are needed. These mechanisms center on the *in situ* formation of a carbene complex. This species adds to the olefin forming a metallacyclobutane intermediate (**14.58**), which then gives the new alkene and the exchanged carbene complex as shown in equation *14.28*.

$$\text{M=C}\overset{H}{\underset{R}{\diagdown}} \underset{\underset{\text{H}_2\text{C}=\text{CH}_2}{}}{\rightleftharpoons} \quad \text{M}\overset{\overset{\displaystyle C-R}{\|}}{\underset{\underset{\text{H}_2\text{C}}{\diagup}}{\searrow \text{CH}_2}} \quad \rightleftharpoons \quad \text{M}\overset{\overset{\displaystyle \overset{\text{H}}{\underset{|}{C}}-R}{}}{\underset{\underset{\text{CH}_2}{}}{\diagdown}}\text{CH}_2$$

**14.58**                     *(14.28)*

$$\text{RHC=CH}_2 + \text{M=CH}_2 \rightleftharpoons \text{M}\overset{\text{RHC}}{\underset{\text{CH}_2}{\diagup \diagdown}}\text{CH}_2$$

The initial carbene complex is generated from the olefin reactants and/or added co-catalysts, although not necessarily by direct homolytic cleavage of an alkene bond.

Considerable effort has been expended in demonstrating the catalytic activity of carbene or metallacyclobutane complexes in olefin metathesis. As shown in equation *14.29*, the carbene complex **14.59** acts as a catalyst when activated by $AlCl_3$. Apparently $AlCl_3$ removes a chloride or phosphorus ligand to generate a vacant coordination site for the alkene. This reaction is an excellent model in that a Group VI

$$\underset{\text{Cl}}{\overset{\text{Cl}_{\prime\prime\prime\prime}}{}}\overset{\overset{\text{L}}{|}}{\underset{\underset{\text{L}}{|}}{\text{W}}}\overset{O}{\underset{C}{\diagdown}}_R^H + \text{H}_2\text{C=C}\overset{Et}{\underset{H}{\diagdown}} \xrightarrow{AlCl_3} \text{L}_2\text{Cl}_2\text{W}\overset{O}{\underset{\text{CH}_2}{\diagdown}} + \text{H}_2\text{C=C}\overset{R}{\underset{H}{\diagdown}} +$$

**14.59**

L = $Et_3P$; R = $CMe_3$

$$\text{L}_2\text{Cl}_2\text{W}\overset{O}{\underset{\underset{\text{H}}{\overset{|}{C-Et}}}{\diagdown}} + \text{H}_2\text{C=CH}_2 + \text{Et(H)C=C(H)Et} \quad (14.29)$$

metal is involved as an oxo complex and a strong Lewis acid is required. A typical commercial catalyst for olefin metathesis is generated from a mixture of $WCl_6$, EtOH, and $EtAlCl_2$.

An example of a metallacyclobutane that effects olefin metathesis is shown in equation *14.30*. The intermediate formation of a carbene complex in this reaction is not detected.

$$Cp_2Ti \underset{H}{\overset{R}{\diamondsuit}} + D_2C{=}C\underset{H}{\overset{CMe_2Et}{<}} \rightleftharpoons Cp_2Ti \underset{H}{\overset{D \quad D}{\diamondsuit}} CMe_2Et + H_2C{=}CHR$$

$$(14.30)$$

$R = CMe_3$

Clearly, organometallic chemistry has contributed significantly to understanding the olefin metathesis reaction and to the development of new catalysts. Future work may yield a catalytically active system in which both carbene and metallacyclobutane species can be detected.

## EPILOGUE

A tremendous amount of research has furthered our understanding of the reaction mechanisms for homogeneous catalytic processes. Similar effort has been applied to other homogeneous processes known to be catalyzed by transition metal organometallic compounds. Furthermore, much basic research in transition metal organometallic chemistry can be justified from a practical standpoint as providing model studies for reaction steps in commercial processes.

The general reactions of organometallic chemistry presented in earlier chapters are viewed differently when they are incorporated into a catalytic cycle. Simple ligand substitution permits the coordination of a substrate molecule to a metal atom. Oxidative-addition might add a second reagent to the same metal complex. Intramolecular reactions, such as 1,2-addition and alkyl migration, or intermolecular reactions, such as nucleophilic attack by an external base, convert the initial ligand into a different type of ligand. Removal of the modified ligand from the metal center is effected by reactions like reductive-elimination, $\beta$-hydrogen elimination or simple ligand substitution. When the original metal complex is regenerated, a catalytic cycle is possible.

Our understanding of these processes is limited by our ability to determine the identity of intermediate species and to measure the reaction rate of each step. The preparation and isolation of new classes of organometallic compounds facilitates the search for reaction intermediates. In some instances, kinetic studies of individual reactions can be conducted independently. Our ability to apply the principles of organometallic chemistry to the development of new catalytic processes is limited only by our knowledge and chemical ingenuity. Only the same limitations apply to the future contributions of organometallic chemistry to stoichiometric reactions. The author hopes that some of the readers will be enticed into joining this scientific adventure. Exciting discoveries will certainly be made.

## SUGGESTED READING

### Shift Reaction, Fischer–Tropsch, and CO Reduction Reactions

Casey, C. P. *et al. Pure Appl. Chem.* **52,** 625 (1980).

Ford, P. C. *Acc. Chem. Res.* **14,** 31 (1981).

Ford, P. C., ed. *Catalytic Activation of Carbon Monoxide.* ACS Symposium Series, No. 152. Washington: American Chemical Society, 1981.

Haggin, J. *Chem. Eng. News.* **1982,** Feb. 8, p. 13 and Feb. 23, p. 39; **1981,** Oct. 26, p. 22.

Herrmann, W. A. *Angew. Chem. International Ed.* **21,** 117 (1982).

Masters, C. *Adv. Organometal. Chem.* **17,** 61 (1979).

Muetterties, E. L., and Stein, J. *Chem. Rev.* **79,** 479 (1979).

Rofer-DePoorter, C. K. *Chem. Rev.* **81,** 447 (1981).

Wolczanski, P. T., and Bercaw, J. E. *Acc. Chem. Res.* **13,** 121 (1980).

### Metal Activation of Carbon-Hydrogen Bonds

Brookhart, M., and Green, M. L. H. *J. Organometal. Chem.* **250,** 395 (1983).

Deeming, A. J., and Rothwell, I. P. *Pure Appl. Chem.* **52,** 649 (1980).

Hoyano, J. K., McMaster, A. D., and Graham, W. A. G. *J. Am. Chem. Soc.* **105,** 7190 (1983).

Janowicz, A. H., and Bergman, R. G. *J. Am. Chem. Soc.* **105,** 3929 (1983).

Parshall, G. W. *Catalysis (London).* **1977,** 1 (1977).

Parshall, G. W. *Acc. Chem. Res.* **8,** 113 (1975).

Shilov, A. E. *Pure Appl. Chem.* **50,** 725 (1978).

Shilov, A. E., and Shteinman, A. A. *Coord. Chem. Rev.* **24,** 97 (1977).

Wax, M. J., Stryker, J. M., Buchanan, J. M., Kovac, C. A., and Bergman, R. G. *J. Am. Chem. Soc.* **106,** 1121 (1984).

Webster, D. E. *Adv. Organometal. Chem.* **15,** 147 (1977).

### Hydrogenation Catalysis and Asymmetric Synthesis

Bosnick, B., and Fryzuk, M. D. *Topics Stereochem.* **12,** 119 (1981).

Brunner, H. *Acc. Chem. Res.* **12,** 250 (1979).

Chan, A. S. C., and Halpern, J. *J. Am. Chem. Soc.* **102,** 838 (1980).

Crabtree, R. *Acc. Chem. Res.* **12,** 331 (1979).

Dolcetti, G., and Hoffman, N. W. *Inorg. Chim. Acta.* **9,** 269 (1974).

Halpern, J. *Inorg. Chim. Acta.* **50,** 11 (1981).

Harmon, R. E., Gupta, S. K., and Brown, D. J. *Chem. Rev.* **73,** 21 (1973).

James, B. R. *Adv. Organometal. Chem.* **17,** 319 (1979).

### Olefin Metathesis

Calderon, N., Lawrence, J. P., and Ofstead, E. A. *Adv. Organometal. Chem.* **17,** 449 (1979).

Calderon, N., Ofstead, E. A., and Judy, W. A. *Angew. Chem., Int. Ed. Engl.* **15,** 401 (1976).

Grubbs, R. H. *Prog. Inorg. Chem.* **24,** 1 (1978).

Grubbs, R. H. *Comprehensive Organometallic Chemistry.* Vol. 8, Chap. 54, G. Wilkinson, F. G. A. Stone, and E. W. Abel, eds. (Oxford: Pergamon Press, 1982).

# Abbreviation List

| | |
|---|---|
| Å | angstrom unit, $10^{-10}$ m |
| Ac | acetyl, $CH_3C(O)$ |
| acac | acetylacetonate anion |
| atm | atmosphere |
| | |
| bipy | 2,2'-bipyridine |
| Bu or But | butyl |
| | |
| chiraphos | $Ph_2PC(H)(Me)C(H)(Me)PPh_2$ |
| $cm^{-1}$ | wave number |
| CN | coordination number (or cyanide ligand) |
| COD | cyclooctadiene (usually 1,5-COD) |
| COT | cyclooctatetraene |
| Cp | cyclopentadiene |
| crypt | $N(CH_2CH_2OCH_2CH_2OCH_2CH_2)_3N$ |
| | |
| DBA | dibenzylideneacetone |
| DCC | dicyclohexylcarbodiimide |
| dec. | decomposition |
| DH | dimethylglyoximato |
| diars | o-phenylenebisdimethylarsine |
| DIBAL | diisobutylaluminum hydride |
| dien | diethylenetriamine |
| diglyme | diethyleneglycoldimethyl ether |
| DIOP | 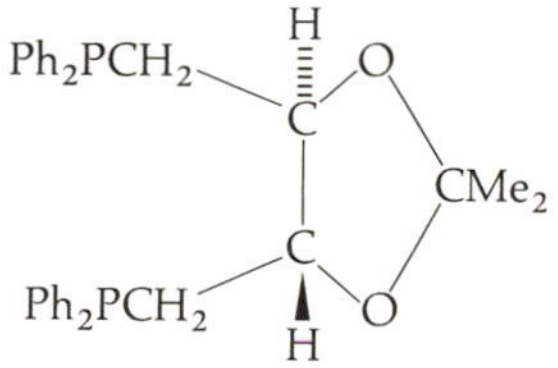 |

| | |
|---|---|
| diphos | 1,2-bis(diphenylphosphino)ethane |
| DME | 1,2-dimethoxyethane |
| DMF | dimethylformamide |
| dmpe | 1,2-bis(dimethylphosphino)ethane |
| DPM or dpm | bis(diphenylphosphino)methane |
| DTO | 2,2,7,7-tetramethyl 1-3,6-dithiaoctane |
| | |
| e$^-$ | electron |
| EAN | effective atomic number |

| | |
|---|---|
| en | ethylenediamine, $H_2NCH_2CH_2NH_2$ |
| Et | ethyl |
| | |
| *hv* | photolytic activation |
| HMPA | hexamethylphosphoramide |
| HOMO | highest occupied molecular orbital |
| | |
| LDA | lithium diisopropylamide |
| LUMO | lowest unoccupied molecular orbital |
| | |
| Me | methyl |
| MO | molecular orbital |
| mp | melting point |
| | |
| NHS | N-hydroxysuccinimide |
| NOR | norbornadiene |
| Np | naphthalenide |
| Nuc | nucleophile |
| | |
| O-A | oxidative-addition |
| | |
| Ph | phenyl |
| phen or *o*-phen | 1,10-phenanthroline |
| $PPN^+$ | $(Ph_3P)_2N^+$ |
| Pr | propyl |
| PTA | phosphatriazaadamantane, 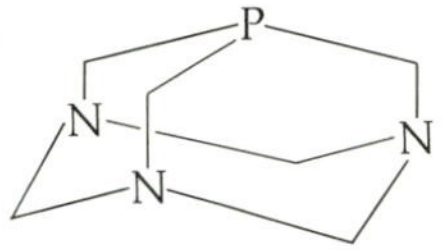 |

| | |
|---|---|
| py | pyridine |
| pz | pyrazoyl |
| | |
| $R_F$ | perfluorocarbon substituent, such as $CF_3$ or $C_2F_5$ |
| $R_p$-CAMP | $(C_6H_{11})PMe(o\text{-MeO-}C_6H_4)$ |
| R-E | reductive-elimination |
| | |
| TFA | trifluoroacetate |
| THF | tetrahydrofuran |
| THP | 2-tetrahydropyranyl |
| TMEDA | N,N,N′,N′-tetramethylethylenediamine |
| TMP | tetramethylpiperidine |
| tmpa | N,N,N′,N′-tetramethylpropylenediamine |
| TRANSPHOS | 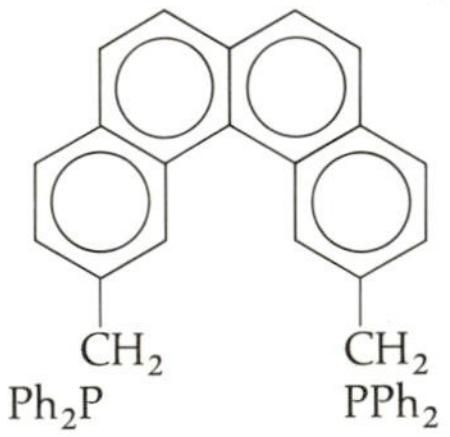 |

| | |
|---|---|
| $T_s$ | tosyl (*p*-toluenesulfonyl) |
| TTP | *meso*-tetra-*p*-tolylporphyrin |
| Δ | thermal activation |

# Index

# Index to Structural Data

This index is an alphabetical listing of formulae of molecules for which some numerical structural data are given in the book. The molecules are listed by metal ion, and the location of the data is indicated by page number, p. xxx, compound number, **x.xx**, equation number, (x.xx), or table number, table x.x. In these structural formulae, the following abbreviations are used: $Cp = \eta^5\text{-}C_5H_5$ and $Cp' = \eta^5\text{-}C_5Me_5$.